计算机组成与设计

硬件/软件接口

（英文版·原书第5版·ARM版）

Computer Organization and Design

The Hardware/Software Interface, ARM Edition

U0186133

［美］ 戴维·A. 帕特森　约翰·L. 亨尼斯　著
　　　David A. Patterson　　John L. Hennessy

机械工业出版社
China Machine Press

图书在版编目（CIP）数据

计算机组成与设计：硬件/软件接口（英文版·原书第5版·ARM版）/（美）戴维·A. 帕特森（David A. Patterson），（美）约翰·L. 亨尼斯（John L. Hennessy）著 . —北京：机械工业出版社，2020.11（经典原版书库）

书名原文：Computer Organization and Design: The Hardware/Software Interface, ARM Edition

ISBN 978-7-111-66835-0

I. 计⋯ II. ①戴⋯ ②约⋯ III. ①计算机体系结构–英文 ②微型计算机–接口设备–英文 IV. ① TP303 ② TP364

中国版本图书馆 CIP 数据核字（2020）第 209152 号

本书版权登记号：图字 01-2020-5373

Computer Organization and Design: The Hardware/Software Interface, ARM Edition
David A. Patterson, John L. Hennessy
ISBN: 9780128017333

Notice

Knowledge and best practice in this field are constantly changing. As new research and experience broaden our understanding, changes in research methods, professional practices, or medical treatment may become necessary. Practitioners and researchers must always rely on their own experience and knowledge in evaluating and using any information, methods, compounds or experiments described herein. Because of rapid advances in the medical sciences, in particular, independent verification of diagnoses and drug dosages should be made. To the fullest extent of the law, no responsibility is assumed by Elsevier, authors, editors or contributors in relation to the adaptation or for any injury and/or damage to persons or property as a matter of products liability, negligence or otherwise, or from any use or operation of any methods, products, instructions, or ideas contained in the material herein.

出版发行：机械工业出版社（北京市西城区百万庄大街 22 号　邮政编码：100037）

责任编辑：曲　熠　　　　　　　　　　　　责任校对：李秋荣

印　　刷：北京市荣盛彩色印刷有限公司　　版　　次：2021 年 1 月第 1 版第 1 次印刷

开　　本：186mm×240mm　1/16　　　　　印　　张：44.75

书　　号：ISBN 978-7-111-66835-0　　　　定　　价：169.00 元

客服电话：（010）88361066　88379833　68326294　　　投稿热线：（010）88379604

华章网站：www.hzbook.com　　　　　　　　　　　　　读者信箱：hzjsj@hzbook.com

版权所有 · 侵权必究

封底无防伪标均为盗版

本书法律顾问：北京大成律师事务所　韩光/邹晓东

出版者的话

文艺复兴以来，源远流长的科学精神和逐步形成的学术规范，使西方国家在自然科学的各个领域取得了垄断性的优势；也正是这样的优势，使美国在信息技术发展的六十多年间名家辈出、独领风骚。在商业化的进程中，美国的产业界与教育界越来越紧密地结合，计算机学科中的许多泰山北斗同时身处科研和教学的最前线，由此而产生的经典科学著作，不仅擘划了研究的范畴，还揭示了学术的源变，既遵循学术规范，又自有学者个性，其价值并不会因年月的流逝而减退。

近年，在全球信息化大潮的推动下，我国的计算机产业发展迅猛，对专业人才的需求日益迫切。这对计算机教育界和出版界都既是机遇，也是挑战；而专业教材的建设在教育战略上显得举足轻重。在我国信息技术发展时间较短的现状下，美国等发达国家在其计算机科学发展的几十年间积淀和发展的经典教材仍有许多值得借鉴之处。因此，引进一批国外优秀计算机教材将对我国计算机教育事业的发展起到积极的推动作用，也是与世界接轨、建设真正的世界一流大学的必由之路。

机械工业出版社华章公司较早意识到"出版要为教育服务"。自1998年开始，我们就将工作重点放在了遴选、移译国外优秀教材上。经过多年的不懈努力，我们与Pearson、McGraw-Hill、Elsevier、MIT、John Wiley & Sons、Cengage等世界著名出版公司建立了良好的合作关系，从它们现有的数百种教材中甄选出Andrew S. Tanenbaum、Bjarne Stroustrup、Brian W. Kernighan、Dennis Ritchie、Jim Gray、Afred V. Aho、John E. Hopcroft、Jeffrey D. Ullman、Abraham Silberschatz、William Stallings、Donald E. Knuth、John L. Hennessy、Larry L. Peterson等大师名家的一批经典作品，以"计算机科学丛书"为总称出版，供读者学习、研究及珍藏。大理石纹理的封面，也正体现了这套丛书的品位和格调。

"计算机科学丛书"的出版工作得到了国内外学者的鼎力相助，国内的专家不仅提供了中肯的选题指导，还不辞劳苦地担任了翻译和审校的工作；而原书的作者也相当关注其作品在中国的传播，有的还专门为其书的中译本作序。迄今，"计算机科学丛书"已经出版了近500个品种，这些书籍在读者中树立了良好的口碑，并被许多高校采用为正式教材和参考书籍。其影印版"经典原版书库"作为姊妹篇也被越来越多实施双语教学的学校所采用。

权威的作者、经典的教材、一流的译者、严格的审校、精细的编辑，这些因素使我们的图书有了质量的保证。随着计算机科学与技术专业学科建设的不断完善和教材改革的逐渐深化，教育界对国外计算机教材的需求和应用都将步入一个新的阶段，我们的目标是尽善尽美，而反馈的意见正是我们达到这一终极目标的重要帮助。华章公司欢迎老师和读者对我们的工作提出建议或给予指正，我们的联系方法如下：

华章网站：www.hzbook.com

电子邮件：hzjsj@hzbook.com

联系电话：（010）88379604

联系地址：北京市西城区百万庄南街1号

邮政编码：100037

华章科技图书出版中心

In Praise of *Computer Organization and Design: The Hardware/Software Interface*, ARM® Edition

"Textbook selection is often a frustrating act of compromise—pedagogy, content coverage, quality of exposition, level of rigor, cost. *Computer Organization and Design* is the rare book that hits all the right notes across the board, without compromise. It is not only the premier computer organization textbook, it is a shining example of what all computer science textbooks could and should be."

—Michael Goldweber, *Xavier University*

"I have been using *Computer Organization and Design* for years, from the very first edition. This new edition is yet another outstanding improvement on an already classic text. The evolution from desktop computing to mobile computing to Big Data brings new coverage of embedded processors such as the ARM, new material on how software and hardware interact to increase performance, and cloud computing. All this without sacrificing the fundamentals."

—Ed Harcourt, *St. Lawrence University*

"To Millennials: *Computer Organization and Design* is the computer architecture book you should keep on your (virtual) bookshelf. The book is both old and new, because it develops venerable principles—Moore's Law, abstraction, common case fast, redundancy, memory hierarchies, parallelism, and pipelining—but illustrates them with contemporary designs."

—Mark D. Hill, *University of Wisconsin-Madison*

"The new edition of *Computer Organization and Design* keeps pace with advances in emerging embedded and many-core (GPU) systems, where tablets and smartphones will/are quickly becoming our new desktops. This text acknowledges these changes, but continues to provide a rich foundation of the fundamentals in computer organization and design which will be needed for the designers of hardware and software that power this new class of devices and systems."

—Dave Kaeli, *Northeastern University*

"*Computer Organization and Design* provides more than an introduction to computer architecture. It prepares the reader for the changes necessary to meet the ever-increasing performance needs of mobile systems and big data processing at a time that difficulties in semiconductor scaling are making all systems power constrained. In this new era for computing, hardware and software must be co-designed and system-level architecture is as critical as component-level optimizations."

—Christos Kozyrakis, *Stanford University*

"Patterson and Hennessy brilliantly address the issues in ever-changing computer hardware architectures, emphasizing on interactions among hardware and software components at various abstraction levels. By interspersing I/O and parallelism concepts with a variety of mechanisms in hardware and software throughout the book, the new edition achieves an excellent holistic presentation of computer architecture for the post-PC era. This book is an essential guide to hardware and software professionals facing energy efficiency and parallelization challenges in Tablet PC to Cloud computing."

—Jae C. Oh, *Syracuse University*

Preface

About This Book

We believe that learning in computer science and engineering should reflect the current state of the field, as well as introduce the principles that are shaping computing. We also feel that readers in every specialty of computing need to appreciate the organizational paradigms that determine the capabilities, performance, energy, and, ultimately, the success of computer systems.

Modern computer technology requires professionals of every computing specialty to understand both hardware and software. The interaction between hardware and software at a variety of levels also offers a framework for understanding the fundamentals of computing. Whether your primary interest is hardware or software, computer science or electrical engineering, the central ideas in computer organization and design are the same. Thus, our emphasis in this book is to show the relationship between hardware and software and to focus on the concepts that are the basis for current computers.

The recent switch from uniprocessor to multicore microprocessors confirmed the soundness of this perspective, given since the first edition. While programmers could ignore the advice and rely on computer architects, compiler writers, and silicon engineers to make their programs run faster or be more energy-efficient without change, that era is over. For programs to run faster, they must become parallel. While the goal of many researchers is to make it possible for programmers to be unaware of the underlying parallel nature of the hardware they are programming, it will take many years to realize this vision. Our view is that for at least the next decade, most programmers are going to have to understand the hardware/software interface if they want programs to run efficiently on parallel computers.

The audience for this book includes those with little experience in assembly language or logic design who need to understand basic computer organization as well as readers with backgrounds in assembly language and/or logic design who want to learn how to design a computer or understand how a system works and why it performs as it does.

About the Other Book

Some readers may be familiar with *Computer Architecture: A Quantitative Approach*, popularly known as Hennessy and Patterson. (This book in turn is often called Patterson and Hennessy.) Our motivation in writing the earlier book was to describe the principles of computer architecture using solid engineering fundamentals and quantitative cost/performance tradeoffs. We used an approach that combined examples and measurements, based on commercial systems, to create realistic design experiences. Our goal was to demonstrate that computer architecture could be learned using quantitative methodologies instead of a descriptive approach. It was intended for the serious computing professional who wanted a detailed understanding of computers.

A majority of the readers for this book do not plan to become computer architects. The performance and energy efficiency of future software systems will be dramatically affected, however, by how well software designers understand the basic hardware techniques at work in a system. Thus, compiler writers, operating system designers, database programmers, and most other software engineers need a firm grounding in the principles presented in this book. Similarly, hardware designers must understand clearly the effects of their work on software applications.

Thus, we knew that this book had to be much more than a subset of the material in *Computer Architecture*, and the material was extensively revised to match the different audience. We were so happy with the result that the subsequent editions of *Computer Architecture* were revised to remove most of the introductory material; hence, there is much less overlap today than with the first editions of both books.

Why ARMv8 for This Edition?

The choice of instruction set architecture is clearly critical to the pedagogy of a computer architecture textbook. We didn't want an instruction set that required describing unnecessary baroque features for someone's first instruction set, no matter how popular it is. Ideally, your initial instruction set should be an exemplar, just like your first love. Surprisingly, you remember both fondly.

Since there were so many choices at the time, for the first edition of *Computer Architecture: A Quantitative Approach* we invented our own RISC-style instruction set. Given the growing popularity and the simple elegance of the MIPS instruction set, we switched to it for the first edition of this book and to later editions of the other book. MIPS has served us and our readers well.

The incredible popularity of the ARM instruction set—14 billion instances were shipped in 2015—led some instructors to ask for a version of the book based on ARM. We even tried a version of it for a subset of chapters for an Asian edition of this book. Alas, as we feared, the baroqueness of the ARMv7 (32-bit address) instruction set was too much for us to bear, so we did not consider making the change permanent.

To our surprise, when ARM offered a 64-bit address instruction set, it made so many significant changes that in our opinion it bore more similarity to MIPS than it did to ARMv7:

- The registers were expanded from 16 to 32;

- The PC is no longer one of these registers;

- The conditional execution option for every instruction was dropped;

- Load multiple and store multiple instructions were dropped;

- PC-relative branches with large address fields were added;

- Addressing modes were made consistent for all data transfer instructions;

- Fewer instructions set condition codes;

and so on. Although ARMv8 is much, much larger than MIPS—the ARMv8 architecture reference manual is 5400 pages long—we found a subset of ARMv8 instructions that is similar in size and nature to the MIPS core used in prior editions, which we call LEGv8 to avoid confusion. Hence, we wrote this ARMv8 edition.

Given that ARMv8 offers both 32-bit address instructions and 64-bit address instructions within essentially the same instruction set, we could have switched instruction sets but kept the address size at 32 bits. Our publisher polled the faculty who used the book and found that 75% either preferred larger addresses or were neutral, so we increased the address space to 64 bits, which may make more sense today than 32 bits.

The only changes for the ARMv8 edition from the MIPS edition are those associated with the change in instruction sets, which primarily affects Chapter 2, Chapter 3, the virtual memory section in Chapter 5, and the short VMIPS example in Chapter 6. In Chapter 4, we switched to ARMv8 instructions, changed several figures, and added a few "Elaboration" sections, but the changes were simpler than we had feared. Chapter 1 and the rest of the appendices are virtually unchanged. The extensive online documentation and combined with the magnitude of ARMv8 make it difficult to come up with a replacement for the MIPS version of Appendix A ("Assemblers, Linkers, and the SPIM Simulator" in the MIPS Fifth Edition). Instead, Chapters 2, 3, and 5 include quick overviews of the hundreds of ARMv8 instructions outside of the core ARMv8 instructions that we cover in detail in the rest of the book. We believe readers of this edition will have a good understanding of ARMv8 without having to plow through thousands of pages of online documentation. And for any reader that adventurous, it would probably be wise to read these surveys first to get a framework on which to hang on the many features of ARMv8.

Note that we are not (yet) saying that we are permanently switching to ARMv8. For example, both ARMv8 and MIPS versions of the fifth edition are available for sale now. One possibility is that there will be a demand for both MIPS and ARMv8 versions for future editions of the book, or there may even be a demand for a third version with yet another instruction set. We'll cross that bridge when we come to it. For now, we look forward to your reaction to and feedback on this effort.

Changes for the Fifth Edition

We had six major goals for the fifth edition of *Computer Organization and Design:* demonstrate the importance of understanding hardware with a running example; highlight main themes across the topics using margin icons that are introduced early; update examples to reflect changeover from PC era to post-PC era; spread the material on I/O throughout the book rather than isolating it into a single chapter; update the technical content to reflect changes in the industry since the publication of the fourth edition in 2009; and put appendices and optional sections online instead of including a CD to lower costs and to make this edition viable as an electronic book.

Before discussing the goals in detail, let's look at the table on the next page. It shows the hardware and software paths through the material. Chapters 1, 4, 5, and 6 are found on both paths, no matter what the experience or the focus. Chapter 1 discusses the importance of energy and how it motivates the switch from single core to multicore microprocessors and introduces the eight great ideas in computer architecture. Chapter 2 is likely to be review material for the hardware-oriented, but it is essential reading for the software-oriented, especially for those readers interested in learning more about compilers and object-oriented programming languages. Chapter 3 is for readers interested in constructing a datapath or in learning more about floating-point arithmetic. Some will skip parts of Chapter 3, either because they don't need them, or because they offer a review. However, we introduce the running example of matrix multiply in this chapter, showing how subword parallels offers a fourfold improvement, so don't skip Sections 3.6 to 3.8. Chapter 4 explains pipelined processors. Sections 4.1, 4.5, and 4.10 give overviews, and Section 4.12 gives the next performance boost for matrix multiply for those with a software focus. Those with a hardware focus, however, will find that this chapter presents core material; they may also, depending on their background, want to read Appendix A on logic design first. The last chapter on multicores, multiprocessors, and clusters, is mostly new content and should be read by everyone. It was significantly reorganized in this edition to make the flow of ideas more natural and to include much more depth on GPUs, warehouse-scale computers, and the hardware–software interface of network interface cards that are key to clusters.

Chapter or Appendix	Sections	Software focus	Hardware focus
1. Computer Abstractions and Technology	1.1 to 1.11	👓	👓
	1.12 (History)	👓	👓
2. Instructions: Language of the Computer	2.1 to 2.14	👓	👓
	2.15 (Compilers & Java)	👓	
	2.16 to 2.21	👓	👓
	2.22 (History)	👓	👓
D. RISC Instruction-Set Architectures	D.1 to D.17	👓	
3. Arithmetic for Computers	3.1 to 3.5	👓	👓
	3.6 to 3.9 (Subword Parallelism)	👓	👓
	3.10 to 3.11 (Fallacies)	👓	👓
	3.12 (History)	👓	👓
A. The Basics of Logic Design	A.1 to A.13		👓
4. The Processor	4.1 (Overview)	👓	👓
	4.2 (Logic Conventions)		👓
	4.3 to 4.4 (Simple Implementation)	👓	👓
	4.5 (Pipelining Overview)	👓	👓
	4.6 (Pipelined Datapath)	👓	👓
	4.7 to 4.9 (Hazards, Exceptions)		👓
	4.10 to 4.12 (Parallel, Real Stuff)	👓	👓
	4.13 (Verilog Pipeline Control)		👓
	4.14 to 4.15 (Fallacies)	👓	👓
	4.16 (History)	👓	👓
C. Mapping Control to Hardware	C.1 to C.6		👓
5. Large and Fast: Exploiting Memory Hierarchy	5.1 to 5.10	👓	👓
	5.11 (Redundant Arrays of Inexpensive Disks)	👓	👓
	5.12 (Verilog Cache Controller)		👓
	5.13 to 5.16	👓	👓
	5.17 (History)	👓	👓
6. Parallel Process from Client to Cloud	6.1 to 6.8	👓	👓
	6.9 (Networks)	👓	👓
	6.10 to 6.14	👓	👓
	6.15 (History)	👓	👓
B. Graphics Processor Units	B.1 to B.13	👓	👓

Read carefully 👓   Read if have time 👓   Reference 👓
Review or read 👓   Read for culture 👓

The first of the six goals for this fifth edition was to demonstrate the importance of understanding modern hardware to get good performance and energy efficiency with a concrete example. As mentioned above, we start with subword parallelism in Chapter 3 to improve matrix multiply by a factor of 4. We double performance in Chapter 4 by unrolling the loop to demonstrate the value of instruction-level parallelism. Chapter 5 doubles performance again by optimizing for caches using blocking. Finally, Chapter 6 demonstrates a speedup of 14 from 16 processors by using thread-level parallelism. All four optimizations in total add just 24 lines of C code to our initial matrix multiply example.

The second goal was to help readers separate the forest from the trees by identifying eight great ideas of computer architecture early and then pointing out all the places they occur throughout the rest of the book. We use (hopefully) easy-to-remember margin icons and highlight the corresponding word in the text to remind readers of these eight themes. There are nearly 100 citations in the book. No chapter has less than seven examples of great ideas, and no idea is cited less than five times. Performance via parallelism, pipelining, and prediction are the three most popular great ideas, followed closely by Moore's Law. The processor chapter (4) is the one with the most examples, which is not a surprise since it probably received the most attention from computer architects. The one great idea found in every chapter is performance via parallelism, which is a pleasant observation given the recent emphasis in parallelism in the field and in editions of this book.

The third goal was to recognize the generation change in computing from the PC era to the post-PC era by this edition with our examples and material. Thus, Chapter 1 dives into the guts of a tablet computer rather than a PC, and Chapter 6 describes the computing infrastructure of the cloud. We also feature the ARM, which is the instruction set of choice in the personal mobile devices of the post-PC era, as well as the x86 instruction set that dominated the PC era and (so far) dominates cloud computing.

The fourth goal was to spread the I/O material throughout the book rather than have it in its own chapter, much as we spread parallelism throughout all the chapters in the fourth edition. Hence, I/O material in this edition can be found in Sections 1.4, 4.9, 5.2, 5.5, 5.11, and 6.9. The thought is that readers (and instructors) are more likely to cover I/O if it's not segregated to its own chapter.

This is a fast-moving field, and, as is always the case for our new editions, an important goal is to update the technical content. The running example is the ARM Cortex A53 and the Intel Core i7, reflecting our post-PC era. Other highlights include a tutorial on GPUs that explains their unique terminology, more depth on the warehouse-scale computers that make up the cloud, and a deep dive into 10 Gigabyte Ethernet cards.

To keep the main book short and compatible with electronic books, we placed the optional material as online appendices instead of on a companion CD as in prior editions.

Finally, we updated all the exercises in the book.

While some elements changed, we have preserved useful book elements from prior editions. To make the book work better as a reference, we still place definitions of new terms in the margins at their first occurrence. The book element called

"Understanding Program Performance" sections helps readers understand the performance of their programs and how to improve it, just as the "Hardware/Software Interface" book element helped readers understand the tradeoffs at this interface. "The Big Picture" section remains so that the reader sees the forest despite all the trees. "Check Yourself" sections help readers to confirm their comprehension of the material on the first time through with answers provided at the end of each chapter. This edition still includes the green ARMv8 reference card, which was inspired by the "Green Card" of the IBM System/360. This card has been updated and should be a handy reference when writing ARMv8 assembly language programs.

Instructor Support[1]

We have collected a great deal of material to help instructors teach courses using this book. Solutions to exercises, figures from the book, lecture slides, and other materials are available to instructors who register with the publisher. In addition, the companion Web site provides links to a free Community Edition of ARM DS-5 professional software suite which contains an ARMv8-A (64-bit) architecture simulator, as well as additional advanced content for further study, appendices, glossary, references, and recommended reading. Check the publisher's Web site for more information:

booksite.elsevier.com/9780128017333

Concluding Remarks

If you read the following acknowledgments section, you will see that we went to great lengths to correct mistakes. Since a book goes through many printings, we have the opportunity to make even more corrections. If you uncover any remaining, resilient bugs, please contact the publisher by electronic mail at *codARMbugs@ mkp.com* or by low-tech mail using the address found on the copyright page.

This edition is the third break in the long-standing collaboration between Hennessy and Patterson, which started in 1989. The demands of running one of the world's great universities meant that President Hennessy could no longer make the substantial commitment to create a new edition. The remaining author felt once again like a tightrope walker without a safety net. Hence, the people in the acknowledgments and Berkeley colleagues played an even larger role in shaping the contents of this book. Nevertheless, this time around there is only one author to blame for the new material in what you are about to read.

Acknowledgments

With every edition of this book, we are very fortunate to receive help from many readers, reviewers, and contributors. Each of these people has helped to make this book better.

We are grateful for the assistance of **Khaled Benkrid** and his colleagues at ARM Ltd., who carefully reviewed the ARM-related material and provided helpful feedback.

---

[1] 关于本书教辅资源，只有使用本书作为教材的教师才可以申请，需要的教师请访问爱思唯尔的教材网站 https://textbooks.elsevier.com/ 进行申请。——编辑注

Chapter 6 was so extensively revised that we did a separate review for ideas and contents, and I made changes based on the feedback from every reviewer. I'd like to thank **Christos Kozyrakis** of Stanford University for suggesting using the network interface for clusters to demonstrate the hardware–software interface of I/O and for suggestions on organizing the rest of the chapter; **Mario Flagsilk** of Stanford University for providing details, diagrams, and performance measurements of the NetFPGA NIC; and the following for suggestions on how to improve the chapter: **David Kaeli** of Northeastern University, **Partha Ranganathan** of HP Labs, **David Wood** of the University of Wisconsin, and my Berkeley colleagues **Siamak Faridani**, **Shoaib Kamil**, **Yunsup Lee**, **Zhangxi Tan**, and **Andrew Waterman**.

Special thanks goes to **Rimas Avizenis** of UC Berkeley, who developed the various versions of matrix multiply and supplied the performance numbers as well. As I worked with his father while I was a graduate student at UCLA, it was a nice symmetry to work with Rimas at UCB.

I also wish to thank my longtime collaborator **Randy Katz** of UC Berkeley, who helped develop the concept of great ideas in computer architecture as part of the extensive revision of an undergraduate class that we did together.

I'd like to thank **David Kirk**, **John Nickolls**, and their colleagues at NVIDIA (Michael Garland, John Montrym, Doug Voorhies, Lars Nyland, Erik Lindholm, Paulius Micikevicius, Massimiliano Fatica, Stuart Oberman, and Vasily Volkov) for writing the first in-depth appendix on GPUs. I'd like to express again my appreciation to **Jim Larus**, recently named Dean of the School of Computer and Communications Science at EPFL, for his willingness in contributing his expertise on assembly language programming, as well as for welcoming readers of this book with regard to using the simulator he developed and maintains.

I am also very grateful to **Zachary Kurmas** of Grand Valley State University, who updated and created new exercises, based on originals created by **Perry Alexander** (The University of Kansas); **Jason Bakos** (University of South Carolina); **Javier Bruguera** (Universidade de Santiago de Compostela); **Matthew Farrens** (University of California, Davis); **David Kaeli** (Northeastern University); **Nicole Kaiyan** (University of Adelaide); **John Oliver** (Cal Poly, San Luis Obispo); **Milos Prvulovic** (Georgia Tech); **Jichuan Chang** (Google); **Jacob Leverich** (Stanford); **Kevin Lim** (Hewlett-Packard); and **Partha Ranganathan** (Google).

Additional thanks goes to **Jason Bakos** for updating the lecture slides.

I am grateful to the many instructors who have answered the publisher's surveys, reviewed our proposals, and attended focus groups to analyze and respond to our plans for this edition. They include the following individuals: Focus Groups: Bruce Barton (Suffolk County Community College), Jeff Braun (Montana Tech), Ed Gehringer (North Carolina State), Michael Goldweber (Xavier University), Ed Harcourt (St. Lawrence University), Mark Hill (University of Wisconsin, Madison), Patrick Homer (University of Arizona), Norm Jouppi (HP Labs), Dave Kaeli (Northeastern University), Christos Kozyrakis (Stanford University), Jae C. Oh (Syracuse University), Lu Peng (LSU), Milos Prvulovic (Georgia Tech), Partha Ranganathan (HP Labs), David Wood (University of Wisconsin), Craig Zilles (University of Illinois at Urbana-Champaign). Surveys and Reviews: Mahmoud Abou-Nasr (Wayne State University), Perry Alexander (The University of Kansas), Behnam Arad (Sacramento State University), Hakan Aydin (George Mason University), Hussein Badr (State University of New York at Stony Brook), Mac Baker (Virginia Military Institute), Ron Barnes (George Mason University), Douglas Blough (Georgia Institute of Technology), Kevin Bolding (Seattle Pacific University), Miodrag Bolic (University of Ottawa), John Bonomo (Westminster College), Jeff Braun (Montana Tech), Tom Briggs (Shippensburg University), Mike Bright (Grove City College), Scott Burgess (Humboldt State University), Fazli Can (Bilkent University), Warren R. Carithers (Rochester Institute of Technology), Bruce Carlton (Mesa Community College), Nicholas Carter (University of Illinois at Urbana-Champaign), Anthony Cocchi (The City University of New York), Don Cooley (Utah State University), Gene Cooperman (Northeastern University), Robert D. Cupper (Allegheny College), Amy Csizmar Dalal (Carleton College), Daniel Dalle (Université de Sherbrooke), Edward W. Davis (North Carolina State University), Nathaniel J. Davis (Air Force Institute of Technology), Molisa Derk (Oklahoma City University), Andrea Di Blas (Stanford University), Derek Eager (University of Saskatchewan), Ata Elahi (Souther Connecticut State University), Ernest Ferguson (Northwest Missouri State University), Rhonda Kay Gaede (The University of Alabama), Etienne M. Gagnon (L'Université du Québec à Montréal), Costa Gerousis (Christopher Newport University), Paul Gillard (Memorial University of Newfoundland), Michael Goldweber (Xavier University), Georgia Grant (College of San Mateo), Paul V. Gratz (Texas A&M University), Merrill Hall (The Master's College), Tyson Hall (Southern Adventist University), Ed Harcourt (St. Lawrence University), Justin E. Harlow (University of South Florida), Paul F. Hemler (Hampden-Sydney College), Jayantha Herath (St. Cloud State University), Martin Herbordt (Boston University), Steve J. Hodges (Cabrillo College), Kenneth Hopkinson (Cornell University), Bill Hsu (San Francisco State University), Dalton Hunkins (St. Bonaventure University), Baback Izadi (State University of New York—New Paltz), Reza Jafari, Robert W. Johnson (Colorado Technical University), Bharat Joshi (University of North Carolina, Charlotte), Nagarajan Kandasamy (Drexel University), Rajiv Kapadia, Ryan Kastner (University of California, Santa Barbara), E.J. Kim (Texas A&M University), Jihong Kim (Seoul National University), Jim Kirk (Union University), Geoffrey S. Knauth (Lycoming College), Manish M. Kochhal (Wayne State), Suzan Koknar-Tezel (Saint Joseph's University), Angkul Kongmunvattana (Columbus State University), April Kontostathis (Ursinus College), Christos Kozyrakis (Stanford University), Danny Krizanc (Wesleyan University), Ashok Kumar, S. Kumar (The University of Texas), Zachary Kurmas (Grand Valley State University), Adrian Lauf (University of Louisville), Robert N. Lea (University of Houston), Alvin Lebeck (Duke University), Baoxin Li (Arizona State University), Li Liao (University of Delaware), Gary Livingston (University of Massachusetts), Michael Lyle, Douglas W. Lynn (Oregon Institute of Technology), Yashwant K Malaiya (Colorado State University), Stephen Mann (University of Waterloo), Bill Mark (University of Texas at Austin), Ananda Mondal (Claflin University), Alvin Moser (Seattle University),

Walid Najjar (University of California, Riverside), Vijaykrishnan Narayanan (Penn State University), Danial J. Neebel (Loras College), Victor Nelson (Auburn University), John Nestor (Lafayette College), Jae C. Oh (Syracuse University), Joe Oldham (Centre College), Timour Paltashev, James Parkerson (University of Arkansas), Shaunak Pawagi (SUNY at Stony Brook), Steve Pearce, Ted Pedersen (University of Minnesota), Lu Peng (Louisiana State University), Gregory D. Peterson (The University of Tennessee), William Pierce (Hood College), Milos Prvulovic (Georgia Tech), Partha Ranganathan (HP Labs), Dejan Raskovic (University of Alaska, Fairbanks) Brad Richards (University of Puget Sound), Roman Rozanov, Louis Rubinfield (Villanova University), Md Abdus Salam (Southern University), Augustine Samba (Kent State University), Robert Schaefer (Daniel Webster College), Carolyn J. C. Schauble (Colorado State University), Keith Schubert (CSU San Bernardino), William L. Schultz, Kelly Shaw (University of Richmond), Shahram Shirani (McMaster University), Scott Sigman (Drury University), Shai Simonson (Stonehill College), Bruce Smith, David Smith, Jeff W. Smith (University of Georgia, Athens), Mark Smotherman (Clemson University), Philip Snyder (Johns Hopkins University), Alex Sprintson (Texas A&M), Timothy D. Stanley (Brigham Young University), Dean Stevens (Morningside College), Nozar Tabrizi (Kettering University), Yuval Tamir (UCLA), Alexander Taubin (Boston University), Will Thacker (Winthrop University), Mithuna Thottethodi (Purdue University), Manghui Tu (Southern Utah University), Dean Tullsen (UC San Diego), Steve VanderLeest (Calvin College), Christopher Vickery (Queens College of CUNY), Rama Viswanathan (Beloit College), Ken Vollmar (Missouri State University), Guoping Wang (Indiana-Purdue University), Patricia Wenner (Bucknell University), Kent Wilken (University of California, Davis), David Wolfe (Gustavus Adolphus College), David Wood (University of Wisconsin, Madison), Ki Hwan Yum (University of Texas, San Antonio), Mohamed Zahran (City College of New York), Amr Zaky (Santa Clara University), Gerald D. Zarnett (Ryerson University), Nian Zhang (South Dakota School of Mines & Technology), Jiling Zhong (Troy University), Huiyang Zhou (North Carolina State University), Weiyu Zhu (Illinois Wesleyan University).

A special thanks also goes to **Mark Smotherman** for making multiple passes to find technical and writing glitches that significantly improved the quality of this edition.

We wish to thank the extended Morgan Kaufmann family for agreeing to publish this book again under the able leadership of **Todd Green**, **Steve Merken** and **Nate McFadden**: I certainly couldn't have completed the book without them. We also want to extend thanks to **Lisa Jones**, who managed the book production process, and **Matthew Limbert**, who did the cover design. The cover cleverly connects the post-PC era content of this edition to the cover of the first edition.

The contributions of the nearly 150 people we mentioned here have helped make this new edition what I hope will be our best book yet. Enjoy!

David A. Patterson

David A. Patterson has been teaching computer architecture at the University of California, Berkeley, since joining the faculty in 1976, where he holds the Pardee Chair of Computer Science. His teaching has been honored by the Distinguished Teaching Award from the University of California, the Karlstrom Award from ACM, and the Mulligan Education Medal and Undergraduate Teaching Award from IEEE. Patterson received the IEEE Technical Achievement Award and the ACM Eckert-Mauchly Award for contributions to RISC, and he shared the IEEE Johnson Information Storage Award for contributions to RAID. He also shared the IEEE John von Neumann Medal and the C & C Prize with John Hennessy. Like his co-author, Patterson is a Fellow of the American Academy of Arts and Sciences, the Computer History Museum, ACM, and IEEE, and he was elected to the National Academy of Engineering, the National Academy of Sciences, and the Silicon Valley Engineering Hall of Fame. He served on the Information Technology Advisory Committee to the U.S. President, as chair of the CS division in the Berkeley EECS department, as chair of the Computing Research Association, and as President of ACM. This record led to Distinguished Service Awards from ACM, CRA, and SIGARCH.

At Berkeley, Patterson led the design and implementation of RISC I, likely the first VLSI reduced instruction set computer, and the foundation of the commercial SPARC architecture. He was a leader of the Redundant Arrays of Inexpensive Disks (RAID) project, which led to dependable storage systems from many companies. He was also involved in the Network of Workstations (NOW) project, which led to cluster technology used by Internet companies and later to cloud computing. These projects earned four dissertation awards from ACM. His current research projects are Algorithm-Machine-People and Algorithms and Specializers for Provably Optimal Implementations with Resilience and Efficiency. The AMP Lab is developing scalable machine learning algorithms, warehouse-scale-computer-friendly programming models, and crowd-sourcing tools to gain valuable insights quickly from big data in the cloud. The ASPIRE Lab uses deep hardware and software co-tuning to achieve the highest possible performance and energy efficiency for mobile and rack computing systems.

John L. Hennessy is the tenth president of Stanford University, where he has been a member of the faculty since 1977 in the departments of electrical engineering and computer science. Hennessy is a Fellow of the IEEE and ACM; a member of the National Academy of Engineering, the National Academy of Science, and the American Philosophical Society; and a Fellow of the American Academy of Arts and Sciences. Among his many awards are the 2001 Eckert-Mauchly Award for his contributions to RISC technology, the 2001 Seymour Cray Computer Engineering Award, and the 2000 John von Neumann Award, which he shared with David Patterson. He has also received seven honorary doctorates.

In 1981, he started the MIPS project at Stanford with a handful of graduate students. After completing the project in 1984, he took a leave from the university to cofound MIPS Computer Systems (now MIPS Technologies), which developed one of the first commercial RISC microprocessors. As of 2006, over 2 billion MIPS microprocessors have been shipped in devices ranging from video games and palmtop computers to laser printers and network switches. Hennessy subsequently led the DASH (Director Architecture for Shared Memory) project, which prototyped the first scalable cache coherent multiprocessor; many of the key ideas have been adopted in modern multiprocessors. In addition to his technical activities and university responsibilities, he has continued to work with numerous start-ups, both as an early-stage advisor and an investor.

Contents

3 Arithmetic for Computers   186

4 The Processor   254

5 Large and Fast: Exploiting Memory Hierarchy 386

6 Parallel Processors from Client to Cloud 514

A P P E N D I X

A   The Basics of Logic Design   A-2

O N L I N E   C O N T E N T

B   Graphics and Computing GPUs   B-2

Mapping Control to Hardware   C-2

A Survey of RISC Architectures for Desktop, Server, and Embedded Computers   D-2

Computer Abstractions and Technology

Civilization advances by extending the number of important operations which we can perform without thinking about them.

Alfred North Whitehead,
An Introduction to Mathematics, 1911

1.1   Introduction

Welcome to this book! We're delighted to have this opportunity to convey the excitement of the world of computer systems. This is not a dry and dreary field, where progress is glacial and where new ideas atrophy from neglect. No! Computers are the product of the incredibly vibrant information technology industry, all aspects of which are responsible for almost 10% of the gross national product of the United States, and whose economy has become dependent in part on the rapid improvements in information technology promised by Moore's Law. This unusual industry embraces innovation at a breath-taking rate. In the last 30 years, there have been a number of new computers whose introduction appeared to revolutionize the computing industry; these revolutions were cut short only because someone else built an even better computer.

This race to innovate has led to unprecedented progress since the inception of electronic computing in the late 1940s. Had the transportation industry kept pace with the computer industry, for example, today we could travel from New York to London in a second for a penny. Take just a moment to contemplate how such an improvement would change society—living in Tahiti while working in San Francisco, going to Moscow for an evening at the Bolshoi Ballet—and you can appreciate the implications of such a change.

Computers have led to a third revolution for civilization, with the information revolution taking its place alongside the agricultural and industrial revolutions. The resulting multiplication of humankind's intellectual strength and reach naturally has affected our everyday lives profoundly and changed the ways in which the search for new knowledge is carried out. There is now a new vein of scientific investigation, with computational scientists joining theoretical and experimental scientists in the exploration of new frontiers in astronomy, biology, chemistry, and physics, among others.

The computer revolution continues. Each time the cost of computing improves by another factor of 10, the opportunities for computers multiply. Applications that were economically infeasible suddenly become practical. In the recent past, the following applications were "computer science fiction."

- *Computers in automobiles*: Until microprocessors improved dramatically in price and performance in the early 1980s, computer control of cars was ludicrous. Today, computers reduce pollution, improve fuel efficiency via engine controls, and increase safety through blind spot warnings, lane departure warnings, moving object detection, and air bag inflation to protect occupants in a crash.

- *Cell phones*: Who would have dreamed that advances in computer systems would lead to more than half of the planet having mobile phones, allowing person-to-person communication to almost anyone anywhere in the world?

- *Human genome project*: The cost of computer equipment to map and analyze human DNA sequences was hundreds of millions of dollars. It's unlikely that anyone would have considered this project had the computer costs been 10 to 100 times higher, as they would have been 15 to 25 years earlier. Moreover, costs continue to drop; you will soon be able to acquire your own genome, allowing medical care to be tailored to you.

- *World Wide Web*: Not in existence at the time of the first edition of this book, the web has transformed our society. For many, the web has replaced libraries and newspapers.

- *Search engines*: As the content of the web grew in size and in value, finding relevant information became increasingly important. Today, many people rely on search engines for such a large part of their lives that it would be a hardship to go without them.

Clearly, advances in this technology now affect almost every aspect of our society. Hardware advances have allowed programmers to create wonderfully useful software, which explains why computers are omnipresent. Today's science fiction suggests tomorrow's killer applications: already on their way are glasses that augment reality, the cashless society, and cars that can drive themselves.

Traditional Classes of Computing Applications and Their Characteristics

Although a common set of hardware technologies (see Sections 1.4 and 1.5) is used in computers ranging from smart home appliances to cell phones to the largest supercomputers, these different applications have distinct design requirements and employ the core hardware technologies in different ways. Broadly speaking, computers are used in three dissimilar classes of applications.

Personal computers (PCs) are possibly the best-known form of computing, which readers of this book have likely used extensively. Personal computers emphasize delivery of good performance to single users at low cost and usually execute third-party software. This class of computing drove the evolution of many computing technologies, which is merely 35 years old!

Servers are the modern form of what were once much larger computers, and are usually accessed only via a network. Servers are oriented to carrying sizable workloads, which may consist of either single complex applications—usually a scientific or engineering application—or handling many small jobs, such as would occur in building a large web server. These applications are usually based on software from another source (such as a database or simulation system), but are often modified or customized for a particular function. Servers are built from the same basic technology as desktop computers, but provide for greater computing, storage, and input/output capacity. In general, servers also place a higher emphasis on dependability, since a crash is usually more costly than it would be on a single-user PC.

Servers span the widest range in cost and capability. At the low end, a server may be little more than a desktop computer without a screen or keyboard and cost a thousand dollars. These low-end servers are typically used for file storage, small business applications, or simple web serving. At the other extreme are supercomputers, which at the present consist of tens of thousands of processors and many terabytes of memory, and cost tens to hundreds of millions of dollars. Supercomputers are usually used for high-end scientific and engineering calculations, such as weather forecasting, oil exploration, protein structure determination, and other large-scale problems. Although such supercomputers represent the peak of computing capability, they represent a relatively small fraction of the servers and thus a proportionally tiny fraction of the overall computer market in terms of total revenue.

Embedded computers are the largest class of computers and span the widest range of applications and performance. Embedded computers include the microprocessors found in your car, the computers in a television set, and the networks of processors that control a modern airplane or cargo ship. Embedded computing systems are designed to run one application or one set of related applications that are normally integrated with the hardware and delivered as a single system; thus, despite the large number of embedded computers, most users never really see that they are using a computer!

personal computer (PC) A computer designed for use by an individual, usually incorporating a graphics display, a keyboard, and a mouse.

server A computer used for running larger programs for multiple users, often simultaneously, and typically accessed only via a network.

supercomputer A class of computers with the highest performance and cost; they are configured as servers and typically cost tens to hundreds of millions of dollars.

terabyte (TB) Originally 1,099,511,627,776 (2^{40}) bytes, although communications and secondary storage systems developers started using the term to mean 1,000,000,000,000 (10^{12}) bytes. To reduce confusion, we now use the term tebibyte (TiB) for 2^{40} bytes, defining terabyte (TB) to mean 10^{12} bytes. Figure 1.1 shows the full range of decimal and binary values and names.

embedded computer A computer inside another device used for running one predetermined application or collection of software.

Decimal term	Abbreviation	Value	Binary term	Abbreviation	Value	% Larger
kilobyte	KB	10^3	kibibyte	KiB	2^{10}	2%
megabyte	MB	10^6	mebibyte	MiB	2^{20}	5%
gigabyte	GB	10^9	gibibyte	GiB	2^{30}	7%
terabyte	TB	10^{12}	tebibyte	TiB	2^{40}	10%
petabyte	PB	10^{15}	pebibyte	PiB	2^{50}	13%
exabyte	EB	10^{18}	exbibyte	EiB	2^{60}	15%
zettabyte	ZB	10^{21}	zebibyte	ZiB	2^{70}	18%
yottabyte	YB	10^{24}	yobibyte	YiB	2^{80}	21%

FIGURE 1.1   The 2^X vs. 10^Y bytes ambiguity was resolved by adding a binary notation for all the common size terms. In the last column we note how much larger the binary term is than its corresponding decimal term, which is compounded as we head down the chart. These prefixes work for bits as well as bytes, so *gigabit* (Gb) is 10^9 bits while *gibibits* (Gib) is 2^{30} bits.

Embedded applications often have unique application requirements that combine a minimum performance with stringent limitations on cost or power. For example, consider a music player: the processor need only to be as fast as necessary to handle its limited function, and beyond that, minimizing cost and power is the most important objective. Despite their low cost, embedded computers often have lower tolerance for failure, since the results can vary from upsetting (when your new television crashes) to devastating (such as might occur when the computer in a plane or cargo ship crashes). In consumer-oriented embedded applications, such as a digital home appliance, dependability is achieved primarily through simplicity—the emphasis is on doing one function as perfectly as possible. In large embedded systems, techniques of redundancy from the server world are often employed. Although this book focuses on general-purpose computers, most concepts apply directly, or with slight modifications, to embedded computers.

Elaboration: Elaborations are short sections used throughout the text to provide more detail on a particular subject that may be of interest. Disinterested readers may skip over an elaboration, since the subsequent material will never depend on the contents of the elaboration.

Many embedded processors are designed using *processor cores*, a version of a processor written in a hardware description language, such as Verilog or VHDL (see Chapter 4). The core allows a designer to integrate other application-specific hardware with the processor core for fabrication on a single chip.

Welcome to the Post-PC Era

The continuing march of technology brings about generational changes in computer hardware that shake up the entire information technology industry. Since the last edition of the book, we have undergone such a change, as significant in the past as the switch starting 30 years ago to personal computers. Replacing the

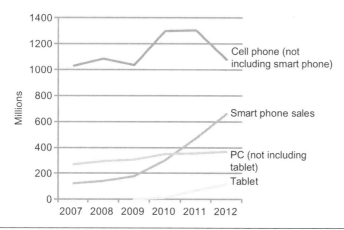

FIGURE 1.2 The number manufactured per year of tablets and smart phones, which reflect the post-PC era, versus personal computers and traditional cell phones. Smart phones represent the recent growth in the cell phone industry, and they passed PCs in 2011. Tablets are the fastest growing category, nearly doubling between 2011 and 2012. Recent PCs and traditional cell phone categories are relatively flat or declining.

PC is the **personal mobile device (PMD)**. PMDs are battery operated with wireless connectivity to the Internet and typically cost hundreds of dollars, and, like PCs, users can download software ("apps") to run on them. Unlike PCs, they no longer have a keyboard and mouse, and are more likely to rely on a touch-sensitive screen or even speech input. Today's PMD is a smart phone or a tablet computer, but tomorrow it may include electronic glasses. Figure 1.2 shows the rapid growth over time of tablets and smart phones versus that of PCs and traditional cell phones.

Taking over from the conventional server is **Cloud Computing**, which relies upon giant datacenters that are now known as *Warehouse Scale Computers* (WSCs). Companies like Amazon and Google build these WSCs containing 100,000 servers and then let companies rent portions of them so that they can provide software services to PMDs without having to build WSCs of their own. Indeed, **Software as a Service (SaaS)** deployed via the Cloud is revolutionizing the software industry just as PMDs and WSCs are revolutionizing the hardware industry. Today's software developers will often have a portion of their application that runs on the PMD and a portion that runs in the Cloud.

What You Can Learn in This Book

Successful programmers have always been concerned about the performance of their programs, because getting results to the user quickly is critical in creating popular software. In the 1960s and 1970s, a primary constraint on computer performance was the size of the computer's memory. Thus, programmers often followed a simple credo: minimize memory space to make programs fast. In the

Personal mobile devices (PMDs) are small wireless devices to connect to the Internet; they rely on batteries for power, and software is installed by downloading apps. Conventional examples are smart phones and tablets.

Cloud Computing refers to large collections of servers that provide services over the Internet; some providers rent dynamically varying numbers of servers as a utility.

Software as a Service (SaaS) delivers software and data as a service over the Internet, usually via a thin program such as a browser that runs on local client devices, instead of binary code that must be installed, and runs wholly on that device. Examples include web search and social networking.

last decade, advances in computer design and memory technology have greatly reduced the importance of small memory size in most applications other than those in embedded computing systems.

Programmers interested in performance now need to understand the issues that have replaced the simple memory model of the 1960s: the parallel nature of processors and the hierarchical nature of memories. We demonstrate the importance of this understanding in Chapters 3 to 6 by showing how to improve performance of a C program by a factor of 200. Moreover, as we explain in Section 1.7, today's programmers need to worry about energy efficiency of their programs running either on the PMD or in the Cloud, which also requires understanding what is below your code. Programmers who seek to build competitive versions of software will therefore need to increase their knowledge of computer organization.

We are honored to have the opportunity to explain what's inside this revolutionary machine, unraveling the software below your program and the hardware under the covers of your computer. By the time you complete this book, we believe you will be able to answer the following questions:

■ How are programs written in a high-level language, such as C or Java, translated into the language of the hardware, and how does the hardware execute the resulting program? Comprehending these concepts forms the basis of understanding the aspects of both the hardware and software that affect program performance.

■ What is the interface between the software and the hardware, and how does software instruct the hardware to perform needed functions? These concepts are vital to understanding how to write many kinds of software.

■ What determines the performance of a program, and how can a programmer improve the performance? As we will see, this depends on the original program, the software translation of that program into the computer's language, and the effectiveness of the hardware in executing the program.

■ What techniques can be used by hardware designers to improve performance? This book will introduce the basic concepts of modern computer design. The interested reader will find much more material on this topic in our advanced book, *Computer Architecture: A Quantitative Approach*.

■ What techniques can be used by hardware designers to improve energy efficiency? What can the programmer do to help or hinder energy efficiency?

multicore microprocessor
A microprocessor containing multiple processors ("cores") in a single integrated circuit.

■ What are the reasons for and the consequences of the recent switch from sequential processing to parallel processing? This book gives the motivation, describes the current hardware mechanisms to support parallelism, and surveys the new generation of "multicore" microprocessors (see Chapter 6).

■ Since the first commercial computer in 1951, what great ideas did computer architects come up with that lay the foundation of modern computing?

Without understanding the answers to these questions, improving the performance of your program on a modern computer or evaluating what features might make one computer better than another for a particular application will be a complex process of trial and error, rather than a scientific procedure driven by insight and analysis.

This first chapter lays the foundation for the rest of the book. It introduces the basic ideas and definitions, places the major components of software and hardware in perspective, shows how to evaluate performance and energy, introduces integrated circuits (the technology that fuels the computer revolution), and explains the shift to multicores.

In this chapter and later ones, you will likely see many new words, or words that you may have heard but are not sure what they mean. Don't panic! Yes, there is a lot of special terminology used in describing modern computers, but the terminology actually helps, since it enables us to describe precisely a function or capability. In addition, computer designers (including your authors) *love* using acronyms, which are *easy* to understand once you know what the letters stand for! To help you remember and locate terms, we have included a highlighted definition of every term in the margins the first time it appears in the text. After a short time of working with the terminology, you will be fluent, and your friends will be impressed as you correctly use acronyms such as BIOS, CPU, DIMM, DRAM, PCIe, SATA, and many others.

To reinforce how the software and hardware systems used to run a program will affect performance, we use a special section, *Understanding Program Performance*, throughout the book to summarize important insights into program performance. The first one appears below.

acronym A word constructed by taking the initial letters of a string of words. For example: RAM is an acronym for Random Access Memory, and CPU is an acronym for Central Processing Unit.

Understanding Program Performance

The performance of a program depends on a combination of the effectiveness of the algorithms used in the program, the software systems used to create and translate the program into machine instructions, and the effectiveness of the computer in executing those instructions, which may include *input/output* (I/O) operations. This table summarizes how the hardware and software affect performance.

Hardware or software component	How this component affects performance	Where is this topic covered?
Algorithm	Determines both the number of source-level statements and the number of I/O operations executed	Other books!
Programming language, compiler, and architecture	Determines the number of computer instructions for each source-level statement	Chapters 2 and 3
Processor and memory system	Determines how fast instructions can be executed	Chapters 4, 5, and 6
I/O system (hardware and operating system)	Determines how fast I/O operations may be executed	Chapters 4, 5, and 6

To demonstrate the impact of the ideas in this book, as mentioned above, we improve the performance of a C program that multiplies a matrix times a vector in a sequence of chapters. Each step leverages understanding how the underlying hardware really works in a modern microprocessor to improve performance by a factor of 200!

- In the category of *data level parallelism*, in Chapter 3 we use *subword parallelism via C intrinsics* to increase performance by a factor of 3.8.

- In the category of *instruction level parallelism*, in Chapter 4 we use *loop unrolling to exploit multiple instruction issue and out-of-order execution hardware* to increase performance by another factor of 2.3.

- In the category of *memory hierarchy optimization*, in Chapter 5 we use *cache blocking* to increase performance on large matrices by another factor of 2.0 to 2.5.

- In the category of *thread level parallelism*, in Chapter 6 we use *parallel for loops in OpenMP to exploit multicore hardware* to increase performance by another factor of 4 to 14.

Check Yourself

Check Yourself sections are designed to help readers assess whether they comprehend the major concepts introduced in a chapter and understand the implications of those concepts. Some *Check Yourself* questions have simple answers; others are for discussion among a group. Answers to the specific questions can be found at the end of the chapter. *Check Yourself* questions appear only at the end of a section, making it easy to skip them if you are sure you understand the material.

1. The number of embedded processors sold every year greatly outnumbers the number of PC and even post-PC processors. Can you confirm or deny this insight based on your own experience? Try to count the number of embedded processors in your home. How does it compare with the number of conventional computers in your home?

2. As mentioned earlier, both the software and hardware affect the performance of a program. Can you think of examples where each of the following is the right place to look for a performance bottleneck?

- The algorithm chosen
- The programming language or compiler
- The operating system
- The processor
- The I/O system and devices

1.2 Eight Great Ideas in Computer Architecture

We now introduce eight great ideas that computer architects have invented in the last 60 years of computer design. These ideas are so powerful they have lasted long after the first computer that used them, with newer architects demonstrating their admiration by imitating their predecessors. These great ideas are themes that we will weave through this and subsequent chapters as examples arise. To point out their influence, in this section we introduce icons and highlighted terms that represent the great ideas and we use them to identify the nearly 100 sections of the book that feature use of the great ideas.

Design for Moore's Law

The one constant for computer designers is rapid change, which is driven largely by **Moore's Law**. It states that integrated circuit resources double every 18–24 months. Moore's Law resulted from a 1965 prediction of such growth in IC capacity made by Gordon Moore, one of the founders of Intel. As computer designs can take years, the resources available per chip can easily double or quadruple between the start and finish of the project. Like a skeet shooter, computer architects must anticipate where the technology will be when the design finishes rather than design for where it starts. We use an "up and to the right" Moore's Law graph to represent designing for rapid change.

MOORE'S LAW

Use Abstraction to Simplify Design

Both computer architects and programmers had to invent techniques to make themselves more productive, for otherwise design time would lengthen as dramatically as resources grew by Moore's Law. A major productivity technique for hardware and software is to use **abstractions** to characterize the design at different levels of representation; lower-level details are hidden to offer a simpler model at higher levels. We'll use the abstract painting icon to represent this second great idea.

ABSTRACTION

Make the Common Case Fast

Making the **common case fast** will tend to enhance performance better than optimizing the rare case. Ironically, the common case is often simpler than the rare case and hence is usually easier to enhance. This common sense advice implies that you know what the common case is, which is only possible with careful experimentation and measurement (see Section 1.6). We use a sports car as the icon for making the common case fast, as the most common trip has one or two passengers, and it's surely easier to make a fast sports car than a fast minivan!

COMMON CASE FAST

PARALLELISM

Performance via Parallelism

Since the dawn of computing, computer architects have offered designs that get more performance by computing operations in parallel. We'll see many examples of parallelism in this book. We use multiple jet engines of a plane as our icon for **parallel performance**.

Performance via Pipelining

A particular pattern of parallelism is so prevalent in computer architecture that it merits its own name: **pipelining**. For example, before fire engines, a "bucket brigade" would respond to a fire, which many cowboy movies show in response to a dastardly act by the villain. The townsfolk form a human chain to carry a water source to fire, as they could much more quickly move buckets up the chain instead of individuals running back and forth. Our pipeline icon is a sequence of pipes, with each section representing one stage of the pipeline.

PIPELINING

Performance via Prediction

Following the saying that it can be better to ask for forgiveness than to ask for permission, the next great idea is **prediction**. In some cases, it can be faster on average to guess and start working rather than wait until you know for sure, assuming that the mechanism to recover from a misprediction is not too expensive and your prediction is relatively accurate. We use the fortune-teller's crystal ball as our prediction icon.

PREDICTION

Hierarchy of Memories

Programmers want the memory to be fast, large, and cheap, as memory speed often shapes performance, capacity limits the size of problems that can be solved, and the cost of memory today is often the majority of computer cost. Architects have found that they can address these conflicting demands with a **hierarchy of memories**, with the fastest, smallest, and the most expensive memory per bit at the top of the hierarchy and the slowest, largest, and cheapest per bit at the bottom. As we shall see in Chapter 5, caches give the programmer the illusion that main memory is almost as fast as the top of the hierarchy and nearly as big and cheap as the bottom of the hierarchy. We use a layered triangle icon to represent the memory hierarchy. The shape indicates speed, cost, and size: the closer to the top, the faster and more expensive per bit the memory; the wider the base of the layer, the bigger the memory.

HIERARCHY

Dependability via Redundancy

Computers not only need to be fast; they need to be dependable. Since any physical device can fail, we make systems **dependable** by including redundant components that can take over when a failure occurs *and* to help detect failures. We use the tractor-trailer as our icon, since the dual tires on each side of its rear axles allow the truck to continue driving even when one tire fails. (Presumably, the truck driver heads immediately to a repair facility so the flat tire can be fixed, thereby restoring redundancy!)

DEPENDABILITY

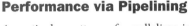

1.3 Below Your Program

A typical application, such as a word processor or a large database system, may consist of millions of lines of code and rely on sophisticated software libraries that implement complex functions in support of the application. As we will see, the hardware in a computer can only execute extremely simple low-level instructions. To go from a complex application to the primitive instructions involves several layers of software that interpret or translate high-level operations into simple computer instructions, an example of the great idea of **abstraction**.

Figure 1.3 shows that these layers of software are organized primarily in a hierarchical fashion, with applications being the outermost ring and a variety of systems software sitting between the hardware and the application software.

There are many types of systems software, but two types of systems software are central to every computer system today: an operating system and a compiler. An operating system interfaces between a user's program and the hardware and provides a variety of services and supervisory functions. Among the most important functions are:

- Handling basic input and output operations

- Allocating storage and memory

- Providing for protected sharing of the computer among multiple applications using it simultaneously

Examples of operating systems in use today are Linux, iOS, and Windows.

A B S T R A C T I O N

In Paris they simply stared when I spoke to them in French; I never did succeed in making those idiots understand their own language.

Mark Twain, *The Innocents Abroad*, 1869

systems software Software that provides services that are commonly useful, including operating systems, compilers, loaders, and assemblers.

operating system Supervising program that manages the resources of a computer for the benefit of the programs that run on that computer.

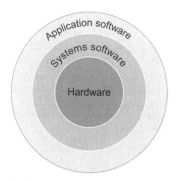

FIGURE 1.3 A simplified view of hardware and software as hierarchical layers, shown as concentric circles with hardware in the center and application software outermost. In complex applications, there are often multiple layers of application software as well. For example, a database system may run on top of the systems software hosting an application, which in turn runs on top of the database.

compiler A program that translates high-level language statements into assembly language statements.

Compilers perform another vital function: the translation of a program written in a high-level language, such as C, C++, Java, or Visual Basic into instructions that the hardware can execute. Given the sophistication of modern programming languages and the simplicity of the instructions executed by the hardware, the translation from a high-level language program to hardware instructions is complex. We give a brief overview of the process here and then go into more depth in Chapter 2.

From a High-Level Language to the Language of Hardware

To speak directly to electronic hardware, you need to send electrical signals. The easiest signals for computers to understand are *on* and *off*, and so the computer alphabet is just two letters. Just as the 26 letters of the English alphabet do not limit how much can be written, the two letters of the computer alphabet do not limit what computers can do. The two symbols for these two letters are the numbers 0 and 1, and we commonly think of the computer language as numbers in base 2, or *binary numbers*. We refer to each "letter" as a binary digit or bit. Computers are slaves to our commands, which are called instructions. Instructions, which are just collections of bits that the computer understands and obeys, can be thought of as numbers. For example, the bits

binary digit Also called a bit. One of the two numbers in base 2 (0 or 1) that are the components of information.

    1000110010100000

instruction A command that computer hardware understands and obeys.

tell one computer to add two numbers. Chapter 2 explains why we use numbers for instructions *and* data; we don't want to steal that chapter's thunder, but using numbers for both instructions and data is a foundation of computing.

The first programmers communicated to computers in binary numbers, but this was so tedious that they quickly invented new notations that were closer to the way humans think. At first, these notations were translated to binary by hand, but this process was still tiresome. Using the computer to help program the computer, the pioneers invented software to translate from symbolic notation to binary. The first of these programs was named an assembler. This program translates a symbolic version of an instruction into the binary version. For example, the programmer would write

assembler A program that translates a symbolic version of instructions into the binary version.

    ADD A,B

and the assembler would translate this notation into

    1000110010100000

assembly language A symbolic representation of machine instructions.

machine language A binary representation of machine instructions.

This instruction tells the computer to add the two numbers A and B. The name coined for this symbolic language, still used today, is assembly language. In contrast, the binary language that the machine understands is the machine language.

Although a tremendous improvement, assembly language is still far from the notations a scientist might like to use to simulate fluid flow or that an accountant might use to balance the books. Assembly language requires the programmer to write one line for every instruction that the computer will follow, forcing the programmer to think like the computer.

The recognition that a program could be written to translate a more powerful language into computer instructions was one of the great breakthroughs in the early days of computing. Programmers today owe their productivity—and their sanity—to the creation of high-level programming languages and compilers that translate programs in such languages into instructions. Figure 1.4 shows the relationships among these programs and languages, which are more examples of the power of **abstraction**.

ABSTRACTION

high-level programming language A portable language such as C, C++, Java, or Visual Basic that is composed of words and algebraic notation that can be translated by a compiler into assembly language.

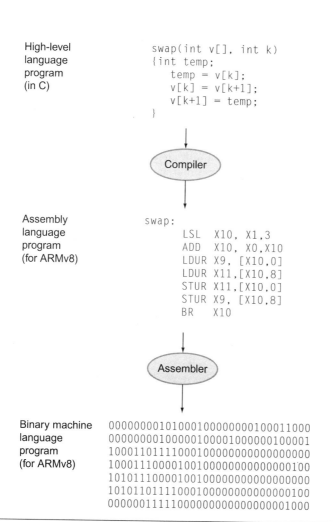

High-level language program (in C)

```
swap(int v[], int k)
{int temp;
    temp = v[k];
    v[k] = v[k+1];
    v[k+1] = temp;
}
```

Compiler

Assembly language program (for ARMv8)

```
swap:
    LSL  X10, X1,3
    ADD  X10, X0,X10
    LDUR X9, [X10,0]
    LDUR X11,[X10,8]
    STUR X11,[X10,0]
    STUR X9, [X10,8]
    BR   X10
```

Assembler

Binary machine language program (for ARMv8)

```
00000000101000100000000100011000
00000000100000100001000000100001
10001101111000100000000000000000
10001110000100100000000000000100
10101110000100100000000000000000
10101101111000100000000000000100
00000011111000000000000000001000
```

FIGURE 1.4 C program compiled into assembly language and then assembled into binary machine language. Although the translation from high-level language to binary machine language is shown in two steps, some compilers cut out the middleman and produce binary machine language directly. These languages and this program are examined in more detail in Chapter 2.

A compiler enables a programmer to write this high-level language expression:

    A + B

The compiler would compile it into this assembly language statement:

    ADD A,B

As shown above, the assembler would translate this statement into the binary instructions that tell the computer to add the two numbers A and B.

High-level programming languages offer several important benefits. First, they allow the programmer to think in a more natural language, using English words and algebraic notation, resulting in programs that look much more like text than like tables of cryptic symbols (see Figure 1.4). Moreover, they allow languages to be designed according to their intended use. Hence, Fortran was designed for scientific computation, Cobol for business data processing, Lisp for symbol manipulation, and so on. There are also domain-specific languages for even narrower groups of users, such as those interested in simulation of fluids, for example.

The second advantage of programming languages is improved programmer productivity. One of the few areas of widespread agreement in software development is that it takes less time to develop programs when they are written in languages that require fewer lines to express an idea. Conciseness is a clear advantage of high-level languages over assembly language.

The final advantage is that programming languages allow programs to be independent of the computer on which they were developed, since compilers and assemblers can translate high-level language programs to the binary instructions of any computer. These three advantages are so strong that today little programming is done in assembly language.

1.4 Under the Covers

Now that we have looked below your program to uncover the underlying software, let's open the covers of your computer to learn about the underlying hardware. The underlying hardware in any computer performs the same basic functions: inputting data, outputting data, processing data, and storing data. How these functions are performed is the primary topic of this book, and subsequent chapters deal with different parts of these four tasks.

input device A mechanism through which the computer is fed information, such as a keyboard.

output device A mechanism that conveys the result of a computation to a user, such as a display, or to another computer.

When we come to an important point in this book, a point so significant that we hope you will remember it forever, we emphasize it by identifying it as a *Big Picture* item. We have about a dozen Big Pictures in this book, the first being the five components of a computer that perform the tasks of inputting, outputting, processing, and storing data.

Two key components of computers are input devices, such as the microphone, and output devices, such as the speaker. As the names suggest, input feeds the computer, and output is the result of computation sent to the user. Some devices, such as wireless networks, provide both input and output to the computer.

Chapters 5 and 6 describe *input/output* (I/O) devices in more detail, but let's take an introductory tour through the computer hardware, starting with the external I/O devices.

The five classic components of a computer are input, output, memory, datapath, and control, with the last two sometimes combined and called the processor. Figure 1.5 shows the standard organization of a computer. This organization is independent of hardware technology: you can place every piece of every computer, past and present, into one of these five categories. To help you keep all this in perspective, the five components of a computer are shown on the front page of each of the following chapters, with the portion of interest to that chapter highlighted.

The **BIG**
Picture

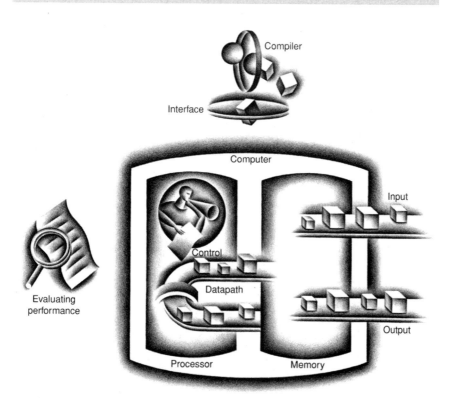

FIGURE 1.5   The organization of a computer, showing the five classic components. The processor gets instructions and data from memory. Input writes data to memory, and output reads data from memory. Control sends the signals that determine the operations of the datapath, memory, input, and output.

Through the Looking Glass

liquid crystal display (LCD) A display technology using a thin layer of liquid polymers that can be used to transmit or block light according to whether a charge is applied.

active matrix display A liquid crystal display using a transistor to control the transmission of light at each individual pixel.

pixel The smallest individual picture element. Screens are composed of hundreds of thousands to millions of pixels, organized in a matrix.

Through computer displays I have landed an airplane on the deck of a moving carrier, observed a nuclear particle hit a potential well, flown in a rocket at nearly the speed of light and watched a computer reveal its innermost workings.

Ivan Sutherland, the "father" of computer graphics, Scientific American, 1984

The most fascinating I/O device is probably the graphics display. Most personal mobile devices use liquid crystal displays (LCDs) to get a thin, low-power display. The LCD is not the source of light; instead, it controls the transmission of light. A typical LCD includes rod-shaped molecules in a liquid that form a twisting helix that bends light entering the display, from either a light source behind the display or less often from reflected light. The rods straighten out when a current is applied and no longer bend the light. Since the liquid crystal material is between two screens polarized at 90 degrees, the light cannot pass through unless it is bent. Today, most LCD displays use an active matrix that has a tiny transistor switch at each pixel to control current precisely and make sharper images. A red-green-blue mask associated with each dot on the display determines the intensity of the three-color components in the final image; in a color active matrix LCD, there are three transistor switches at each point.

The image is composed of a matrix of picture elements, or pixels, which can be represented as a matrix of bits, called a *bit map*. Depending on the size of the screen and the resolution, the display matrix in a typical tablet ranges in size from 1024×768 to 2048×1536. A color display might use 8 bits for each of the three colors (red, blue, and green), for 24 bits per pixel, permitting millions of different colors to be displayed.

The computer hardware support for graphics consists mainly of a *raster refresh buffer*, or *frame buffer*, to store the bit map. The image to be represented onscreen is stored in the frame buffer, and the bit pattern per pixel is read out to the graphics display at the refresh rate. Figure 1.6 shows a frame buffer with a simplified design of just 4 bits per pixel.

The goal of the bit map is to represent faithfully what is on the screen. The challenges in graphics systems arise because the human eye is very good at detecting even subtle changes on the screen.

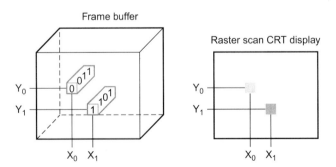

FIGURE 1.6 Each coordinate in the frame buffer on the left determines the shade of the corresponding coordinate for the raster scan CRT display on the right. Pixel (X_0, Y_0) contains the bit pattern 0011, which is a lighter shade on the screen than the bit pattern 1101 in pixel (X_1, Y_1).

Touchscreen

While PCs also use LCD displays, the tablets and smartphones of the post-PC era have replaced the keyboard and mouse with touch-sensitive displays, which has the wonderful user interface advantage of users pointing directly at what they are interested in rather than indirectly with a mouse.

While there are a variety of ways to implement a touch screen, many tablets today use capacitive sensing. Since people are electrical conductors, if an insulator like glass is covered with a transparent conductor, touching distorts the electrostatic field of the screen, which results in a change in capacitance. This technology can allow multiple touches simultaneously, which recognizes gestures that can lead to attractive user interfaces.

Opening the Box

Figure 1.7 shows the contents of the Apple iPad 2 tablet computer. Unsurprisingly, of the five classic components of the computer, I/O dominates this reading device. The list of I/O devices includes a capacitive multitouch LCD display, front-facing camera, rear-facing camera, microphone, headphone jack, speakers, accelerometer, gyroscope, Wi-Fi network, and Bluetooth network. The datapath, control, and memory are a tiny portion of the components.

The small rectangles in Figure 1.8 contain the devices that drive our advancing technology, called integrated circuits and nicknamed chips. The A5 package seen in the middle of Figure 1.8 contains two ARM processors that operate at a clock rate of 1 GHz. The *processor* is the active part of the computer, following the instructions of a program to the letter. It adds numbers, tests numbers, signals I/O devices to activate, and so on. Occasionally, people call the processor the CPU, for the more bureaucratic-sounding central processor unit.

Descending even lower into the hardware, Figure 1.9 reveals details of a microprocessor. The processor logically comprises two main components: datapath and control, the respective brawn and brain of the processor. The datapath performs the arithmetic operations, and control tells the datapath, memory, and I/O devices what to do according to the wishes of the instructions of the program. Chapter 4 explains the datapath and control for a higher-performance design.

The A5 package in Figure 1.8 also includes two memory chips, each with 2 gibibits of capacity, thereby supplying 512 MiB. The memory is where the programs are kept when they are running; it also contains the data needed by the running programs. The memory is built from DRAM chips. *DRAM* stands for dynamic random access memory. Multiple DRAMs are used together to contain the instructions and data of a program. In contrast to sequential access memories, such as magnetic tapes, the *RAM* portion of the term DRAM means that memory accesses take basically the same amount of time no matter what portion of the memory is read.

Descending into the depths of any component of the hardware reveals insights into the computer. Inside the processor is another type of memory—cache memory.

integrated circuit Also called a chip. A device combining dozens to millions of transistors.

central processor unit (CPU) Also called processor. The active part of the computer, which contains the datapath and control and which adds numbers, tests numbers, signals I/O devices to activate, and so on.

datapath The component of the processor that performs arithmetic operations.

control The component of the processor that commands the datapath, memory, and I/O devices according to the instructions of the program.

memory The storage area in which programs are kept when they are running and that contains the data needed by the running programs.

dynamic random access memory (DRAM) Memory built as an integrated circuit; it provides random access to any location. Access times are 50 nanoseconds and cost per gigabyte in 2012 was $5 to $10.

FIGURE 1.7   Components of the Apple iPad 2 A1395. The metal back of the iPad (with the reversed Apple logo in the middle) is in the center. At the top is the capacitive multitouch screen and LCD display. To the far right is the 3.8 V, 25 watt-hour, polymer battery, which consists of three Li-ion cell cases and offers 10 hours of battery life. To the far left is the metal frame that attaches the LCD to the back of the iPad. The small components surrounding the metal back in the center are what we think of as the computer; they are often L-shaped to fit compactly inside the case next to the battery. Figure 1.8 shows a close-up of the L-shaped board to the lower left of the metal case, which is the logic printed circuit board that contains the processor and the memory. The tiny rectangle below the logic board contains a chip that provides wireless communication: Wi-Fi, Bluetooth, and FM tuner. It fits into a small slot in the lower left corner of the logic board. Near the upper left corner of the case is another L-shaped component, which is a front-facing camera assembly that includes the camera, headphone jack, and microphone. Near the right upper corner of the case is the board containing the volume control and silent/screen rotation lock button along with a gyroscope and accelerometer. These last two chips combine to allow the iPad to recognize six-axis motion. The tiny rectangle next to it is the rear-facing camera. Near the bottom right of the case is the L-shaped speaker assembly. The cable at the bottom is the connector between the logic board and the camera/volume control board. The board between the cable and the speaker assembly is the controller for the capacitive touchscreen. (Courtesy iFixit, www.ifixit.com)

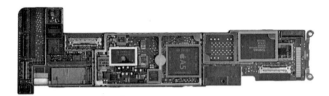

FIGURE 1.8   The logic board of Apple iPad 2 in Figure 1.7. The photo highlights five integrated circuits. The large integrated circuit in the middle is the Apple A5 chip, which contains dual ARM processor cores that run at 1 GHz as well as 512 MB of main memory inside the package. Figure 1.9 shows a photograph of the processor chip inside the A5 package. The similar-sized chip to the left is the 32 GB flash memory chip for non-volatile storage. There is an empty space between the two chips where a second flash chip can be installed to double storage capacity of the iPad. The chips to the right of the A5 include power controller and I/O controller chips. (Courtesy iFixit, www.ifixit.com)

FIGURE 1.9 The processor integrated circuit inside the A5 package. The size of chip is 12.1 by 10.1 mm, and it was manufactured originally in a 45-nm process (see Section 1.5). It has two identical ARM processors or cores in the middle left of the chip and a PowerVR *graphical processor unit* (GPU) with four datapaths in the upper left quadrant. To the left and bottom side of the ARM cores are interfaces to main memory (DRAM). (Courtesy Chipworks, www.chipworks.com)

cache memory A small, fast memory that acts as a buffer for a slower, larger memory.

static random access memory (SRAM) Also memory built as an integrated circuit, but faster and less dense than DRAM.

Cache memory consists of a small, fast memory that acts as a buffer for the DRAM memory. (The nontechnical definition of *cache* is a safe place for hiding things.) Cache is built using a different memory technology, static random access memory (SRAM). SRAM is faster but less dense, and hence more expensive, than DRAM (see Chapter 5). SRAM and DRAM are two layers of the **memory hierarchy**.

HIERARCHY

A B S T R A C T I O N

instruction set architecture Also called architecture. An abstract interface between the hardware and the lowest-level software that encompasses all the information necessary to write a machine language program that will run correctly, including instructions, registers, memory access, I/O, and so on.

application binary interface (ABI) The user portion of the instruction set plus the operating system interfaces used by application programmers. It defines a standard for binary portability across computers.

The **BIG** Picture

implementation Hardware that obeys the architecture abstraction.

volatile memory Storage, such as DRAM, that retains data only if it is receiving power.

nonvolatile memory A form of memory that retains data even in the absence of a power source and that is used to store programs between runs. A DVD disk is nonvolatile.

As mentioned above, one of the great ideas to improve design is abstraction. One of the most important **abstractions** is the interface between the hardware and the lowest-level software. Because of its importance, it is given a special name: the instruction set architecture, or simply architecture, of a computer. The instruction set architecture includes anything programmers need to know to make a binary machine language program work correctly, including instructions, I/O devices, and so on. Typically, the operating system will encapsulate the details of doing I/O, allocating memory, and other low-level system functions so that application programmers do not need to worry about such details. The combination of the basic instruction set and the operating system interface provided for application programmers is called the application binary interface (ABI).

An instruction set architecture allows computer designers to talk about functions independently from the hardware that performs them. For example, we can talk about the functions of a digital clock (keeping time, displaying the time, setting the alarm) separately from the clock hardware (quartz crystal, LED displays, plastic buttons). Computer designers distinguish architecture from an implementation of an architecture along the same lines: an implementation is hardware that obeys the architecture abstraction. These ideas bring us to another Big Picture.

Both hardware and software consist of hierarchical layers using abstraction, with each lower layer hiding details from the level above. One key interface between the levels of abstraction is the *instruction set architecture*—the interface between the hardware and low-level software. This abstract interface enables many *implementations* of varying cost and performance to run identical software.

A Safe Place for Data

Thus far, we have seen how to input data, compute using the data, and display data. If we were to lose power to the computer, however, everything would be lost because the memory inside the computer is volatile—that is, when it loses power, it forgets. In contrast, a DVD disk doesn't forget the movie when you turn off the power to the DVD player, and is therefore a nonvolatile memory technology.

To distinguish between the volatile memory used to hold data and programs while they are running and this nonvolatile memory used to store data and programs between runs, the term main memory or primary memory is used for the former, and secondary memory for the latter. Secondary memory forms the next lower layer of the **memory hierarchy**. DRAMs have dominated main memory since 1975, but magnetic disks dominated secondary memory starting even earlier. Because of their size and form factor, personal Mobile Devices use flash memory, a nonvolatile semiconductor memory, instead of disks. Figure 1.8 shows the chip containing the flash memory of the iPad 2. While slower than DRAM, it is much cheaper than DRAM in addition to being nonvolatile. Although costing more per bit than disks, it is smaller, it comes in much smaller capacities, it is more rugged, and it is more power efficient than disks. Hence, flash memory is the standard secondary memory for PMDs. Alas, unlike disks and DRAM, flash memory bits wear out after 100,000 to 1,000,000 writes. Thus, file systems must keep track of the number of writes and have a strategy to avoid wearing out storage, such as by moving popular data. Chapter 5 describes disks and flash memory in more detail.

Communicating with Other Computers

We've explained how we can input, compute, display, and save data, but there is still one missing item found in today's computers: computer networks. Just as the processor shown in Figure 1.5 is connected to memory and I/O devices, networks interconnect whole computers, allowing computer users to extend the power of computing by including communication. Networks have become so popular that they are the backbone of current computer systems; a new personal mobile device or server without a network interface would be ridiculed. Networked computers have several major advantages:

- *Communication*: Information is exchanged between computers at high speeds.

- *Resource sharing*: Rather than each computer having its own I/O devices, computers on the network can share I/O devices.

- *Nonlocal access*: By connecting computers over long distances, users need not be near the computer they are using.

Networks vary in length and performance, with the cost of communication increasing according to both the speed of communication and the distance that information travels. Perhaps the most popular type of network is *Ethernet*. It can be up to a kilometer long and transfer at up to 40 gigabits per second. Its length and speed make Ethernet useful to connect computers on the same floor of a building;

HIERARCHY

main memory Also called **primary memory**. Memory used to hold programs while they are running; typically consists of DRAM in today's computers.

secondary memory Nonvolatile memory used to store programs and data between runs; typically consists of flash memory in PMDs and magnetic disks in servers.

magnetic disk Also called **hard disk**. A form of nonvolatile secondary memory composed of rotating platters coated with a magnetic recording material. Because they are rotating mechanical devices, access times are about 5 to 20 milliseconds and cost per gigabyte in 2012 was $0.05 to $0.10.

flash memory A nonvolatile semi-conductor memory. It is cheaper and slower than DRAM but more expensive per bit and faster than magnetic disks. Access times are about 5 to 50 microseconds and cost per gigabyte in 2012 was $0.75 to $1.00.

hence, it is an example of what is generically called a local area network. Local area networks are interconnected with switches that can also provide routing services and security. Wide area networks cross continents and are the backbone of the Internet, which supports the web. They are typically based on optical fibers and are leased from telecommunication companies.

Networks have changed the face of computing in the last 30 years, both by becoming much more ubiquitous and by making dramatic increases in performance. In the 1970s, very few individuals had access to electronic mail, the Internet and web did not exist, and physically mailing magnetic tapes was the primary way to transfer large amounts of data between two locations. Local area networks were almost nonexistent, and the few existing wide area networks had limited capacity and restricted access.

As networking technology improved, it became considerably cheaper and had a significantly higher capacity. For example, the first standardized local area network technology, developed about 30 years ago, was a version of Ethernet that had a maximum capacity (also called bandwidth) of 10 million bits per second, typically shared by tens of, if not a hundred, computers. Today, local area network technology offers a capacity of from 1 to 40 gigabits per second, usually shared by at most a few computers. Optical communications technology has allowed similar growth in the capacity of wide area networks, from hundreds of kilobits to gigabits and from hundreds of computers connected to a worldwide network to millions of computers connected. This dramatic rise in deployment of networking combined with increases in capacity have made network technology central to the information revolution of the last 30 years.

For the last decade another innovation in networking is reshaping the way computers communicate. Wireless technology is widespread, which enabled the post-PC era. The ability to make a radio in the same low-cost semiconductor technology (CMOS) used for memory and microprocessors enabled a significant improvement in price, leading to an explosion in deployment. Currently available wireless technologies, called by the IEEE standard name 802.11, allow for transmission rates from 1 to nearly 100 million bits per second. Wireless technology is quite a bit different from wire-based networks, since all users in an immediate area share the airwaves.

**Check
Yourself**

- Semiconductor DRAM memory, flash memory, and disk storage differ significantly. For each technology, list its volatility, approximate relative access time, and approximate relative cost compared to DRAM.

1.5 Technologies for Building Processors and Memory

Processors and memory have improved at an incredible rate, because computer designers have long embraced the latest in electronic technology to try to win the race to design a better computer. Figure 1.10 shows the technologies that have

Year	Technology used in computers	Relative performance/unit cost
1951	Vacuum tube	1
1965	Transistor	35
1975	Integrated circuit	900
1995	Very large-scale integrated circuit	2,400,000
2013	Ultra large-scale integrated circuit	250,000,000,000

FIGURE 1.10 Relative performance per unit cost of technologies used in computers over time. Source: Computer Museum, Boston, with 2013 extrapolated by the authors. See ▦ Section 1.12.

been used over time, with an estimate of the relative performance per unit cost for each technology. Since this technology shapes what computers will be able to do and how quickly they will evolve, we believe all computer professionals should be familiar with the basics of integrated circuits.

A transistor is simply an on/off switch controlled by electricity. The *integrated circuit* (IC) combined dozens to hundreds of transistors into a single chip. When Gordon Moore predicted the continuous doubling of resources, he was forecasting the growth rate of the number of transistors per chip. To describe the tremendous increase in the number of transistors from hundreds to millions, the adjective *very large scale* is added to the term, creating the abbreviation *VLSI*, for very large-scale integrated circuit.

This rate of increasing integration has been remarkably stable. Figure 1.11 shows the growth in DRAM capacity since 1977. For 35 years, the industry has consistently quadrupled capacity every 3 years, resulting in an increase in excess of 16,000 times!

To understand how to manufacture integrated circuits, we start at the beginning. The manufacture of a chip begins with silicon, a substance found in sand. Because silicon does not conduct electricity well, it is called a semiconductor. With a special chemical process, it is possible to add materials to silicon that allow tiny areas to transform into one of three devices:

- Excellent conductors of electricity (using either microscopic copper or aluminum wire)

transistor An on/off switch controlled by an electric signal.

very large-scale integrated (VLSI) circuit A device containing hundreds of thousands to millions of transistors.

silicon A natural element that is a semiconductor.

semiconductor A substance that does not conduct electricity well.

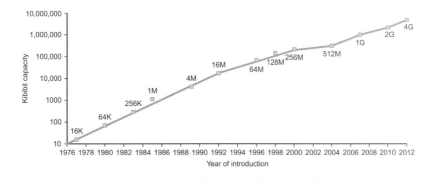

FIGURE 1.11 Growth of capacity per DRAM chip over time. The *y*-axis is measured in kibibits (2^{10} bits). The DRAM industry quadrupled capacity almost every three years, a 60% increase per year, for 20 years. In recent years, the rate has slowed down and is somewhat closer to doubling every two to three years.

- Excellent insulators from electricity (like plastic sheathing or glass)

- Areas that can conduct or insulate under specific conditions (as a switch)

Transistors fall into the last category. A VLSI circuit, then, is just billions of combinations of conductors, insulators, and switches manufactured in a single small package.

The manufacturing process for integrated circuits is critical to the cost of the chips and hence important to computer designers. Figure 1.12 shows that process. The process starts with a **silicon crystal ingot**, which looks like a giant sausage. Today, ingots are 8–12 inches in diameter and about 12–24 inches long. An ingot is finely sliced into **wafers** no more than 0.1 inches thick. These wafers then go through a series of processing steps, during which patterns of chemicals are placed on each wafer, creating the transistors, conductors, and insulators discussed earlier. Today's integrated circuits contain only one layer of transistors but may have from two to eight levels of metal conductor, separated by layers of insulators.

silicon crystal ingot
A rod composed of a silicon crystal that is between 8 and 12 inches in diameter and about 12 to 24 inches long.

wafer A slice from a silicon ingot no more than 0.1 inches thick, used to create chips.

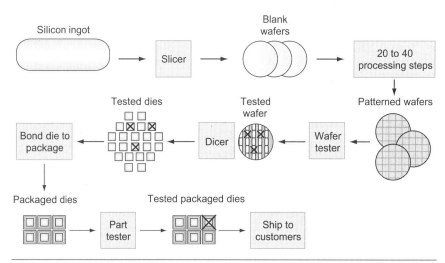

FIGURE 1.12   The chip manufacturing process. After being sliced from the silicon ingot, blank wafers are put through 20 to 40 steps to create patterned wafers (see Figure 1.13). These patterned wafers are then tested with a wafer tester, and a map of the good parts is made. Next, the wafers are diced into dies (see Figure 1.9). In this figure, one wafer produced 20 dies, of which 17 passed testing. (X means the die is bad.) The yield of good dies in this case was 17/20, or 85%. These good dies are then bonded into packages and tested one more time before shipping the packaged parts to customers. One bad packaged part was found in this final test.

defect A microscopic flaw in a wafer or in patterning steps that can result in the failure of the die containing that defect.

A single microscopic flaw in the wafer itself or in one of the dozens of patterning steps can result in that area of the wafer failing. These **defects**, as they are called, make it virtually impossible to manufacture a perfect wafer. The simplest way to cope with imperfection is to place many independent components on a single wafer. The patterned wafer is then chopped up, or *diced*, into these components,

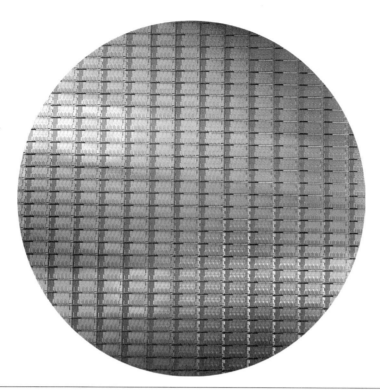

FIGURE 1.13 A 12-inch (300 mm) wafer of Intel Core i7 (Courtesy Intel). The number of dies on this 300 mm (12 inch) wafer at 100% yield is 280, each 20.7 by 10.5 mm. The several dozen partially rounded chips at the boundaries of the wafer are useless; they are included because it's easier to create the masks used to pattern the silicon. This die uses a 32-nanometer technology, which means that the smallest features are approximately 32 nm in size, although they are typically somewhat smaller than the actual feature size, which refers to the size of the transistors as "drawn" versus the final manufactured size.

called **dies** and more informally known as **chips**. Figure 1.13 shows a photograph of a wafer containing microprocessors before they have been diced; earlier, Figure 1.9 shows an individual microprocessor die.

Dicing enables you to discard only those dies that were unlucky enough to contain the flaws, rather than the whole wafer. This concept is quantified by the **yield** of a process, which is defined as the percentage of good dies from the total number of dies on the wafer.

The cost of an integrated circuit rises quickly as the die size increases, due both to the lower yield and to the fewer dies that fit on a wafer. To reduce the cost, using the next generation process shrinks a large die as it uses smaller sizes for both transistors and wires. This improves the yield and the die count per wafer. A 32-nanometer (nm) process was typical in 2012, which means essentially that the smallest feature size on the die is 32 nm.

die The individual rectangular sections that are cut from a wafer, more informally known as **chips**.

yield The percentage of good dies from the total number of dies on the wafer.

Once you've found good dies, they are connected to the input/output pins of a package, using a process called *bonding*. These packaged parts are tested a final time, since mistakes can occur in packaging, and then they are shipped to customers.

Elaboration: The cost of an integrated circuit can be expressed in three simple equations:

$$\text{Cost per die} = \frac{\text{Cost per wafer}}{\text{Dies per wafer} \times \text{yield}}$$

$$\text{Dies per wafer} \approx \frac{\text{Wafer area}}{\text{Die area}}$$

$$\text{Yield} = \frac{1}{(1 + (\text{Defects per area} \times \text{Die area}/2))^2}$$

The first equation is straightforward to derive. The second is an approximation, since it does not subtract the area near the border of the round wafer that cannot accommodate the rectangular dies (see Figure 1.13). The final equation is based on empirical observations of yields at integrated circuit factories, with the exponent related to the number of critical processing steps.

Hence, depending on the defect rate and the size of the die and wafer, costs are generally not linear in the die area.

Check Yourself

A key factor in determining the cost of an integrated circuit is volume. Which of the following are reasons why a chip made in high volume should cost less?

1. With high volumes, the manufacturing process can be tuned to a particular design, increasing the yield.

2. It is less work to design a high-volume part than a low-volume part.

3. The masks used to make the chip are expensive, so the cost per chip is lower for higher volumes.

4. Engineering development costs are high and largely independent of volume; thus, the development cost per die is lower with high-volume parts.

5. High-volume parts usually have smaller die sizes than low-volume parts and therefore, have higher yield per wafer.

1.6 Performance

Assessing the performance of computers can be quite challenging. The scale and intricacy of modern software systems, together with the wide range of performance improvement techniques employed by hardware designers, have made performance assessment much more difficult.

When trying to choose among different computers, performance is an important attribute. Accurately measuring and comparing different computers is critical to purchasers and therefore, to designers. The people selling computers know this as well. Often, salespeople would like you to see their computer in the best possible light, whether or not this light accurately reflects the needs of the purchaser's application. Hence, understanding how best to measure performance and the limitations of those measurements is important in selecting a computer.

The rest of this section describes different ways in which performance can be determined; then, we describe the metrics for measuring performance from the viewpoint of both a computer user and a designer. We also look at how these metrics are related and present the classical processor performance equation, which we will use throughout the text.

Defining Performance

When we say one computer has better performance than another, what do we mean? Although this question might seem simple, an analogy with passenger airplanes shows how subtle the question of performance can be. Figure 1.14 lists some typical passenger airplanes, together with their cruising speed, range, and capacity. If we wanted to know which of the planes in this table had the best performance, we would first need to define performance. For example, considering different measures of performance, we see that the plane with the highest cruising speed was the Concorde (retired from service in 2003), the plane with the longest range is the DC-8, and the plane with the largest capacity is the 747.

Airplane	Passenger capacity	Cruising range (miles)	Cruising speed (m.p.h.)	Passenger throughput (passengers × m.p.h.)
Boeing 777	375	4630	610	228,750
Boeing 747	470	4150	610	286,700
BAC/Sud Concorde	132	4000	1350	178,200
Douglas DC-8-50	146	8720	544	79,424

FIGURE 1.14 The capacity, range, and speed for a number of commercial airplanes. The last column shows the rate at which the airplane transports passengers, which is the capacity times the cruising speed (ignoring range and takeoff and landing times).

Let's suppose we define performance in terms of speed. This still leaves two possible definitions. You could define the fastest plane as the one with the highest cruising speed, taking a single passenger from one point to another in the least time. If you were interested in transporting 450 passengers from one point to another, however, the 747 would clearly be the fastest, as the last column of the figure shows. Similarly, we can define computer performance in several distinct ways.

If you were running a program on two different desktop computers, you'd say that the faster one is the desktop computer that gets the job done first. If you were running a datacenter that had several servers running jobs submitted by many users, you'd say that the faster computer was the one that completed the most jobs during a day. As an individual computer user, you are interested in reducing response time—the time between the start and completion of a task—also referred to as

response time Also called **execution time**. The total time required for the computer to complete a task, including disk accesses, memory accesses, I/O activities, operating system overhead, CPU execution time, and so on.

throughput Also called bandwidth. Another measure of performance, it is the number of tasks completed per unit time.

execution time. Datacenter managers often care about increasing throughput or bandwidth—the total amount of work done in a given time. Hence, in most cases, we will need different performance metrics as well as different sets of applications to benchmark personal mobile devices, which are more focused on response time, versus servers, which are more focused on throughput.

EXAMPLE

Throughput and Response Time

Do the following changes to a computer system increase throughput, decrease response time, or both?

1. Replacing the processor in a computer with a faster version

2. Adding additional processors to a system that uses multiple processors for separate tasks—for example, searching the web

ANSWER

Decreasing response time almost always improves throughput. Hence, in case 1, both response time and throughput are improved. In case 2, no one task gets work done faster, so only throughput increases.

If, however, the demand for processing in the second case was almost as large as the throughput, the system might force requests to queue up. In this case, increasing the throughput could also improve response time, since it would reduce the waiting time in the queue. Thus, in many real computer systems, changing either execution time or throughput often affects the other.

In discussing the performance of computers, we will be primarily concerned with response time for the first few chapters. To maximize performance, we want to minimize response time or execution time for some task. Thus, we can relate performance and execution time for a computer X:

$$\text{Performance}_X = \frac{1}{\text{Execution time}_X}$$

This means that for two computers X and Y, if the performance of X is greater than the performance of Y, we have

$$\text{Performance}_X > \text{Performance}_Y$$

$$\frac{1}{\text{Execution time}_X} > \frac{1}{\text{Execution time}_Y}$$

$$\text{Execution time}_Y > \text{Execution time}_X$$

That is, the execution time on Y is longer than that on X, if X is faster than Y.

In discussing a computer design, we often want to relate the performance of two different computers quantitatively. We will use the phrase "X is n times faster than Y"—or equivalently "X is n times as fast as Y"—to mean

$$\frac{\text{Performance}_X}{\text{Performance}_Y} = n$$

If X is n times as fast as Y, then the execution time on Y is n times as long as it is on X:

$$\frac{\text{Performance}_X}{\text{Performance}_Y} = \frac{\text{Execution time}_Y}{\text{Execution time}_X} = n$$

Relative Performance

If computer A runs a program in 10 seconds and computer B runs the same program in 15 seconds, how much faster is A than B?

EXAMPLE

We know that A is n times as fast as B if

$$\frac{\text{Performance}_A}{\text{Performance}_B} = \frac{\text{Execution time}_B}{\text{Execution time}_A} = n$$

ANSWER

Thus the performance ratio is

$$\frac{15}{10} = 1.5$$

and A is therefore 1.5 times as fast as B.

In the above example, we could also say that computer B is 1.5 times *slower than* computer A, since

$$\frac{\text{Performance}_A}{\text{Performance}_B} = 1.5$$

means that

$$\frac{\text{Performance}_A}{1.5} = \text{Performance}_B$$

For simplicity, we will normally use the terminology *as fast as* when we try to compare computers quantitatively. Because performance and execution time are reciprocals, increasing performance requires decreasing execution time. To avoid the potential confusion between the terms *increasing* and *decreasing*, we usually say "improve performance" or "improve execution time" when we mean "increase performance" and "decrease execution time."

Measuring Performance

Time is the measure of computer performance: the computer that performs the same amount of work in the least time is the fastest. Program *execution time* is measured in seconds per program. However, time can be defined in different ways, depending on what we count. The most straightforward definition of time is called *wall clock time*, *response time*, or *elapsed time*. These terms mean the total time to complete a task, including disk accesses, memory accesses, *input/output* (I/O) activities, operating system overhead—everything.

Computers are often shared, however, and a processor may work on several programs simultaneously. In such cases, the system may try to optimize throughput rather than attempt to minimize the elapsed time for one program. Hence, we often want to distinguish between the elapsed time and the time over which the processor is working on our behalf. CPU execution time or simply CPU time, which recognizes this distinction, is the time the CPU spends computing for this task and does not include time spent waiting for I/O or running other programs. (Remember, though, that the response time experienced by the user will be the elapsed time of the program, not the CPU time.) CPU time can be further divided into the CPU time spent in the program, called user CPU time, and the CPU time spent in the operating system performing tasks on behalf of the program, called system CPU time. Differentiating between system and user CPU time is difficult to do accurately, because it is often hard to assign responsibility for operating system activities to one user program rather than another and because of the functionality differences between operating systems.

For consistency, we maintain a distinction between performance based on elapsed time and that based on CPU execution time. We will use the term *system performance* to refer to elapsed time on an unloaded system and *CPU performance* to refer to user CPU time. We will focus on CPU performance in this chapter, although our discussions of how to summarize performance can be applied to either elapsed time or CPU time measurements.

CPU execution time Also called CPU time. The actual time the CPU spends computing for a specific task.

user CPU time The CPU time spent in a program itself.

system CPU time The CPU time spent in the operating system performing tasks on behalf of the program.

Understanding Program Performance

Different applications are sensitive to different aspects of the performance of a computer system. Many applications, especially those running on servers, depend as much on I/O performance, which, in turn, relies on both hardware and software. Total elapsed time measured by a wall clock is the measurement of interest. In some application environments, the user may care about throughput, response time, or a complex combination of the two (e.g., maximum throughput with a worst-case response time). To improve the performance of a program, one must have a clear definition of what performance metric matters and then proceed to find performance bottlenecks by measuring program execution and looking for the likely bottlenecks. In the following chapters, we will describe how to search for bottlenecks and improve performance in various parts of the system.

Although as computer users we care about time, when we examine the details of a computer it's convenient to think about performance in other metrics. In particular, computer designers may want to think about a computer by using a measure that relates to how fast the hardware can perform basic functions. Almost all computers are constructed using a clock that determines when events take place in the hardware. These discrete time intervals are called clock cycles (or ticks, clock ticks, clock periods, clocks, cycles). Designers refer to the length of a clock period both as the time for a complete *clock cycle* (e.g., 250 picoseconds, or 250 ps) and as the *clock rate* (e.g., 4 gigahertz, or 4 GHz), which is the inverse of the clock period. In the next subsection, we will formalize the relationship between the clock cycles of the hardware designer and the seconds of the computer user.

clock cycle Also called tick, clock tick, clock period, clock, or cycle. The time for one clock period, usually of the processor clock, which runs at a constant rate.

clock period The length of each clock cycle.

1. Suppose we know that an application that uses both personal mobile devices and the Cloud is limited by network performance. For the following changes, state whether only the throughput improves, both response time and throughput improve, or neither improves.

   a. An extra network channel is added between the PMD and the Cloud, increasing the total network throughput and reducing the delay to obtain network access (since there are now two channels).

   b. The networking software is improved, thereby reducing the network communication delay, but not increasing throughput.

   c. More memory is added to the computer.

2. Computer C's performance is four times as fast as the performance of computer B, which runs a given application in 28 seconds. How long will computer C take to run that application?

Check Yourself

CPU Performance and Its Factors

Users and designers often examine performance using different metrics. If we could relate these different metrics, we could determine the effect of a design change on the performance as experienced by the user. Since we are confining ourselves to CPU performance at this point, the bottom-line performance measure is CPU execution time. A simple formula relates the most basic metrics (clock cycles and clock cycle time) to CPU time:

$$\begin{array}{ll} \text{CPU execution time} \\ \text{for a program} \end{array} = \begin{array}{ll} \text{CPU clock cycles} \\ \text{for a program} \end{array} \times \text{Clock cycle time}$$

Alternatively, because clock rate and clock cycle time are inverses,

$$\begin{array}{ll} \text{CPU execution time} \\ \text{for a program} \end{array} = \frac{\text{CPU clock cycles for a program}}{\text{Clock rate}}$$

This formula makes it clear that the hardware designer can improve performance by reducing the number of clock cycles required for a program or the length of the clock cycle. As we will see in later chapters, the designer often faces a trade-off between the number of clock cycles needed for a program and the length of each cycle. Many techniques that decrease the number of clock cycles may also increase the clock cycle time.

Improving Performance

EXAMPLE

Our favorite program runs in 10 seconds on computer A, which has a 2 GHz clock. We are trying to help a computer designer build a computer, B, which will run this program in 6 seconds. The designer has determined that a substantial increase in the clock rate is possible, but this increase will affect the rest of the CPU design, causing computer B to require 1.2 times as many clock cycles as computer A for this program. What clock rate should we tell the designer to target?

ANSWER

Let's first find the number of clock cycles required for the program on A:

$$\text{CPU time}_A = \frac{\text{CPU clock cycles}_A}{\text{Clock rate}_A}$$

$$10 \, \text{seconds} = \frac{\text{CPU clock cycles}_A}{2 \times 10^9 \, \dfrac{\text{cycles}}{\text{second}}}$$

$$\text{CPU clock cycles}_A = 10 \, \text{seconds} \times 2 \times 10^9 \, \frac{\text{cycles}}{\text{second}} = 20 \times 10^9 \, \text{cycles}$$

CPU time for B can be found using this equation:

$$\text{CPU time}_B = \frac{1.2 \times \text{CPU clock cycles}_A}{\text{Clock rate}_B}$$

$$6 \text{ seconds} = \frac{1.2 \times 20 \times 10^9 \text{ cycles}}{\text{Clock rate}_B}$$

$$\text{Clock rate}_B = \frac{1.2 \times 20 \times 10^9 \text{ cycles}}{6 \text{ seconds}} = \frac{0.2 \times 20 \times 10^9 \text{ cycles}}{\text{second}} = \frac{4 \times 10^9 \text{ cycles}}{\text{second}} = 4 \text{ GHz}$$

To run the program in 6 seconds, B must have twice the clock rate of A.

Instruction Performance

The performance equations above did not include any reference to the number of instructions needed for the program. However, since the compiler clearly generated instructions to execute, and the computer had to execute the instructions to run the program, the execution time must depend on the number of instructions in a program. One way to think about execution time is that it equals the number of instructions executed multiplied by the average time per instruction. Therefore, the number of clock cycles required for a program can be written as

$$\text{CPU clock cycles} = \text{Instructions for a program} \times \frac{\text{Average clock cycles}}{\text{per instruction}}$$

The term clock cycles per instruction, which is the average number of clock cycles each instruction takes to execute, is often abbreviated as CPI. Since different instructions may take different amounts of time depending on what they do, CPI is an average of all the instructions executed in the program. CPI provides one way of comparing two different implementations of the identical instruction set architecture, since the number of instructions executed for a program will, of course, be the same.

clock cycles per instruction (CPI) Average number of clock cycles per instruction for a program or program fragment.

Using the Performance Equation

Suppose we have two implementations of the same instruction set architecture. Computer A has a clock cycle time of 250 ps and a CPI of 2.0 for some program, and computer B has a clock cycle time of 500 ps and a CPI of 1.2 for the same program. Which computer is faster for this program and by how much?

EXAMPLE

ANSWER

We know that each computer executes the same number of instructions for the program; let's call this number I. First, find the number of processor clock cycles for each computer:

$$\text{CPU clock cycles}_A = I \times 2.0$$
$$\text{CPU clock cycles}_B = I \times 1.2$$

Now we can compute the CPU time for each computer:

$$\text{CPU time}_A = \text{CPU clock cycles}_A \times \text{Clock cycle time}$$
$$= I \times 2.0 \times 250 \text{ ps} = 500 \times I \text{ ps}$$

Likewise, for B:

$$\text{CPU time}_B = I \times 1.2 \times 500 \text{ ps} = 600 \times I \text{ ps}$$

Clearly, computer A is faster. The amount faster is given by the ratio of the execution times:

$$\frac{\text{CPU performance}_A}{\text{CPU performance}_B} = \frac{\text{Execution time}_B}{\text{Execution time}_A} = \frac{600 \times I \text{ ps}}{500 \times I \text{ ps}} = 1.2$$

We can conclude that computer A is 1.2 times as fast as computer B for this program.

The Classic CPU Performance Equation

instruction count The number of instructions executed by the program.

We can now write this basic performance equation in terms of instruction count (the number of instructions executed by the program), CPI, and clock cycle time:

$$\text{CPU time} = \text{Instruction count} \times \text{CPI} \times \text{Clock cycle time}$$

or, since the clock rate is the inverse of clock cycle time:

$$\text{CPU time} = \frac{\text{Instruction count} \times \text{CPI}}{\text{Clock rate}}$$

These formulas are particularly useful because they separate the three key factors that affect performance. We can use these formulas to compare two different implementations or to evaluate a design alternative if we know its impact on these three parameters.

Comparing Code Segments

A compiler designer is trying to decide between two code sequences for a computer. The hardware designers have supplied the following facts:

	CPI for each instruction class		
	A	B	C
CPI	1	2	3

For a particular high-level language statement, the compiler writer is considering two code sequences that require the following instruction counts:

Code sequence	Instruction counts for each instruction class		
	A	B	C
1	2	1	2
2	4	1	1

Which code sequence executes the most instructions? Which will be faster? What is the CPI for each sequence?

Sequence 1 executes $2 + 1 + 2 = 5$ instructions. Sequence 2 executes $4 + 1 + 1 = 6$ instructions. Therefore, sequence 1 executes fewer instructions.

We can use the equation for CPU clock cycles based on instruction count and CPI to find the total number of clock cycles for each sequence:

$$\text{CPU clock cycles} = \sum_{i=1}^{n}(\text{CPI}_i \times \text{C}_i)$$

This yields

$$\text{CPU clock cycles}_1 = (2 \times 1) + (1 \times 2) + (2 \times 3) = 2 + 2 + 6 = 10 \text{ cycles}$$
$$\text{CPU clock cycles}_2 = (4 \times 1) + (1 \times 2) + (1 \times 3) = 4 + 2 + 3 = 9 \text{ cycles}$$

So code sequence 2 is faster, even though it executes one extra instruction. Since code sequence 2 takes fewer overall clock cycles but has more instructions, it must have a lower CPI. The CPI values can be computed by

$$\text{CPI} = \frac{\text{CPU clock cycles}}{\text{Instruction count}}$$

$$\text{CPI}_1 = \frac{\text{CPU clock cycles}_1}{\text{Instruction count}_1} = \frac{10}{5} = 2.0$$

$$\text{CPI}_2 = \frac{\text{CPU clock cycles}_2}{\text{Instruction count}_2} = \frac{9}{6} = 1.5$$

Figure 1.15 shows the basic measurements at different levels in the computer and what is being measured in each case. We can see how these factors are combined to yield execution time measured in seconds per program:

$$\text{Time} = \text{Seconds/Program} = \frac{\text{Instructions}}{\text{Program}} \times \frac{\text{Clock cycles}}{\text{Instruction}} \times \frac{\text{Seconds}}{\text{Clock cycle}}$$

The **BIG** Picture

Always bear in mind that the only complete and reliable measure of computer performance is time. For example, changing the instruction set to lower the instruction count may lead to an organization with a slower clock cycle time or higher CPI that offsets the improvement in instruction count. Similarly, because CPI depends on the type of instructions executed, the code that executes the fewest number of instructions may not be the fastest.

Components of performance	Units of measure
CPU execution time for a program	Seconds for the program
Instruction count	Instructions executed for the program
Clock cycles per instruction (CPI)	Average number of clock cycles per instruction
Clock cycle time	Seconds per clock cycle

FIGURE 1.15 The basic components of performance and how each is measured.

How can we determine the value of these factors in the performance equation? We can measure the CPU execution time by running the program, and the clock cycle time is usually published as part of the documentation for a computer. The instruction count and CPI can be more difficult to obtain. Of course, if we know the clock rate and CPU execution time, we need only one of the instruction count or the CPI to determine the other.

We can measure the instruction count by using software tools that profile the execution or by using a simulator of the architecture. Alternatively, we can use hardware counters, which are included in most processors, to record a variety of measurements, including the number of instructions executed, the average CPI, and often, the sources of performance loss. Since the instruction count depends on the architecture, but not on the exact implementation, we can measure the instruction count without knowing all the details of the implementation. The CPI, however, depends on a wide variety of design details in the computer, including both the memory system and the processor structure (as we will see in Chapter 4 and Chapter 5), as well as on the mix of instruction types executed in an application. Thus, CPI varies by application, as well as among implementations with the same instruction set.

The above example shows the danger of using only one factor (instruction count) to assess performance. When comparing two computers, you must look at all three components, which combine to form execution time. If some of the factors are identical, like the clock rate in the above example, performance can be determined by comparing all the nonidentical factors. Since CPI varies by instruction mix, both instruction count and CPI must be compared, even if clock rates are equal. Several exercises at the end of this chapter ask you to evaluate a series of computer and compiler enhancements that affect clock rate, CPI, and instruction count. In Section 1.10, we'll examine a common performance measurement that does not incorporate all the terms and can thus be misleading.

instruction mix
A measure of the dynamic frequency of instructions across one or many programs.

The performance of a program depends on the algorithm, the language, the compiler, the architecture, and the actual hardware. The following table summarizes how these components affect the factors in the CPU performance equation.

Understanding Program Performance

Hardware or software component	Affects what?	How?
Algorithm	Instruction count, possibly CPI	The algorithm determines the number of source program instructions executed and hence the number of processor instructions executed. The algorithm may also affect the CPI, by favoring slower or faster instructions. For example, if the algorithm uses more divides, it will tend to have a higher CPI.
Programming language	Instruction count, CPI	The programming language certainly affects the instruction count, since statements in the language are translated to processor instructions, which determine instruction count. The language may also affect the CPI because of its features; for example, a language with heavy support for data abstraction (e.g., Java) will require indirect calls, which will use higher CPI instructions.
Compiler	Instruction count, CPI	The efficiency of the compiler affects both the instruction count and average cycles per instruction, since the compiler determines the translation of the source language instructions into computer instructions. The compiler's role can be very complex and affect the CPI in varied ways.
Instruction set architecture	Instruction count, clock rate, CPI	The instruction set architecture affects all three aspects of CPU performance, since it affects the instructions needed for a function, the cost in cycles of each instruction, and the overall clock rate of the processor.

Elaboration: Although you might expect that the minimum CPI is 1.0, as we'll see in Chapter 4, some processors fetch and execute multiple instructions per clock cycle. To reflect that approach, some designers invert CPI to talk about *IPC*, or *instructions per clock cycle*. If a processor executes on average two instructions per clock cycle, then it has an IPC of 2 and hence a CPI of 0.5.

Elaboration: Although clock cycle time has traditionally been fixed, to save energy or temporarily boost performance, today's processors can vary their clock rates, so we would need to use the *average* clock rate for a program. For example, the Intel Core i7 will temporarily increase clock rate by about 10% until the chip gets too warm. Intel calls this *Turbo mode*.

Check Yourself

A given application written in Java runs 15 seconds on a desktop processor. A new Java compiler is released that requires only 0.6 as many instructions as the old compiler. Unfortunately, it increases the CPI by 1.1. How fast can we expect the application to run using this new compiler? Pick the right answer from the three choices below:

a. $\dfrac{15 \times 0.6}{1.1} = 8.2\,\mathrm{sec}$

b. $15 \times 0.6 \times 1.1 = 9.9\,\mathrm{sec}$

c. $\dfrac{1.5 \times 1.1}{0.6} = 27.5\,\mathrm{sec}$

1.7   The Power Wall

Figure 1.16 shows the increase in clock rate and power of eight generations of Intel microprocessors over 30 years. Both clock rate and power increased rapidly for decades and then flattened off recently. The reason they grew together is that they are correlated, and the reason for their recent slowing is that we have run into the practical power limit for cooling commodity microprocessors.

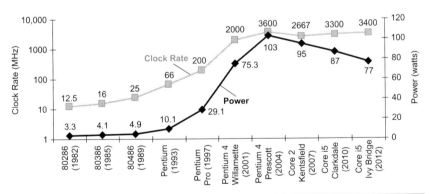

FIGURE 1.16   Clock rate and Power for Intel x86 microprocessors over eight generations and 30 years. The Pentium 4 made a dramatic jump in clock rate and power but less so in performance. The Prescott thermal problems led to the abandonment of the Pentium 4 line. The Core 2 line reverts to a simpler pipeline with lower clock rates and multiple processors per chip. The Core i5 pipelines follow in its footsteps.

Although power provides a limit to what we can cool, in the post-PC era the really valuable resource is energy. Battery life can trump performance in the personal mobile device, and the architects of warehouse scale computers try to reduce the costs of powering and cooling 100,000 servers as the costs are high at this scale. Just as measuring time in seconds is a safer evaluation of program performance than a rate like MIPS (see Section 1.10), the energy metric joules is a better measure than a power rate like watts, which is just joules/second.

The dominant technology for integrated circuits is called CMOS (*complementary metal oxide semiconductor*). For CMOS, the primary source of energy consumption is so-called dynamic energy—that is, energy that is consumed when transistors switch states from 0 to 1 and vice versa. The dynamic energy depends on the capacitive loading of each transistor and the voltage applied:

$$Energy \propto Capacitive\ load \times Voltage^2$$

This equation is the energy of a pulse during the logic transition of $0 \rightarrow 1 \rightarrow 0$ or $1 \rightarrow 0 \rightarrow 1$. The energy of a single transition is then

$$Energy \propto 1/2 \times Capacitive\ load \times Voltage^2$$

The power required per transistor is just the product of energy of a transition and the frequency of transitions:

$$Power \propto 1/2 \times Capacitive\ load \times Voltage^2 \times Frequency\ switched$$

Frequency switched is a function of the clock rate. The capacitive load per transistor is a function of both the number of transistors connected to an output (called the *fanout*) and the technology, which determines the capacitance of both wires and transistors.

With regard to Figure 1.16, how could clock rates grow by a factor of 1000 while power increased by only a factor of 30? Energy and thus power can be reduced by lowering the voltage, which occurred with each new generation of technology, and power is a function of the voltage squared. Typically, the voltage was reduced about 15% per generation. In 20 years, voltages have gone from 5 V to 1 V, which is why the increase in power is only 30 times.

Relative Power

EXAMPLE

Suppose we developed a new, simpler processor that has 85% of the capacitive load of the more complex older processor. Further, assume that it can adjust voltage so that it can reduce voltage 15% compared to processor B, which results in a 15% shrink in frequency. What is the impact on dynamic power?

ANSWER

$$\frac{\text{Power}_{\text{new}}}{\text{Power}_{\text{old}}} = \frac{\langle\text{Capacitive load} \times 0.85\rangle \times \langle\text{Voltage} \times 0.85\rangle^2 \times \langle\text{Frequency switched} \times 0.85\rangle}{\text{Capacitive load} \times \text{Voltage}^2 \times \text{Frequency switched}}$$

Thus the power ratio is

$$0.85^4 = 0.52$$

Hence, the new processor uses about half the power of the old processor.

The modern problem is that further lowering of the voltage appears to make the transistors too leaky, like water faucets that cannot be completely shut off. Even today about 40% of the power consumption in server chips is due to leakage. If transistors started leaking more, the whole process could become unwieldy.

To try to address the power problem, designers have already attached large devices to increase cooling, and they turn off parts of the chip that are not used in a given clock cycle. Although there are many more expensive ways to cool chips and thereby raise their power to, say, 300 watts, these techniques are generally too costly for personal computers and even servers, not to mention personal mobile devices.

Since computer designers slammed into a power wall, they needed a new way forward. They chose a different path from the way they designed microprocessors for their first 30 years.

Elaboration: Although dynamic energy is the primary source of energy consumption in CMOS, static energy consumption occurs because of leakage current that flows even when a transistor is off. In servers, leakage is typically responsible for 40% of the energy consumption. Thus, increasing the number of transistors increases power dissipation, even if the transistors are always off. A variety of design techniques and technology innovations are being deployed to control leakage, but it's hard to lower voltage further.

Elaboration: Power is a challenge for integrated circuits for two reasons. First, power must be brought in and distributed around the chip; modern microprocessors use hundreds of pins just for power and ground! Similarly, multiple levels of chip interconnect are used solely for power and ground distribution to portions of the chip. Second, power is dissipated as heat and must be removed. Server chips can burn more than 100 watts, and cooling the chip and the surrounding system is a major expense in warehouse scale computers (see Chapter 6).

1.8 The Sea Change: The Switch from Uniprocessors to Multiprocessors

The power limit has forced a dramatic change in the design of microprocessors. Figure 1.17 shows the improvement in response time of programs for desktop microprocessors over time. Since 2002, the rate has slowed from a factor of 1.5 per year to a factor of 1.2 per year.

Rather than continuing to decrease the response time of one program running on the single processor, as of 2006 all desktop and server companies are shipping microprocessors with multiple processors per chip, where the benefit is often more on throughput than on response time. To reduce confusion between the words processor and microprocessor, companies refer to processors as "cores," and such microprocessors are generically called multicore microprocessors. Hence, a "quadcore" microprocessor is a chip that contains four processors or four cores.

In the past, programmers could rely on innovations in hardware, architecture, and compilers to double performance of their programs every 18 months without having to change a line of code. Today, for programmers to get significant improvement in response time, they need to rewrite their programs to take advantage of multiple processors. Moreover, to get the historic benefit of running faster on new microprocessors, programmers will have to continue to improve the performance of their code as the number of cores increases.

To reinforce how the software and hardware systems work together, we use a special section, *Hardware/Software Interface*, throughout the book, with the first one appearing below. These elements summarize important insights at this critical interface.

Up to now, most software has been like music written for a solo performer; with the current generation of chips we're getting a little experience with duets and quartets and other small ensembles; but scoring a work for large orchestra and chorus is a different kind of challenge.

Brian Hayes, *Computing in a Parallel Universe*, 2007.

PARALLELISM

Parallelism has always been crucial to performance in computing, but it was often hidden. Chapter 4 will explain **pipelining**, an elegant technique that runs programs faster by overlapping the execution of instructions. This optimization is one example of *instruction-level parallelism*, where the parallel nature of the hardware is abstracted away so the programmer and compiler can think of the hardware as executing instructions sequentially.

Forcing programmers to be aware of the parallel hardware and to rewrite their programs to be parallel had been the "third rail" of computer architecture, for companies in the past that depended on such a change in behavior failed (see ⊞ Section 6.15). From this historical perspective, it's startling that the whole IT industry has bet its future that programmers will finally successfully switch to explicitly parallel programming.

Hardware/ Software Interface

PIPELINING

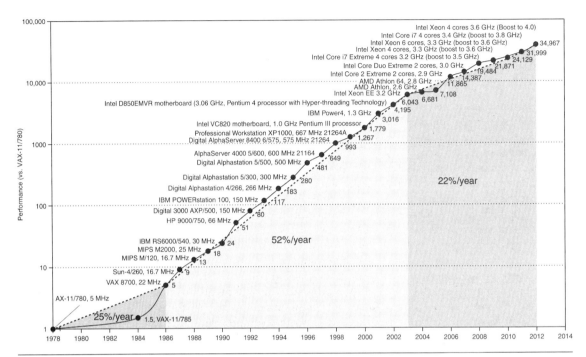

FIGURE 1.17 Growth in processor performance since the mid-1980s. This chart plots performance relative to the VAX 11/780 as measured by the SPECint benchmarks (see Section 1.10). Prior to the mid-1980s, processor performance growth was largely technology-driven and averaged about 25% per year. The increase in growth to about 52% since then is attributable to more advanced architectural and organizational ideas. The higher annual performance improvement of 52% since the mid-1980s meant performance was about a factor of seven larger in 2002 than it would have been had it stayed at 25%. Since 2002, the limits of power, available instruction-level parallelism, and long memory latency have slowed uniprocessor performance recently, to about 22% per year.

Why has it been so hard for programmers to write explicitly parallel programs? The first reason is that parallel programming is by definition performance programming, which increases the difficulty of programming. Not only does the program need to be correct, solve an important problem, and provide a useful interface to the people or other programs that invoke it; the program must also be fast. Otherwise, if you don't need performance, just write a sequential program.

The second reason is that fast for parallel hardware means that the programmer must divide an application so that each processor has roughly the same amount to do at the same time, and that the overhead of scheduling and coordination doesn't fritter away the potential performance benefits of parallelism.

As an analogy, suppose the task was to write a newspaper story. Eight reporters working on the same story could potentially write a story eight times faster. To achieve this increased speed, one would need to break up the task so that each reporter had something to do at the same time. Thus, we must *schedule* the sub-tasks. If anything went wrong and just one reporter took longer than the seven others did, then the benefits of having eight writers would be diminished. Thus, we must *balance the load* evenly to get the desired speedup. Another danger would be if reporters had to spend a lot of time talking to each other to write their sections. You would also fall short if one part of the story, such as the conclusion, couldn't be written until all the other parts were completed. Thus, care must be taken to *reduce communication and synchronization overhead.* For both this analogy and parallel programming, the challenges include scheduling, load balancing, time for synchronization, and overhead for communication between the parties. As you might guess, the challenge is stiffer with more reporters for a newspaper story and more processors for parallel programming.

To reflect this sea change in the industry, the next five chapters in this edition of the book each has a section on the implications of the parallel revolution to that chapter:

MOORE'S LAW

PIPELINING

- *Chapter 2, Section 2.11: Parallelism and Instructions: Synchronization.* Usually independent parallel tasks need to coordinate at times, such as to say when they have completed their work. This chapter explains the instructions used by multicore processors to synchronize tasks.

- *Chapter 3, Section 3.6: Parallelism and Computer Arithmetic: Subword Parallelism.* Perhaps the simplest form of parallelism to build involves computing on elements in parallel, such as when multiplying two vectors. Subword parallelism takes advantage of the resources supplied by **Moore's Law** to provider wider arithmetic units that can operate on many operands simultaneously.

- *Chapter 4, Section 4.10: Parallelism via Instructions.* Given the difficulty of explicitly parallel programming, tremendous effort was invested in the 1990s in having the hardware and the compiler uncover implicit parallelism, initially via **pipelining**. This chapter describes some of these aggressive techniques, including fetching and executing multiple instructions concurrently and guessing on the outcomes of decisions, and executing instructions speculatively using **prediction**.

PREDICTION

HIERARCHY

- *Chapter 5, Section 5.10: Parallelism and **Memory Hierarchies**: Cache Coherence.* One way to lower the cost of communication is to have all processors use the same address space, so that any processor can read or write any data. Given that all processors today use caches to keep a temporary copy of the data in faster memory near the processor, it's easy to imagine that parallel programming would be even more difficult if the caches associated with each processor had inconsistent values of the shared data. This chapter describes the mechanisms that keep the data in all caches consistent.

- *Chapter 5, Section 5.11: Parallelism and Memory Hierarchy: Redundant Arrays of Inexpensive Disks.* This section describes how using many disks in conjunction can offer much higher throughput, which was the original inspiration of *Redundant Arrays of Inexpensive Disks* (RAID). The real popularity of RAID proved to be the much greater dependability offered by including a modest number of redundant disks. The section explains the differences in performance, cost, and dependability between the various RAID levels.

I thought [computers] would be a universally applicable idea, like a book is. But I didn't think it would develop as fast as it did, because I didn't envision we'd be able to get as many parts on a chip as we finally got. The transistor came along unexpectedly. It all happened much faster than we expected.

J. Presper Eckert, coinventor of ENIAC, speaking in 1991

workload A set of programs run on a computer that is either the actual collection of applications run by a user or constructed from real programs to approximate such a mix. A typical workload specifies both the programs and the relative frequencies.

COMMON CASE FAST

benchmark A program selected for use in comparing computer performance.

In addition to these sections, there is a full chapter on parallel processing. Chapter 6 goes into more detail on the challenges of parallel programming; presents the two contrasting approaches to communication of shared addressing and explicit message passing; describes a restricted model of parallelism that is easier to program; discusses the difficulty of benchmarking parallel processors; introduces a new simple performance model for multicore microprocessors; and, finally, describes and evaluates four examples of multicore microprocessors using this model.

As mentioned above, Chapters 3 to 6 use matrix vector multiply as a running example to show how each type of parallelism can significantly increase performance.

🌐 Appendix B describes an increasingly popular hardware component that is included with desktop computers, the *graphics processing unit* (GPU). Invented to accelerate graphics, GPUs are becoming programming platforms in their own right. As you might expect, given these times, GPUs rely on **parallelism**.

🌐 Appendix B describes the NVIDIA GPU and highlights parts of its parallel programming environment.

1.9   Real Stuff: Benchmarking the Intel Core i7

Each chapter has a section entitled "Real Stuff" that ties the concepts in the book with a computer you may use every day. These sections cover the technology underlying modern computers. For this first "Real Stuff" section, we look at how integrated circuits are manufactured and how performance and power are measured, with the Intel Core i7 as the example.

SPEC CPU Benchmark

A computer user who runs the same programs day in and day out would be the perfect candidate to evaluate a new computer. The set of programs run would form a workload. To evaluate two computer systems, a user would simply compare the execution time of the workload on the two computers. Most users, however, are not in this situation. Instead, they must rely on other methods that measure the performance of a candidate computer, hoping that the methods will reflect how well the computer will perform with the user's workload. This alternative is usually followed by evaluating the computer using a set of benchmarks—programs specifically chosen to measure performance. The benchmarks form a workload that the user hopes will predict the performance of the actual workload. As we noted above, to make the common case fast, you first need to know accurately which case is common, so benchmarks play a critical role in computer architecture.

SPEC (*System Performance Evaluation Cooperative*) is an effort funded and supported by a number of computer vendors to create standard sets of benchmarks for modern computer systems. In 1989, SPEC originally created a benchmark

Description	Name	Instruction Count x 10⁹	CPI	Clock cycle time (seconds x 10⁻⁹)	Execution Time (seconds)	Reference Time (seconds)	SPECratio
Interpreted string processing	perl	2252	0.60	0.376	508	9770	19.2
Block-sorting compression	bzip2	2390	0.70	0.376	629	9650	15.4
GNU C compiler	gcc	794	1.20	0.376	358	8050	22.5
Combinatorial optimization	mcf	221	2.66	0.376	221	9120	41.2
Go game (AI)	go	1274	1.10	0.376	527	10490	19.9
Search gene sequence	hmmer	2616	0.60	0.376	590	9330	15.8
Chess game (AI)	sjeng	1948	0.80	0.376	586	12100	20.7
Quantum computer simulation	libquantum	659	0.44	0.376	109	20720	190.0
Video compression	h264avc	3793	0.50	0.376	713	22130	31.0
Discrete event simulation library	omnetpp	367	2.10	0.376	290	6250	21.5
Games/path finding	astar	1250	1.00	0.376	470	7020	14.9
XML parsing	xalancbmk	1045	0.70	0.376	275	6900	25.1
Geometric mean	–	–	–	–	–	–	25.7

FIGURE 1.18 SPECINTC2006 benchmarks running on a 2.66 GHz Intel Core i7 920. As the equation on page 36 explains, execution time is the product of the three factors in this table: instruction count in billions, *clocks per instruction* (CPI), and clock cycle time in nanoseconds. SPECratio is simply the reference time, which is supplied by SPEC, divided by the measured execution time. The single number quoted as SPECINTC2006 is the geometric mean of the SPECratios.

set focusing on processor performance (now called SPEC89), which has evolved through five generations. The latest is SPEC CPU2006, which consists of a set of 12 integer benchmarks (CINT2006) and 17 floating-point benchmarks (CFP2006). The integer benchmarks vary from part of a C compiler to a chess program to a quantum computer simulation. The floating-point benchmarks include structured grid codes for finite element modeling, particle method codes for molecular dynamics, and sparse linear algebra codes for fluid dynamics.

Figure 1.18 describes the SPEC integer benchmarks and their execution time on the Intel Core i7 and shows the factors that explain execution time: instruction count, CPI, and clock cycle time. Note that CPI varies by more than a factor of 5.

To simplify the marketing of computers, SPEC decided to report a single number summarizing all 12 integer benchmarks. Dividing the execution time of a reference processor by the execution time of the evaluated computer normalizes the execution time measurements; this normalization yields a measure, called the *SPECratio*, which has the advantage that bigger numeric results indicate faster performance. That is, the SPECratio is the inverse of execution time. A CINT2006 or CFP2006 summary measurement is obtained by taking the geometric mean of the SPECratios.

Elaboration: When comparing two computers using SPECratios, apply the geometric mean so that it gives the same relative answer no matter what computer is used to normalize the results. If we averaged the normalized execution time values with an arithmetic mean, the results would vary depending on the computer we choose as the reference.

The formula for the geometric mean is

$$\sqrt[n]{\prod_{i=1}^{n} \text{Execution time ratio}_i}$$

where Execution time ratio$_i$ is the execution time, normalized to the reference computer, for the ith program of a total of n in the workload, and

$$\prod_{i=1}^{n} a_i \text{ means the product } a_1 \times a_2 \times \ldots \times a_n$$

SPEC Power Benchmark

Given the increasing importance of energy and power, SPEC added a benchmark to measure power. It reports power consumption of servers at different workload levels, divided into 10% increments, over a period of time. Figure 1.19 shows the results for a server using Intel Nehalem processors similar to the above.

Target Load %	Performance (ssj_ops)	Average Power (watts)
100%	865,618	258
90%	786,688	242
80%	698,051	224
70%	607,826	204
60%	521,391	185
50%	436,757	170
40%	345,919	157
30%	262,071	146
20%	176,061	135
10%	86,784	121
0%	0	80
Overall Sum	4,787,166	1922
$\sum$ssj_ops / $\sum$power =		2490

FIGURE 1.19   SPECpower_ssj2008 running on a dual socket 2.66 GHz Intel Xeon X5650 with 16 GB of DRAM and one 100 GB SSD disk.

SPECpower started with another SPEC benchmark for Java business applications (SPECJBB2005), which exercises the processors, caches, and main memory as well as the Java virtual machine, compiler, garbage collector, and pieces of the operating system. Performance is measured in throughput, and the units are business operations per second. Once again, to simplify the marketing of computers, SPEC boils these numbers down to one number, called "overall ssj_ops per watt." The formula for this single summarizing metric is

$$\text{overall ssj\_ops per watt} = \left(\sum_{i=0}^{10} \text{ssj\_ops}_i\right) \bigg/ \left(\sum_{i=0}^{10} \text{power}_i\right)$$

where ssj_ops_i is performance at each 10% increment and power_i is power consumed at each performance level.

1.10 Fallacies and Pitfalls

The purpose of a section on fallacies and pitfalls, which will be found in every chapter, is to explain some commonly held misconceptions that you might encounter. We call them *fallacies*. When discussing a fallacy, we try to give a counterexample. We also discuss *pitfalls*, or easily made mistakes. Often pitfalls are generalizations of principles that are true in a limited context. The purpose of these sections is to help you avoid making these mistakes in the computers you may design or use. Cost/performance fallacies and pitfalls have ensnared many a computer architect, including us. Accordingly, this section suffers no shortage of relevant examples. We start with a pitfall that traps many designers and reveals an important relationship in computer design.

Science must begin with myths, and the criticism of myths.

Sir Karl Popper, *The Philosophy of Science*, 1957

Pitfall: Expecting the improvement of one aspect of a computer to increase overall performance by an amount proportional to the size of the improvement.

The great idea of making the **common case fast** has a demoralizing corollary that has plagued designers of both hardware and software. It reminds us that the opportunity for improvement is affected by how much time the event consumes.

A simple design problem illustrates it well. Suppose a program runs in 100 seconds on a computer, with multiply operations responsible for 80 seconds of this time. How much do I have to improve the speed of multiplication if I want my program to run five times faster?

The execution time of the program after making the improvement is given by the following simple equation known as Amdahl's Law:

COMMON CASE FAST

$$\text{Execution time after improvement}$$
$$= \frac{\text{Execution time affected by improvement}}{\text{Amount of improvement}} + \text{Execution time unaffected}$$

For this problem:

$$\text{Execution time after improvement} = \frac{80\,\text{seconds}}{n} + (100 - 80\,\text{seconds})$$

Amdahl's Law
A rule stating that the performance enhancement possible with a given improvement is limited by the amount that the improved feature is used. It is a quantitative version of the law of diminishing returns.

Since we want the performance to be five times faster, the new execution time should be 20 seconds, giving

$$20 \text{ seconds} = \frac{80 \text{ seconds}}{n} + 20 \text{ seconds}$$

$$0 = \frac{80 \text{ seconds}}{n}$$

That is, there is *no amount* by which we can enhance-multiply to achieve a fivefold increase in performance, if multiply accounts for only 80% of the workload. The performance enhancement possible with a given improvement is limited by the amount that the improved feature is used. In everyday life this concept also yields what we call the law of diminishing returns.

We can use Amdahl's Law to estimate performance improvements when we know the time consumed for some function and its potential speedup. Amdahl's Law, together with the CPU performance equation, is a handy tool for evaluating possible enhancements. Amdahl's Law is explored in more detail in the exercises.

Amdahl's Law is also used to argue for practical limits to the number of parallel processors. We examine this argument in the Fallacies and Pitfalls section of Chapter 6.

Fallacy: Computers at low utilization use little power.

Power efficiency matters at low utilizations because server workloads vary. Utilization of servers in Google's warehouse scale computer, for example, is between 10% and 50% most of the time and at 100% less than 1% of the time. Even given 5 years to learn how to run the SPECpower benchmark well, the specially configured computer with the best results in 2012 still uses 33% of the peak power at 10% of the load. Systems in the field that are not configured for the SPECpower benchmark are surely worse.

Since servers' workloads vary but use a large fraction of peak power, Luiz Barroso and Urs Hölzle [2007] argue that we should redesign hardware to achieve "energy-proportional computing." If future servers used, say, 10% of peak power at 10% workload, we could reduce the electricity bill of datacenters and become good corporate citizens in an era of increasing concern about CO_2 emissions.

Fallacy: Designing for performance and designing for energy efficiency are unrelated goals.

Since energy is power over time, it is often the case that hardware or software optimizations that take less time save energy overall even if the optimization takes a bit more energy when it is used. One reason is that all the rest of the computer is consuming energy while the program is running, so even if the optimized portion uses a little more energy, the reduced time can save the energy of the whole system.

Pitfall: Using a subset of the performance equation as a performance metric.

We have already warned about the danger of predicting performance based on simply one of the clock rate, instruction count, or CPI. Another common mistake is to use only two of the three factors to compare performance. Although using two of the three factors may be valid in a limited context, the concept is also easily misused. Indeed, nearly all proposed alternatives to the use of time as the performance metric have led eventually to misleading claims, distorted results, or incorrect interpretations.

One alternative to time is MIPS (million instructions per second). For a given program, MIPS is simply

$$MIPS = \frac{Instruction\,count}{Execution\,time \times 10^6}$$

Since MIPS is an instruction execution rate, MIPS specifies performance inversely to execution time; faster computers have a higher MIPS rating. The good news about MIPS is that it is easy to understand, and quicker computers mean bigger MIPS, which matches intuition.

There are three problems with using MIPS as a measure for comparing computers. First, MIPS specifies the instruction execution rate but does not take into account the capabilities of the instructions. We cannot compare computers with different instruction sets using MIPS, since the instruction counts will certainly differ. Second, MIPS varies between programs on the same computer; thus, a computer cannot have a single MIPS rating. For example, by substituting for execution time, we see the relationship between MIPS, clock rate, and CPI:

$$MIPS = \frac{Instruction\,count}{\dfrac{Instruction\,count \times CPI}{Clock\,rate} \times 10^6} = \frac{Clock\,rate}{CPI \times 10^6}$$

The CPI varied by a factor of 5 for SPEC CPU2006 on an Intel Core i7 computer in Figure 1.18, so MIPS does as well. Finally, and most importantly, if a new program executes more instructions but each instruction is faster, MIPS can vary independently from performance!

Consider the following performance measurements for a program:

Measurement	Computer A	Computer B
Instruction count	10 billion	8 billion
Clock rate	4 GHz	4 GHz
CPI	1.0	1.1

a. Which computer has the higher MIPS rating?

b. Which computer is faster?

million instructions per second (MIPS) A measurement of program execution speed based on the number of millions of instructions. MIPS is computed as the instruction count divided by the product of the execution time and 10^6.

Check Yourself

1.11   Concluding Remarks

Where ... the ENIAC is equipped with 18,000 vacuum tubes and weighs 30 tons, computers in the future may have 1,000 vacuum tubes and perhaps weigh just 1½ tons.

Popular Mechanics,
March 1949

ABSTRACTION

Although it is difficult to predict exactly what level of cost/performance computers will have in the future, it's a safe bet that they will be much better than they are today. To participate in these advances, computer designers and programmers must understand a wider variety of issues.

Both hardware and software designers construct computer systems in hierarchical layers, with each lower layer hiding details from the level above. This great idea of **abstraction** is fundamental to understanding today's computer systems, but it does not mean that designers can limit themselves to knowing a single abstraction. Perhaps the most important example of abstraction is the interface between hardware and low-level software, called the *instruction set architecture*. Maintaining the instruction set architecture as a constant enables many implementations of that architecture—presumably varying in cost and performance—to run identical software. On the downside, the architecture may preclude introducing innovations that require the interface to change.

There is a reliable method of determining and reporting performance by using the execution time of real programs as the metric. This execution time is related to other important measurements we can make by the following equation:

$$\frac{\text{Seconds}}{\text{Program}} = \frac{\text{Instructions}}{\text{Program}} \times \frac{\text{Clock cycles}}{\text{Instruction}} \times \frac{\text{Seconds}}{\text{Clock cycle}}$$

We will use this equation and its constituent factors many times. Remember, though, that individually the factors do not determine performance: only the product, which equals execution time, is a reliable measure of performance.

The **BIG** Picture

Execution time is the only valid and unimpeachable measure of performance. Many other metrics have been proposed and found wanting. Sometimes these metrics are flawed from the start by not reflecting execution time; other times a metric that is sound in a limited context is extended and used beyond that context or without the additional clarification needed to make it valid.

The key hardware technology for modern processors is silicon. Equal in importance to an understanding of integrated circuit technology is an understanding of the expected rates of technological change, as predicted by **Moore's Law**. While silicon fuels the rapid advance of hardware, new ideas in the organization of computers have improved price/performance. Two of the key ideas are exploiting parallelism in the program, normally today via multiple processors, and exploiting locality of accesses to a **memory hierarchy**, typically via caches.

MOORE'S LAW

Energy efficiency has replaced die area as the most critical resource of microprocessor design. Conserving power while trying to increase performance has forced the hardware industry to switch to multicore microprocessors, thereby requiring the software industry to switch to programming parallel hardware. **Parallelism** is now required for performance.

HIERARCHY

Computer designs have always been measured by cost and performance, as well as other important factors such as energy, dependability, cost of ownership, and scalability. Although this chapter has focused on cost, performance, and energy, the best designs will strike the appropriate balance for a given market among all the factors.

Road Map for This Book

At the bottom of these abstractions is the five classic components of a computer: datapath, control, memory, input, and output (refer to Figure 1.5). These five components also serve as the framework for the rest of the chapters in this book:

PARALLELISM

- *Datapath*: Chapter 3, Chapter 4, Chapter 6, and ▦ **Appendix B**

- *Control*: Chapter 4, Chapter 6, and ▦ **Appendix B**

- *Memory*: Chapter 5

- *Input*: Chapters 5 and 6

- *Output*: Chapters 5 and 6

As mentioned above, Chapter 4 describes how processors exploit implicit parallelism, Chapter 6 describes the explicitly parallel multicore microprocessors that are at the heart of the parallel revolution, and ▦ **Appendix B** describes the highly parallel graphics processor chip. Chapter 5 describes how a memory hierarchy exploits locality. Chapter 2 describes instruction sets—the interface between compilers and the computer—and emphasizes the role of compilers and programming languages in using the features of the instruction set. Chapter 3 describes how computers handle arithmetic data. Appendix A introduces logic design.

Historical Perspective and Further Reading

An active field of science is like an immense anthill; the individual almost vanishes into the mass of minds tumbling over each other, carrying information from place to place, passing it around at the speed of light.

Lewis Thomas, "Natural Science," in *The Lives of a Cell*, 1974

For each chapter in the text, a section devoted to a historical perspective can be found online on a site that accompanies this book. We may trace the development of an idea through a series of computers or describe some important projects, and we provide references in case you are interested in probing further.

The historical perspective for this chapter provides a background for some of the key ideas presented in this opening chapter. Its purpose is to give you the human story behind the technological advances and to place achievements in their historical context. By studying the past, you may be better able to understand the forces that will shape computing in the future. Each Historical Perspective section online ends with suggestions for further reading, which are also collected separately online under the section "Further Reading." The rest of 🖥 Section 1.12 is found online.

1.13   Exercises

The relative time ratings of exercises are shown in square brackets after each exercise number. On average, an exercise rated [10] will take you twice as long as one rated [5]. Sections of the text that should be read before attempting an exercise will be given in angled brackets; for example, <§1.4> means you should have read Section 1.4, Under the Covers, to help you solve this exercise.

1.1 [2] <§1.1> Aside from the smart cell phones used by a billion people, list and describe four other types of computers.

1.2 [5] <§1.2> The eight great ideas in computer architecture are similar to ideas from other fields. Match the eight ideas from computer architecture, "Design for Moore's Law," "Use Abstraction to Simplify Design," "Make the Common Case Fast," "Performance via Parallelism," "Performance via Pipelining," "Performance via Prediction," "Hierarchy of Memories," and "Dependability via Redundancy" to the following ideas from other fields:

a. Assembly lines in automobile manufacturing

b. Suspension bridge cables

c. Aircraft and marine navigation systems that incorporate wind information

d. Express elevators in buildings

e. Library reserve desk

f. Increasing the gate area on a CMOS transistor to decrease its switching time

g. Adding electromagnetic aircraft catapults (which are electrically powered as opposed to current steam-powered models), allowed by the increased power generation offered by the new reactor technology

h. Building self-driving cars whose control systems partially rely on existing sensor systems already installed into the base vehicle, such as lane departure systems and smart cruise control systems

1.3 [2] <§1.3> Describe the steps that transform a program written in a high-level language such as C into a representation that is directly executed by a computer processor.

1.4 [2] <§1.4> Assume a color display using 8 bits for each of the primary colors (red, green, blue) per pixel and a frame size of 1280 × 1024.

a. What is the minimum size in bytes of the frame buffer to store a frame?

b. How long would it take, at a minimum, for the frame to be sent over a 100 Mbit/s network?

1.5 [4] <§1.6> Consider three different processors P1, P2, and P3 executing the same instruction set. P1 has a 3 GHz clock rate and a CPI of 1.5. P2 has a 2.5 GHz clock rate and a CPI of 1.0. P3 has a 4.0 GHz clock rate and has a CPI of 2.2.

a. Which processor has the highest performance expressed in instructions per second?

b. If the processors each execute a program in 10 seconds, find the number of cycles and the number of instructions.

c. We are trying to reduce the execution time by 30%, but this leads to an increase of 20% in the CPI. What clock rate should we have to get this time reduction?

1.6 [20] <§1.6> Consider two different implementations of the same instruction set architecture. The instructions can be divided into four classes according to their CPI (classes A, B, C, and D). P1 with a clock rate of 2.5 GHz and CPIs of 1, 2, 3, and 3, and P2 with a clock rate of 3 GHz and CPIs of 2, 2, 2, and 2.

Given a program with a dynamic instruction count of 1.0E6 instructions divided into classes as follows: 10% class A, 20% class B, 50% class C, and 20% class D, which is faster: P1 or P2?

a. What is the global CPI for each implementation?

b. Find the clock cycles required in both cases.

1.7 [15] <§1.6> Compilers can have a profound impact on the performance of an application. Assume that for a program, compiler A results in a dynamic instruction count of 1.0E9 and has an execution time of 1.1 s, while compiler B results in a dynamic instruction count of 1.2E9 and an execution time of 1.5 s.

a. Find the average CPI for each program given that the processor has a clock cycle time of 1 ns.

b. Assume the compiled programs run on two different processors. If the execution times on the two processors are the same, how much faster is the clock of the processor running compiler A's code versus the clock of the processor running compiler B's code?

c. A new compiler is developed that uses only 6.0E8 instructions and has an average CPI of 1.1. What is the speedup of using this new compiler versus using compiler A or B on the original processor?

1.8 The Pentium 4 Prescott processor, released in 2004, had a clock rate of 3.6 GHz and voltage of 1.25 V. Assume that, on average, it consumed 10 W of static power and 90 W of dynamic power.

The Core i5 Ivy Bridge, released in 2012, has a clock rate of 3.4 GHz and voltage of 0.9 V. Assume that, on average, it consumed 30 W of static power and 40 W of dynamic power.

1.8.1 [5] < §1.7> For each processor find the average capacitive loads.

1.8.2 [5] < §1.7> Find the percentage of the total dissipated power comprised by static power and the ratio of static power to dynamic power for each technology.

1.8.3 [15] < §1.7> If the total dissipated power is to be reduced by 10%, how much should the voltage be reduced to maintain the same leakage current? Note: power is defined as the product of voltage and current.

1.9 Assume for arithmetic, load/store, and branch instructions, a processor has CPIs of 1, 12, and 5, respectively. Also assume that on a single processor a program requires the execution of 2.56E9 arithmetic instructions, 1.28E9 load/store instructions, and 256 million branch instructions. Assume that each processor has a 2 GHz clock frequency.

Assume that, as the program is parallelized to run over multiple cores, the number of arithmetic and load/store instructions per processor is divided by $0.7 \times p$ (where p is the number of processors) but the number of branch instructions per processor remains the same.

1.9.1 [5] < §1.7> Find the total execution time for this program on 1, 2, 4, and 8 processors, and show the relative speedup of the 2, 4, and 8 processors result relative to the single processor result.

1.9.2 [10] <§§1.6, 1.8> If the CPI of the arithmetic instructions was doubled, what would the impact be on the execution time of the program on 1, 2, 4, or 8 processors?

1.9.3 [10] <§§1.6, 1.8> To what should the CPI of load/store instructions be reduced in order for a single processor to match the performance of four processors using the original CPI values?

1.10 Assume a 15 cm diameter wafer has a cost of 12, contains 84 dies, and has 0.020 defects/cm². Assume a 20 cm diameter wafer has a cost of 15, contains 100 dies, and has 0.031 defects/cm².

1.10.1 [10] <§1.5> Find the yield for both wafers.

1.10.2 [5] <§1.5> Find the cost per die for both wafers.

1.10.3 [5] <§1.5> If the number of dies per wafer is increased by 10% and the defects per area unit increases by 15%, find the die area and yield.

1.10.4 [5] <§1.5> Assume a fabrication process improves the yield from 0.92 to 0.95. Find the defects per area unit for each version of the technology given a die area of 200 mm².

1.11 The results of the SPEC CPU2006 bzip2 benchmark running on an AMD Barcelona has an instruction count of 2.389E12, an execution time of 750 s, and a reference time of 9650 s.

1.11.1 [5] <§§1.6, 1.9> Find the CPI if the clock cycle time is 0.333 ns.

1.11.2 [5] <§1.9> Find the SPECratio.

1.11.3 [5] <§§1.6, 1.9> Find the increase in CPU time if the number of instructions of the benchmark is increased by 10% without affecting the CPI.

1.11.4 [5] <§§1.6, 1.9> Find the increase in CPU time if the number of instructions of the benchmark is increased by 10% and the CPI is increased by 5%.

1.11.5 [5] <§§1.6, 1.9> Find the change in the SPECratio for this change.

1.11.6 [10] <§1.6> Suppose that we are developing a new version of the AMD Barcelona processor with a 4 GHz clock rate. We have added some additional instructions to the instruction set in such a way that the number of instructions has been reduced by 15%. The execution time is reduced to 700 s and the new SPECratio is 13.7. Find the new CPI.

1.11.7 [10] <§1.6> This CPI value is larger than obtained in 1.11.1 as the clock rate was increased from 3 GHz to 4 GHz. Determine whether the increase in the CPI is similar to that of the clock rate. If they are dissimilar, why?

1.11.8 [5] <§1.6> By how much has the CPU time been reduced?

1.11.9 [10] <§1.6> For a second benchmark, libquantum, assume an execution time of 960 ns, CPI of 1.61, and clock rate of 3 GHz. If the execution time is reduced by an additional 10% without affecting the CPI and with a clock rate of 4 GHz, determine the number of instructions.

1.11.10 [10] <§1.6> Determine the clock rate required to give a further 10% reduction in CPU time while maintaining the number of instructions and with the CPI unchanged.

1.11.11 [10] <§1.6> Determine the clock rate if the CPI is reduced by 15% and the CPU time by 20% while the number of instructions is unchanged.

1.12 Section 1.10 cites as a pitfall the utilization of a subset of the performance equation as a performance metric. To illustrate this, consider the following two processors. P1 has a clock rate of 4 GHz, average CPI of 0.9, and requires the execution of 5.0E9 instructions. P2 has a clock rate of 3 GHz, an average CPI of 0.75, and requires the execution of 1.0E9 instructions.

1.12.1 [5] <§§1.6, 1.10> One usual fallacy is to consider the computer with the largest clock rate as having the highest performance. Check if this is true for P1 and P2.

1.12.2 [10] <§§1.6, 1.10> Another fallacy is to consider that the processor executing the largest number of instructions will need a larger CPU time. Considering that processor P1 is executing a sequence of 1.0E9 instructions and that the CPI of processors P1 and P2 do not change, determine the number of instructions that P2 can execute in the same time that P1 needs to execute 1.0E9 instructions.

1.12.3 [10] <§§1.6, 1.10> A common fallacy is to use MIPS (*millions of instructions per second*) to compare the performance of two different processors, and consider that the processor with the largest MIPS has the largest performance. Check if this is true for P1 and P2.

1.12.4 [10] <§1.10> Another common performance figure is MFLOPS (millions of floating-point operations per second), defined as

MFLOPS = No. FP operations /(execution time × 1E6)

but this figure has the same problems as MIPS. Assume that 40% of the instructions executed on both P1 and P2 are floating-point instructions. Find the MFLOPS figures for the processors.

1.13 Another pitfall cited in Section 1.10 is expecting to improve the overall performance of a computer by improving only one aspect of the computer. Consider a computer running a program that requires 250 s, with 70 s spent executing FP instructions, 85 s executed L/S instructions, and 40 s spent executing branch instructions.

1.13.1 [5] <§1.10> By how much is the total time reduced if the time for FP operations is reduced by 20%?

1.13.2 [5] <§1.10> By how much is the time for INT operations reduced if the total time is reduced by 20%?

1.13.3 [5] <§1.10> Can the total time can be reduced by 20% by reducing only the time for branch instructions?

1.14 Assume a program requires the execution of 50×10^6 FP instructions, 110×10^6 INT instructions, 80×10^6 L/S instructions, and 16×10^6 branch instructions. The CPI for each type of instruction is 1, 1, 4, and 2, respectively. Assume that the processor has a 2 GHz clock rate.

1.14.1 [10] <§1.10> By how much must we improve the CPI of FP instructions if we want the program to run two times faster?

1.14.2 [10] <§1.10> By how much must we improve the CPI of L/S instructions if we want the program to run two times faster?

1.14.3 [5] <§1.10> By how much is the execution time of the program improved if the CPI of INT and FP instructions is reduced by 40% and the CPI of L/S and Branch is reduced by 30%?

1.15 [5] <§1.8> When a program is adapted to run on multiple processors in a multiprocessor system, the execution time on each processor is comprised of computing time and the overhead time required for locked critical sections and/or to send data from one processor to another.

Assume a program requires $t = 100$ s of execution time on one processor. When run p processors, each processor requires t/p s, as well as an additional 4 s of overhead, irrespective of the number of processors. Compute the per-processor execution time for 2, 4, 8, 16, 32, 64, and 128 processors. For each case, list the corresponding speedup relative to a single processor and the ratio between actual speedup versus ideal speedup (speedup if there was no overhead).

§1.1, page 10: Discussion questions: many answers are acceptable.
§1.4, page 24: DRAM memory: volatile, short access time of 50 to 70 nanoseconds, and cost per GB is $5 to $10. Disk memory: nonvolatile, access times are 100,000 to 400,000 times slower than DRAM, and cost per GB is 100 times cheaper than DRAM. Flash memory: nonvolatile, access times are 100 to 1000 times slower than DRAM, and cost per GB is 7 to 10 times cheaper than DRAM.
§1.5, page 28: 1, 3, and 4 are valid reasons. Answer 5 can be generally true because high volume can make the extra investment to reduce die size by, say, 10% a good economic decision, but it doesn't have to be true.
§1.6, page 33: 1. a: both, b: latency, c: neither. 7 seconds.
§1.6, page 40: b.
§1.10, page 51: a. Computer A has the higher MIPS rating. b. Computer B is faster.

Answers to Check Yourself

Instructions: Language of the Computer

I speak Spanish to God, Italian to women, French to men, and German to my horse.

Charles V, Holy Roman Emperor
(1500–1558)

The Five Classic Components of a Computer

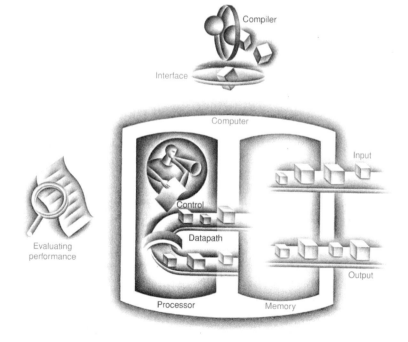

2.1 Introduction

To command a computer's hardware, you must speak its language. The words of a computer's language are called *instructions*, and its vocabulary is called an *instruction set*. In this chapter, you will see the instruction set of a real computer, both in the form written by people and in the form read by the computer. We introduce instructions in a top-down fashion. Starting from a notation that looks like a restricted programming language, we refine it step-by-step until you see the actual language of a real computer. Chapter 3 continues our downward descent, unveiling the hardware for arithmetic and the representation of floating-point numbers.

instruction set The vocabulary of commands understood by a given architecture.

You might think that the languages of computers would be as diverse as those of people, but in reality, computer languages are quite similar, more like regional dialects than independent languages. Hence, once you learn one, it is easy to pick up others.

The chosen instruction set is ARMv8, which comes from ARM Holdings plc and was announced in 2011. For pedagogic reasons, we will use a subset of ARMv8 instructions in this book. We will use the term ARMv8 when talking about the original full instruction set, and LEGv8 when referring to the teaching subset, which is of course based on ARM's ARMv8 instruction set. (LEGv8 is intended as a pun on ARMv8, but it is also a backronym for "Lessen Extrinsic Garrulity.") We'll identify differences between the two in the elaborations. Note that this chapter and several others have a section to give an overview of the rest of the features in ARMv8 that are not in LEGv8 (see Sections 2.19, 3.8, and 5.14).

To demonstrate how easy it is to pick up other instruction sets, we will take a quick look at three other popular instruction sets.

1. MIPS is an elegant example of the instruction sets designed since the 1980s.

2. ARMv7 is an older instruction set also from ARM Holdings plc, but with 32-bit addresses instead of ARMv8's 64 bits. More than 14 billion chips with ARM processors were manufactured in 2015, making them the most popular instruction sets in the world. Ironically, in the authors' opinion, and as shall be seen, ARMv8 is closer to MIPS than it is to ARMv7.

3. The final example is the Intel x86, which powers both the PC and the Cloud of the post-PC era.

This similarity of instruction sets occurs because all computers are constructed from hardware technologies based on similar underlying principles and because there are a few basic operations that all computers must provide. Moreover, computer designers have a common goal: to find a language that makes it easy to build the hardware and the compiler while maximizing performance and minimizing cost and energy. This goal is time-honored; the following quote was written before you could buy a computer, and it is as true today as it was in 1947:

It is easy to see by formal-logical methods that there exist certain [instruction sets] that are in abstract adequate to control and cause the execution of any sequence of operations.... The really decisive considerations from the present point of view, in selecting an [instruction set], are more of a practical nature: simplicity of the equipment demanded by the [instruction set], and the clarity of its application to the actually important problems together with the speed of its handling of those problems.

Burks, Goldstine, and von Neumann, 1947

The "simplicity of the equipment" is as valuable a consideration for today's computers as it was for those of the 1950s. The goal of this chapter is to teach an instruction set that follows this advice, showing both how it is represented in hardware and the relationship between high-level programming languages and this more primitive one. Our examples are in the C programming language; ▦ Section 2.15 shows how these would change for an object-oriented language like Java.

By learning how to represent instructions, you will also discover the secret of computing: the stored-program concept. Moreover, you will exercise your "foreign language" skills by writing programs in the language of the computer and running them on the simulator that comes with this book. You will also see the impact of programming languages and compiler optimization on performance. We conclude with a look at the historical evolution of instruction sets and an overview of other computer dialects.

We reveal our first instruction set a piece at a time, giving the rationale along with the computer structures. This top-down, step-by-step tutorial weaves the components with their explanations, making the computer's language more palatable. Figure 2.1 gives a sneak preview of the instruction set covered in this chapter.

stored-program concept The idea that instructions and data of many types can be stored in memory as numbers and thus be easy to change, leading to the stored-program computer.

2.2 Operations of the Computer Hardware

Every computer must be able to perform arithmetic. The LEGv8 assembly language notation

```
ADD a, b, c
```

instructs a computer to add the two variables b and c and to put their sum in a.

This notation is rigid in that each LEGv8 arithmetic instruction performs only one operation and must always have exactly three variables. For example, suppose we want to place the sum of four variables b, c, d, and e into variable a. (In this section, we are being deliberately vague about what a "variable" is; in the next section, we'll explain in detail.)

The following sequence of instructions adds the four variables:

```
ADD a, b, c     // The sum of b and c is placed in a
ADD a, a, d     // The sum of b, c, and d is now in a
ADD a, a, e     // The sum of b, c, d, and e is now in a
```

Thus, it takes three instructions to sum the four variables.

There must certainly be instructions for performing the fundamental arithmetic operations.

Burks, Goldstine, and von Neumann, 1947

LEGv8 operands

Name	Example	Comments
32 registers	X0-X30, XZR	Fast locations for data. In LEGv8, data must be in registers to perform arithmetic, register XZR always equals 0.
2^{62} memory words	Memory[0], Memory[4], . . . , Memory[4,611,686,018,427,387, 904]	Accessed only by data transfer instructions. LEGv8 uses byte addresses, so sequential doubleword addresses differ by 8. Memory holds data structures, arrays, and spilled registers.

LEGv8 assembly language

Category	Instruction	Example	Meaning	Comments
Arithmetic	add	ADD X1, X2, X3	X1 = X2 + X3	Three register operands
	subtract	SUB X1, X2, X3	X1 = X2 − X3	Three register operands
	add immediate	ADDI X1, X2, 20	X1 = X2 + 20	Used to add constants
	subtract immediate	SUBI X1, X2, 20	X1 = X2 − 20	Used to subtract constants
	add and set flags	ADDS X1, X2, X3	X1 = X2 + X3	Add, set condition codes
	subtract and set flags	SUBS X1, X2, X3	X1 = X2 − X3	Subtract, set condition codes
	add immediate and set flags	ADDIS X1, X2, 20	X1 = X2 + 20	Add constant, set condition codes
	subtract immediate and set flags	SUBIS X1, X2, 20	X1 = X2 − 20	Subtract constant, set condition codes
Data transfer	load register	LDUR X1, [X2,40]	X1 = Memory[X2 + 40]	Doubleword from memory to register
	store register	STUR X1, [X2,40]	Memory[X2 + 40] = X1	Doubleword from register to memory
	load signed word	LDURSW X1,[X2,40]	X1 = Memory[X2 + 40]	Word from memory to register
	store word	STURW X1, [X2,40]	Memory[X2 + 40] = X1	Word from register to memory
	load half	LDURH X1, [X2,40]	X1 = Memory[X2 + 40]	Halfword memory to register
	store half	STURH X1, [X2,40]	Memory[X2 + 40] = X1	Halfword register to memory
	load byte	LDURB X1, [X2,40]	X1 = Memory[X2 + 40]	Byte from memory to register
	store byte	STURB X1, [X2,40]	Memory[X2 + 40] = X1	Byte from register to memory
	load exclusive register	LDXR X1, [X2,0]	X1 = Memory[X2]	Load; 1st half of atomic swap
	store exclusive register	STXR X1, X3 [X2]	Memory[X2]=X1;X3=0 or 1	Store; 2nd half of atomic swap
	move wide with zero	MOVZ X1,20, LSL 0	X1 = 20 or 20 $*$ 2^{16} or 20 $*$ 2^{32} or 20 $*$ 2^{48}	Loads 16-bit constant, rest zeros
	move wide with keep	MOVK X1,20, LSL 0	X1 = 20 or 20 $*$ 2^{16} or 20 $*$ 2^{32} or 20 $*$ 2^{48}	Loads 16-bit constant, rest unchanged

FIGURE 2.1 LEGv8 assembly language revealed in this chapter. This information is also found in Column 1 of the LEGv8 Reference Data Card at the front of this book.

Logical	and	AND X1, X2, X3	X1 = X2 & X3	Three reg. operands; bit-by-bit AND
	inclusive or	ORR X1, X2, X3	X1 = X2 | X3	Three reg. operands; bit-by-bit OR
	exclusive or	EOR X1, X2, X3	X1 = X2 ^ X3	Three reg. operands; bit-by-bit XOR
	and immediate	ANDI X1, X2, 20	X1 = X2 & 20	Bit-by-bit AND reg. with constant
	inclusive or immediate	ORRI X1, X2, 20	X1 = X2 | 20	Bit-by-bit OR reg. with constant
	exclusive or immediate	EORI X1, X2, 20	X1 = X2 ^ 20	Bit-by-bit XOR reg. with constant
	logical shift left	LSL X1, X2, 10	X1 = X2 << 10	Shift left by constant
	logical shift right	LSR X1, X2, 10	X1 = X2 >> 10	Shift right by constant
Conditional branch	compare and branch on equal 0	CBZ X1, 25	if (X1 == 0) go to PC + 100	Equal 0 test; PC-relative branch
	compare and branch on not equal 0	CBNZ X1, 25	if (X1 != 0) go to PC + 100	Not equal 0 test; PC-relative branch
	branch conditionally	B.cond 25	if (condition true) go to PC + 100	Test condition codes; if true, branch
Unconditional branch	branch	B 2500	go to PC + 10000	Branch to target address; PC-relative
	branch to register	BR X30	go to X30	For switch, procedure return
	branch with link	BL 2500	X30 = PC + 4; PC + 10000	For procedure call PC-relative

FIGURE 2.1 (Continued).

The words to the right of the double slashes (//) on each line above are *comments* for the human reader, so the computer ignores them. Note that unlike other programming languages, each line of this language can contain at most one instruction. Another difference from C is that comments always terminate at the end of a line.

The natural number of operands for an operation like addition is three: the two numbers being added together and a place to put the sum. Requiring every instruction to have exactly three operands, no more and no less, conforms to the philosophy of keeping the hardware simple: hardware for a variable number of operands is more complicated than hardware for a fixed number. This situation illustrates the first of three underlying principles of hardware design:

Design Principle 1: Simplicity favors regularity.

We can now show, in the two examples that follow, the relationship of programs written in higher-level programming languages to programs in this more primitive notation.

Compiling Two C Assignment Statements into LEGv8

EXAMPLE

This segment of a C program contains the five variables a, b, c, d, and e. Since Java evolved from C, this example and the next few work for either high-level programming language:

```
a = b + c;
d = a - e;
```

The *compiler* translates from C to LEGv8 assembly language instructions. Show the LEGv8 code produced by a compiler.

ANSWER

An LEGv8 instruction operates on two source operands and places the result in one destination operand. Hence, the two simple statements above compile directly into these two LEGv8 assembly language instructions:

```
ADD a, b, c
SUB d, a, e
```

Compiling a Complex C Assignment into LEGv8

EXAMPLE

A somewhat complicated statement contains the five variables f, g, h, i, and j:

```
f = (g + h) - (i + j);
```

What might a C compiler produce?

ANSWER

The compiler must break this statement into several assembly instructions, since only one operation is performed per LEGv8 instruction. The first LEGv8 instruction calculates the sum of g and h. We must place the result somewhere, so the compiler creates a temporary variable, called t0:

```
ADD t0,g,h // temporary variable t0 contains g + h
```

Although the next operation is subtract, we need to calculate the sum of i and j before we can subtract. Thus, the second instruction places the sum of i and j in another temporary variable created by the compiler, called t1:

```
ADD t1,i,j // temporary variable t1 contains i + j
```

Finally, the subtract instruction subtracts the second sum from the first and places the difference in the variable f, completing the compiled code:

```
SUB f,t0,t1 // f gets t0 - t1, which is (g + h) - (i + j)
```

For a given function, which programming language likely takes the most lines of code? Put the three representations below in order.

Check Yourself

1. Java

2. C

3. LEGv8 assembly language

Elaboration: To increase portability, Java was originally envisioned as relying on a software interpreter. The instruction set of this interpreter is called *Java bytecodes* (see ⊕ Section 2.15), which is quite different from the LEGv8 instruction set. To get performance close to the equivalent C program, Java systems today typically compile Java bytecodes into the native instruction sets like LEGv8. Because this compilation is normally done much later than for C programs, such Java compilers are often called *Just In Time* (JIT) compilers. Section 2.12 shows how JITs are used later than C compilers in the start-up process, and Section 2.13 shows the performance consequences of compiling versus interpreting Java programs.

2.3 Operands of the Computer Hardware

Unlike programs in high-level languages, the operands of arithmetic instructions are restricted; they must be from a limited number of special locations built directly in hardware called *registers*. Registers are primitives used in hardware design that are also visible to the programmer when the computer is completed, so you can think of registers as the bricks of computer construction. The size of a register in the LEGv8 architecture is 64 bits; groups of 64 bits occur so frequently that they are given the name **doubleword** in the LEGv8 architecture. (Another popular size is a group of 32 bits, called a **word** in the LEGv8 architecture.)

One major difference between the variables of a programming language and registers is the limited number of registers, typically 32 on current computers, like LEGv8. (See ⊕ Section 2.22 for the history of the number of registers.) Thus, continuing in our top-down, stepwise evolution of the symbolic representation of the LEGv8 language, in this section we have added the restriction that the three operands of LEGv8 arithmetic instructions must each be chosen from one of the 32 64-bit registers.

The reason for the limit of 32 registers may be found in the second of our three underlying design principles of hardware technology:

Design Principle 2: Smaller is faster.

A very large number of registers may increase the clock cycle time simply because it takes electronic signals longer when they must travel farther.

Guidelines such as "smaller is faster" are not absolutes; 31 registers may not be faster than 32. Even so, the truth behind such observations causes computer designers to take them seriously. In this case, the designer must balance the craving of

doubleword Another natural unit of access in a computer, usually a group of 64 bits; corresponds to the size of a register in the LEGv8 architecture.

word A natural unit of access in a computer, usually a group of 32 bits.

programs for more registers with the designer's desire to keep the clock cycle fast. Another reason for not using more than 32 is the number of bits it would take in the instruction format, as Section 2.5 demonstrates.

Chapter 4 shows the central role that registers play in hardware construction; as we shall see in that chapter, effective use of registers is critical to program performance.

Although we could simply write instructions using numbers for registers, from 0 to 31, the LEGv8 convention is X followed by the number of the register, except for a few register names that we will cover later.

EXAMPLE

Compiling a C Assignment Using Registers

It is the compiler's job to associate program variables with registers. Take, for instance, the assignment statement from our earlier example:

```
f = (g + h) — (i + j);
```

The variables f, g, h, i, and j are assigned to the registers X19, X20, X21, X22, and X23, respectively. What is the compiled LEGv8 code?

ANSWER

The compiled program is very similar to the prior example, except we replace the variables with the register names mentioned above plus two temporary registers, X9 and X10, which correspond to the temporary variables above:

```
ADD X9,X20,X21   // register X9 contains g + h
ADD X10,X22,X23  // register X10 contains i + j
SUB X19,X9,X10   // f gets X9 — X10, which is (g + h) — (i + j)
```

Memory Operands

Programming languages have simple variables that contain single data elements, as in these examples, but they also have more complex data structures—arrays and structures. These composite data structures can contain many more data elements than there are registers in a computer. How can a computer represent and access such large structures?

Recall the five components of a computer introduced in Chapter 1 and repeated on page 61. The processor can keep only a small amount of data in registers, but computer memory contains billions of data elements. Hence, data structures (arrays and structures) are kept in memory.

data transfer instruction A command that moves data between memory and registers.

As explained above, arithmetic operations occur only on registers in LEGv8 instructions; thus, LEGv8 must include instructions that transfer data between memory and registers. Such instructions are called data transfer instructions.

To access a word or doubleword in memory, the instruction must supply the memory address. Memory is just a large, single-dimensional array, with the address acting as the index to that array, starting at 0. For example, in Figure 2.2, the address of the third data element is 2, and the value of memory [2] is 10.

address A value used to delineate the location of a specific data element within a memory array.

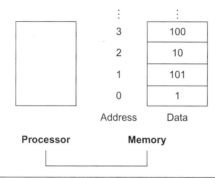

Address	Data
3	100
2	10
1	101
0	1

Processor   Memory

FIGURE 2.2 Memory addresses and contents of memory at those locations. If these elements were doublewords, these addresses would be incorrect, since LEGv8 actually uses byte addressing, with each doubleword representing 8 bytes. Figure 2.3 shows the correct memory addressing for sequential doubleword addresses.

The data transfer instruction that copies data from memory to a register is traditionally called *load*. The format of the load instruction is the name of the operation followed by the register to be loaded, then register and a constant used to access memory. The sum of the constant portion of the instruction and the contents of the second register forms the memory address. The real LEGv8 name for this instruction is LDUR, standing for *load register*.

Elaboration: The U in LDUR stands for *unscaled* immediate as opposite to *scaled* immediate, which we explain in Section 2.19.

Compiling an Assignment When an Operand Is in Memory

EXAMPLE

Let's assume that A is an array of 100 doublewords and that the compiler has associated the variables g and h with the registers X20 and X21 as before. Let's also assume that the starting address, or *base address*, of the array is in X22. Compile this C assignment statement:

```
g = h + A[8];
```

Although there is a single operation in this assignment statement, one of the operands is in memory, so we must first transfer A[8] to a register. The address

ANSWER

of this array element is the sum of the base of the array A, found in register X22, plus the number to select element 8. The data should be placed in a temporary register for use in the next instruction. Based on Figure 2.2, the first compiled instruction is

```
LDUR      X9,[X22,#8] // Temporary reg X9 gets A[8]
```

(We'll be making a slight adjustment to this instruction, but we'll use this simplified version for now.) The following instruction can operate on the value in X9 (which equals A[8]) since it is in a register. The instruction must add h (contained in X21) to A[8] (contained in X9) and put the sum in the register corresponding to g (associated with X20):

```
ADD      X20,X21,X9 // g = h + A[8]
```

The register added to form the address (X22) is called the *base register*, and the constant in a data transfer instruction (8) is called the *offset*.

Hardware/ Software Interface

In addition to associating variables with registers, the compiler allocates data structures like arrays and structures to locations in memory. The compiler can then place the proper starting address into the data transfer instructions.

Since 8-bit *bytes* are useful in many programs, virtually all architectures today address individual bytes. Therefore, the address of a doubleword matches the address of one of the 8 bytes within the doubleword, and addresses of sequential doublewords differ by 8. For example, Figure 2.3 shows the actual LEGv8 addresses for the doublewords in Figure 2.2; the byte address of the third doubleword is 16.

Computers divide into those that use the address of the leftmost or "big end" byte as the doubleword address versus those that use the rightmost or "little end" byte. LEGv8 can work either as *big-endian* or *little-endian*. Since the order matters only if you access the identical data both as a doubleword and as eight bytes, few need to be aware of the "endianess".

Byte addressing also affects the array index. To get the proper byte address in the code above, *the offset to be added to the base register* X22 *must be 8 × 8, or 64*, so that the load address will select A[8] and not A[8/8]. (See the related *Pitfall* on page 171 of Section 2.20.)

The instruction complementary to load is traditionally called *store*; it copies data from a register to memory. The format of a store is similar to that of a load: the name of the operation, followed by the register to be stored, then the base register, and finally the offset to select the array element. Once again, the LEGv8 address is specified in part by a constant and in part by the contents of a register. The actual LEGv8 name is STUR, standing for *store register*.

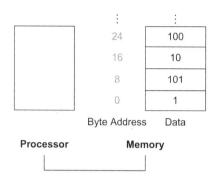

Byte Address   Data

Processor   Memory

FIGURE 2.3 Actual LEGv8 memory addresses and contents of memory for those doublewords. The changed addresses are highlighted to contrast with Figure 2.2. Since LEGv8 addresses each byte, doubleword addresses are multiples of 8: there are 8 bytes in a doubleword.

Elaboration: In many architectures, words must start at addresses that are multiples of 4 and doublewords must start at addresses that are multiples of 8. This requirement is called an alignment restriction. (Chapter 4 suggests why alignment leads to faster data transfers.) ARMv8 and Intel x86 do *not* have alignment restrictions, but ARMv7 and MIPS do.

alignment restriction A requirement that data be aligned in memory on natural boundaries.

Elaboration: It is not quite true that ARMv8 has no alignment restrictions. While it does support unaligned access to normal memory for most data transfer instructions, stack accesses and instruction fetches do have alignment restrictions.

As the addresses in loads and stores are binary numbers, we can see why the DRAM for main memory comes in binary sizes rather than in decimal sizes. That is, in gibibytes (2^{30}) or tebibytes (2^{40}), not in gigabytes (10^9) or terabytes (10^{12}); see Figure 1.1.

Hardware/ Software Interface

Compiling Using Load and Store

Assume variable h is associated with register X21 and the base address of the array A is in X22. What is the LEGv8 assembly code for the C assignment statement below?

```
A[12] = h + A[8];
```

EXAMPLE

Although there is a single operation in the C statement, now two of the operands are in memory, so we need even more LEGv8 instructions. The first two instructions are the same as in the prior example, except this time we use

ANSWER

the proper offset for byte addressing in the load register instruction to select A[8], and the ADD instruction places the sum in X9:

```
LDUR  X9, [X22,#64]  // Temporary reg X9 gets A[8]
ADD   X9,X21,X9       // Temporary reg X9 gets h + A[8]
```

The final instruction stores the sum into A[12], using 96 (8 × 12) as the offset and register X22 as the base register.

```
STUR X9, [X22,#96] // Stores h + A[8] back into A[12]
```

Load register and store register are the instructions that copy doublewords between memory and registers in the ARMv8 architecture. Some brands of computers use other instructions along with load and store to transfer data. An architecture with such alternatives is the Intel x86, described in Section 2.18.

Hardware/ Software Interface

Many programs have more variables than computers have registers. Consequently, the compiler tries to keep the most frequently used variables in registers and places the rest in memory, using loads and stores to move variables between registers and memory. The process of putting less frequently used variables (or those needed later) into memory is called *spilling* registers.

The hardware principle relating size and speed suggests that memory must be slower than registers, since there are fewer registers. This suggestion is indeed the case; data accesses are faster if data are in registers instead of memory.

Moreover, data are more useful when in a register. A LEGv8 arithmetic instruction can read two registers, operate on them, and write the result. A LEGv8 data transfer instruction only reads one operand or writes one operand, without operating on it.

Thus, registers take less time to access *and* have higher throughput than memory, making data in registers both considerably faster to access and simpler to use. Accessing registers also uses much less energy than accessing memory. To achieve the highest performance and conserve energy, an instruction set architecture must have enough registers, and compilers must use registers efficiently.

Elaboration: Let's put the energy and performance of registers versus memory into perspective. Assuming 64-bit data, registers are roughly 200 times faster (0.25 vs. 50 nanoseconds) and are 10,000 times more energy efficient (0.1 vs. 1000 picoJoules) than DRAM in 2015. These large differences led to caches, which reduce the performance and energy penalties of going to memory (see Chapter 5).

Constant or Immediate Operands

Many times a program will use a constant in an operation—for example, incrementing an index to point to the next element of an array. In fact, more than half of the LEGv8 arithmetic instructions have a constant as an operand when running the SPEC CPU2006 benchmarks.

Using only the instructions we have seen so far, we would have to load a constant from memory to use one. (The constants would have been placed in memory when the program was loaded.) For example, to add the constant 4 to register X22, we could use the code

```
LDUR X9, [X20, AddrConstant4]      // X9 = constant 4
ADD  X22,X22,X9                    // X22 = X22 + X9 (X9 == 4)
```

assuming that X20 + AddrConstant4 is the memory address of the constant 4.

An alternative that avoids the load instruction is to offer versions of the arithmetic instructions in which one operand is a constant. This quick add instruction with one constant operand is called *add immediate* or ADDI. To add 4 to register X22, we just write

```
ADDI       X22,X22,#4             // X22 = X22 + 4
```

Constant operands occur frequently, and by including constants inside arithmetic instructions, operations are much faster and use less energy than if constants were loaded from memory.

The constant zero has another role, which is to simplify the instruction set by offering useful variations. For example, the move operation is just an add instruction where one operand is zero. Hence, LEGv8 dedicates a register XZR to be hard-wired to the value zero. (It corresponds to register number 31.) Using frequency to justify the inclusions of constants is another example of the great idea from Chapter 1 of making the **common case fast**.

COMMON CASE FAST

Given the importance of registers, what is the rate of increase in the number of registers in a chip over time?

Check Yourself

1. Very fast: They increase as fast as **Moore's Law**, which predicts doubling the number of transistors on a chip every 18 months.

2. Very slow: Since programs are usually distributed in the language of the computer, there is inertia in instruction set architecture, and so the number of registers increases only as fast as new instruction sets become viable.

Elaboration: Although the LEGv8 registers in this book are 64 bits wide, the full ARMv8 instruction set has two execution states: *AArch32*, in which registers are 32 bits wide (see Section 2.19), and *AArch64*, which has a 64-bit wide register. The former supports the A32 and T32 instruction sets and the latter supports A64. In this chapter, we use a subset of A64 for LEGv8.

MOORE'S LAW

Elaboration: The LEGv8 offset plus base register addressing is an excellent match to structures as well as arrays, since the register can point to the beginning of the structure and the offset can select the desired element. We'll see such an example in Section 2.13.

Elaboration: The register in the data transfer instructions was originally invented to hold an index of an array with the offset used for the starting address of an array. Thus, the base register is also called the *index register*. Today's memories are much larger, and the software model of data allocation is more sophisticated, so the base address of the array is normally passed in a register since it won't fit in the offset, as we shall see.

Elaboration: The migration from 32-bit address computers to 64-bit address computers left compiler writers a choice of the size of data types in C. Clearly, pointers should be 64 bits, but what about integers? Moreover, C has the data types `int`, `long int`, and `long long int`. The problems come from converting from one data type to another and having an unexpected overflow in C code that is not fully standard compliant, which unfortunately is not rare code. The table below shows the two popular options:

Operating System	pointers	int	long int	long long int
Microsoft Windows	64 bits	32 bits	32 bits	64 bits
Linux, Most Unix	64 bits	32 bits	64 bits	64 bits

While each compiler could have different choices, generally the compilers associated with each operating system make the same decision. To keep the examples simple, in this book we'll assume pointers are all 64 bits and declare all C integers as `long long int` to keep them the same size. We also follow C99 standard and declare variables used as indexes to arrays to be `size_t`, which guarantees they are the right size no matter how big the array. They typically declared the same as `long int`.

Elaboration: In the full ARMv8 instruction set, register 31 is `XZR` in most instructions but the stack point (`SP`) in others. We think it is confusing, so register 31 is always `XZR` in LEGv8 and `SP` is always register 28. Besides confusing the reader, it would also complicate datapath design in Chapter 4 if register 31 meant 0 for some instructions and `SP` for others.

Elaboration: The full ARMv8 instruction set does not use the mnemonic `ADDI` when one of the operands is an immediate; it just uses `ADD`, and lets the assembler pick the proper opcode. We worry that it might be confusing to use the same mnemonic for both opcodes, so for teaching purposes LEGv8 distinguishes the two cases with different mnemonics.

Signed and Unsigned Numbers

First, let's quickly review how a computer represents numbers. Humans are taught to think in base 10, but numbers may be represented in any base. For example, 123 base 10 = 1111011 base 2.

Numbers are kept in computer hardware as a series of high and low electronic signals, and so they are considered base 2 numbers. (Just as base 10 numbers are called *decimal* numbers, base 2 numbers are called *binary* numbers.)

A single digit of a binary number is thus the "atom" of computing, since all information is composed of binary digits or *bits*. This fundamental building block can be one of two values, which can be thought of as several alternatives: high or low, on or off, true or false, or 1 or 0.

Generalizing the point, in any number base, the value of ith digit d is

$$d \times \text{Base}^i$$

where i starts at 0 and increases from right to left. This representation leads to an obvious way to number the bits in the doubleword: simply use the power of the base for that bit. We subscript decimal numbers with *ten* and binary numbers with *two*. For example,

$$1011_{two}$$

represents

$$\begin{aligned}
&(1 \times 2^3) + (0 \times 2^2) + (1 \times 2^1) + (1 \times 2^0)_{ten} \\
=~&(1 \times 8) + (0 \times 4) + (1 \times 2) + (1 \times 1)_{ten} \\
=~&\quad 8 \quad + \quad 0 \quad + \quad 2 \quad + \quad 1_{ten} \\
=~&11_{ten}
\end{aligned}$$

We number the bits 0, 1, 2, 3, … from *right to left* in a doubleword. The drawing below shows the numbering of bits within an LEGv8 doubleword and the placement of the number 1011_{two}, (which we must unfortunately split in half to fit on the page of the book):

63 62 61 60	59 58 57 56	55 54 53 52	51 50 49 48	47 46 45 44	43 42 41 40	39 38 37 36	35 34 33 32
0 0 0 0	0 0 0 0	0 0 0 0	0 0 0 0	0 0 0 0	0 0 0 0	0 0 0 0	0 0 0 0

31 30 29 28	27 26 25 24	23 22 21 20	19 18 17 16	15 14 13 12	11 10 9 8	7 6 5 4	3 2 1 0
0 0 0 0	0 0 0 0	0 0 0 0	0 0 0 0	0 0 0 0	0 0 0 0	0 0 0 0	1 0 1 1

(64 bits wide, split into two 32-bit rows)

Since doublewords are drawn vertically as well as horizontally, leftmost and rightmost may be unclear. Hence, the phrase least significant bit is used to refer to the rightmost bit (bit 0 above) and most significant bit to the leftmost bit (bit 63).

binary digit Also called binary bit. One of the two numbers in base 2, 0 or 1, that are the components of information.

least significant bit The rightmost bit in an LEGv8 doubleword.

most significant bit The leftmost bit in an LEGv8 doubleword.

The LEGv8 doubleword is 64 bits long, so we can represent 2^{64} different 64-bit patterns. It is natural to let these combinations represent the numbers from 0 to 2^{64} -1 ($18,446,774,073,709,551,615_{ten}$):

```
00000000 00000000 00000000 00000000 00000000 00000000 00000000 00000000_two = 0_ten
00000000 00000000 00000000 00000000 00000000 00000000 00000000 00000001_two = 1_ten
00000000 00000000 00000000 00000000 00000000 00000000 00000000 00000010_two = 2_ten
. . .                        . . .
11111111 11111111 11111111 11111111 11111111 11111111 11111111 11111101_two = 18,446,774,073,709,551,613_ten
11111111 11111111 11111111 11111111 11111111 11111111 11111111 11111110_two = 18,446,744,073,709,551,614_ten
11111111 11111111 11111111 11111111 11111111 11111111 11111111 11111111_two = 18,446,744,073,709,551,615_ten
```

That is, 64-bit binary numbers can be represented in terms of the bit value times a power of 2 (here x_i means the ith bit of x):

$$(x_{63} \times 2^{63}) + (x_{62} \times 2^{62}) + (x_{61} \times 2^{61}) + \cdots + (x_1 \times 2^1) + (x_0 \times 2^0)$$

For reasons we will shortly see, these positive numbers are called unsigned numbers.

Hardware/ Software Interface

Base 2 is not natural to human beings; we have 10 fingers and so find base 10 natural. Why didn't computers use decimal? In fact, the first commercial computer *did* offer decimal arithmetic. The problem was that the computer still used on and off signals, so a decimal digit was simply represented by several binary digits. Decimal proved so inefficient that subsequent computers reverted to all binary, converting to base 10 only for the relatively infrequent input/output events.

Keep in mind that the binary bit patterns above are simply *representatives* of numbers. Numbers really have an infinite number of digits, with almost all being 0 except for a few of the rightmost digits. We just don't normally show leading 0s.

Hardware can be designed to add, subtract, multiply, and divide these binary bit patterns. If the number that is the proper result of such operations cannot be represented by these rightmost hardware bits, *overflow* is said to have occurred. It's up to the programming language, the operating system, and the program to determine what to do if overflow occurs.

Computer programs calculate both positive and negative numbers, so we need a representation that distinguishes the positive from the negative. The most obvious solution is to add a separate sign, which conveniently can be represented in a single bit; the name for this representation is *sign and magnitude*.

Alas, sign and magnitude representation has several shortcomings. First, it's not obvious where to put the sign bit. To the right? To the left? Early computers tried both. Second, adders for sign and magnitude may need an extra step to set the sign because we can't know in advance what the proper sign will be. Finally, a separate sign bit means that sign and magnitude has both a positive and a negative zero, which can lead to problems for inattentive programmers. Because of these shortcomings, sign and magnitude representation was soon abandoned.

In the search for a more attractive alternative, the question arose as to what would be the result for unsigned numbers if we tried to subtract a large number from a small one. The answer is that it would try to borrow from a string of leading 0s, so the result would have a string of leading 1s.

Given that there was no obvious better alternative, the final solution was to pick the representation that made the hardware simple: leading 0s mean positive, and leading 1s mean negative. This convention for representing signed binary numbers is called *two's complement* representation:

```
00000000 00000000 00000000 00000000 00000000 00000000 00000000_two = 0_ten
00000000 00000000 00000000 00000000 00000000 00000000 00000001_two = 1_ten
00000000 00000000 00000000 00000000 00000000 00000000 00000010_two = 2_ten
...                          ...
01111111 11111111 11111111 11111111 11111111 11111111 11111101_two = 9,223,372,036,854,775,805_ten
01111111 11111111 11111111 11111111 11111111 11111111 11111110_two = 9,223,372,036,854,775,806_ten
01111111 11111111 11111111 11111111 11111111 11111111 11111111_two = 9,223,372,036,854,775,807_ten
10000000 00000000 00000000 00000000 00000000 00000000 00000000_two = - 9,223,372,036,854,775,808_ten
10000000 00000000 00000000 00000000 00000000 00000000 00000001_two = - 9,223,372,036,854,775,807_ten
10000000 00000000 00000000 00000000 00000000 00000000 00000010_two = - 9,223,372,036,854,775,806_ten
...                          ...
11111111 11111111 11111111 11111111 11111111 11111111 11111101_two = - 3_ten
11111111 11111111 11111111 11111111 11111111 11111111 11111110_two = - 2_ten
11111111 11111111 11111111 11111111 11111111 11111111 11111111_two = - 1_ten
```

The positive half of the numbers, from 0 to $9{,}223{,}372{,}036{,}854{,}775{,}807_{ten}$ ($2^{63}-1$), use the same representation as before. The following bit pattern ($1000 \ldots 0000_{two}$) represents the most negative number $-9{,}223{,}372{,}036{,}854{,}775{,}808_{ten}$ (-2^{63}). It is followed by a declining set of negative numbers: $-9{,}223{,}372{,}036{,}854{,}775{,}807_{ten}$ ($1000 \ldots 0001_{two}$) down to -1_{ten} ($1111 \ldots 1111_{two}$).

Two's complement does have one negative number that has no corresponding positive number: $-9{,}223{,}372{,}036{,}854{,}775{,}808_{ten}$. Such imbalance was also a worry to the inattentive programmer, but sign and magnitude had problems for both the programmer *and* the hardware designer. Consequently, every computer today uses two's complement binary representations for signed numbers.

Two's complement representation has the advantage that all negative numbers have a 1 in the most significant bit. Thus, hardware needs to test only this bit to see if a number is positive or negative (with the number 0 is considered positive). This bit is often called the *sign bit*. By recognizing the role of the sign bit, we can represent positive and negative 64-bit numbers in terms of the bit value times a power of 2:

$$(x_{63} \times -2^{63}) + (x_{62} \times 2^{62}) + (x_{61} \times 2^{61}) + \cdots + (x_1 \times 2^1) + (x_0 \times 2^0)$$

The sign bit is multiplied by -2^{63}, and the rest of the bits are then multiplied by positive versions of their respective base values.

Binary to Decimal Conversion

EXAMPLE

What is the decimal value of this 64-bit two's complement number?

11111111 11111111 11111111 11111111 11111111 11111111 11111111 11111100$_{two}$

ANSWER

Substituting the number's bit values into the formula above:

$$(1 \times -2^{63}) + (1 \times 2^{62}) + (1 \times 2^{61}) + \cdots + (1 \times 2^1) + (0 \times 2^1) + (0 \times 2^0)$$
$$= -2^{63} + 2^{62} + 2^{61} + \cdots + 2^2 + 0 + 0$$
$$= -9{,}223{,}372{,}036{,}854{,}775{,}808_{ten} + 9{,}223{,}372{,}036{,}854{,}775{,}804_{ten}$$
$$= -4_{ten}$$

We'll see a shortcut to simplify conversion from negative to positive soon.

Just as an operation on unsigned numbers can overflow the capacity of hardware to represent the result, so can an operation on two's complement numbers. Overflow occurs when the leftmost retained bit of the binary bit pattern is not the same as the infinite number of digits to the left (the sign bit is incorrect): a 0 on the left of the bit pattern when the number is negative or a 1 when the number is positive.

Signed versus unsigned applies to loads as well as to arithmetic. The *function* of a signed load is to copy the sign repeatedly to fill the rest of the register—called *sign extension*—but its *purpose* is to place a correct representation of the number within that register. Unsigned loads simply fill with 0s to the left of the data, since the number represented by the bit pattern is unsigned.

When loading a 64-bit doubleword into a 64-bit register, the point is moot; signed and unsigned loads are identical. ARMv8 does offer two flavors of byte loads: *load byte* (LDURB) treats the byte as an unsigned number and thus zero-extends to fill the leftmost bits of the register, while *load byte signed* (LDURSB) works with signed integers. Since C programs almost always use bytes to represent characters rather than consider bytes as very short signed integers, LDURB is used practically exclusively for byte loads.

Hardware/ Software Interface

Unlike the signed numbers discussed above, memory addresses naturally start at 0 and continue to the largest address. Put another way, negative addresses make no sense. Thus, programs want to deal sometimes with numbers that can be positive or negative and sometimes with numbers that can be only positive. Some programming languages reflect this distinction. C, for example, names the former *integers* (declared as long long int in the program) and the latter *unsigned integers* (unsigned long long int). Some C style guides even recommend declaring the former as signed long long int to keep the distinction clear.

Hardware/ Software Interface

Let's examine two useful shortcuts when working with two's complement numbers. The first shortcut is a quick way to negate a two's complement binary number. Simply invert every 0 to 1 and every 1 to 0, then add one to the result. This shortcut is based on the observation that the sum of a number and its inverted representation must be $111 \ldots 111_{two}$, which represents -1. Since $x + \bar{x} = -1$, therefore $x + \bar{x} + 1 = 0$ or $\bar{x} + 1 = -x$. (We use the notation $\bar{x}$ to mean invert every bit in x from 0 to 1 and vice versa.)

EXAMPLE

Negation Shortcut

Negate 2_{ten}, and then check the result by negating -2_{ten}.

ANSWER

2_{ten} = 00000000 00000000 00000000 00000000 00000000 00000000 00000000 00000010_{two}

Negating this number by inverting the bits and adding one,

$$11111111\ 11111111\ 11111111\ 11111111\ 11111111\ 11111111\ 11111111\ 11111101_{two}$$
$$+\ 1_{two}$$
$$=\ 11111111\ 11111111\ 11111111\ 11111111\ 11111111\ 11111111\ 11111111\ 11111110_{two}$$
$$=\ -2_{ten}$$

Going the other direction,

$$11111111\ 11111111\ 11111111\ 11111111\ 11111111\ 11111111\ 11111111\ 11111110_{two}$$

is first inverted and then incremented:

$$00000000\ 00000000\ 00000000\ 00000000\ 00000000\ 00000000\ 00000000\ 00000001_{two}$$
$$+\ 1_{two}$$
$$=\ 00000000\ 00000000\ 00000000\ 00000000\ 00000000\ 00000000\ 00000000\ 00000010_{two}$$
$$=\ 2_{ten}$$

Our next shortcut tells us how to convert a binary number represented in n bits to a number represented with more than n bits. The shortcut is to take the most significant bit from the smaller quantity—the sign bit—and replicate it to fill the new bits of the larger quantity. The old nonsign bits are simply copied into the right portion of the new doubleword. This shortcut is commonly called *sign extension*.

EXAMPLE

ANSWER

Sign Extension Shortcut

Convert 16-bit binary versions of 2_{ten} and -2_{ten} to 64-bit binary numbers.

The 16-bit binary version of the number 2 is

$$00000000\ 00000010_{two}\ =\ 2_{ten}$$

It is converted to a 64-bit number by making 48 copies of the value in the most significant bit (0) and placing that in the left of the doubleword. The right part gets the old value:

$$00000000\ 00000000\ 00000000\ 00000000\ 00000000\ 00000000\ 00000000\ 00000010_{two}\ =\ 2_{ten}$$

Let's negate the 16-bit version of 2 using the earlier shortcut. Thus,

$$0000\ 0000\ 0000\ 0010_{two}$$

becomes

$$1111\ 1111\ 1111\ 1101_{two}$$
$$+\ 1_{two}$$
$$=\ 1111\ 1111\ 1111\ 1110_{two}$$

Creating a 64-bit version of the negative number means copying the sign bit 48 times and placing it on the left:

$$11111111\ 11111111\ 11111111\ 11111111\ 11111111\ 11111111\ 11111111\ 11111110_{two} = -2_{ten}$$

This trick works because positive two's complement numbers really have an infinite number of 0s on the left and negative two's complement numbers have an infinite number of 1s. The binary bit pattern representing a number hides leading bits to fit the width of the hardware; sign extension simply restores some of them.

Summary

The main point of this section is that we need to represent both positive and negative integers within a computer, and although there are pros and cons to any option, the unanimous choice since 1965 has been two's complement.

Elaboration: For signed decimal numbers, we used "−" to represent negative because there are no limits to the size of a decimal number. Given a fixed data size, binary and hexadecimal (see Figure 2.4) bit strings can encode the sign; therefore, we do not normally use "+" or "−" with binary or hexadecimal notation.

What is the decimal value of this 64-bit two's complement number?

Check Yourself

$$11111111\ 11111111\ 11111111\ 11111111\ 11111111\ 11111111\ 11111111\ 11111000_{two}$$

1) -4_{ten}

2) -8_{ten}

3) -16_{ten}

4) $18,446,744,073,709,551,609_{ten}$

Elaboration: Two's complement gets its name from the rule that the unsigned sum of an n-bit number and its n-bit negative is 2^n; hence, the negation or complement of a number x is $2^n - x$, or its "two's complement."

one's complement A notation that represents the most negative value by 10 … 000$_{two}$ and the most positive value by 01 … 11$_{two}$, leaving an equal number of negatives and positives but ending up with two zeros, one positive (00 … 00$_{two}$) and one negative (11 … 11$_{two}$). The term is also used to mean the inversion of every bit in a pattern: 0 to 1 and 1 to 0.

biased notation A notation that represents the most negative value by 00 … 000$_{two}$ and the most positive value by 11 … 11$_{two}$, with 0 typically having the value 10 … 00$_{two}$, thereby biasing the number such that the number plus the bias has a non-negative representation.

A third alternative representation to two's complement and sign and magnitude is called **one's complement**. The negative of a one's complement is found by inverting each bit, from 0 to 1 and from 1 to 0, or $\overline{x}$. This relation helps explain its name since the complement of x is $2^n - x - 1$. It was also an attempt to be a better solution than sign and magnitude, and several early scientific computers did use the notation. This representation is similar to two's complement except that it also has two 0s: 00 … 00$_{two}$ is positive 0 and 11 … 11$_{two}$ is negative 0. The most negative number, 10 … 000$_{two}$, represents $-2{,}147{,}483{,}647_{ten}$, and so the positives and negatives are balanced. One's complement adders did need an extra step to subtract a number, and hence two's complement dominates today.

A final notation, which we will look at when we discuss floating point in Chapter 3, is to represent the most negative value by 00 … 000$_{two}$ and the most positive value by 11 … 11$_{two}$, with 0 typically having the value 10 … 00$_{two}$. This representation is called a **biased notation**, since it biases the number such that the number plus the bias has a non-negative representation.

2.5 Representing Instructions in the Computer

We are now ready to explain the difference between the way humans instruct computers and the way computers see instructions.

Instructions are kept in the computer as a series of high and low electronic signals and may be represented as numbers. In fact, each piece of an instruction can be considered as an individual number, and placing these numbers side by side forms the instruction. The 32 registers of LEGv8 are just referred to by their number, from 0 to 31.

EXAMPLE

Translating a LEGv8 Assembly Instruction into a Machine Instruction

Let's do the next step in the refinement of the LEGv8 language as an example. We'll show the real LEGv8 language version of the instruction represented symbolically as

```
ADD X9,X20,X21
```

first as a combination of decimal numbers and then of binary numbers.

The decimal representation is

ANSWER

| 1112 | 21 | 0 | 20 | 9 |

Each of these segments of an instruction is called a *field*. The first field (containing 1112 in this case) tells the LEGv8 computer that this instruction performs addition. The second field gives the number of the register that is the second source operand of the addition operation (21 for X21), and the fourth field gives the other source operand for the addition (20 for X20). The fifth field contains the number of the register that is to receive the sum (9 for X9). (The third field is unused in this instruction, so it is set to 0.) Thus, this instruction adds register X20 to register X21 and places the sum in register X9.

This instruction can also be represented as fields of binary numbers instead of decimal:

10001011000	10101	000000	10100	01001
11 bits	5 bits	6 bits	5 bits	5 bits

This layout of the instruction is called the instruction format. As you can see from counting the number of bits, this LEGv8 instruction takes exactly 32 bits—a word, or one half of a doubleword. In keeping with our design principle that simplicity favors regularity, all LEGv8 instructions are 32 bits long.

To distinguish it from assembly language, we call the numeric version of instructions machine language and a sequence of such instructions *machine code*.

It would appear that you would now be reading and writing long, tiresome strings of binary numbers. We avoid that tedium by using a higher base than binary that converts easily into binary. Since almost all computer data sizes are multiples of 4, hexadecimal (base 16) numbers are popular. As base 16 is a power of 2, we can trivially convert by replacing each group of four binary digits by a single hexadecimal digit, and vice versa. Figure 2.4 converts between hexadecimal and binary.

instruction format A form of representation of an instruction composed of fields of binary numbers.

machine language Binary representation used for communication within a computer system.

hexadecimal Numbers in base 16.

Hexadecimal	Binary	Hexadecimal	Binary	Hexadecimal	Binary	Hexadecimal	Binary
0_{hex}	0000_{two}	4_{hex}	0100_{two}	8_{hex}	1000_{two}	c_{hex}	1100_{two}
1_{hex}	0001_{two}	5_{hex}	0101_{two}	9_{hex}	1001_{two}	d_{hex}	1101_{two}
2_{hex}	0010_{two}	6_{hex}	0110_{two}	a_{hex}	1010_{two}	e_{hex}	1110_{two}
3_{hex}	0011_{two}	7_{hex}	0111_{two}	b_{hex}	1011_{two}	f_{hex}	1111_{two}

FIGURE 2.4 The hexadecimal–binary conversion table. Just replace one hexadecimal digit by the corresponding four binary digits, and vice versa. If the length of the binary number is not a multiple of 4, go from right to left.

Because we frequently deal with different number bases, to avoid confusion, we will subscript decimal numbers with *ten*, binary numbers with *two*, and hexadecimal numbers with *hex*. (If there is no subscript, the default is base 10.) By the way, C and Java use the notation 0x*nnnn* for hexadecimal numbers.

EXAMPLE

Binary to Hexadecimal and Back

Convert the following 8-digit hexadecimal and 32-bit binary numbers into the other base:

eca8 6420$_{hex}$
0001 0011 0101 0111 1001 1011 1101 1111$_{two}$

ANSWER

Using Figure 2.4, the answer is just a table lookup one way:

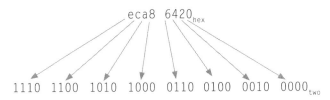

And then the other direction:

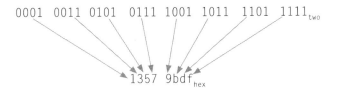

LEGv8 Fields

LEGv8 fields are given names to make them easier to discuss:

opcode	Rm	shamt	Rn	Rd
11 bits	5 bits	6 bits	5 bits	5 bits

Here is the meaning of each name of the fields in LEGv8 instructions:

opcode The field that denotes the operation and format of an instruction.

- *opcode:* Basic operation of the instruction, and this abbreviation is its traditional name.

- *Rm:* The second register source operand.

- *shamt:* Shift amount. (Section 2.6 explains shift instructions and this term; it will not be used until then, and hence the field contains zero in this section.)

- *Rn:* The first register source operand.

- *Rd:* The register destination operand. It gets the result of the operation.

A problem occurs when an instruction needs longer fields than those shown above. For example, the load register instruction must specify two registers and a constant. If the address were to use one of the 5-bit fields in the format above, the largest constant within the load register instruction would be limited to only 2^5-1 or 31. This constant is used to select elements from arrays or data structures, and it often needs to be much larger than 31. This 5-bit field is too small to be useful.

Hence, we have a conflict between the desire to keep all instructions the same length and the desire to have a single instruction format. This conflict leads us to the final hardware design principle:

Design Principle 3: Good design demands good compromises.

The compromise chosen by the LEGv8 designers is to keep all instructions the same length, thereby requiring distinct instruction formats for different kinds of instructions. For example, the format above is called *R-type* (for register) or *R-format*. A second type of instruction format is *D-type* or *D-format* and is used by the data transfer instructions (loads and stores). The fields of D-format are

opcode	address	op2	Rn	Rt
11 bits	9 bits	2 bits	5 bits	5 bits

The 9-bit address means a load register instruction can load any doubleword within a region of $\pm 2^8$ or 256 bytes ($\pm 2^5$ or 32 doublewords) of the address in the base register Rn. We see that more than 32 registers would be difficult in this format, as the Rn and Rt fields would each need another bit, making it harder to fit everything in one word. (The last field of D-type is called Rt instead of Rd because for store instructions, the field indicates a data source and not a data destination.)

Let's look at the load register instruction from page 72:

```
LDUR X9, [X22,#64] // Temporary reg X9 gets A[8]
```

Here, 22 (for X22) is placed in the Rn field, 64 is placed in the address field, and 9 (for X9) is placed in the Rt field. Note that in a load register instruction, the Rt field specifies the *destination* register, which receives the result of the load.

We also need a format for the immediate instructions ADDI, SUBI, and immediate instructions that we will introduce later. While we could have used the D-format instruction since it has a 9-bit field holding a constant, the ARMv8 architects decided it would be useful to have a larger immediate field for these instructions, even shaving a bit from the opcode field to make a 12-bit immediate. The fields of *immediate* or *I-type* format are

opcode	immediate	Rn	Rd
10 bits	12 bits	5 bits	5 bits

Although multiple formats complicate the hardware, we can reduce the complexity by keeping the formats similar. For example, the last two fields of all three formats are the identical size and almost the same names, and the opcode field is the same size in two of the three formats.

In case you were wondering, the formats are distinguished by the values in the first field: each format is assigned a distinct set of values in the first field (opcode) so that the hardware knows how to treat the rest of the instruction. Figure 2.5 shows the numbers used in each field for the LEGv8 instructions covered so far.

Instruction	Format	opcode	Rm	shamt	address	op2	Rn	Rd
ADD (add)	R	1112_{ten}	reg	0	n.a.	n.a.	reg	reg
SUB (subtract)	R	1624_{ten}	reg	0	n.a.	n.a.	reg	reg
ADDI (add immediate)	I	580_{ten}	reg	n.a.	constant	n.a.	reg	n.a.
SUBI (sub immediate)	I	836_{ten}	reg	n.a.	constant	n.a.	reg	n.a.
LDUR (load word)	D	1986_{ten}	reg	n.a.	address	0	reg	n.a.
STUR (store word)	D	1984_{ten}	reg	n.a.	address	0	reg	n.a.

FIGURE 2.5   LEGv8 instruction encoding. In the table above, "reg" means a register number between 0 and 31, "address" means a 9-bit address or 12-bit constant, and "n.a." (not applicable) means this field does not appear in this format. The op2 field expands the opcode field.

EXAMPLE

Translating LEGv8 Assembly Language into Machine Language

We can now take an example all the way from what the programmer writes to what the computer executes. If X10 has the base of the array A and X21 corresponds to h, the assignment statement

```
A[30] = h + A[30] + 1;
```

is compiled into

```
LDUR X9, [X10,#240] // Temporary reg X9 gets A[30]
ADD  X9,X21,X9       // Temporary reg X9 gets h+A[30]
ADDI X9,X9,#1        // Temporary reg X9 gets h+A[30]+1
STUR X9, [X10,#240] // Stores h+A[30]+1 back into A[30]
```

What is the LEGv8 machine language code for these three instructions?

ANSWER

For convenience, let's first represent the machine language instructions using decimal numbers. From Figure 2.5, we can determine the three machine language instructions:

opcode	Rm/address	shamt/op2	Rn	Rd/Rt
1986	240	0	10	9
1112	9	0	21	9
580	1		9	9
1984	240	0	10	9

The LDUR instruction is identified by 1986 (see Figure 2.5) in the first field (opcode). The base register 10 is specified in the fourth field (Rn), and the destination register 9 is specified in the last field (Rt). The offset to select A[30] (240 = 30 × 8) is found in the second field (address).

The ADD instruction that follows is specified with 1112 in the first field (opcode). The three register operands (9, 21, and 9) are found in the second, fourth, and fifth fields, with 0 in the third field (shamt).

The following ADDI instruction is specified with 580 in the first field (opcode), the immediate value 1 in the second, and the register operands (9 in both cases) in the last two fields.

The STUR instruction is identified with 1984 in the first field. The rest of this final instruction is identical to the LDUR instruction.

Since $240_{ten} = 0\ 1111\ 0000_{two}$, the binary equivalent to the decimal form is:

11111000010	011110000	00	01010	01001
10001011000	01001	000000	10101	01001
1001000100	000000000001		01001	01001
11111000000	011110000	00	01010	01001

Note the similarity of the binary representations of the first and last instructions. The only difference is in the tenth bit from the left, which is highlighted here.

Elaboration: ARMv8 assembly language programmers aren't forced to use ADDI when working with constants. The programmer simply writes ADD, and the assembler generates the proper opcode and the proper instruction format depending on whether the operands are all registers (R-format) or if one is a constant (I-format). We use the explicit names in LEGv8 for the different opcodes and formats as we think it is less confusing when introducing assembly language versus machine language.

Elaboration: Note that unlike MIPS, the LEGv8 immediate field in I-format is zero-extended. Thus, LEGv8 includes both ADDI and SUBI instructions, while MIPS has just ADDI and both positive and negative immediates.

The desire to keep all instructions the same size conflicts with the desire to have as many registers as possible. Any increase in the number of registers uses up at least one more bit in every register field of the instruction format. Given these constraints and the design principle that smaller is faster, most instruction sets today have 16 or 32 general-purpose registers.

Hardware/ Software Interface

Figure 2.6 summarizes the portions of LEGv8 machine language described in this section. As we shall see in Chapter 4, the similarity of the binary representations of related instructions simplifies hardware design. These similarities are another example of regularity in the LEGv8 architecture.

LEGv8

Name	Format	Example						Comments
ADD	R	1112	3		0	2	1	ADD X1, X2, X3
SUB	R	1624	3		0	2	1	SUB X1, X2, X3
ADDI	I	580	100			2	1	ADDI X1, X2, #100
SUBI	I	836	100			2	1	SUBI X1, X2, #100
LDUR	D	1986	100		0	2	1	LDUR X1, [X2, #100]
STUR	D	1984	100		0	2	1	STUR X1, [X2, #100]
Field size		11 or 10 bits	5 bits	5 or 4 bits	2 bits	5 bits	5 bits	All ARM instructions are 32 bits long
R-format	R	opcode	Rm	shamt		Rn	Rd	Arithmetic instruction format
I-format	I	opcode	immediate			Rn	Rd	Immediate format
D-format	D	opcode	address		op2	Rn	Rt	Data transfer format

FIGURE 2.6   LEGv8 architecture revealed through Section 2.5. The three LEGv8 instruction formats so far are R, I and D. The last 10 bits contain a *Rn* field, giving one of the sources; and the *Rd* or *Rt* field, which specifies the destination register, except for store register, where it specifies the value to be stored. R-format divides the rest into an 11-bit opcode; a 5-bit *Rm* field, specifying the other source operand; and a 6-bit *shamt* field, which Section 2.6 explains. I-format combines 12 bits into a single *immediate* field, which requires shrinking the opcode field to 10 bits. The D-format uses a full 11-bit opcode like the R-format, plus a 9-bit *address* field, and a 2-bit *op2* field. The op2 field is logically an extension of the opcode field.

The **BIG**
Picture

Today's computers are built on two key principles:

1.   Instructions are represented as numbers.

2.   Programs are stored in memory to be read or written, just like data.

These principles lead to the *stored-program* concept; its invention let the computing genie out of its bottle. Figure 2.7 shows the power of the concept; specifically, memory can contain the source code for an editor program, the corresponding compiled machine code, the text that the compiled program is using, and even the compiler that generated the machine code.

One consequence of instructions as numbers is that programs are often shipped as files of binary numbers. The commercial implication is that computers can inherit ready-made software provided they are compatible with an existing instruction set. Such "binary compatibility" often leads industry to align around a small number of instruction set architectures.

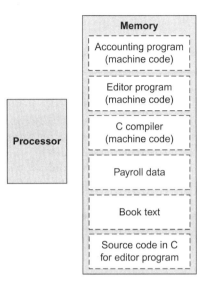

FIGURE 2.7 The stored-program concept. Stored programs allow a computer that performs accounting to become, in the blink of an eye, a computer that helps an author write a book. The switch happens simply by loading memory with programs and data and then telling the computer to begin executing at a given location in memory. Treating instructions in the same way as data greatly simplifies both the memory hardware and the software of computer systems. Specifically, the memory technology needed for data can also be used for programs, and programs like compilers, for instance, can translate code written in a notation far more convenient for humans into code that the computer can understand.

What LEGv8 instruction does this represent? Choose from one of the four options below.

Check Yourself

opcode	Rm	shamt	Rn	Rd
1624	9	0	10	11

1. SUB X9, X10, X11

2. ADD X11, X9, X10

3. SUB X11, X10, X9

4. SUB X11, X9, X10

Elaboration: You might be asking yourself why the LEGv8 opcode field is so big given the modest number of instructions in Figure 2.1? The main reason is that the full ARMv8 instruction set is very large; depending how you count, it is on the order of 1000 instructions. We'll survey the full ARMv8 instruction set in some of the last sections of this chapter and Chapters 3 and 5.

"Contrariwise,"
continued Tweedledee,
"if it was so, it might
be; and if it were so, it
would be; but as
it isn't, it ain't.
That's logic."

Lewis Carroll,
Alice's Adventures in
Wonderland, 1865

2.6   Logical Operations

Although the first computers operated on full words, it soon became clear that it was useful to operate on fields of bits within a word or even on individual bits. Examining characters within a word, each of which is stored as 8 bits, is one example of such an operation (see Section 2.9). It follows that operations were added to programming languages and instruction set architectures to simplify, among other things, the packing and unpacking of bits into words. These instructions are called *logical operations.* Figure 2.8 shows logical operations in C, Java, and LEGv8.

Logical operations	C operators	Java operators	LEGv8 instructions
Shift left	<<	<<	LSL
Shift right	>>	>>>	LSR
Bit-by-bit AND	&	&	AND,ANDI
Bit-by-bit OR	|	|	OR,ORI
Bit-by-bit NOT	~	~	EOR,EORI

FIGURE 2.8   C and Java logical operators and their corresponding LEGv8 instructions. One way to implement NOT is to use EOR with one operand being all ones (FFFF FFFF FFFF FFFF$_{hex}$).

The first class of such operations is called *shifts.* They move all the bits in a doubleword to the left or right, filling the emptied bits with 0s. For example, if register X19 contained

00000000 00000000 00000000 00000000 00000000 00000000 00000000 00001001$_{two}$ = 9$_{ten}$

and the instruction to shift left by 4 was executed, the new value would be:

00000000 00000000 00000000 00000000 00000000 00000000 00000000 10010000$_{two}$ = 144$_{ten}$

The dual of a shift left is a shift right. The actual names of the two LEGv8 shift instructions are *logical shift left* (LSL) and *logical shift right* (LSR). The following instruction performs the operation above, if the original value was in register X19 and the result should go in register X11:

```
LSL X11,X19,#4 // reg X11 = reg X19 << 4 bits
```

We delayed explaining the *shamt* field in the R-format. Used in shift instructions, it stands for *shift amount.* Hence, the machine language version of the instruction above is:

opcode	Rm	shamt	Rn	Rd
1691	0	4	19	11

The encoding of LSL is 1691 in the opcode field, Rd contains 11, Rn contains 19, and shamt contains 4. The Rm field is unused and thus is set to 0.

Shift left logical provides a bonus benefit. Shifting left by i bits gives the identical result as multiplying by 2^i, just as shifting a decimal number by i digits is equivalent to multiplying by 10^i. For example, the above LSL shifts by 4, which gives the same result as multiplying by 2^4 or 16. The first bit pattern above represents 9, and $9 \times 16 = 144$, the value of the second bit pattern.

Another useful operation that isolates fields is AND. (We capitalize the word to avoid confusion between the operation and the English conjunction.) AND is a bit-by-bit operation that leaves a 1 in the result only if both bits of the operands are 1. For example, if register X11 contains

> **AND** A logical bit-by-bit operation with two operands that calculates a 1 only if there is a 1 in *both* operands.

```
00000000 00000000 00000000 00000000 00000000 00000000 00001101 11000000₂
```

and register X10 contains

```
00000000 00000000 00000000 00000000 00000000 00000000 00111100 00000000₂
```

then, after executing the LEGv8 instruction

```
    AND X9,X10,X11      // reg X9 = reg X10 & reg X11
```

the value of register X9 would be

```
00000000 00000000 00000000 00000000 00000000 00000000 00001100 00000000₂
```

As you can see, AND can apply a bit pattern to a set of bits to force 0s where there is a 0 in the bit pattern. Such a bit pattern in conjunction with AND is traditionally called a *mask*, since the mask "conceals" some bits.

To place a value into one of these seas of 0s, there is the dual to AND, called OR. It is a bit-by-bit operation that places a 1 in the result if *either* operand bit is a 1. To elaborate, if the registers X10 and X11 are unchanged from the preceding example, the result of the LEGv8 instruction

> **OR** A logical bit-by-bit operation with two operands that calculates a 1 if there is a 1 in *either* operand.

```
    ORR X9,X10,X11 // reg X9 = reg X10 | reg X11
```

is this value in register X9:

```
00000000 00000000 00000000 00000000 00000000 00000000 00111101 11000000₂
```

The final logical operation is a contrarian. NOT takes one operand and places a 1 in the result if one operand bit is a 0, and vice versa. Using our prior notation, it calculates $\bar{x}$.

> **NOT** A logical bit-by-bit operation with one operand that inverts the bits; that is, it replaces every 1 with a 0, and every 0 with a 1.

EOR A logical bit-by-bit operation with two operands that calculates the Exclusive OR of the two operands. That is, it calculates a 1 only if the values are different in the two operands.

In keeping with the three-operand format, the designers of ARMv8 decided to include the instruction EOR (Exclusive OR) instead of NOT. Since exclusive OR creates a 0 when bits are the same and a 1 if they are different, the equivalent to NOT is an EOR 111…111.

If the register X10 is unchanged from the preceding example and register X12 has the value 0, the result of the LEGv8 instruction

```
EOR X9,X10,X12 // reg X9 = reg X10 | reg X12
```

is this value in register X9:

```
00000000 00000000 00000000 00000000 00000000 00000000 00110001 11000000_two
```

Figure 2.8 above shows the relationship between the C and Java operators and the LEGv8 instructions. Constants are useful in logical operations as well as in arithmetic operations, so LEGv8 also provides the instructions *and immediate* (ANDI), *or immediate* (ORRI), and *exclusive or immediate* (EORI).

Elaboration: C allows *bit fields* or *fields* to be defined within doublewords, both allowing objects to be packed within a doubleword and to match an externally enforced interface such as an I/O device. All fields must fit within a single doubleword. Fields are unsigned integers that can be as short as 1 bit. C compilers insert and extract fields using logical instructions in LEGv8: AND, ORR, LSL, and LSR.

Elaboration: The immediate fields for ANDI, ORRI, and EORI of the full ARMv8 instruction set are not simple 12-bit immediates. Once again, like ARMv7, it has the unusual feature of using a complex algorithm for encoding immediate values; ARMv8 does it with repeating patterns. This means that some small constants (e.g., 1, 2, 3, 4, and 6) are valid, while others (e.g., −1, 0, 5) are not. LEGv8 simply uses normal 12-bit immediates as found in ADDI. This difference means EORI X1,X1,#5 is legal for LEGv8 but not ARMv8. Once again, in addition to its rarity among other instruction sets, immediate encoding is omitted because it would complicate the datapaths in Chapter 4 significantly.

Check Yourself

Which operations can isolate a field in a doubleword?

1. AND

2. A shift left followed by a shift right

Elaboration: Unlike almost all other computer architectures, ARMv8 (and ARMv7) allows a register to be shifted as part of an arithmetic or logical instruction: add an optionally shifted register, subtract an optionally shifted register, AND an optionally shifted register, and so on. Since this combination is unusual in computer architectures and not frequently generated by compilers—and since supporting it would make the data path in Chapter 4 much more complicated and unlike other computer datapaths—we decided to treat shifts as separate instructions, as it is in virtually every other computer architecture. While you can synthesize a shift using either ADD with XZR or OR with XZR, it would be confusing to use the same opcode for shifts as ADD or OR. Thus, we follow the ARMv8 recommendation of using an UBFM (unsigned bitfield move) instruction and its opcode. We simplify the values put into the Rm and shamt fields to be 0 and the actual immediate shift amount, which is what it looks like in ARMv8 assembly language. The actual values in the Rm and shamt fields of UBFM should be (–shift amount MOD 64) and (63 – shift amount) for LSL and shift amount and 63 for LSR. The opcode field includes part of immediate field in ARMv8, so we make the two opcodes 1691 and 1690, respectively, to distinguish them.

2.7 Instructions for Making Decisions

What distinguishes a computer from a simple calculator is its ability to make decisions. Based on the input data and the values created during computation, different instructions execute. Decision making is commonly represented in programming languages using the *if* statement, sometimes combined with *go to* statements and labels. LEGv8 assembly language includes two decision-making instructions, similar to an *if* statement with a *go to*. The first instruction is

```
CBZ register, L1
```

This instruction means go to the statement labeled L1 if the value in register equals zero. The mnemonic CBZ stands for *compare and branch if zero*. The second instruction is

```
CBNZ register, L1
```

It means go to the statement labeled L1 if the value in register does *not* equal zero. The mnemonic CBNZ stands for *compare and branch if not zero*. These two instructions are traditionally called conditional branches.

The utility of an automatic computer lies in the possibility of using a given sequence of instructions repeatedly, the number of times it is iterated being dependent upon the results of the computation.... This choice can be made to depend upon the sign of a number (zero being reckoned as plus for machine purposes). Consequently, we introduce an [instruction] (the conditional transfer [instruction]) which will, depending on the sign of a given number, cause the proper one of two routines to be executed.

Burks, Goldstine, and von Neumann, 1947

Compiling *if-then-else* into Conditional Branches

In the following code segment, f, g, h, i, and j are variables. If the five variables f through j correspond to the five registers X19 through X23, what is the compiled LEGv8 code for this C *if* statement?

```
if (i == j) f = g + h; else f = g - h;
```

Figure 2.9 shows a flowchart of what the LEGv8 code should do. The first expression compares for equality between two variables in registers. Given that the instructions above can only test to see if a register is zero, the first step is to subtract j from i to test if the difference is zero. It would seem that we would next want to branch if the difference is zero (CBZ). In general, the code will be more efficient if we test for the opposite condition to branch over the code that branches if the difference is *not* equal to zero (CBNZ). Here are the two instructions, using register X9 to hold the result of subtracting j from i:

```
SUB  X9,X22,X23 // X9 = i - j
CBNZ X9, Else   // go to Else if i ≠ j (X9 ≠ 0)
```

conditional branch An instruction that tests a value and that allows for a subsequent transfer of control to a new address in the program based on the outcome of the test.

The next assignment statement performs a single operation, and if all the operands are allocated to registers, it is just one instruction:

```
ADD X19,X20,X21        // f = g + h (skipped if i ≠ j)
```

We now need to go to the end of the *if* statement. This example introduces another kind of branch, often called an *unconditional branch*. This instruction says that the processor always follows the branch. To distinguish between conditional and unconditional branches, the LEGv8 name for this type of instruction is *branch*, abbreviated as B (the label Exit is defined below).

```
B Exit       // go to Exit
```

The assignment statement in the *else* portion of the *if* statement can again be compiled into a single instruction. We just need to append the label Else to this instruction. We also show the label Exit that is after this instruction, showing the end of the *if-then-else* compiled code:

```
Else:SUB X19,X20,X21    // f = g - h (skipped if i = j)
Exit:
```

Notice that the assembler relieves the compiler and the assembly language programmer from the tedium of calculating addresses for branches, just as it does for calculating data addresses for loads and stores (see Section 2.12).

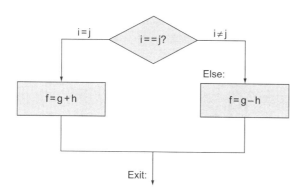

FIGURE 2.9 Illustration of the options in the *if* statement above. The left box corresponds to the *then* part of the *if* statement, and the right box corresponds to the *else* part.

Compilers frequently create branches and labels where they do not appear in the programming language. Avoiding the burden of writing explicit labels and branches is one benefit of writing in high-level programming languages and is a reason coding is faster at that level.

Hardware/ Software Interface

Loops

Decisions are important both for choosing between two alternatives—found in *if* statements—and for iterating a computation—found in loops. The same assembly instructions are the building blocks for both cases.

Compiling a *while* Loop in C

Here is a traditional loop in C:

```
while (save[i] == k)
      i += 1;
```

Assume that i and k correspond to registers X22 and X24 and the base of the array save is in X25. What is the LEGv8 assembly code corresponding to this C code?

EXAMPLE

The first step is to load save[i] into a temporary register. Before we can load save[i] into a temporary register, we need to have its address. Before we can add i to the base of array save to form the address, we must multiply the index i by 8 due to the byte addressing issue. Fortunately, we can use shift

ANSWER

left, since shifting left by 3 bits multiplies by 2^3 or 8 (see page 91 in the prior section). We need to add the label Loop to it so that we can branch back to that instruction at the end of the loop:

```
Loop: LSL X10,X22,#3      // Temp reg X10 = i * 8
```

To get the address of save[i], we need to add X10 and the base of save in X25:

```
ADD X10,X10,X25       // X10 = address of save[i]
```

Now we can use that address to load save[i] into a temporary register:

```
LDUR X9, [X10,#0]      // Temp reg X9 = save[i]
```

The next instruction subtracts k from save[i] and puts the difference into X11 to set up the loop test. If X11 is not 0, then they are unequal (save[i] ≠ k):

```
SUB X11,X9,X24        // X11 = save[i] − k
```

The next instruction performs the loop test, exiting if save[i] ≠ k:

```
CBNZ X11, Exit        // go to Exit if save[i] ≠ k(X11 ≠ 0)
```

The following instruction adds 1 to i:

```
ADDI X22,X22,#1       // i = i + 1
```

The end of the loop branches back to the *while* test at the top of the loop. We just add the Exit label after it, and we're done:

```
B       Loop        // go to Loop

Exit:
```

(See the exercises for an optimization of this sequence.)

Hardware/ Software Interface

Such sequences of instructions that end in a branch are so fundamental to compiling that they are given their own buzzword: a **basic block** is a sequence of instructions without branches, except possibly at the end, and without branch targets or branch labels, except possibly at the beginning. One of the first early phases of compilation is breaking the program into basic blocks.

basic block A sequence of instructions without branches (except possibly at the end) and without branch targets or branch labels (except possibly at the beginning).

The test for equality or inequality is probably the most popular test, but there are many other relationships between two numbers. For example, a *for* loop may want to test to see if the index variable is less than 0. The full set of comparisons is less than (<), less than or equal (≤), greater than (>), greater than or equal (≥), equal (=), and not equal (≠).

Comparison of bit patterns must also deal with the dichotomy between signed and unsigned numbers. Sometimes a bit pattern with a 1 in the most significant bit represents a negative number and, of course, is less than any positive number, which must have a 0 in the most significant bit. With unsigned integers, on the other hand, a 1 in the most significant bit represents a number that is *larger* than any that begins with a 0. (We'll soon take advantage of this dual meaning of the most significant bit to reduce the cost of the array bounds checking.)

Architects long ago figured out how to handle all these cases by keeping just four extra bits that record what occurred during an instruction. These four added bits, called *condition codes* or *flags*, are named:

- *negative (N)* – the result that set the condition code had a 1 in the most significant bit;

- *zero (Z)* – the result that set the condition code was 0;

- *overflow (V)* – the result that set the condition code overflowed; and

- *carry (C)* – the result that set the condition code had a carry out of the most significant bit or a borrow into the most significant bit.

Conditional branches then use combinations of these condition codes to perform the desired sets. In LEGv8, this conditional branch instruction is B.cond. cond can be used for any of the signed comparison instructions: EQ (= or Equal), NE (≠ or Not Equal), LT (< or Less Than), LE (≤ or Less than or Equal), GT (> or Greater Than), or GE (≥ or Greater than or Equal). It can also be used for the unsigned comparison instruction: LO (< or Lower), LS (≤ or Lower or Same), HI (> or Higher), or HS (≥ or Higher or Same). If the instruction that set the condition codes was a subtract (A-B), Figure 2.10 shows the LEGv8 instructions and the values of the condition codes that perform the full set of comparisons for signed and unsigned numbers.

In addition to the 10 conditional branch instructions in Figure 2.10, LEGv8 includes these four branches to complete the testing of the individual condition code bits:

- Branch on minus (B.MI): N= 1;

- Branch on plus (B.PL): N= 0;

- Branch on overflow set (B.VS): V= 1;

- Branch on overflow clear (B.VC): V= 0.

One alternative to condition codes is to have instructions that compare two registers and then branch based on the result. A second option is to compare two registers and set a third register to a result indicating the success of the comparison,

	Signed numbers		Unsigned numbers	
Comparison	Instruction	CC Test	Instruction	CC Test
=	B.EQ	Z=1	B.EQ	Z=1
≠	B.NE	Z=0	B.NE	Z=0
<	B.LT	N!=V	B.LO	C=0
≤	B.LE	~(Z=0 & N=V)	B.LS	~(Z=0 & C=1)
>	B.GT	(Z=0 & N=V)	B.HI	(Z=0 & C=1)
≥	B.GE	N=V	B.HS	C=1

FIGURE 2.10   How to do all comparisons if the instruction that set the condition codes was a subtract. If it was an ADD or AND, the test is simply on the result of the operation as compared to zero. For AND, C and V are always set to 0.

which a subsequent conditional branch instruction then tests to see if register is non-zero (condition is true) or zero (condition is false). Conditional branches in MIPS follow the latter approach (see Section 2.16).

One downside to condition codes is that if many instructions always set them, it will create dependencies that will make it difficult for pipelined execution (see Chapter 4). Hence, LEGv8 limits condition code (flag) setting to just a few instructions—ADD, ADDI, AND, ANDI, SUB, and SUBI—and even then condition code setting is optional. In LEGv8 assembly language, you simply append an S to the end of one of these instructions if you want to set condition codes: ADDS, ADDIS, ANDS, ANDIS, SUBS, and SUBIS. The instruction name actually uses the term flag, so the proper name of ADDS is "add and set flags."

Bounds Check Shortcut

Treating signed numbers as if they were unsigned gives us a low-cost way of checking if $0 \leq x < y$, which matches the index out-of-bounds check for arrays. The key is that negative integers in two's complement notation look like large numbers in unsigned notation; that is, the most significant bit is a sign bit in the former notation but a large part of the number in the latter. Thus, an unsigned comparison of $x < y$ checks if x is negative as well as if x is less than y.

EXAMPLE

Use this shortcut to reduce an index-out-of-bounds check: branch to IndexOutOfBounds if X20 ≥ X11 or if X20 is negative.

ANSWER

The checking code just uses unsigned greater than or equal to do both checks:

```
SUBS XZR,X20,X11 // Test if X20 >= length or X20 < 0
B.HS IndexOutOfBounds //if bad, goto Error
```

Case/Switch Statement

Most programming languages have a *case* or *switch* statement that allows the programmer to select one of many alternatives depending on a single value. The simplest way to implement *switch* is via a sequence of conditional tests, turning the *switch* statement into a chain of *if-then-else* statements.

Sometimes the alternatives may be more efficiently encoded as a table of addresses of alternative instruction sequences, called a branch address table or branch table, and the program needs only to index into the table and then branch to the appropriate sequence. The branch table is therefore just an array of double-words containing addresses that correspond to labels in the code. The program loads the appropriate entry from the branch table into a register. It then needs to branch using the address in the register. To support such situations, computers like LEGv8 include a *branch register* instruction (BR), meaning an unconditional branch to the address specified in a register. Then it branches to the proper address using this instruction. We'll see an even more popular use of BR in the next section.

branch address table Also called **branch table**. A table of addresses of alternative instruction sequences.

Although there are many statements for decisions and loops in programming languages like C and Java, the bedrock statement that implements them at the instruction set level is the conditional branch.

Hardware/ Software Interface

I. C has many statements for decisions and loops, while LEGv8 has few. Which of the following does or does not explain this imbalance? Why?

Check Yourself

1. More decision statements make code easier to read and understand.

2. Fewer decision statements simplify the task of the underlying layer that is responsible for execution.

3. More decision statements mean fewer lines of code, which generally reduces coding time.

4. More decision statements mean fewer lines of code, which generally results in the execution of fewer operations.

II. Why does C provide two sets of operators for AND (& and &&) and two sets of operators for OR (| and ||), while LEGv8 doesn't?

1. Logical operations AND and ORR implement & and |, while conditional branches implement && and ||.

2. The previous statement has it backwards: && and || correspond to logical operations, while & and | map to conditional branches.

3. They are redundant and mean the same thing: && and || are simply inherited from the programming language B, the predecessor of C.

2.8 Supporting Procedures in Computer Hardware

procedure A stored subroutine that performs a specific task based on the parameters with which it is provided.

A procedure or function is one tool programmers use to structure programs, both to make them easier to understand and to allow code to be reused. Procedures allow the programmer to concentrate on just one portion of the task at a time; parameters act as an interface between the procedure and the rest of the program and data, since they can pass values and return results. We describe the equivalent to procedures in Java in 🖥 Section 2.15, but Java needs everything from a computer that C needs. Procedures are one way to implement **abstraction** in software.

ABSTRACTION

You can think of a procedure like a spy who leaves with a secret plan, acquires resources, performs the task, covers his or her tracks, and then returns to the point of origin with the desired result. Nothing else should be perturbed once the mission is complete. Moreover, a spy operates on only a "need to know" basis, so the spy can't make assumptions about the spymaster.

Similarly, in the execution of a procedure, the program must follow these six steps:

1. Put parameters in a place where the procedure can access them.

2. Transfer control to the procedure.

3. Acquire the storage resources needed for the procedure.

4. Perform the desired task.

5. Put the result value in a place where the calling program can access it.

6. Return control to the point of origin, since a procedure can be called from several points in a program.

As mentioned above, registers are the fastest place to hold data in a computer, so we want to use them as much as possible. LEGv8 software follows the following convention for procedure calling in allocating its 32 registers:

■ X0-X7: eight parameter registers in which to pass parameters or return values.

■ LR (X30): one return address register to return to the point of origin.

branch-and-link instruction An instruction that branches to an address and simultaneously saves the address of the following instruction in a register (LR or X30 in LEGv8).

In addition to allocating these registers, LEGv8 assembly language includes an instruction just for the procedures: it branches to an address and simultaneously saves the address of the following instruction in register LR (X30). The branch-and-link instruction (BL) is simply written

```
BL ProcedureAddress
```

The *link* portion of the name means that an address or link is formed that points to the calling site to allow the procedure to return to the proper address. This "link," stored in register LR (register 30), is called the return address. The return address is needed because the same procedure could be called from several parts of the program.

To support the return from a procedure, computers like LEGv8 use the *branch register* instruction (BR), introduced above to help with case statements, meaning an unconditional branch to the address specified in a register:

```
BR    LR
```

The branch register instruction branches to the address stored in register LR—which is just what we want. Thus, the calling program, or caller, puts the parameter values in X0-X7 and uses BL X to branch to procedure X (sometimes named the callee). The callee then performs the calculations, places the results in the same parameter registers, and returns control to the caller using BR LR.

Implicit in the stored-program idea is the need to have a register to hold the address of the current instruction being executed. For historical reasons, this register is almost always called the program counter, abbreviated *PC* in the LEGv8 architecture, although a more sensible name would have been *instruction address register*. The BL instruction actually saves PC + 4 in register LR to link to the byte address of the following instruction to set up the procedure return.

Using More Registers

Suppose a compiler needs more registers for a procedure than the eight argument registers. Since we must cover our tracks after our mission is complete, any registers needed by the caller must be restored to the values that they contained *before* the procedure was invoked. This situation is an example in which we need to spill registers to memory, as mentioned in the *Hardware/Software Interface* section on page 72.

The ideal data structure for spilling registers is a stack—a last-in-first-out queue. A stack needs a pointer to the most recently allocated address in the stack to show where the next procedure should place the registers to be spilled or where old register values are found. The stack pointer (SP), which is just one of the 32 registers, is adjusted by one doubleword for each register that is saved or restored. Stacks are so popular that they have their own buzzwords for transferring data to and from the stack: placing data onto the stack is called a push, and removing data from the stack is called a pop.

By historical precedent, stacks "grow" from higher addresses to lower addresses. This convention means that you push values onto the stack by subtracting from the stack pointer. Adding to the stack pointer shrinks the stack, thereby popping values off the stack.

return address A link to the calling site that allows a procedure to return to the proper address; in LEGv8 it is stored in register LR (X30).

caller The program that instigates a procedure and provides the necessary parameter values.

callee A procedure that executes a series of stored instructions based on parameters provided by the caller and then returns control to the caller.

program counter (PC) The register containing the address of the instruction in the program being executed.

stack A data structure for spilling registers organized as a last-in-first-out queue.

stack pointer A value denoting the most recently allocated address in a stack that shows where registers should be spilled or where old register values can be found. In LEGv8, it is register SP.

push Add element to stack.

pop Remove element from stack.

Elaboration: As mentioned above, in the full ARMv8 instruction set, the stack pointer is folded into register 31. In some instructions—data transfers and arithmetic immediates that don't set flags when it is the destination register or first source register—register 31 indicates SP but in the rest, such as arithmetic register instructions or in flag setting instructions, it indicates the *zero register* (XZR). Given that this trick only saves one register, would complicate the datapath in Chapter 4, and it is a bit confusing, LEGv8 just assumes SP is one of the other 31 general purpose registers; we use X28 for SP.

EXAMPLE

Compiling a C Procedure That Doesn't Call Another Procedure

Let's turn the example on page 66 from Section 2.2 into a C procedure:

```
long long int leaf_example (long long int g, long long
int h, long long int i, long long int j)
{
      long long int f;

      f = (g + h) − (i + j);
      return f;
}
```

What is the compiled LEGv8 assembly code?

ANSWER

The parameter variables g, h, i, and j correspond to the argument registers X0, X1, X2, and X3, and f corresponds to X19. The compiled program starts with the label of the procedure:

```
leaf_example:
```

The next step is to save the registers used by the procedure. The C assignment statement in the procedure body is identical to the example on page 68, which uses two temporary registers (X9 and X10). Thus, we need to save three registers: X19, X9, and X10. We "push" the old values onto the stack by creating space for three doublewords (24 bytes) on the stack and then store them:

```
SUBI SP,  SP, #24     // adjust stack to make room for 3 items
STUR X10, [SP,#16]    // save register X10 for use afterwards
STUR X9,  [SP,#8]     // save register X9 for use afterwards
STUR X19, [SP,#0]     // save register X19 for use afterwards
```

Figure 2.11 shows the stack before, during, and after the procedure call.

The next three statements correspond to the body of the procedure, which follows the example on page 68:

```
ADD X9,X0,X1   // register X9 contains g + h
ADD X10,X2,X3  // register X10 contains i + j
SUB X19,X9,X10 // f = X9 - X10, which is (g + h) - (i + j)
```

To return the value of f, we copy it into a parameter register:

```
ADD X0,X19,XZR // returns f (X0 = X19 + 0)
```

Before returning, we restore the three old values of the registers we saved by "popping" them from the stack:

```
LDUR X19, [SP,#0]   // restore register X19 for caller
LDUR X9, [SP,#8]    // restore register X9 for caller
LDUR X10, [SP,#16]  // restore register X10 for caller
ADDI SP,SP,#24      // adjust stack to delete 3 items
```

The procedure ends with a branch register using the return address:

```
BR    LR    // branch back to calling routine
```

In the previous example, we used temporary registers and assumed their old values must be saved and restored. To avoid saving and restoring a register whose value is never used, which might happen with a temporary register, LEGv8 software separates 19 of the registers into two groups:

- X9-X17: temporary registers that are *not* preserved by the callee (called procedure) on a procedure call

- X19-X28: saved registers that must be preserved on a procedure call (if used, the callee saves and restores them)

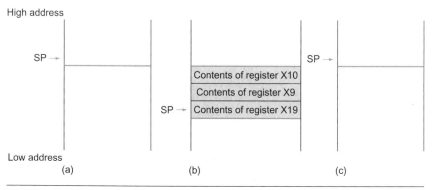

FIGURE 2.11 The values of the stack pointer and the stack (a) before, (b) during, and (c) after the procedure call. The stack pointer always points to the "top" of the stack, or the last doubleword in the stack in this drawing.

This simple convention reduces register spilling. In the example above, since the caller does not expect registers X9 and X10 to be preserved across a procedure call, we can drop two stores and two loads from the code. We still must save and restore X19, since the callee must assume that the caller needs its value.

Nested Procedures

Procedures that do not call others are called *leaf* procedures. Life would be simple if all procedures were leaf procedures, but they aren't. Just as a spy might employ other spies as part of a mission, who in turn might use even more spies, so do procedures invoke other procedures. Moreover, recursive procedures even invoke "clones" of themselves. Just as we need to be careful when using registers in procedures, attention must be paid when invoking nonleaf procedures.

For example, suppose that the main program calls procedure A with an argument of 3, by placing the value 3 into register X0 and then using BL A. Then suppose that procedure A calls procedure B via BL B with an argument of 7, also placed in X0. Since A hasn't finished its task yet, there is a conflict over the use of register X0. Similarly, there is a conflict over the return address in register LR, since it now has the return address for B. Unless we take steps to prevent the problem, this conflict will eliminate procedure A's ability to return to its caller.

One solution is to push all the other registers that must be preserved on the stack, just as we did with the saved registers. The caller pushes any argument registers (X0-X7) or temporary registers (X9-X17) that are needed after the call. The callee pushes the return address register LR and any saved registers (X19-X25) used by the callee. The stack pointer SP is adjusted to account for the number of registers placed on the stack. Upon the return, the registers are restored from memory, and the stack pointer is readjusted.

EXAMPLE

Compiling a Recursive C Procedure, Showing Nested Procedure Linking

Let's tackle a recursive procedure that calculates factorial:

```
long long int fact (long long int n)
{
        if (n < 1) return (1);
                else return (n * fact(n − 1));
}
```

What is the LEGv8 assembly code?

The parameter variable n corresponds to the argument register X0. The compiled program starts with the label of the procedure and then saves two registers on the stack, the return address and X0:

```
fact:
    SUBI  SP, SP, #16 // adjust stack for 2 items
    STUR  LR, [SP,#8] // save the return address
    STUR  X0, [SP,#0] // save the argument n
```

The first time fact is called, STUR saves an address in the program that called fact. The next two instructions test whether n is less than 1, going to L1 if n ≥ 1.

```
    SUBIS   ZXR,X0, #1  // test for n < 1
    B.GE    L1          // if n >= 1, go to L1
```

If n is less than 1, fact returns 1 by putting 1 into a value register: it adds 1 to 0 and places that sum in X1. It then pops the two saved values off the stack and branches to the return address:

```
    ADDI    X1,XZR, #1  // return 1
    ADDI    SP,SP,#16   // pop 2 items off stack
    BR      LR          // return to caller
```

Before popping two items off the stack, we could have loaded X0 and LR. Since X0 and LR don't change when n is less than 1, we skip those instructions.

If n is not less than 1, the argument n is decremented and then fact is called again with the decremented value:

```
L1: SUBI X0,X0,#1  // n >= 1: argument gets (n − 1)
    BL fact        // call fact with (n − 1)
```

The next instruction is where fact returns. Now the old return address and old argument are restored, along with the stack pointer:

```
    LDUR    X0, [SP,#0]  // return from BL: restore argument n
    LDUR    LR, [SP,#8]  // restore the return address
    ADDI    SP, SP, #16  // adjust stack pointer to pop 2 items
```

Next, the value register X1 gets the product of old argument X0 and the current value of the value register. We assume a multiply instruction is available, even though it is not covered until Chapter 3:

```
    MUL     X1,X0,X1     // return n * fact (n − 1)
```

Finally, fact branches again to the return address:

```
    BR      LR       // return to the caller
```

Hardware/
Software
Interface

global pointer The
register that is reserved to
point to the static area.

A C variable is generally a location in storage, and its interpretation depends both on its *type* and *storage class*. Example types include integers and characters (see Section 2.9). C has two storage classes: *automatic* and *static*. Automatic variables are local to a procedure and are discarded when the procedure exits. Static variables exist across exits from and entries to procedures. C variables declared outside all procedures are considered static, as are any variables declared using the keyword *static*. The rest are automatic. To simplify access to static data, some LEGv8 compilers reserve a register, called the global pointer, or GP. For example X27 could be reserved for GP.

Figure 2.12 summarizes what is preserved across a procedure call. Note that several schemes preserve the stack, guaranteeing that the caller will get the same data back on a load from the stack as it stored onto the stack. The stack above SP is preserved simply by making sure the callee does not write above SP; SP is itself preserved by the callee adding exactly the same amount that was subtracted from it; and the other registers are preserved by saving them on the stack (if they are used) and restoring them from there.

Preserved	Not preserved
Saved registers: X19-X27	Temporary registers: X9-X15
Stack pointer register: X28(SP)	Argument/Result registers: X0-X7
Frame pointer register: X29(FP)	
Link Register (return address): X30(LR)	
Stack above the stack pointer	Stack below the stack pointer

FIGURE 2.12   What is and what is not preserved across a procedure call. If the software relies on the global pointer register, discussed in the following subsections, it is also preserved.

procedure frame Also
called activation record.
The segment of the stack
containing a procedure's
saved registers and local
variables.

Allocating Space for New Data on the Stack

The final complexity is that the stack is also used to store variables that are local to the procedure but do not fit in registers, such as local arrays or structures. The segment of the stack containing a procedure's saved registers and local variables is called a procedure frame or activation record. Figure 2.13 shows the state of the stack before, during, and after the procedure call.

frame pointer A value
denoting the location of
the saved registers and
local variables for a given
procedure.

Some ARMv8 compilers use a frame pointer (FP) to point to the first doubleword of the frame of a procedure. A stack pointer might change during the procedure, and so references to a local variable in memory might have different offsets depending on where they are in the procedure, making the procedure harder to understand. Alternatively, a frame pointer offers a stable base register within a procedure for local memory-references. Note that an activation record appears on the stack whether or not an explicit frame pointer is used. We've been avoiding using FP by avoiding changes to SP within a procedure: in our examples, the stack is adjusted only on entry to and exit from the procedure.

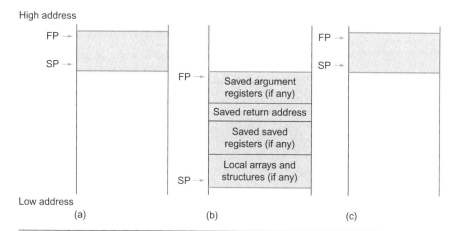

FIGURE 2.13 Illustration of the stack allocation (a) before, (b) during, and (c) after the procedure call. The frame pointer (FP or X29) points to the first doubleword of the frame, often a saved argument register, and the stack pointer (SP) points to the top of the stack. The stack is adjusted to make room for all the saved registers and any memory-resident local variables. Since the stack pointer may change during program execution, it's easier for programmers to reference variables via the stable frame pointer, although it could be done just with the stack pointer and a little address arithmetic. If there are no local variables on the stack within a procedure, the compiler will save time by *not* setting and restoring the frame pointer. When a frame pointer is used, it is initialized using the address in SP on a call, and SP is restored using FP. This information is also found in Column 4 of the LEGv8 Reference Data Card at the front of this book.

Allocating Space for New Data on the Heap

In addition to automatic variables that are local to procedures, C programmers need space in memory for static variables and for dynamic data structures. Figure 2.14 shows the LEGv8 convention for allocation of memory when running the Linux operating system. The stack starts in the high end of the user addresses space (see Chapter 5) and grows down. The first part of the low end of memory is reserved, followed by the home of the LEGv8 machine code, traditionally called the text segment. Above the code is the *static data segment*, which is the place for constants and other static variables. Although arrays tend to be a fixed length and thus are a good match to the static data segment, data structures like linked lists tend to grow and shrink during their lifetimes. The segment for such data structures is traditionally called the *heap*, and it is placed next in memory. Note that this allocation allows the stack and heap to grow toward each other, thereby allowing the efficient use of memory as the two segments wax and wane.

C allocates and frees space on the heap with explicit functions. malloc() allocates space on the heap and returns a pointer to it, and free() releases space on the heap to which the pointer points. C programs control memory allocation, which is the source of many common and difficult bugs. Forgetting to free space leads to a "memory leak," which ultimately uses up so much memory that the operating system may crash. Freeing space too early leads to "dangling pointers," which can cause pointers to point to things that the program never intended. Java uses automatic memory allocation and garbage collection just to avoid such bugs.

text segment The segment of a UNIX object file that contains the machine language code for routines in the source file.

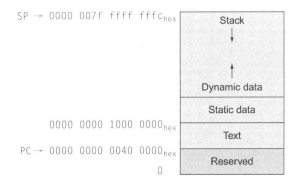

SP → 0000 007f ffff fffc_{hex}

0000 0000 1000 0000_{hex}

PC → 0000 0000 0040 0000_{hex}

0

FIGURE 2.14 The LEGv8 memory allocation for program and data. These addresses are only a software convention, and not part of the LEGv8 architecture. The user address space is set to 2^{39} of the potential 2^{64} total address space given a 64-bit architecture (see Chapter 5). The stack pointer is initialized to $0000\ 007f$ $ffff\ fffc_{hex}$ and grows down toward the data segment. At the other end, the program code ("text") starts at $0000\ 0000\ 0040\ 0000_{hex}$. The static data starts immediately after the end of the text segment; in this example, we assume that address is $0000\ 0000\ 1000\ 0000_{hex}$. Dynamic data, allocated by malloc in C and by new in Java, is next. It grows up toward the stack in an area called the *heap*. This information is also found in Column 4 of the LEGv8 Reference Data Card at the front of this book.

Figure 2.15 summarizes the register conventions for the LEGv8 assembly language. This convention is another example of making the **common case fast**: most procedures can be satisfied with up to eight argument registers, nine saved registers, and seven temporary registers without ever going to memory.

Elaboration: What if there are more than eight parameters? The LEGv8 convention is to place the extra parameters on the stack just above the frame pointer. The procedure then expects the first eight parameters to be in registers X0 through X7 and the rest in memory, addressable via the frame pointer.

As mentioned in the caption of Figure 2.13, the frame pointer is convenient because all references to variables in the stack within a procedure will have the same offset. The frame pointer is not necessary, however. The ARMv8 C compiler uses a frame pointer, but some C compilers do not; they treat register 29 as another save register.

Elaboration: Some recursive procedures can be implemented iteratively without using recursion. Iteration can significantly improve performance by removing the overhead associated with recursive procedure calls. For example, consider a procedure used to accumulate a sum:

```
long long int sum (long long int n, long long int acc) {
    if (n >0)
        return sum(n - 1, acc + n);
    else
        return acc;
}
```

Name	Register number	Usage	Preserved on call?
X0–X7	0–7	Arguments/Results	no
X8	8	Indirect result location register	no
X9–X15	9–15	Temporaries	no
X16 (IP0)	16	May be used by linker as a scratch register; other times used as temporary register	no
X17 (IP1)	17	May be used by linker as a scratch register; other times used as temporary register	no
X18	18	Platform register for platform independent code; otherwise a temporary register	no
X19–X27	19–27	Saved	yes
X28 (SP)	28	Stack Pointer	yes
X29 (FP)	29	Frame Pointer	yes
X30 (LR)	30	Link Register (return address)	yes
XZR	31	The constant value 0	n.a.

FIGURE 2.15 LEGv8 register conventions. This information is also found in Column 2 of the LEGv8 Reference Data Card at the front of this book. X8 is used by procedures that return a result via a pointer. ARM discourages the use of registers X16 to X18 as X16 and X17 may be used by the linker (see Section 2.12), and X18 may be used to create platform independent code, which would be specified by the platforms Application Binary Interface.

Consider the procedure call sum(3,0). This will result in recursive calls to sum(2,3), sum(1,5), and sum(0,6), and then the result 6 will be returned four times. This recursive call of sum is referred to as a *tail call*, and this example use of tail recursion can be implemented very efficiently (assume X0 = n, X1 = acc, and the result goes into X2):

```
sum: SUBS XZR, X0, XZR  // compare n to 0
     B.LE sum_exit      // go to sum_exit if n <= 0
     ADD X1, X1, X0     // add n to acc
     SUBI X0, X0, #1    // subtract 1 from n
     B sum              // go to sum
sum_exit:
     ADD X2, X1, XZR    // return value acc
     BR LR              // return to caller
```

Which of the following statements about C and Java is generally true?

1. C programmers manage data explicitly, while it's automatic in Java.

2. C leads to more pointer bugs and memory leak bugs than does Java.

!(@ | = > (wow open tab at bar is great)

Fourth line of the keyboard poem "Hatless Atlas," 1991 (some give names to ASCII characters: "!" is "wow," "(" is open, "|" is bar, and so on).

2.9   Communicating with People

Computers were invented to crunch numbers, but as soon as they became commercially viable they were used to process text. Most computers today offer 8-bit bytes to represent characters, with the *American Standard Code for Information Interchange* (ASCII) being the representation that nearly everyone follows. Figure 2.16 summarizes ASCII.

ASCII value	Char-acter	ASCII value	Char-acter	ASCII value	Char-acter	ASCII value	Char-acter	ASCII value	Char-acter	ASCII value	Char-acter
32	space	48	0	64	@	80	P	96	`	112	p
33	!	49	1	65	A	81	Q	97	a	113	q
34	"	50	2	66	B	82	R	98	b	114	r
35	#	51	3	67	C	83	S	99	c	115	s
36	$	52	4	68	D	84	T	100	d	116	t
37	%	53	5	69	E	85	U	101	e	117	u
38	&	54	6	70	F	86	V	102	f	118	v
39	'	55	7	71	G	87	W	103	g	119	w
40	(	56	8	72	H	88	X	104	h	120	x
41	)	57	9	73	I	89	Y	105	i	121	y
42	*	58	:	74	J	90	Z	106	j	122	z
43	+	59	;	75	K	91	[	107	k	123	{
44	,	60	<	76	L	92	\	108	l	124		
45	-	61	=	77	M	93	]	109	m	125	}
46	.	62	>	78	N	94	^	110	n	126	~
47	/	63	?	79	O	95	_	111	o	127	DEL

FIGURE 2.16   ASCII representation of characters. Note that upper- and lowercase letters differ by exactly 32; this observation can lead to shortcuts in checking or changing upper- and lowercase. Values not shown include formatting characters. For example, 8 represents a backspace, 9 represents a tab character, and 13 a carriage return. Another useful value is 0 for null, the value the programming language C uses to mark the end of a string.

ASCII versus Binary Numbers

We could represent numbers as strings of ASCII digits instead of as integers. How much does storage increase if the number 1 billion is represented in ASCII versus a 32-bit integer?

One billion is 1,000,000,000, so it would take 10 ASCII digits, each 8 bits long. Thus the storage expansion would be (10 × 8)/32 or 2.5. Beyond the expansion in storage, the hardware to add, subtract, multiply, and divide such decimal numbers is difficult and would consume more energy. Such difficulties explain why computing professionals are raised to believe that binary is natural and that the occasional decimal computer is bizarre.

A series of instructions can extract a byte from a doubleword, so load register and store register are sufficient for transferring bytes as well as words. Because of the popularity of text in some programs, however, LEGv8 provides instructions to move bytes. *Load byte* (LDURB) loads a byte from memory, placing it in the rightmost 8 bits of a register. *Store byte* (STURB) takes a byte from the rightmost 8 bits of a register and writes it to memory. Thus, we copy a byte with the sequence

```
LDURB X9,[X0,#0]    // Read byte from source
STURB X9,[X1,#0]    // Write byte to destination
```

Characters are normally combined into strings, which have a variable number of characters. There are three choices for representing a string: (1) the first position of the string is reserved to give the length of a string, (2) an accompanying variable has the length of the string (as in a structure), or (3) the last position of a string is indicated by a character used to mark the end of a string. C uses the third choice, terminating a string with a byte whose value is 0 (named null in ASCII). Thus, the string "Cal" is represented in C by the following 4 bytes, shown as decimal numbers: 67, 97, 108, and 0. (As we shall see, Java uses the first option.)

EXAMPLE

Compiling a String Copy Procedure, Showing How to Use C Strings

The procedure strcpy copies string y to string x using the null byte termination convention of C:

```
void strcpy (char x[], char y[])
{
    size t i;
    i = 0;
    while ((x[i] = y[i]) != '\0') /* copy & test byte */
    i += 1;
}
```

What is the LEGv8 assembly code?

ANSWER

Below is the basic LEGv8 assembly code segment. Assume that base addresses for arrays x and y are found in X0 and X1, while i is in X19. strcpy adjusts the stack pointer and then saves the saved register X19 on the stack:

```
strcpy:
    SUBI  SP,SP,#8      // adjust stack for 1 more item
    STUR  X19, [SP,#0]  // save X19
```

To initialize i to 0, the next instruction sets X19 to 0 by adding 0 to 0 and placing that sum in X19:

```
ADD  X19,XZR,XZR // i = 0 + 0
```

This is the beginning of the loop. The address of y[i] is first formed by adding i to y[]:

```
L1: ADD  X10,X19,X1  // address of y[i] in X10
```

Note that we don't have to multiply i by 8 since y is an array of *bytes* and not of doublewords, as in prior examples.

To load the character in y[i], we use load byte unsigned, which puts the character into X11:

```
LDURB  X11, [X10,#0]  // X11 = y[i]
```

A similar address calculation puts the address of x[i] in X12, and then the character in X11 is stored at that address.

```
ADD    X12,X19,X0      // address of x[i] in X12
STURB  X11, [X12,#0]   // x[i] = y[i]
```

Next, we exit the loop if the character was 0. That is, we exit if it is the last character of the string:

```
CBZ  X11,L2  // if y[i] == 0, go to L2
```

If not, we increment i and loop back:

```
ADDI  X19, X19,#1  // i = i + 1
B     L1           // go to L1
```

If we don't loop back, it was the last character of the string; we restore X19 and the stack pointer, and then return.

```
L2: LDUR  X19, [SP,#0]  // y[i] == 0: end of string.
                        // Restore old X19
    ADDI  SP,SP,#8      // pop 1 doubleword off stack
    BR    LR            // return
```

String copies usually use pointers instead of arrays in C to avoid the operations on i in the code above. See Section 2.14 for an explanation of arrays versus pointers.

Since the procedure strcpy above is a leaf procedure, the compiler could allocate i to a temporary register and avoid saving and restoring X19. Hence, instead of thinking of these registers as being just for temporaries, we can think of them as registers that the callee should use whenever convenient. When a compiler finds a leaf procedure, it exhausts all temporary registers before using registers it must save.

Characters and Strings in Java

Unicode is a universal encoding of the alphabets of most human languages. Figure 2.17 gives a list of Unicode alphabets; there are almost as many *alphabets* in Unicode as there are useful *symbols* in ASCII. To be more inclusive, Java uses Unicode for characters. By default, it uses 16 bits to represent a character.

Latin	Malayalam	Tagbanwa	General Punctuation
Greek	Sinhala	Khmer	Spacing Modifier Letters
Cyrillic	Thai	Mongolian	Currency Symbols
Armenian	Lao	Limbu	Combining Diacritical Marks
Hebrew	Tibetan	Tai Le	Combining Marks for Symbols
Arabic	Myanmar	Kangxi Radicals	Superscripts and Subscripts
Syriac	Georgian	Hiragana	Number Forms
Thaana	Hangul Jamo	Katakana	Mathematical Operators
Devanagari	Ethiopic	Bopomofo	Mathematical Alphanumeric Symbols
Bengali	Cherokee	Kanbun	Braille Patterns
Gurmukhi	Unified Canadian Aboriginal Syllabic	Shavian	Optical Character Recognition
Gujarati	Ogham	Osmanya	Byzantine Musical Symbols
Oriya	Runic	Cypriot Syllabary	Musical Symbols
Tamil	Tagalog	Tai Xuan Jing Symbols	Arrows
Telugu	Hanunoo	Yijing Hexagram Symbols	Box Drawing
Kannada	Buhid	Aegean Numbers	Geometric Shapes

FIGURE 2.17 Example alphabets in Unicode. Unicode version 4.0 has more than 160 "blocks," which is their name for a collection of symbols. Each block is a multiple of 16. For example, Greek starts at 0370_{hex}, and Cyrillic at 0400_{hex}. The first three columns show 48 blocks that correspond to human languages in roughly Unicode numerical order. The last column has 16 blocks that are multilingual and are not in order. A 16-bit encoding, called UTF-16, is the default. A variable-length encoding, called UTF-8, keeps the ASCII subset as eight bits and uses 16 or 32 bits for the other characters. UTF-32 uses 32 bits per character. To learn more, see www.unicode.org.

The LEGv8 instruction set has explicit instructions to load and store such 16-bit quantities, called *halfwords*. *Load half* (LDURH) loads a halfword from memory, placing it in the rightmost 16 bits of a register. Like load byte, *load half* (LDURH) treats the halfword as a signed number and thus sign-extends to fill the 48 leftmost bits of the register. *Store half* (STURH) takes a halfword from the rightmost 16 bits of a register and writes it to memory. We copy a halfword with the sequence

```
LDURH X19,[X0,#0] // Read halfword (16 bits) from source
STURH X9,[X1,#0]  // Write halfword (16 bits) to dest.
```

Strings are a standard Java class with special built-in support and predefined methods for concatenation, comparison, and conversion. Unlike C, Java includes a word that gives the length of the string, similar to Java arrays.

Elaboration: ARMv8 software is required to keep the stack aligned to "quadword" (16 byte) addresses to get better performance. This convention means that a char variable allocated on the stack occupies 16 bytes, even though it needs less. However, a C string variable or an array of bytes *will* pack 16 bytes per quadword, and a Java string variable or array of shorts packs 8 halfwords per quadword.

Elaboration: Reflecting the international nature of the web, most web pages today use Unicode instead of ASCII. Hence, Unicode may be even more popular than ASCII today.

Elaboration: LEGv8 keeps everything 64 bits vs. providing both 32-bit and 64-bit address instructions as in ARMv8, which means it needs to include STURW (store word) as an instruction even though it is not specified in ARMv8 in assembly language. ARMv8 just uses STUR with a W register name (32-bit register) instead of X register name (64-bit register).

I. Which of the following statements about characters and strings in C and Java is true? **Check Yourself**

   1. A string in C takes about half the memory as the same string in Java.

   2. Strings are just an informal name for single-dimension arrays of characters in C and Java.

   3. Strings in C and Java use null (0) to mark the end of a string.

   4. Operations on strings, like length, are faster in C than in Java.

II. Which type of variable that can contain $1,000,000,000_{ten}$ takes the most memory space?

   1. `long long int` in C

   2. `string` in C

   3. `string` in Java

2.10 LEGv8 Addressing for Wide Immediates and Addresses

Although keeping all LEGv8 instructions 32 bits long simplifies the hardware, there are times where it would be convenient to have 32-bit or larger constants or addresses. This section starts with the general solution for large constants, and then shows the optimizations for instruction addresses used in branches.

Wide Immediate Operands

Although constants are frequently short and fit into the 12-bit fields, sometimes they are bigger. The LEGv8 instruction set includes the instruction *move wide with zeros* (MOVZ) and *move wide with keep* (MOVK) specifically to set any 16 bits of a constant in a register. The former instruction zeros the rest of the bits of the register and the latter leaves the remaining bits unchanged. The 16-bit field to be loaded is specified by adding LSL and then the number 0, 16, 32, or 48 depending on which quadrant of the 64-bit double word is desired. Tese instructions allow, for example, a 32-bit constant to be created from two 32-bit instructions. Figure 2.18 shows the operation of MOVZ and MOVK.

The machine language version of MOVZ X9, 255, LSL 16:

| 110100101 | 01 | 0000 0000 1111 1111 | 01001 |

Contents of register X9 after executing MOVZ X9, 255, LSL 16:

| 0000 0000 0000 0000 | 0000 0000 0000 0000 | 0000 0000 1111 1111 | 0000 0000 0000 0000 |

The machine language version of MOVK X9, 255, LSL 0:

| 111100101 | 00 | 0000 0000 1111 1111 | 01001 |

Given value of X9 above, new contents of X9 after executing MOVK X9, 255, LSL 0:

| 0000 0000 0000 0000 | 0000 0000 0000 0000 | 0000 0000 1111 1111 | 0000 0000 1111 1111 |

FIGURE 2.18   The effect of the MOVZ **and** MOVK **instructions.** The instruction MOVZ transfers a 16-bit immediate constant field value into one of the four quadrants leftmost of a 64-bit register, filling the other 48 bits with 0s. The instruction MOVK only changes 16 bits of the register, keeping the other bits the same.

EXAMPLE

Loading a 32-Bit Constant

What is the LEGv8 assembly code to load this 64-bit constant into register X19?

00000000 00000000 00000000 00000000 00000000 00111101 00001001 00000000

First, we would load bits 16 to 31 with that bit pattern, which is 61 in decimal, using MOVZ:

ANSWER

```
MOVZ    X19, 61, LSL 16 // 61 decimal = 0000 0000 0011 1101 binary
```

The value of register X19 afterward is:

00000000 00000000 00000000 00000000 00000000 00111101 00000000 00000000

The next step is to insert the lowest 16 bits, whose decimal value is 2304:

```
MOVK    X19, 2304, LSL 0 // 2304 decimal = 00001001 00000000
```

The final value in register X19 is the desired value:

00000000 00000000 00000000 00000000 00000000 00111101 00001001 00000000

Either the compiler or the assembler must break large constants into pieces and then reassemble them into a register. As you might expect, the immediate field's size restriction may be a problem for memory addresses in loads and stores as well as for constants in immediate instructions.

Hence, the symbolic representation of the LEGv8 machine language is no longer limited by the hardware, but by whatever the creator of an assembler chooses to include (see Section 2.12). We stick close to the hardware to explain the architecture of the computer, noting when we use the enhanced language of the assembler that is not found in the processor.

**Hardware/
Software
Interface**

Addressing in Branches

The LEGv8 branch instructions have the simplest addressing. They use the LEGv8 instruction format, called the *B-type*, which consists of 6 bits for the operation field and the rest of the bits for the address field. Thus,

```
B    10000    // go to location 10000ten
```

could be assembled into this format (it's actually a bit more complicated, as we will see):

5	10000_{ten}
6 bits	26 bits

where the value of the branch opcode is 5 and the branch address is 10000_{ten}.

Unlike the branch instruction, a conditional branch instruction can specify one operand in addition to the branch address. Thus,

```
CBNZ  X19, Exit  // go to Exit if X19 ≠ 0
```

is assembled into this instruction, leaving only 19 bits for the branch address:

181	Exit	19
8 bits	19 bits	5 bits

This format is called *CB-type*, for conditional branch. (The conditional branch instructions that rely on condition codes also use the CB-type format, but they use the final field to select among the many possible branch conditions.)

If addresses of the program had to fit in this 19-bit field, it would mean that no program could be bigger than 2^{19}, which is far too small to be a realistic option today. An alternative would be to specify a register that would always be added to the branch offset, so that a branch instruction would calculate the following:

$$\text{Program counter} = \text{Register} + \text{Branch offset}$$

This sum allows the program to be as large as 2^{64} and still be able to use conditional branches, solving the branch address size problem. Then the question is, which register?

The answer comes from seeing how conditional branches are used. Conditional branches are found in loops and in *if* statements, so they tend to branch to a nearby instruction. For example, about half of all conditional branches in SPEC benchmarks go to locations less than 16 instructions away. Since the *program counter* (PC) contains the address of the current instruction, we can branch within $\pm 2^{18}$ words of the current instruction if we use the PC as the register to be added to the address. Almost all loops and *if* statements are much smaller than 2^{18} words, so the PC is the ideal choice. This form of branch addressing is called PC-relative addressing.

Like most recent computers, LEGv8 uses PC-relative addressing for all conditional branches, because the destination of these instructions is likely to be close to the branch. On the other hand, branch-and-link instructions invoke procedures that have no reason to be near the call, so they normally use other forms of addressing. Hence, the LEGv8 architecture offers long addresses for procedure calls by using the B-type format for both branch and branch-and-link instructions.

Since all LEGv8 instructions are 4 bytes long, LEGv8 stretches the distance of the branch by having PC-relative addressing refer to the number of *words* to the next instruction instead of the number of bytes. Thus, the 19-bit field can branch four times as far by interpreting the field as a relative word address rather than as a relative byte address: ± 1 MB from the current PC. Similarly, the 26-bit field in branch instructions is also a word address, meaning that it represents a 28-bit byte address.

The unconditional branch is also PC-relative, which means it can branch ± 128 MB from the current PC.

PC-relative addressing An addressing regime in which the address is the sum of the *program counter* (PC) and a constant in the instruction.

Showing Branch Offset in Machine Language

The *while* loop on page 95 was compiled into this LEGv8 assembler code:

```
Loop:LSL X10,X22,#3      // Temp reg X10 = 8 * i
     ADD  X10,X10,X25     // X10 = address of save[i]
     LDUR X9,[X10,#0]     // Temp reg X9 = save[i]
     SUB  X11,X9,X24      // X11 = save[i] − k
     CBNZ X11, Exit       // go to Exit if save[i] ≠ k (X11≠0)
     ADDI X22,X22,#1      // i = i + 1
     B    Loop            // go to Loop
Exit:
```

If we assume we place the loop starting at location 80000 in memory, what is the LEGv8 machine code for this loop?

The assembled instructions and their addresses are:

80000	1691	0	3	22	10
80004	1112	25	0	10	10
80008	1896	0	0	10	9
80012	1624	24	0	9	11
80016	181	3			11
80020	580	1		22	22
80024	5	−6			
80028	...				

Remember that LEGv8 instructions have byte addresses, so addresses of sequential words differ by 4, the number of bytes in a word, and that branches multiply their address fields by 4, the size of LEGv8 instructions in bytes. The CBNZ instruction on the fifth line adds 3 words or 12 bytes to the address of the instruction, specifying the branch destination relative to the branch instruction (12 + 80016) and not using the full destination address (80028). The branch instruction on the last line does a similar calculation for a backwards branch (−24 + 80024), corresponding to the label Loop.

Most conditional branches are to a nearby location, but occasionally they branch far away, farther than can be represented in the 19 bits of the conditional branch instruction. The assembler comes to the rescue just as it did with large addresses or constants: it inserts an unconditional branch to the branch target, and inverts the condition so that the conditional branch decides whether to skip the unconditional branch.

EXAMPLE

Branching Far Away

Given a branch on register X19 being equal to register zero,

```
CBZ    X19, L1
```

replace it by a pair of instructions that offers a much greater branching distance. These instructions replace the short-address conditional branch:

ANSWER

```
        CBNZ    X19, L2
        B       L1
L2:
```

LEGv8 Addressing Mode Summary

addressing mode One of several addressing regimes delimited by their varied use of operands and/or addresses.

Multiple forms of addressing are generically called addressing modes. Figure 2.19 shows how operands are identified for each addressing mode. The addressing modes of the LEGv8 instructions are the following:

1. *Immediate addressing*, where the operand is a constant within the instruction itself.

2. *Register addressing*, where the operand is a register.

3. *Base* or *displacement addressing*, where the operand is at the memory location whose address is the sum of a register and a constant in the instruction.

4. *PC-relative addressing*, where the branch address is the sum of the PC and a constant in the instruction.

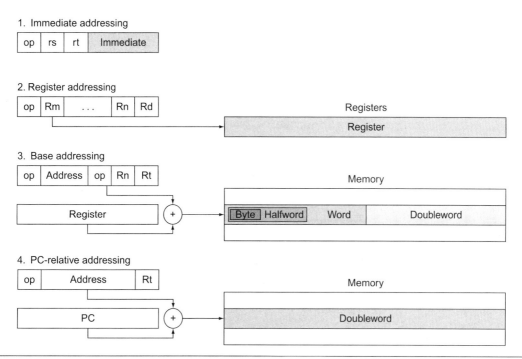

1. Immediate addressing

2. Register addressing

3. Base addressing

4. PC-relative addressing

FIGURE 2.19 Illustration of four LEGv8 addressing modes. The operands are shaded in color. The operand of mode 3 is in memory, whereas the operand for mode 2 is a register. Note that versions of load and store access bytes, halfwords, words, or doublewords. For mode 1, the operand is part of the instruction itself. Mode 4 addresses instructions in memory, with mode 4 adding a long address shifted left 2 bits to the PC. Note that a single operation can use more than one addressing mode. Add, for example, uses both immediate (ADDI) and register (ADD) addressing.

Decoding Machine Language

Sometimes you are forced to reverse-engineer machine language to create the original assembly language. One example is when looking at "core dump." Figure 2.20 shows the LEGv8 encoding of the opcodes for the LEGv8 machine language. This figure helps when translating by hand between assembly language and machine language.

Instruction	Opcode	Opcode Size	11-bit opcode range		Instruction Format
			Start	End	
B	000101	6	160	191	B - format
STURB	00111000000	11	448		D - format
LDURB	00111000010	11	450		D - format
B.cond	01010100	8	672	679	CB - format
ORRI	1011001000	10	712	713	I - format
EORI	1101001000	10	840	841	I - format
STURH	01111000000	11	960		D - format
LDURH	01111000010	11	962		D - format
AND	10001010000	11	1104		R - format
ADD	10001011000	11	1112		R - format
ADDI	1001000100	10	1160	1161	I - format
ANDI	1001001000	10	1168	1169	I - format
BL	100101	6	1184	1215	B - format
ORR	10101010000	11	1360		R - format
ADDS	10101011000	11	1368		R - format
ADDIS	1011000100	10	1416	1417	I - format
CBZ	10110100	8	1440	1447	CB - format
CBNZ	10110101	8	1448	1455	CB - format
STURW	10111000000	11	1472		D - format
LDURSW	10111000100	11	1476		D - format
STXR	11001000000	11	1600		D - format
LDXR	11001000010	11	1602		D - format
EOR	11101010000	11	1616		R - format
SUB	11001011000	11	1624		R - format
SUBI	1101000100	10	1672	1673	I - format
MOVZ	110100101	9	1684	1687	IM - format
LSR	11010011010	11	1690		R - format
LSL	11010011011	11	1691		R - format
BR	11010110000	11	1712		R - format
ANDS	11101010000	11	1872		R - format
SUBS	11101011000	11	1880		R - format
SUBIS	1111000100	10	1928	1929	I - format
ANDIS	1111001000	10	1936	1937	I - format
MOVK	111100101	9	1940	1943	IM - format
STUR	11111000000	11	1984		D - format
LDUR	11111000010	11	1986		D - format

FIGURE 2.20 LEGv8 instruction encoding. The varying size opcode values can be mapped into the space they occupy in the widest opcodes. By looking at the first 11 bits of the instruction and looking up the value, you can see which instruction it refers to.

Decoding Machine Code

What is the assembly language statement corresponding to this machine instruction?

8b0f0013$_{hex}$

The first step is converting hexadecimal to binary:

1000 1011 0000 1111 0000 0000 0001 0011

To know how to interpret the bits, we need to determine the instruction format, and to do that we first need to find the opcode field. The problem is that the opcode varies from 6 bits to 11 bits depending on the format. Since opcodes must be unique, one way to identify them is to see how many 11-bit opcodes do the shorter opcodes correspond.

For example, the branch instruction B use a 6-bit opcode with the value:

00 0101

Measured in 11-bit opcodes, it occupies all the opcode values from

00 0101 00000

to

00 0101 11111

That is, if any 11-bit opcode had a value in that range, such as

00 0101 00100

it would conflict with the 6-bit opcode of branch.

Figure 2.20 lists the instructions in LEGv8 in numerical order by opcode, showing the range of the 11-bit opcode space they occupy. For example, branch goes from 160 to 191, which is the decimal version of the bit patterns above. To determine the opcode, you just take the first 11 bits of the instruction, covert it to decimal representation, and then look up the table to find the instruction and its format.

In this example, the tentative opcode field is then 10001011000_{two}, which is 1112_{ten}. When we search Figure 2.20, we see that opcode corresponds to the ADD instruction, which uses the R-format. Thus, we can parse the binary format into fields listed in Figure 2.21:

opcode	Rm	shamt	Rn	Rd
10001011000	00101	000000	01111	10000

We decode the rest of the instruction by looking at the field values. The decimal values are 5 for the Rm field, 15 for Rn, and 16 for Rd (shamt is unused). These numbers represent registers X5, X15, and X16. Now we can reveal the assembly instruction:

```
ADD X16,X15,X5
```

Figure 2.21 shows all the LEGv8 instruction formats. Figure 2.1 on pages 64–65 shows the LEGv8 assembly language revealed in this chapter. The next chapter covers LEGv8 instructions for multiply, divide, and arithmetic for real numbers.

Name		Fields						Comments
Field size		6 to 11 bits	5 to 10 bits	5 or 4 bits	2 bits	5 bits	5 bits	All LEGv8 instructions are 32 bits long
R-format	R	opcode	Rm	shamt		Rn	Rd	Arithmetic instruction format
I-format	I	opcode	immediate			Rn	Rd	Immediate format
D-format	D	opcode	address		op2	Rn	Rt	Data transfer format
B-format	B	opcode	address					Unconditional Branch format
CB-format	CB	opcode	address				Rt	Conditional Branch format
IW-format	IW	opcode	immediate				Rd	Wide Immediate format

FIGURE 2.21   LEGv8 instruction formats.

Check Yourself

I. What is the range of addresses for conditional branches in LEGv8 (K = 1024)?

1. Addresses between 0 and 512K − 1
2. Addresses between 0 and 2048K − 1
3. Addresses up to about 256K before the branch to about 256K after
4. Addresses up to about 1024K before the branch to about 1024K after

II. What is the range of addresses for branch and branch and link in LEGv8 (M = 1024K)?

1. Addresses between 0 and 64M − 1
2. Addresses between 0 and 256M − 1
3. Addresses up to about 32M before the branch to about 32M after
4. Addresses up to about 128M before the branch to about 128M after

Elaboration: An easy-to-understand way for hardware to figure out the format of the instruction in parallel is to think of the hardware as having a read-only memory whose address size matches the largest opcode and whose content tells the hardware what to do for the specific instruction. Thus, instructions like *add* (ADD) that have an 11-bit opcode have a single entry in the memory, but instructions like *branch* (B) with a 6-bit opcode have many redundant copies. In fact, B has $2^{11}/2^6=2^5$ or 32 entries. A more efficient hardware structure than a read-only memory that will accomplish the same task is a *programmable-logic array* (PLA), which essentially modifies the address decoder so that there is a single entry for every opcode, no matter its size (see Appendix B).

2.11 Parallelism and Instructions: Synchronization

PARALLELISM

Parallel execution is easier when tasks are independent, but often they need to cooperate. Cooperation usually means some tasks are writing new values that others must read. To know when a task is finished writing so that it is safe for another to read, the tasks need to synchronize. If they don't synchronize, there is a danger of a data race, where the results of the program can change depending on how events happen to occur.

data race Two memory accesses form a data race if they are from different threads to the same location, at least one is a write, and they occur one after another.

For example, recall the analogy of the eight reporters writing a story on pages 44–45 of Chapter 1. Suppose one reporter needs to read all the prior sections before writing a conclusion. Hence, he or she must know when the other reporters have finished their sections, so that there is no danger of sections being changed afterwards. That is, they had better synchronize the writing and reading of each section so that the conclusion will be consistent with what is printed in the prior sections.

In computing, synchronization mechanisms are typically built with user-level software routines that rely on hardware-supplied synchronization instructions. In this section, we focus on the implementation of *lock* and *unlock* synchronization operations. Lock and unlock can be used straightforwardly to create regions where only a single processor can operate, called a *mutual exclusion*, as well as to implement more complex synchronization mechanisms.

The critical ability we require to implement synchronization in a multiprocessor is a set of hardware primitives with the ability to *atomically* read and modify a memory location. That is, nothing else can interpose itself between the read and the write of the memory location. Without such a capability, the cost of building basic synchronization primitives will be high and will increase unreasonably as the processor count increases.

There are a number of alternative formulations of the basic hardware primitives, all of which provide the ability to atomically read and modify a location, together with some way to tell if the read and write were performed atomically. In general, architects do not expect users to employ the basic hardware primitives, but instead expect system programmers will use the primitives to build a synchronization library, a process that is often complex and tricky.

Let's start with one such hardware primitive and show how it can be used to build a basic synchronization primitive. One typical operation for building synchronization operations is the *atomic exchange* or *atomic swap*, which interchanges a value in a register for a value in memory.

To see how to use this to build a basic synchronization primitive, assume that we want to build a simple lock where the value 0 is used to indicate that the lock is free and 1 is used to indicate that the lock is unavailable. A processor tries to set the lock by doing an exchange of 1, which is in a register, with the memory address corresponding to the lock. The value returned from the exchange instruction is 1 if some other processor had already claimed access, and 0 otherwise. In the latter case, the value is also changed to 1, preventing any competing exchange in another processor from also retrieving a 0.

For example, consider two processors that each try to do the exchange simultaneously: this race is prevented, since exactly one of the processors will perform the exchange first, returning 0, and the second processor will return 1 when it does the exchange. The key to using the exchange primitive to implement synchronization is that the operation is atomic: the exchange is indivisible, and two simultaneous exchanges will be ordered by the hardware. It is impossible for two processors trying to set the synchronization variable in this manner to both think they have simultaneously set the variable.

Implementing a single atomic memory operation introduces some challenges in the design of the processor, since it requires both a memory read and a write in a single, uninterruptible instruction.

An alternative is to have a pair of instructions in which the second instruction returns a value showing whether the pair of instructions was executed as if the pair was atomic. The pair of instructions is effectively atomic if it appears as if all other operations executed by any processor occurred before or after the pair. Thus, when an instruction pair is effectively atomic, no other processor can change the value between the pair of instructions.

In LEGv8 this pair of instructions includes a special load called a *load exclusive register* (LDXR) and a special store called a *store exclusive register* (STXR). These instructions are used in sequence: if the contents of the memory location specified by the load exclusive are changed before the store exclusive to the same address occurs, then the store exclusive fails and does not write the value to memory. The store exclusive is defined to both store the value of a (presumably different) register in memory *and* to change the value of another register to a 0 if it succeeds and to a 1 if it fails. Thus, STXR specifies three registers: one to hold the address, one to indicate whether the atomic operation failed or succeeded, and one to hold the value to be stored in memory if it succeeded. Since the load exclusive returns the initial value, and the store exclusive returns 0 only if it succeeds, the following sequence implements an atomic exchange on the memory location specified by the contents of X20:

```
again:LDXR X10,[X20,#0]      // load exclusive
      STXR X23, X9, [X20]    // store exclusive
      CBNZ X9,again          // branch if store fails
      ADD  X23,XZR,X10       // put loaded value in X23
```

Any time a processor intervenes and modifies the value in memory between the LDXR and STXR instructions, the STXR returns 1 in X9, causing the code sequence to try again. At the end of this sequence, the contents of X23 and the memory location specified by X20 have been atomically exchanged.

Elaboration: Although it was presented for multiprocessor synchronization, atomic exchange is also useful for the operating system in dealing with multiple processes in a single processor. To make sure nothing interferes in a single processor, the store exclusive also fails if the processor does a context switch between the two instructions (see Chapter 5).

Elaboration: An advantage of the load/store exclusive mechanism is that it can be used to build other synchronization primitives, such as *atomic compare and swap* or *atomic fetch-and-increment*, which are used in some parallel programming models. These involve more instructions between the LDXR and the STXR, but not too many.

Since the store exclusive will fail after either another attempted store to the load exclusive address or any exception, care must be taken in choosing which instructions are inserted between the two instructions. In particular, only register–register instructions can safely be permitted; otherwise, it is possible to create deadlock situations where the processor can never complete the STXR because of repeated page faults. In addition, the number of instructions between the load exclusive and the store exclusive should be small to minimize the probability that either an unrelated event or a competing processor causes the store exclusive to fail frequently.

Elaboration: While the code above implemented an atomic exchange, the following code would more efficiently acquire a lock at the location in register X20, where the value of 0 means the lock was free and 1 to mean lock was acquired:

```
        ADDI  X11,XZR,#1       // copy locked value
again:  LDXR  X10,[X20,#0]     // load exclusive to read lock
        CBNZ  X10, again       // check if it is 0 yet
        STXR  X11, X9, [X20]   // attempt to store new value
        BNEZ  X9,again         // branch if store fails
```

We release the lock just using a regular store to write 0 into the location:

```
        STUR  XZR, [X20,#0]    // free lock by writing 0
```

When do you use primitives like load exclusive and store exclusive?

Check Yourself

1. When cooperating threads of a parallel program need to synchronize to get proper behavior for reading and writing shared data.

2. When cooperating processes on a uniprocessor need to synchronize for reading and writing shared data.

2.12 Translating and Starting a Program

This section describes the four steps in transforming a C program in a file from storage (disk or flash memory) into a program running on a computer. Figure 2.22 shows the translation hierarchy. Some systems combine these steps to reduce translation time, but programs go through these four logical phases. This section follows this translation hierarchy.

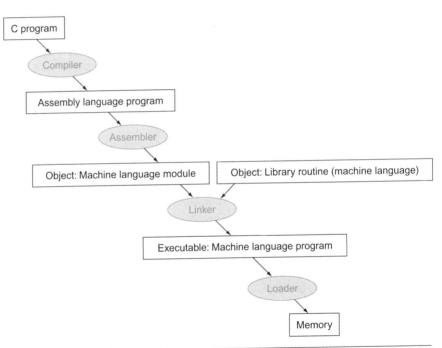

FIGURE 2.22 A translation hierarchy for C. A high-level language program is first compiled into an assembly language program and then assembled into an object module in machine language. The linker combines multiple modules with library routines to resolve all references. The loader then places the machine code into the proper memory locations for execution by the processor. To speed up the translation process, some steps are skipped or combined. Some compilers produce object modules directly, and some systems use linking loaders that perform the last two steps. To identify the type of file, UNIX follows a suffix convention for files: C source files are named x.c, assembly files are x.s, object files are named x.o, statically linked library routines are x.a, dynamically linked library routes are x.so, and executable files by default are called a.out. MS-DOS uses the suffixes .C, .ASM, .OBJ, .LIB, .DLL, and .EXE to the same effect.

Compiler

The compiler transforms the C program into an *assembly language program,* a symbolic form of what the machine understands. High-level language programs take many fewer lines of code than assembly language, so programmer productivity is much higher.

In 1975, many operating systems and assemblers were written in assembly language because memories were small and compilers were inefficient. The million-fold increase in memory capacity per single DRAM chip has reduced program size concerns, and optimizing compilers today can produce assembly language programs nearly as well as an assembly language expert, and sometimes even better for large programs.

assembly language A symbolic language that can be translated into binary machine language.

Assembler

Since assembly language is an interface to higher-level software, the assembler can also treat common variations of machine language instructions as if they were instructions in their own right. The hardware need not implement these instructions; however, their appearance in assembly language simplifies translation and programming. Such instructions are called pseudoinstructions.

As mentioned above, the LEGv8 hardware makes sure that register XZR (X31) always has the value 0. That is, whenever register XZR is used, it supplies a 0, and the programmer cannot change the value of register XZR. Register XZR is used to create the assembly language instruction that copies the contents of one register to another. Thus, the LEGv8 assembler accepts the following instruction even though it is not found in the LEGv8 machine language:

pseudoinstruction A common variation of assembly language instructions often treated as if it were an instruction in its own right.

```
MOV X9,X10        // register X9 gets register X10
```

The assembler converts this assembly language instruction into the machine language equivalent of the following instruction:

```
ORR X9,XZR,X10 // register X9 gets 0 OR register X10
```

The LEGv8 assembler also converts CMP (compare) into a subtract instruction that sets the condition codes and has XZR as the destination. Thus

```
CMP X9,X10  // compare X9 to X10 and set condition codes
```

becomes

```
SUBS XZR,X9,X10  // use X9 - X10 to set condition codes
```

It also converts branches to faraway locations into two branches. As mentioned above, the LEGv8 assembler allows large constants to be loaded into a register despite the limited size of the immediate instructions. Thus, the assembler can accept *load address* (LDA) and turn it into the necessary instruction sequence. Finally, it can simplify the instruction set by determining which variation of an instruction the programmer wants. For example, the LEGv8 assembler does not require the programmer to specify the immediate version of the instruction when using a constant for arithmetic and logical instructions; it just generates the proper opcode. Thus

```
AND X9,X10,#15 // register X9 gets X10 AND 15
```

becomes

```
ANDI X9,X10,#15 // register X9 gets X10 AND 15
```

We include the "I" on the instructions to remind the reader that this instruction produces a different opcode in a different instruction format than the AND instruction with no immediate operands.

In summary, pseudoinstructions give LEGv8 a richer set of assembly language instructions than those implemented by the hardware. If you are going to write assembly programs, use pseudoinstructions to simplify your task. To understand the LEGv8 architecture and be sure to get best performance, however, study the real LEGv8 instructions found in Figures 2.1 and 2.20.

Assemblers will also accept numbers in a variety of bases. In addition to binary and decimal, they usually accept a base that is more succinct than binary yet converts easily to a bit pattern. LEGv8 assemblers use hexadecimal.

Such features are convenient, but the primary task of an assembler is assembly into machine code. The assembler turns the assembly language program into an *object file*, which is a combination of machine language instructions, data, and information needed to place instructions properly in memory.

To produce the binary version of each instruction in the assembly language program, the assembler must determine the addresses corresponding to all labels. Assemblers keep track of labels used in branches and data transfer instructions in a **symbol table**. As you might expect, the table contains pairs of symbols and addresses.

symbol table A table that matches names of labels to the addresses of the memory words that instructions occupy.

The object file for UNIX systems typically contains six distinct pieces:

- The *object file header* describes the size and position of the other pieces of the object file.

- The *text segment* contains the machine language code.

- The *static data segment* contains data allocated for the life of the program. (UNIX allows programs to use both *static data*, which is allocated throughout the program, and *dynamic data*, which can grow or shrink as needed by the program. See Figure 2.14.)

- The *relocation information* identifies instructions and data words that depend on absolute addresses when the program is loaded into memory.

- The *symbol table* contains the remaining labels that are not defined, such as external references.

- The *debugging information* contains a concise description of how the modules were compiled so that a debugger can associate machine instructions with C source files and make data structures readable.

The next subsection shows how to attach such routines that have already been assembled, such as library routines.

Elaboration: Similar to the case of ADD and ADDI mentioned above, the full ARMv8 instruction set does not use ANDI when one of the operands is an immediate; it just uses AND, and lets the assembler pick the proper opcode. For teaching purposes, LEGv8 again distinguishes the two cases with different mnemonics.

Linker

What we have presented so far suggests that a single change to one line of one procedure requires compiling and assembling the whole program. Complete retranslation is a terrible waste of computing resources. This repetition is particularly wasteful for standard library routines, because programmers would be compiling and assembling routines that by definition almost never change. An alternative is to compile and assemble each procedure independently, so that a change to one line would require compiling and assembling only one procedure. This alternative requires a new systems program, called a link editor or linker, which takes all the independently assembled machine language programs and "stitches" them together. The reason a linker is useful is that it is much faster to patch code than it is to recompile and reassemble.

linker Also called **link editor**. A systems program that combines independently assembled machine language programs and resolves all undefined labels into an executable file.

There are three steps for the linker:

1. Place code and data modules symbolically in memory.
2. Determine the addresses of data and instruction labels.
3. Patch both the internal and external references.

The linker uses the relocation information and symbol table in each object module to resolve all undefined labels. Such references occur in branch instructions and data addresses, so the job of this program is much like that of an editor: it finds the old addresses and replaces them with the new addresses. Editing is the origin of the name "link editor," or linker for short.

If all external references are resolved, the linker next determines the memory locations each module will occupy. Recall that Figure 2.14 on page 108 shows the LEGv8 convention for allocation of program and data to memory. Since the files were assembled in isolation, the assembler could not know where a module's instructions and data would be placed relative to other modules. When the linker places a module in memory, all *absolute* references, that is, memory addresses that are not relative to a register, must be *relocated* to reflect its true location.

The linker produces an executable file that can be run on a computer. Typically, this file has the same format as an object file, except that it contains no unresolved references. It is possible to have partially linked files, such as library routines, that still have unresolved addresses and hence result in object files.

executable file A functional program in the format of an object file that contains no unresolved references. It can contain symbol tables and debugging information. A "stripped executable" does not contain that information. Relocation information may be included for the loader.

EXAMPLE

Linking Object Files

Link the two object files below. Show updated addresses of the first few instructions of the completed executable file. We show the instructions in assembly language just to make the example understandable; in reality, the instructions would be numbers.

Note that in the object files we have highlighted the addresses and symbols that must be updated in the link process: the instructions that refer to the addresses of procedures A and B and the instructions that refer to the addresses of data doublewords X and Y.

Object file header			
	Name	Procedure A	
	Text size	100_{hex}	
	Data size	20_{hex}	
Text segment	Address	Instruction	
	0	LDUR X0, [X27,#0]	
	4	BL 0	
	. . .	. . .	
Data segment	0	(X)	
	. . .	. . .	
Relocation information	Address	Instruction type	Dependency
	0	LDUR	X
	4	BL	B
Symbol table	Label	Address	
	X	–	
	B	–	
	Name	Procedure B	
	Text size	200_{hex}	
	Data size	30_{hex}	
Text segment	Address	Instruction	
	0	STUR X1, [X27,#0]	
	4	BL 0	
	. . .	. . .	
Data segment	0	(Y)	
	. . .	. . .	
Relocation information	Address	Instruction type	Dependency
	0	STUR	Y
	4	BL	A
Symbol table	Label	Address	
	Y	–	
	A	–	

Procedure A needs to find the address for the variable labeled X to put in the load instruction and to find the address of procedure B to place in the BL instruction. Procedure B needs the address of the variable labeled Y for the store instruction and the address of procedure A for its BL instruction.

From Figure 2.14 on page 108, we know that the text segment starts at address $0000\ 0000\ 0040\ 0000_{hex}$ and the data segment at $0000\ 0000\ 1000\ 0000_{hex}$. The text of procedure A is placed at the first address and its data at the second. The object file header for procedure A says that its text is 100_{hex} bytes and its data is 20_{hex} bytes, so the starting address for procedure B text is $40\ 0100_{hex}$, and its data starts at $1000\ 0020_{hex}$.

ANSWER

Executable file header			
	Text size		300_{hex}
	Data size		50_{hex}
Text segment	Address		Instruction
	$0000\ 0000\ 0040\ 0000_{hex}$		LDUR X0, [X27,#0_{hex}]
	$0000\ 0000\ 0040\ 0004_{hex}$		BL 000 00FC$_{hex}$
	. . .		. . .
	$0000\ 0000\ 0040\ 0100_{hex}$		STUR X1, [X27,#20_{hex}]
	$0000\ 0000\ 0040\ 0104_{hex}$		BL 3FF FEFC$_{hex}$
	. . .		. . .
Data segment	Address		
	$0000\ 0000\ 1000\ 0000_{hex}$		(X)
	. . .		. . .
	$0000\ 0000\ 1000\ 0020_{hex}$		(Y)
	. . .		. . .

Now the linker updates the address fields of the instructions. It uses the instruction type field to know the format of the address to be edited. We have two types here:

1. The branch and link instructions use PC-relative addressing. Thus, for the BL at address $40\ 0004_{hex}$ to go to $40\ 0100_{hex}$ (the address of procedure B), it must put ($40\ 0100_{hex} - 40\ 0004_{hex}$) or $000\ 00FC_{hex}$ in its address field. Similarly, since $40\ 0000_{hex}$ is the address of procedure A, the BL at $40\ 0104_{hex}$ gets the negative number $3FF\ FEFC_{hex}$ ($40\ 0000_{hex} - 40\ 0104_{hex}$) in its address field.

2. The load and store addresses are harder because they are relative to a base register. This example uses X27 as the base register, assuming it is initialized to $0000\ 0000\ 1000\ 0000_{hex}$. To get the address $0000\ 0000\ 1000\ 0000_{hex}$ (the address of doubleword X), we place 0_{hex} in the address field of LDUR at address $40\ 0000_{hex}$. Similarly, we place 20_{hex} in the address field of STUR at address $40\ 0100_{hex}$ to get the address $0000\ 0000\ 1000\ 0020_{hex}$ (the address of doubleword Y).

Elaboration: Recall that LEGv8 instructions are word aligned, so BL drops the right two bits to increase the instruction's address range. Thus, it uses 26 bits to create a 28-bit byte address. Hence, the actual address in the lower 26 bits of the first BL instruction in this example is $000\ 003F_{hex}$, rather than $000\ 00FC_{hex}$.

Loader

loader A systems program that places an object program in main memory so that it is ready to execute.

Now that the executable file is on disk, the operating system reads it to memory and starts it. The loader follows these steps in UNIX systems:

1. Reads the executable file header to determine size of the text and data segments.

2. Creates an address space large enough for the text and data.

3. Copies the instructions and data from the executable file into memory.

4. Copies the parameters (if any) to the main program onto the stack.

5. Initializes the processor registers and sets the stack pointer to the first free location.

6. Branches to a start-up routine that copies the parameters into the argument registers and calls the main routine of the program. When the main routine returns, the start-up routine terminates the program with an exit system call.

Dynamically Linked Libraries

Virtually every problem in computer science can be solved by another level of indirection.

David Wheeler

The first part of this section describes the traditional approach to linking libraries before the program is run. Although this static approach is the fastest way to call library routines, it has a few disadvantages:

- The library routines become part of the executable code. If a new version of the library is released that fixes bugs or supports new hardware devices, the statically linked program keeps using the old version.

- It loads all routines in the library that are called anywhere in the executable, even if those calls are not executed. The library can be large relative to the program; for example, the standard C library is 2.5 MB.

dynamically linked libraries (DLLs) Library routines that are linked to a program during execution.

These disadvantages lead to dynamically linked libraries (DLLs), where the library routines are not linked and loaded until the program is run. Both the program and library routines keep extra information on the location of nonlocal procedures and their names. In the original version of DLLs, the loader ran a dynamic linker, using the extra information in the file to find the appropriate libraries and to update all external references.

The downside of the initial version of DLLs was that it still linked all routines of the library that might be called, versus just those that are called during the running of the program. This observation led to the lazy procedure linkage version of DLLs, where each routine is linked only *after* it is called.

Like many innovations in our field, this trick relies on a level of indirection. Figure 2.23 shows the technique. It starts with the nonlocal routines calling a set of dummy routines at the end of the program, with one entry per nonlocal routine. These dummy entries each contain an indirect branch.

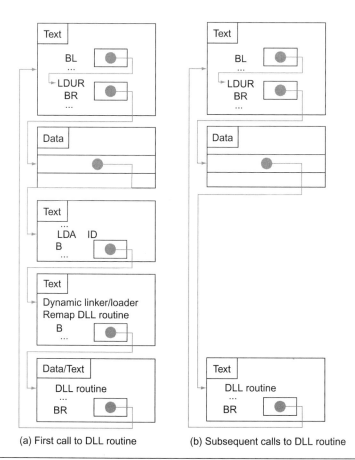

(a) First call to DLL routine   (b) Subsequent calls to DLL routine

FIGURE 2.23 Dynamically linked library via lazy procedure linkage. (a) Steps for the first time a call is made to the DLL routine. (b) The steps to find the routine, remap it, and link it are skipped on subsequent calls. As we will see in Chapter 5, the operating system may avoid copying the desired routine by remapping it using virtual memory management.

The first time the library routine is called, the program calls the dummy entry and follows the indirect branch. It points to code that puts a number in a register to identify the desired library routine and then branches to the dynamic linker/loader. The linker/loader finds the wanted routine, remaps it, and changes the address in the indirect branch location to point to that routine. It then branches to it. When the routine completes, it returns to the original calling site. Thereafter, the call to the library routine branches indirectly to the routine without the extra hops.

In summary, DLLs require additional space for the information needed for dynamic linking, but do not require that whole libraries be copied or linked. They pay a good deal of overhead the first time a routine is called, but only a single indirect branch thereafter. Note that the return from the library pays no extra overhead. Microsoft's Windows relies extensively on dynamically linked libraries, and it is also the default when executing programs on UNIX systems today.

Starting a Java Program

The discussion above captures the traditional model of executing a program, where the emphasis is on fast execution time for a program targeted to a specific instruction set architecture, or even a particular implementation of that architecture. Indeed, it is possible to execute Java programs just like C. Java was invented with a different set of goals, however. One was to run safely on any computer, even if it might slow execution time.

Java bytecode
Instruction from an instruction set designed to interpret Java programs.

Figure 2.24 shows the typical translation and execution steps for Java. Rather than compile to the assembly language of a target computer, Java is compiled first to instructions that are easy to interpret: the Java bytecode instruction set (see Section 2.15). This instruction set is designed to be close to the Java language so that this compilation step is trivial. Virtually no optimizations are performed. Like the C compiler, the Java compiler checks the types of data and produces the proper operation for each type. Java programs are distributed in the binary version of these bytecodes.

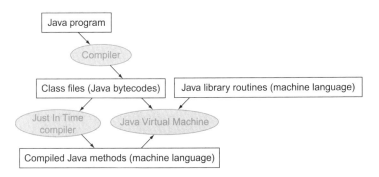

FIGURE 2.24 A translation hierarchy for Java. A Java program is first compiled into a binary version of Java bytecodes, with all addresses defined by the compiler. The Java program is now ready to run on the interpreter, called the *Java Virtual Machine* (JVM). The JVM links to desired methods in the Java library while the program is running. To achieve greater performance, the JVM can invoke the JIT compiler, which selectively compiles methods into the native machine language of the machine on which it is running.

A software interpreter, called a Java Virtual Machine (JVM), can execute Java bytecodes. An interpreter is a program that simulates an instruction set architecture. For example, the ARMv8 simulator used with this book is an interpreter. There is no need for a separate assembly step since either the translation is so simple that the compiler fills in the addresses or JVM finds them at runtime.

The upside of interpretation is portability. The availability of software Java virtual machines meant that most people could write and run Java programs shortly after Java was announced. Today, Java virtual machines are found in billions of devices, in everything from cell phones to Internet browsers.

The downside of interpretation is lower performance. The incredible advances in performance of the 1980s and 1990s made interpretation viable for many important applications, but the factor of 10 slowdown when compared to traditionally compiled C programs made Java unattractive for some applications.

To preserve portability and improve execution speed, the next phase of Java's development was compilers that translated *while* the program was running. Such Just In Time compilers (JIT) typically profile the running program to find where the "hot" methods are and then compile them into the native instruction set on which the virtual machine is running. The compiled portion is saved for the next time the program is run, so that it can run faster each time it is run. This balance of interpretation and compilation evolves over time, so that frequently run Java programs suffer little of the overhead of interpretation.

As computers get faster so that compilers can do more, and as researchers invent betters ways to compile Java on the fly, the performance gap between Java and C or C++ is closing. Section 2.15 goes into much greater depth on the implementation of Java, Java bytecodes, JVM, and JIT compilers.

> **Java Virtual Machine (JVM)** The program that interprets Java bytecodes.

> **Just In Time compiler (JIT)** The name commonly given to a compiler that operates at runtime, translating the interpreted code segments into the native code of the computer.

Which of the advantages of an interpreter over a translator was the most important for the designers of Java?

1. Ease of writing an interpreter

2. Better error messages

3. Smaller object code

4. Machine independence

Check Yourself

2.13 A C Sort Example to Put it All Together

One danger of showing assembly language code in snippets is that you will have no idea what a full assembly language program looks like. In this section, we derive the LEGv8 code from two procedures written in C: one to swap array elements and one to sort them.

```
void swap(long long int v[], size_t k)
{
  long long int temp;
  temp = v[k];
  v[k] = v[k+1];
  v[k+1] = temp;
}
```

FIGURE 2.25   A C procedure that swaps two locations in memory. This subsection uses this procedure in a sorting example.

The Procedure swap

Let's start with the code for the procedure swap in Figure 2.25. This procedure simply swaps two locations in memory. When translating from C to assembly language by hand, we follow these general steps:

1. Allocate registers to program variables.

2. Produce code for the body of the procedure.

3. Preserve registers across the procedure invocation.

This section describes the swap procedure in these three pieces, concluding by putting all the pieces together.

Register Allocation for swap

As mentioned on page 100, the LEGv8 convention on parameter passing is to use registers X0 to X7. Since swap has just two parameters, v and k, they will be found in registers X0 and X1. The only other variable is temp, which we associate with register X9 since swap is a leaf procedure (see page 113). This register allocation corresponds to the variable declarations in the first part of the swap procedure in Figure 2.25.

Code for the Body of the Procedure swap

The remaining lines of C code in swap are

```
temp = v[k];
v[k] = v[k+1];
v[k+1] = temp;
```

Recall that the memory address for LEGv8 refers to the *byte* address, and so doublewords are really 8 bytes apart. Hence, we need to multiply the index k by 8 before adding it to the address. *Forgetting that sequential doubleword addresses differ by 8 instead of by 1 is a common mistake in assembly language programming.*

Hence, the first step is to get the address of v[k] by multiplying k by 8 via a shift left by 3:

```
LSL     X10, X1,#3      // reg X10 = k * 8
ADD     X10, X0,X10     // reg X10 = v + (k * 8)
                        // reg X10 has the address of v[k]
```

Now we load v[k] using X10, and then v[k+1] by adding 8 to X10:

```
LDUR    X9,  [X10,#0]   // reg X9 (temp) = v[k]
LDUR    X11, [X10,#8]   // reg X11 = v[k + 1]
                        // refers to next element of v
```

Next we store X9 and X11 to the swapped addresses:

```
STUR    X11, [X10,#0]   // v[k] = reg X11
STUR    X9,  [X10,#8]   // v[k+1] = reg X9 (temp)
```

Now we have allocated registers and written the code to perform the operations of the procedure. What is missing is the code for preserving the saved registers used within swap. Since we are not using saved registers in this leaf procedure, there is nothing to preserve.

The Full swap Procedure

We are now ready for the whole routine, which includes the procedure label and the return branch. To make it easier to follow, we identify in Figure 2.26 each block of code with its purpose in the procedure.

Procedure body			
swap:	LSL	X10, X1,#3	# reg X10 = k * 8
	ADD	X10, X0,X10	# reg X10 = v + (k * 8)
			# reg X10 has the address of v[k]
	LDUR	X9, [X10,#0]	# reg X9 (temp) = v[k]
	LDUR	X11,[X10,#8]	# reg X11 = v[k + 1]
			# refers to next element of v
	STUR	X11,[X10,#0]	# v[k] = reg X11
	STUR	X9, [X10,#8]	# v[k+1] = reg X9 (temp)

Procedure return		
BR	LR	# return to calling routine

FIGURE 2.26   LEGv8 assembly code of the procedure swap **in Figure 2.25.**

```
void sort (long long int v[], size_t int n)
{
   size_t, j;
   for (i = 0; i < n; i += 1) {
       for (j = i - 1; j >= 0 && v[j] > v[j + 1]; j +=1) {
           swap(v,j);
       }
   }
}
```

FIGURE 2.27   A C procedure that performs a sort on the array v.

The Procedure sort

To ensure that you appreciate the rigor of programming in assembly language, we'll try a second, longer example. In this case, we'll build a routine that calls the swap procedure. This program sorts an array of integers, using bubble or exchange sort, which is one of the simplest if not the fastest sorts. Figure 2.27 shows the C version of the program. Once again, we present this procedure in several steps, concluding with the full procedure.

Register Allocation for sort

The two parameters of the procedure sort, v and n, are in the parameter registers X0 and X1, and we assign register X19 to i and register X20 to j.

Code for the Body of the Procedure sort

The procedure body consists of two nested *for* loops and a call to swap that includes parameters. Let's unwrap the code from the outside to the middle.
    The first translation step is the first *for* loop:

```
for (i = 0; i <n; i += 1) {
```

Recall that the C *for* statement has three parts: initialization, loop test, and iteration increment. It takes just one instruction to initialize i to 0, the first part of the *for* statement:

```
MOV   X19, XZR  // i = 0
```

(Remember that MOV is a pseudoinstruction provided by the assembler for the convenience of the assembly language programmer; see page 129.) It also takes just one instruction to increment i, the last part of the *for* statement:

```
ADDI  X19, X19, #1  // i += 1
```

The loop should be exited if i < n is *not* true or, said another way, should be exited if i ≥ n. This test takes two instructions:

```
for1tst: CMP X19, X1      // compare X19 to X1(i to n)
         B.GE exit1       // go to exit1 if X19 ≥ X1 (i≥n)
```

The bottom of the loop just branches back to the loop test:

```
         B    for1tst     // branch to test of outer loop
exit1:
```

The skeleton code of the first *for* loop is then

```
         MOV X19, XZR      // i = 0
for1tst:CMP X19, X1        // compare X19 to X1 (i to n)
         B.GE exit1        // go to exit1 if X19 ≥ X1 (i≥n)
         ...
         (body of first for loop)
         ...
         ADDI X19, X19, #1 // i += 1
         B    for1tst      // branch to test of outer loop
exit1:
```

Voila! (The exercises explore writing faster code for similar loops.)

The second *for* loop looks like this in C:

```
for (j = i − 1; j >= 0 && v[j] > v[j + 1]; j − = 1) {
```

The initialization portion of this loop is again one instruction:

```
SUBI        X20, X19, #1 // j = i − 1
```

The decrement of j at the end of the loop is also one instruction:

```
SUBI        X20, X20, #1 // j − = 1
```

The loop test has two parts. We exit the loop if either condition fails, so the first test must exit the loop if it fails (j < 0):

```
for2tst: CMP X20,XZR      // compare X20 to 0 (j to 0)
         B.LT exit2       // go to exit2 if X20 < 0 (j < 0)
```

This branch will skip over the second condition test. If it doesn't skip, then j ≥ 0.

The second test exits if v[j] > v[j + 1] is *not* true, or exits if v[j] ≤ v[j + 1]. First we create the address by multiplying j by 8 (since we need a byte address) and add it to the base address of v:

```
LSL      X10, X20, #3      // reg X10 = j * 8
ADD      X11, X0, X10      // reg X11 = v + (j * 8)
```

Now we load v[j]:

```
LDUR     X12, [X11,#0]        // reg X12 = v[j]
```

Since we know that the second element is just the following doubleword, we add 8 to the address in register X11 to get v[j + 1]:

```
LDUR     X13, [X11,#8]        // reg X13 = v[j + 1]
```

We test v[j] ≤ v[j + 1] to exit the loop

```
CMP      X12, X13     // compare X12 to X13
B.LE     exit2        // go to exit2 if X12 ≤ X13
```

The bottom of the loop branches back to the inner loop test:

```
B for2tst // branch to test of inner loop
```

Combining the pieces, the skeleton of the second *for* loop looks like this:

```
          SUBI X20, X19, #1       // j = i - 1
for2tst:  CMP X20,XZR             // compare X20 to 0 (j to 0)
          B.LT exit2              // go to exit2 if X20 < 0 (j < 0)
          LSL X10, X20, #3        // reg X10 = j * 8
          ADD X11, X0, X10        // reg X11 = v + (j * 8)
          LDUR X12, [X11,#0]      // reg X12 = v[j]
          LDUR X13, [X11,#8]      // reg X13 = v[j + 1]
          CMP    X12, X13         // compare X12 to X13
          B.LE    exit2           // go to exit2 if X12 ≤ X13

               . . .
          (body of second for loop)
               . . .
          SUBI X20, X20, #1       // j -= 1
          B        for2tst        // branch to test of inner loop
exit2:
```

The Procedure Call in sort

The next step is the body of the second *for* loop:

```
swap(v,j);
```

Calling swap is easy enough:

```
BL swap
```

Passing Parameters in sort

The problem comes when we want to pass parameters because the sort procedure needs the values in registers X0 and X1, yet the swap procedure needs to have its parameters placed in those same registers. One solution is to copy the parameters for sort into other registers earlier in the procedure, making registers X0 and X1 available for the call of swap. (This copy is faster than saving and restoring on the stack.) We first copy X0 and X1 into X21 and X22 during the procedure:

```
MOV X21, X0        // copy parameter X0 into X21
MOV X22, X1        // copy parameter X1 into X22
```

Then we pass the parameters to swap with these two instructions:

```
MOV X0, X21        // first swap parameter is v
MOV X1, X20        // second swap parameter is j
```

Preserving Registers in sort

The only remaining code is the saving and restoring of registers. Clearly, we must save the return address in register LR, since sort is a procedure and is called itself. The sort procedure also uses the callee-saved registers X19, X20, X21, and X22, so they must be saved. The prologue of the sort procedure is then

```
SUBI   SP,SP,#40      // make room on stack for 5 regs
STUR   LR,[SP,#32]    // save LR on stack
STUR   X22,[SP,#24]   // save X22 on stack
STUR   X21,[SP,#16]   // save X21 on stack
STUR   X20,[SP,#8]    // save X20 on stack
STUR   X19,[SP,#0]    // save X19 on stack
```

The tail of the procedure simply reverses all these instructions, and then adds a BR to return.

The Full Procedure sort

Now we put all the pieces together in Figure 2.28, being careful to replace references to registers X0 and X1 in the *for* loops with references to registers X21 and X22.

Saving registers				
	sort:	SUBI	SP,SP,#40	// make room on stack for 5 registers
		STUR	X30,[SP,#32]	// save LR on stack
		STUR	X22,[SP,#24]	// save X22 on stack
		STUR	X21,[SP,#16]	// save X21 on stack
		STUR	X20, [SP,#8]	// save X20 on stack
		STUR	X19, [SP,#0]	// save X19 on stack

Procedure body				
Move parameters		MOV	X21, X0	# copy parameter X0 into X21
		MOV	X22, X1	# copy parameter X1 into X22
Outer loop		MOV	X19, XZR	# i = 0
	for1tst:CMP		X19, X1	# compare X19 to X1 (i to n)
		B.GE	exit1	# go to exit1 if X19 ≥ X1 (i≥n)
Inner loop		SUBI	X20, X19, #1	# j = i - 1
	for2tst:CMP		X20,XZR	# compare X20 to 0 (j to 0)
		B.LT	exit2	# go to exit2 if X20 < 0 (j < 0)
		LSL	X10, X20, #3	# reg X10 = j * 8
		ADD	X11, X0, X10	# reg X11 = v + (j * 8)
		LDUR	X12, [X11,#0]	# reg X12 = v[j]
		LDUR	X13, [X11,#8]	# reg X13 = v[j + 1]
		CMP	X12, X13	# compare X12 to X13
		B.LE	exit2	# go to exit2 if X12 ≤ X13
Pass parameters and call		MOV	X0, X21	# first swap parameter is v
		MOV	X1, X20	# second swap parameter is j
		BL	swap	
Inner loop		SUBI	X20, X20, #1	# j -= 1
		B	for2tst	# branch to test of inner loop
Outer loop	exit2:	ADDI	X19, X19, #1	# i += 1
		B	for1tst	# branch to test of outer loop

Restoring registers				
	exit1:	STUR	X19, [SP,#0]	# restore X19 from stack
		STUR	X20, [SP,#8]	# restore X20 from stack
		STUR	X21,[SP,#16]	# restore X21 from stack
		STUR	X22,[SP,#24]	# restore X22 from stack
		STUR	X30,[SP,#32]	# restore LR from stack
		SUBI	SP,SP,#40	# restore stack pointer

Procedure return				
		BR	LR	# return to calling routine

FIGURE 2.28 LEGv8 assembly version of procedure sort **in Figure 2.27.**

Once again, to make the code easier to follow, we identify each block of code with its purpose in the procedure. In this example, nine lines of the sort procedure in C became 34 lines in the LEGv8 assembly language.

Elaboration: One optimization that works with this example is *procedure inlining.* Instead of passing arguments in parameters and invoking the code with a BL instruction, the compiler would copy the code from the body of the swap procedure where the call to swap appears in the code. Inlining would avoid four instructions in this example. The downside of the inlining optimization is that the compiled code would be bigger if the inlined procedure is called from several locations. Such a code expansion might turn into *lower* performance if it increased the cache miss rate; see Chapter 5.

Figure 2.29 shows the impact of compiler optimization on sort program performance, compile time, clock cycles, instruction count, and CPI. Note that unoptimized code has the best CPI, and O1 optimization has the lowest instruction count, but O3 is the fastest, reminding us that time is the only accurate measure of program performance.

Understanding Program Performance

Figure 2.30 compares the impact of programming languages, compilation versus interpretation, and algorithms on performance of sorts. The fourth column shows that the unoptimized C program is 8.3 times faster than the interpreted Java code for Bubble Sort. Using the JIT compiler makes Java 2.1 times *faster* than the unoptimized C and within a factor of 1.13 of the highest optimized C code. (🖳 Section 2.15 gives more details on interpretation versus compilation of Java and the Java and LEGv8 code for Bubble Sort.) The ratios aren't as close for Quicksort in Column 5, presumably because it is harder to amortize the cost of runtime compilation over the shorter execution time. The last column demonstrates the impact of a better algorithm, offering three orders of magnitude a performance increase by when sorting 100,000 items. Even comparing interpreted Java in Column 5 to the C compiler at highest optimization in Column 4, Quicksort beats Bubble Sort by a factor of 50 (0.05 × 2468, or 123 times faster than the unoptimized C code versus 2.41 times faster).

Elaboration: The ARMv8 compilers always save room on the stack for the arguments in case they need to be stored, so in reality they always decrement SP by 64 to make room for all eight argument registers (64 bytes). One reason is that C provides a vararg option that allows a pointer to pick, say, the third argument to a procedure. When the compiler encounters the rare vararg, it copies the eight argument registers onto the stack into the eight reserved locations.

gcc optimization	Relative performance	Clock cycles (millions)	Instruction count (millions)	CPI
None	1.00	158,615	114,938	1.38
O1 (medium)	2.37	66,990	37,470	1.79
O2 (full)	2.38	66,521	39,993	1.66
O3 (procedure integration)	2.41	65,747	44,993	1.46

FIGURE 2.29 Comparing performance, instruction count, and CPI using compiler optimization for Bubble Sort. The programs sorted 100,000 32-bit words with the array initialized to random values. These programs were run on a Pentium 4 with a clock rate of 3.06 GHz and a 533 MHz system bus with 2 GB of PC2100 DDR SDRAM. It used Linux version 2.4.20.

Language	Execution method	Optimization	Bubble Sort relative performance	Quicksort relative performance	Speedup Quicksort vs. Bubble Sort
C	Compiler	None	1.00	1.00	2468
	Compiler	O1	2.37	1.50	1562
	Compiler	O2	2.38	1.50	1555
	Compiler	O3	2.41	1.91	1955
Java	Interpreter	–	0.12	0.05	1050
	JIT compiler	–	2.13	0.29	338

FIGURE 2.30 Performance of two sort algorithms in C and Java using interpretation and optimizing compilers relative to unoptimized C version. The last column shows the advantage in performance of Quicksort over Bubble Sort for each language and execution option. These programs were run on the same system as in Figure 2.29. The JVM is Sun version 1.3.1, and the JIT is Sun Hotspot version 1.3.1.

2.14   Arrays versus Pointers

A challenge for any new C programmer is understanding pointers. Comparing assembly code that uses arrays and array indices to the assembly code that uses pointers offers insights about pointers. This section shows C and LEGv8 assembly versions of two procedures to clear a sequence of doublewords in memory: one using array indices and one with pointers. Figure 2.31 shows the two C procedures.

The purpose of this section is to show how pointers map into LEGv8 instructions, and not to endorse a dated programming style. We'll see the impact of modern compiler optimization on these two procedures at the end of the section.

Array Version of Clear

Let's start with the array version, clear1, focusing on the body of the loop and ignoring the procedure linkage code. We assume that the two parameters array and size are found in the registers X0 and X1, and that i is allocated to register X9.

```
clear1(long long int array[], size_t int size)
{
    size_t i;
    for (i = 0; i < size; i += 1)
        array[i] = 0;
}
clear2(long long int *array, size_t int size)
{
    long long int *p;
    for (p = &array[0]; p < &array[size]; p = p + 1)
        *p = 0;
}
```

FIGURE 2.31   Two C procedures for setting an array to all zeros. clear1 uses indices, while clear2 uses pointers. The second procedure needs some explanation for those unfamiliar with C. The address of a variable is indicated by &, and the object pointed to by a pointer is indicated by *. The declarations declare that array and p are pointers to integers. The first part of the *for* loop in clear2 assigns the address of the first element of array to the pointer p. The second part of the *for* loop tests to see if the pointer is pointing beyond the last element of array. Incrementing a pointer by one, in the bottom part of the *for* loop, means moving the pointer to the next sequential object of its declared size. Since p is a pointer to integers, the compiler will generate LEGv8 instructions to increment p by eight, the number of bytes in an LEGv8 integer. The assignment in the loop places 0 in the object pointed to by p.

The initialization of i, the first part of the *for* loop, is straightforward:

```
MOV     X9,XZR      // i = 0 (register X9 = 0)
```

To set array[i] to 0 we must first get its address. Start by multiplying i by 8 to get the byte address:

```
loop1: LSL     X10,X9,#3     // X10 = i * 8
```

Since the starting address of the array is in a register, we must add it to the index to get the address of array[i] using an add instruction:

```
ADD     X11,X0,X10     // X11 = address of array[i]
```

Finally, we can store 0 in that address:

```
STUR     XZR, [X11,#0]     // array[i] = 0
```

This instruction is the end of the body of the loop, so the next step is to increment i:

```
ADDI X9,X9,#1      // i = i + 1
```

The loop test checks if i is less than size:

```
CMP     X9,X1       // compare i to size
B.LT    loop1       // if (i < size) go to loop1
```

We have now seen all the pieces of the procedure. Here is the LEGv8 code for clearing an array using indices:

```
        MOV    X9,XZR          // i = 0
loop1:  LSL    X10,X9,#3       // X10 = i * 8
        ADD    X11,X0,X10      // X11 = address of array[i]
        STUR   XZR,[X11,#0]    // array[i] = 0
        ADDI   X9,X9,#1        // i = i + 1
        CMP    X9,X1           // compare i to size
        B.LT   loop1           // if (i < size) go to loop1
```

(This code works as long as size is greater than 0; ANSI C requires a test of size before the loop, but we'll skip that legality here.)

Pointer Version of Clear

The second procedure that uses pointers allocates the two parameters array and size to the registers X0 and X1 and allocates p to register X9. The code for the second procedure starts with assigning the pointer p to the address of the first element of the array:

```
    MOV    X9,X0       // p = address of array[0]
```

The next code is the body of the *for* loop, which simply stores 0 into p:

```
    loop2: STUR      XZR,[X9,#0]      // Memory[p] = 0
```

This instruction implements the body of the loop, so the next code is the iteration increment, which changes p to point to the next doubleword:

```
    ADDI      X9,X9,#8      // p = p + 8
```

Incrementing a pointer by 1 means moving the pointer to the next sequential object in C. Since p is a pointer to integers declared as long long int, each of which uses 8 bytes, the compiler increments p by 8.

The loop test is next. The first step is calculating the address of the last element of array. Start with multiplying size by 8 to get its byte address:

```
    LSL      X10,X1,#3      // X10 = size * 8
```

and then we add the product to the starting address of the array to get the address of the first doubleword *after* the array:

```
    ADD    X11,X0,X10      // X11 = address of array[size]
```

The loop test is simply to see if p is less than the last element of array:

```
    CMP    X9,X11      // compare p to &array[size]
    B.LT   loop2       // if (p<&array[size]) go to loop2
```

With all the pieces completed, we can show a pointer version of the code to zero an array:

```
        MOV   X9,X0          // p = address of array[0]
loop2:  STUR  XZR,[X9,#0]    // Memory[p] = 0
        ADDI  X9,X9,#8       // p = p + 8
        LSL   X10,X1,#3      // X10 = size * 8
        ADD   X11,X0,X10     // X11 = address of array[size]
        CMP   X9,X11         // compare p to < &array[size]
        B.LT  loop2          // if (p < &array[size]) go to loop2
```

As in the first example, this code assumes size is greater than 0.

Note that this program calculates the address of the end of the array in every iteration of the loop, even though it does not change. A faster version of the code moves this calculation outside the loop:

```
        MOV   X9,X0          // p = address of array[0]
        LSL   X10,X1,#3      // X10 = size * 8
        ADD   X11,X0,X10     // X11 = address of array[size]
loop2:  STUR  XZR,0[X9,#0]   // Memory[p] = 0
        ADDI  X9,X9,#8       // p = p + 8
        CMP   X9,X11         // compare p to < &array[size]
        B.LT  loop2          // if (p < &array[size]) go to loop2
```

Comparing the Two Versions of Clear

Comparing the two code sequences side by side illustrates the difference between array indices and pointers (the changes introduced by the pointer version are highlighted):

```
        MOV   X9,XZR        // i = 0              MOV   X9,X0         // p = & array[0]
loop1:  LSL   X10,X9,#3     // X10 = i * 8        LSL   X10,X1,#3     // X10 = size * 8
        ADD   X11,X0,X10    // X11 = &array[i]    ADD   X11,X0,X10    // X11 = &array[size]
        STUR  XZR,[X11,#0]  // array[i] = 0   loop2: STUR XZR,[X9,#0]) // Memory[p] = 0
        ADDI  X9,X9,#1      // i = i + 1          ADDI  X9,X9,#8      // p = p + 8
        CMP   X9,X1         // compare i to size  CMP   X9,X11        // compare p to &array[size]
        B.LT  loop1         // if () go to loop1  B.LT  loop2  // if (p < &array[size]) go to loop2
```

The version on the left must have the "multiply" and add inside the loop because i is incremented and each address must be recalculated from the new index. The memory pointer version on the right increments the pointer p directly. The pointer version moves the scaling shift and the array bound addition outside the loop, thereby reducing the instructions executed per iteration from six to four. This manual optimization corresponds to the compiler optimization of strength reduction (shift instead of multiply) and induction variable elimination (eliminating array address calculations within loops). Section 2.15 describes these two and many other optimizations.

Elaboration: As mentioned earlier, a C compiler would add a test to be sure that s i ze is greater than 0. One way would be to add a branch just before the first instruction of the loop to the CMP instruction.

Understanding Program Performance

People were once taught to use pointers in C to get greater efficiency than that available with arrays: "Use pointers, even if you can't understand the code." Modern optimizing compilers can produce code for the array version that is just as good. Most programmers today prefer that the compiler do the heavy lifting.

Advanced Material: Compiling C and Interpreting Java

This section gives a brief overview of how the C compiler works and how Java is executed. Because the compiler will significantly affect the performance of a computer, understanding compiler technology today is critical to understanding performance. Keep in mind that the subject of compiler construction is usually taught in a one- or two-semester course, so our introduction will necessarily only touch on the basics.

object-oriented language A programming language that is oriented around objects rather than actions, or data versus logic.

The second part of this section is for readers interested in seeing how an object-oriented language like Java executes on an LEGv8 architecture. It shows the Java byte-codes used for interpretation and the LEGv8 code for the Java version of some of the C segments in prior sections, including Bubble Sort. It covers both the Java Virtual Machine and JIT compilers.

The rest of Section 2.15 can be found online.

2.16 Real Stuff: MIPS Instructions

The instruction set closest to ARMv8 comes from another company. MIPS and ARMv8 share the same design philosophy, despite MIPS being 25 years more

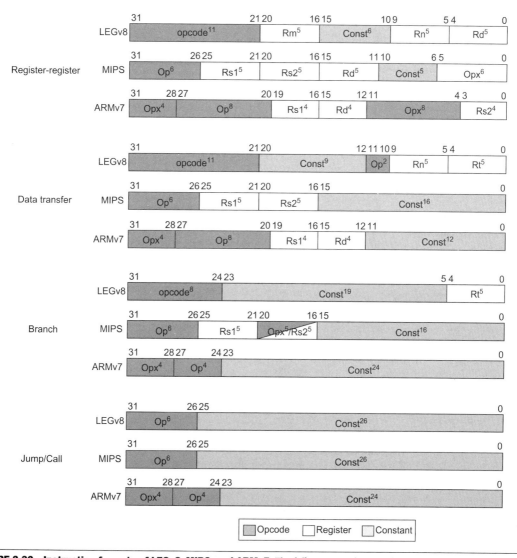

FIGURE 2.32   Instruction formats of LEGv8, MIPS, and ARMv7. The differences result in part from whether the architecture has 16 registers (ARMv7) or 32 registers (ARMv8 and MIPS).

senior than ARMv8. The good news is that if you know ARMv8, it will be very easy to pick up MIPS. To show their similarity, Figure 2.32 compares instruction formats for ARMv8 and MIPS, as well as ARMv7.

The MIPS ISA has both 32-bit address and 64-bit address versions, sensibly called MIPS-32 and MIPS-64. These instruction sets are virtually identical except for the larger address size needing 64-bit registers instead of 32-bit registers. Here are the common features between ARMv8 and MIPS:

- All instructions are 32 bits wide for both architectures.

- Both have 32 general-purpose registers, with one register being hardwired to 0.

- The only way to access memory is via load and store instructions on both architectures.

- Unlike some architectures, there are no instructions that can load or store many registers in MIPS or ARMv8.

- Both have instructions that branch if a register is equal to zero and branch if a register is not equal to zero.

- Both have 32 floating-point registers, as we shall see in Chapter 3.

- Both sets of addressing modes work for all word sizes.

One of the main differences between ARMv8 and MIPS is for conditional branches other than equal or not equal. ARMv8 and many other architectures use condition codes. MIPS instead relies on a comparison instruction that sets a register to 0 or 1 depending on whether the comparison is true. Programmers then follow that comparison instruction with a branch on equal to or not equal to zero depending on the desired outcome of the comparison. Keeping with its minimalist philosophy, MIPS only performs less than comparisons, leaving it up to the programmer to switch order of operands or to switch the condition being tested by the branch to get all the desired outcomes. MIPS has both signed and unsigned versions of the set on less than instructions: SLT and SLTU.

When we look beyond the core instructions that are most commonly used, the other main difference is that the full ARMv8 is a much larger instruction set than the complete MIPS instruction set, as we shall see in Section 2.19. That size difference means many more instruction formats, many more addressing modes, and many more operations.

2.17 Real Stuff: ARMv7 (32-bit) Instructions

Standing originally for the Acorn RISC Machine, later changed to Advanced RISC Machine, ARMv1 came out the same year as MIPS. Both architectures used 32-bit addresses, which was just fine in 1985. Many versions of the 32-bit address ARM instruction set came out over the years, culminating with ARMv7 in 2005.

ARM architects could see the writing on the wall of its 32-bit address computer and began design of the 64-bit address version of ARM in 2007. Of the many potential problems in an instruction set, the one that is almost impossible to overcome is having too small a memory address. For example, the 16-bit address MOStek 6502 powered the Apple II, but even given this head start with the first commercially successful personal computer, its lack of address bits condemned it to the dustbin of history.

ARMv8 was announced in 2011 with 64-bit addresses, and was finally revealed in 2013. Rather than some minor cosmetic changes to make all the registers 64 bits wide, which is basically what MIPS did earlier to create MIPS-64, ARM did a complete overhaul. Perhaps surprisingly, in our opinion the philosophy of the ARMv8 instruction ended up being much closer to MIPS than it is to ARMv7.

Here are the similarities between ARMv7 and ARMv8:

- All instructions are 32 bits wide for both architectures.

- The only way to access memory is via load and store instructions on both architectures.

Here are some of the differences:

- ARMv7 and the earlier ARM instruction sets have just 15 general-purpose registers, which can't make compiler writers happy.

- No register is hardwired to 0, so ARMv7 and its predecessors need extra instructions to perform some operations that ARMv8 can do with XZR.

- The missing 16th register in ARMv7 and its predecessors is the *program counter* (PC). Thus, programmers get unexpected branches if that register is altered as part of an arithmetic logic instruction. PC is not one of the 32 ARMv8 registers.

- ARMv7 addressing modes do not work for all data sizes, which is not the case in ARMv8.

- ARMv7 has Load Multiple and Store Multiple instructions, which allow several registers to be moved to and from memory. ARMv8 does not.

- Virtually every one of the ARMv5 (and earlier) instructions had a 4-bit conditional execution field, which turns it into a non-operating instruction (NOP) if the condition is false. Indeed, the conditional branch instruction is really just a special case of the regular branch instruction. This field is missing in many ARMv7 instructions, in part to make room for new instructions. ARMv8 has no such field.

- Rather than the immediate field simply being a constant, it is essentially an input to a function that produces a constant. The eight least-significant bits of ARMv7's 12-bit immediate field are zero-extended to a 32-bit value and then rotated right the number of bits specified in the first four bits of the field multiplied by two. The goal was to encode more useful constants in fewer bits. The full ARMv8 has its own complicated encoding for immediate fields of logical instructions. Whether this technique actually catches many more immediates than a simple constant field—which is what MIPS uses—would be an interesting study.

- Unlike ARMv8, the early ARM instruction sets omitted a divide instruction (see Chapter 3). It is included in ARMv7, but it is optional in a compliant ARMv7 core.

2.18   Real Stuff: x86 Instructions

Beauty is altogether in the eye of the beholder.

Margaret Wolfe Hungerford, *Molly Bawn*, 1877

Designers of instruction sets sometimes provide more powerful operations than those found in ARMv8 and MIPS. The goal is generally to reduce the number of instructions executed by a program. The danger is that this reduction can occur at the cost of simplicity, increasing the time a program takes to execute because the instructions are slower. This slowness may be the result of a slower clock cycle time or of requiring more clock cycles than a simpler sequence.

The path toward operation complexity is thus fraught with peril. Section 2.20 demonstrates the pitfalls of complexity.

Evolution of the Intel x86

ARMv8 and MIPS were the vision of single groups working at the same time; the pieces of these architectures fit nicely together. Such is not the case for the x86; it is the product of several independent groups who evolved the architecture over 35 years, adding new features to the original instruction set as someone might add clothing to a packed bag. Here are important x86 milestones.

general-purpose register (GPR) A register that can be used for addresses or for data with virtually any instruction.

- **1978:** The Intel 8086 architecture was announced as an assembly language-compatible extension of the then-successful Intel 8080, an 8-bit microprocessor. The 8086 is a 16-bit architecture, with all internal registers 16 bits wide. Unlike ARMv8, the registers have dedicated uses, and hence the 8086 is not considered a general-purpose register (GPR) architecture.

- **1980:** The Intel 8087 floating-point coprocessor is announced. This architecture extends the 8086 with about 60 floating-point instructions. Instead of using registers, it relies on a stack (see Section 2.22 and Section 3.7).

- **1982:** The 80286 extended the 8086 architecture by increasing the address space to 24 bits, by creating an elaborate memory-mapping and protection model (see Chapter 5), and by adding a few instructions to round out the instruction set and to manipulate the protection model.

- **1985:** The 80386 extended the 80286 architecture to 32 bits. In addition to a 32-bit architecture with 32-bit registers and a 32-bit address space, the 80386 added new addressing modes and additional operations. The expanded instructions make the 80386 nearly a general-purpose register machine. The

80386 also added paging support in addition to segmented addressing (see Chapter 5). Like the 80286, the 80386 has a mode to execute 8086 programs without change.

■ **1989–95**: The subsequent 80486 in 1989, Pentium in 1992, and Pentium Pro in 1995 were aimed at higher performance, with only four instructions added to the user-visible instruction set: three to help with multiprocessing (see Chapter 6) and a conditional move instruction.

■ **1997**: After the Pentium and Pentium Pro were shipping, Intel announced that it would expand the Pentium and the Pentium Pro architectures with MMX (*Multi Media Extensions*). This new set of 57 instructions uses the floating-point stack to accelerate multimedia and communication applications. MMX instructions typically operate on multiple short data elements at a time, in the tradition of *single instruction, multiple data* (SIMD) architectures (see Chapter 6). Pentium II did not introduce any new instructions.

■ **1999**: Intel added another 70 instructions, labeled SSE (*Streaming SIMD Extensions*) as part of Pentium III. The primary changes were to add eight separate registers, double their width to 128 bits, and add a single precision floating-point data type. Hence, four 32-bit floating-point operations can be performed in parallel. To improve memory performance, SSE includes cache prefetch instructions plus streaming store instructions that bypass the caches and write directly to memory.

■ **2001**: Intel added yet another 144 instructions, this time labeled SSE2. The new data type is double precision arithmetic, which allows pairs of 64-bit floating-point operations in parallel. Almost all of these 144 instructions are versions of existing MMX and SSE instructions that operate on 64 bits of data in parallel. Not only does this change enable more multimedia operations; it gives the compiler a different target for floating-point operations than the unique stack architecture. Compilers can choose to use the eight SSE registers as floating-point registers like those found in other computers. This change boosted the floating-point performance of the Pentium 4, the first microprocessor to include SSE2 instructions.

■ **2003**: A company other than Intel enhanced the x86 architecture this time. AMD announced a set of architectural extensions to increase the address space from 32 to 64 bits. Similar to the transition from a 16- to 32-bit address space in 1985 with the 80386, AMD64 widens all registers to 64 bits. It also increases the number of registers to 16 and increases the number of 128-bit SSE registers to 16. The primary ISA change comes from adding a new mode called *long mode* that redefines the execution of all x86 instructions with 64-bit addresses and data. To address the larger number of registers, it adds a new prefix to instructions. Depending how you count, long mode also adds four to 10 new instructions and drops 27 old ones. PC-relative data addressing is another extension. AMD64 still has a mode that is identical to x86 (*legacy mode*) plus a mode that restricts user programs to x86 but allows operating systems to use AMD64 (*compatibility mode*). These modes allow a more graceful transition to 64-bit addressing than the HP/Intel IA-64 architecture.

- **2004**: Intel capitulates and embraces AMD64, relabeling it *Extended Memory 64 Technology* (EM64T). The major difference is that Intel added a 128-bit atomic compare and swap instruction, which probably should have been included in AMD64. At the same time, Intel announced another generation of media extensions. SSE3 adds 13 instructions to support complex arithmetic, graphics operations on arrays of structures, video encoding, floating-point conversion, and thread synchronization (see Section 2.11). AMD added SSE3 in subsequent chips and the missing atomic swap instruction to AMD64 to maintain binary compatibility with Intel.

- **2006**: Intel announces 54 new instructions as part of the SSE4 instruction set extensions. These extensions perform tweaks like sum of absolute differences, dot products for arrays of structures, sign or zero extension of narrow data to wider sizes, population count, and so on. They also added support for virtual machines (see Chapter 5).

- **2007**: AMD announces 170 instructions as part of SSE5, including 46 instructions of the base instruction set that adds three operand instructions like ARMv8.

- **2011**: Intel ships the Advanced Vector Extension that expands the SSE register width from 128 to 256 bits, thereby redefining about 250 instructions and adding 128 new instructions.

This history illustrates the impact of the "golden handcuffs" of compatibility on the x86, as the existing software base at each step was too important to jeopardize with significant architectural changes.

Whatever the artistic failures of the x86, keep in mind that this instruction set largely drove the PC generation of computers and still dominates the Cloud portion of the post-PC era. Manufacturing 350M x86 chips per year may seem small compared to 14 billion ARM chips, but many companies would love to control such a market. Nevertheless, this checkered ancestry has led to an architecture that is difficult to explain and impossible to love.

Brace yourself for what you are about to see! Do *not* try to read this section with the care you would need to write x86 programs; the goal instead is to give you familiarity with the strengths and weaknesses of the world's most popular desktop architecture.

Rather than show the entire 16-bit, 32-bit, and 64-bit instruction set, in this section we concentrate on the 32-bit subset that originated with the 80386. We start our explanation with the registers and addressing modes, move on to the integer operations, and conclude with an examination of instruction encoding.

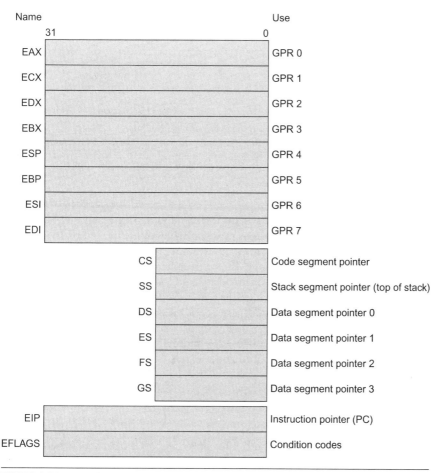

FIGURE 2.33 The 80386 register set. Starting with the 80386, the top eight registers were extended to 32 bits and could also be used as general-purpose registers.

x86 Registers and Data Addressing Modes

The registers of the 80386 show the evolution of the instruction set (Figure 2.33). The 80386 extended all 16-bit registers (except the segment registers) to 32 bits, prefixing an *E* to their name to indicate the 32-bit version. We'll refer to them generically as GPRs (*general-purpose registers*). The 80386 contains only eight GPRs. This means ARMv8 and MIPS programs can use four times as many and ARMv7 twice as many.

Figure 2.34 shows the arithmetic, logical, and data transfer instructions are two-operand instructions. There are two important differences here. The x86 arithmetic and logical instructions must have one operand act as both a source

Source/destination operand type	Second source operand
Register	Register
Register	Immediate
Register	Memory
Memory	Register
Memory	Immediate

FIGURE 2.34 Instruction types for the arithmetic, logical, and data transfer instructions. The x86 allows the combinations shown. The only restriction is the absence of a memory-memory mode. Immediates may be 8, 16, or 32 bits in length; a register is any one of the 14 major registers in Figure 2.33 (not EIP or EFLAGS).

and a destination; ARMv7, ARMv8 and MIPS allow separate registers for source and destination. This restriction puts more pressure on the limited registers, since one source register must be modified. The second important difference is that one of the operands can be in memory. Thus, virtually any instruction may have one operand in memory, unlike ARMv7, ARMv8 and MIPS.

Data memory-addressing modes, described in detail below, offer two sizes of addresses within the instruction. These so-called *displacements* can be 8 bits or 32 bits.

Although a memory operand can use any addressing mode, there are restrictions on which *registers* can be used in a mode. Figure 2.35 shows the x86 addressing modes and which GPRs cannot be used with each mode, as well as how to get the same effect using ARMv8 instructions.

x86 Integer Operations

The 8086 provides support for both 8-bit (*byte*) and 16-bit (*word*) data types. The 80386 adds 32-bit addresses and data (*doublewords*) in the x86. (AMD64 adds 64-bit addresses and data, called *quad words*; we'll stick to the 80386 in this section.) The data type distinctions apply to register operations as well as memory accesses.

Almost every operation works on both 8-bit data and on one longer data size. That size is determined by the mode and is either 16 bits or 32 bits.

Clearly, some programs want to operate on data of all three sizes, so the 80386 architects provided a convenient way to specify each version without expanding code size significantly. They decided that either 16-bit or 32-bit data dominate most programs, and so it made sense to be able to set a default large size. This default data size is set by a bit in the code segment register. To override the default data size, an 8-bit *prefix* is attached to the instruction to tell the machine to use the other large size for this instruction.

The prefix solution was borrowed from the 8086, which allows multiple prefixes to modify instruction behavior. The three original prefixes override the default segment register, lock the bus to support synchronization (see Section 2.11), or repeat the following instruction until the register ECX counts down to 0. This last

Mode	Description	Register restrictions	ARMv8 equivalent
Register indirect	Address is in a register.	Not ESP or EBP	`LDUR X1, [X2,#40]`
Based mode with 8- or 32-bit displacement	Address is contents of base register plus displacement.	Not ESP	`LDUR X1, [X2,#40]` `# <= 9-bit displacement`

Base plus scaled index	The address is
	Base + (2^Scale x Index)
	where Scale has the value 0, 1, 2, or 3.
		Base: any GPR
		Index: not ESP
			`MUL X0, X2, #8`
			`ADD X0, X0, X1`
			`LDUR X1, [X0,0]`

Base plus scaled index with 8- or 32-bit displacement	The address is
	Base + (2^Scale x Index) + displacement
	where Scale has the value 0, 1, 2, or 3.
		Base: any GPR
		Index: not ESP
			`MUL X0, X2, #8`
			`ADD X0, X0, X1`
			`LDUR X1, [X0,#40] // <= 9-bit`
			`displacement`

FIGURE 2.35 x86 32-bit addressing modes with register restrictions and the equivalent ARMv8 code. The Base plus Scaled Index addressing mode, not found in LEGv8 or MIPS, is included to avoid the multiplies by 8 (scale factor of 3) to turn an index in a register into a byte address (see Figures 2.26 and 2.28). A scale factor of 1 is used for 16-bit data, and a scale factor of 2 for 32-bit data. A scale factor of 0 means the address is not scaled. If the displacement is longer than 9 bits in the second or fourth modes, then the ARMv8 equivalent mode would need more instructions. Either a MOVZ to load 16 bits of the displacement and an ADDI to sum the upper address with the base register X2, or a MOVZ followed by a MOVK to get 32-bits of address plus an ADDI to add this address to the base register X2. (Intel gives two different names to what is called Based addressing mode—Based and Indexed—but they are essentially identical and we combine them here.)

prefix was intended to be paired with a byte move instruction to move a variable number of bytes. The 80386 also added a prefix to override the default address size.

The x86 integer operations can be divided into four major classes:

1. Data movement instructions, including move, push, and pop.

2. Arithmetic and logic instructions, including test, integer, and decimal arithmetic operations.

3. Control flow, including conditional branches, unconditional branches, calls, and returns.

4. String instructions, including string move and string compare.

The first two categories are unremarkable, except that the arithmetic and logic instruction operations allow the destination to be either a register or a memory location. Figure 2.36 shows some typical x86 instructions and their functions.

Conditional branches on the x86 are based on *condition codes* or *flags*, like ARMv7. Condition codes are set as a side effect of an operation; most are used to compare the value of a result to 0. Branches then test the condition codes. PC-relative branch addresses must be specified in the number of bytes, since unlike ARMv7, ARMv8 and MIPS, 80386 instructions are not all 4 bytes in length.

String instructions are part of the 8080 ancestry of the x86 and are not commonly executed in most programs. They are often slower than equivalent software routines (see the *Fallacy* on page 169).

Figure 2.37 lists some of the integer x86 instructions. Many of the instructions are available in both byte and word formats.

Instruction	Function
`je name`	`if equal(condition code) {EIP=name};` `EIP-128 <= name < EIP+128`
`jmp name`	`EIP=name`
`call name`	`SP=SP-4; M[SP]=EIP+5; EIP=name;`
`movw EBX,[EDI+45]`	`EBX=M[EDI+45]`
`push ESI`	`SP=SP-4; M[SP]=ESI`
`pop EDI`	`EDI=M[SP]; SP=SP+4`
`add EAX,#6765`	`EAX= EAX+6765`
`test EDX,#42`	Set condition code (flags) with EDX and 42
`movsl`	`M[EDI]=M[ESI];` `EDI=EDI+4; ESI=ESI+4`

FIGURE 2.36   Some typical x86 instructions and their functions. A list of frequent operations appears in Figure 2.37. The CALL saves the EIP of the next instruction on the stack. (EIP is the Intel PC.)

Instruction	Meaning
Control	**Conditional and unconditional branches**
`jnz, jz`	Jump if condition to EIP + 8-bit offset; JNE (for JNZ), JE (for JZ) are alternative names
`jmp`	Unconditional jump—8-bit or 16-bit offset
`call`	Subroutine call—16-bit offset; return address pushed onto stack
`ret`	Pops return address from stack and jumps to it
`loop`	Loop branch—decrement ECX; jump to EIP + 8-bit displacement if ECX $\neq$ 0
Data transfer	**Move data between registers or between register and memory**
`move`	Move between two registers or between register and memory
`push, pop`	Push source operand on stack; pop operand from stack top to a register
`les`	Load ES and one of the GPRs from memory
Arithmetic, logical	**Arithmetic and logical operations using the data registers and memory**
`add, sub`	Add source to destination; subtract source from destination; register-memory format
`cmp`	Compare source and destination; register-memory format
`shl, shr, rcr`	Shift left; shift logical right; rotate right with carry condition code as fill
`cbw`	Convert byte in eight rightmost bits of EAX to 16-bit word in right of EAX
`test`	Logical AND of source and destination sets condition codes
`inc, dec`	Increment destination, decrement destination
`or, xor`	Logical OR; exclusive OR; register-memory format
String	**Move between string operands; length given by a repeat prefix**
`movs`	Copies from string source to destination by incrementing ESI and EDI; may be repeated
`lods`	Loads a byte, word, or doubleword of a string into the EAX register

FIGURE 2.37   Some typical operations on the x86. Many operations use register-memory format, where either the source or the destination may be memory and the other may be a register or immediate operand.

x86 Instruction Encoding

Saving the worst for last, the encoding of instructions in the 80386 is complex, with many different instruction formats. Instructions for the 80386 may vary from 1 byte, when there are no operands, up to 15 bytes.

Figure 2.38 shows the instruction format for several of the example instructions in Figure 2.36. The opcode byte usually contains a bit saying whether the operand is 8 bits or 32 bits. For some instructions, the opcode may include the addressing mode and the register; this is true in many instructions that have the form "register = register op

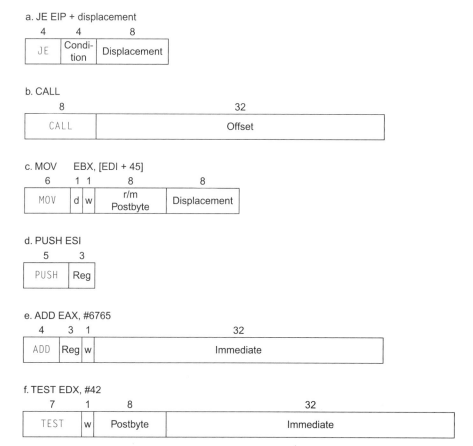

FIGURE 2.38 **Typical x86 instruction formats.** Figure 2.39 shows the encoding of the postbyte. Many instructions contain the 1-bit field *w*, which says whether the operation is a byte or a doubleword. The *d* field in MOV is used in instructions that may move to or from memory and shows the direction of the move. The ADD instruction requires 32 bits for the immediate field, because in 32-bit mode, the immediates are either 8 bits or 32 bits. The immediate field in the TEST is 32 bits long because there is no 8-bit immediate for test in 32-bit mode. Overall, instructions may vary from 1 to 15 bytes in length. The long length comes from extra 1-byte prefixes, having both a 4-byte immediate and a 4-byte displacement address, using an opcode of 2 bytes, and using the scaled index mode specifier, which adds another byte.

reg	w = 0	w = 1		r/m	mod = 0		mod = 1		mod = 2		mod = 3
		16b	32b		16b	32b	16b	32b	16b	32b	
0	AL	AX	EAX	0	addr=BX+SI	=EAX	*same*	*same*	*same*	*same*	*same*
1	CL	CX	ECX	1	addr=BX+DI	=ECX	*addr as*	*addr as*	*addr as*	*addr as*	*as*
2	DL	DX	EDX	2	addr=BP+SI	=EDX	*mod=0*	*mod=0*	*mod=0*	*mod=0*	*reg*
3	BL	BX	EBX	3	addr=BP+SI	=EBX	*+ disp8*	*+ disp8*	*+ disp16*	*+ disp32*	*field*
4	AH	SP	ESP	4	addr=SI	=(sib)	SI+disp8	(sib)+disp8	SI+disp8	(sib)+disp32	"
5	CH	BP	EBP	5	addr=DI	=disp32	DI+disp8	EBP+disp8	DI+disp16	EBP+disp32	"
6	DH	SI	ESI	6	addr=disp16	=ESI	BP+disp8	ESI+disp8	BP+disp16	ESI+disp32	"
7	BH	DI	EDI	7	addr=BX	=EDI	BX+disp8	EDI+disp8	BX+disp16	EDI+disp32	"

FIGURE 2.39   The encoding of the first address specifier of the x86: *mod, reg, r/m*. The first four columns show the encoding of the 3-bit reg field, which depends on the *w* bit from the opcode and whether the machine is in 16-bit mode (8086) or 32-bit mode (80386). The remaining columns explain the *mod* and *r/m* fields. The meaning of the 3-bit *r/m* field depends on the value in the 2-bit *mod* field and the address size. Basically, the registers used in the address calculation are listed in the sixth and seventh columns, under *mod* = 0, with *mod* = 1 adding an 8-bit displacement and *mod* = 2 adding a 16-bit or 32-bit displacement, depending on the address mode. The exceptions are 1) *r/m* = 6 when *mod* = 1 or *mod* = 2 in 16-bit mode selects BP plus the displacement; 2) *r/m* = 5 when *mod* = 1 or *mod* = 2 in 32-bit mode selects EBP plus displacement; and 3) *r/m* = 4 in 32-bit mode when *mod* does not equal 3, where (sib) means use the scaled index mode shown in Figure 2.35. When *mod* = 3, the *r/m* field indicates a register, using the same encoding as the reg field combined with the *w* bit.

immediate." Other instructions use a "postbyte" or extra opcode byte, labeled "mod, reg, r/m," which contains the addressing mode information. This postbyte is used for many of the instructions that address memory. The base plus scaled index mode uses a second postbyte, labeled "sc, index, base."

Figure 2.39 shows the encoding of the two postbyte address specifiers for both 16-bit and 32-bit modes. Unfortunately, to understand fully which registers and which addressing modes are available, you need to see the encoding of all addressing modes and sometimes even the encoding of the instructions.

x86 Conclusion

Intel had a 16-bit microprocessor two years before its competitors' more elegant architectures, such as the Motorola 68000, and this head start led to the selection of the 8086 as the CPU for the IBM PC. Intel engineers generally acknowledge that the x86 is more difficult to build than computers like ARMv7 and MIPS, but the large market meant in the PC era that AMD and Intel could afford more resources to help overcome the added complexity. What the x86 lacks in style, it rectifies with market size, making it beautiful from the right perspective.

Its saving grace is that the most frequently used x86 architectural components are not too difficult to implement, as AMD and Intel have demonstrated by rapidly improving performance of integer programs since 1978. To get that performance, compilers must avoid the portions of the architecture that are hard to implement fast.

In the post-PC era, however, despite considerable architectural and manufacturing expertise, x86 has not yet been competitive in the personal mobile device.

2.19 Real Stuff: The Rest of the ARMv8 Instruction set

Figure 2.40 suggests the size of the full ARMv8 architecture by listing the number of assembly language instructions and machine language instructions in each class of instruction types. Chapter 3 describes the floating point, multiply–divide, and SIMD portion of the architecture and Chapter 5 covers the system instructions. In this section, we highlight the integer operations and branches.

One reason for the much larger number of machine instructions than assembly instructions is that ARMv8 includes both 32-bit and 64-bit versions of instructions within the same architecture. In assembly language, programmers use registers named W0, W1, ... instead of the X0, X1, ... to specify 32-bit operations. Thus, this 64-bit operation

```
ADD    X9,X21,X9
```

can be turned into a 32-bit operation by writing instead

```
ADD    W9,W21,W9
```

Figure 2.40 counts this as one assembly language instruction but two machine language instructions since they have different opcodes.

Our goal for the rest of this section is to explain any remaining instructions in ARMv8. It is also to give insight into how the 36 LEGv8 instructions in Figure 2.1 increase to 146 in the full ARMv8 instruction set: 123 assembly language instructions in the first row in Figure 2.40 plus 23 branch instructions in the second to last row.

Class	Loads/Stores		Operations		Branches		Total	
	AL	ML	AL	ML	AL	ML	AL	ML
Integer	49	145	74	105	—	—	123	250
Floating Point & Int Mul/Div	0	18	63	156	—	—	63	174
SIMD/Vector	16	166	229	371	—	—	245	537
System/Special	11	55	52	40	—	—	63	95
—	—	—	—	—	23	14	23	14
Total	76	384	418	672	23	14	517	1070

FIGURE 2.40 The number of instructions of each type and for each function, counted separately for assembly language instructions (AL) and machine language instructions (ML). The counts were based on evaluating the *ARM Architecture Reference Manual: ARMv8, for ARMv8-A architecture profile*, beta edition, 2014. There were also a few minor adjustments made to match the version of the assembly language instructions in this book. For example, this book shows different assembly language instructions for single and precision floating-point arithmetic (FADDS, FADDD, ...) but ARMv8 actually uses just has one version (FADD), using the register names to select the opcode.

Full ARMv8 Integer Arithmetic Logic Instructions

A unique feature of the ARMv8 instruction set is that the second register of all arithmetic and logical processing operations has the option of being shifted before being operated on. The shift options are shift left logical, shift right logical, shift right arithmetic, and rotate right. (For the 12-bit immediate instructions, the only shift option is a logical left shift by 12 bits.) Although the assembler has explicit instructions with these names (LSL, LSR, SRA, and ROR), these are really just pseudoinstructions of a regular instruction that only performs the shift. We picked versions of the instructions that put 0 in the shift amount field so that the second operand is unchanged in the LEGv8, but the full instruction set allows shifts as part of arithmetic and logic instructions.

Figure 2.41 lists all 74 assembly language instructions for the integer operations of the full ARMv8 instruction set. We start with the 20 integer operations in the LEGv8 in Figure 2.1. First, the total count of assembly language instructions includes pseudoinstructions from above—CMP, CMPI, MOV—so that bumps the count to 23. There are also five pseudoinstructions based on the instructions in the LEGv8—negate, negate and set flags, compare negative, compare negative immediate, and test immediate—which brings us to 28.

To support arithmetic on narrower data types, there are another six instructions that let you mix data sizes of the second operand by either sign extending it or zero extending it to the full width. The *extended-register* instructions work with bytes, halfwords, or words, which gets us to 34 assembly language instructions.

To support add and subtract operations on operands larger than one doubleword, ARM includes six more assembly instructions to add or subtract the carry from a previous operation. Thus, the first instruction does the regular add of the registers with the right half of the operands and sets the condition codes (ADDS), and the second one does add with the carry out from that operation when adding the registers with the left half of the operands (ADC). These six bring the total to 40 instructions.

The logical instructions (AND, ANDS, ORR, EOR) also have versions where the second operand is complemented: that is, every 0 becomes 1 and vice versa. Including new pseudoinstructions, this variation leads to six more instructions, or a total now of 46.

In addition to the logical shift left and logical shift right instructions, ARMv8 includes two more, as mentioned above. ASR does arithmetic shift right, which replicates the sign bit during the shift, and ROR rotates the bits to the right; that is, the bits shift off to the right are inserted on the left. Although we didn't list them in LEGv8, there are versions of all the shift instructions that determine the amount to be shifted based on a value in a register rather than as an immediate within the instruction. These shifts add six more instructions, bringing us to 52.

There is also a version of the MOV wide instructions (MOVN) that complements all 64 bits that are created from the 16-bit constant; that is, the other 48 bits are ones instead of zeros and the 16-bit immediate field is complemented too, which bumps the count to 53.

To manipulate fields of bits, the full ARMv8 instruction set includes 10 instructions that can extract a bit field from a register and insert it into another.

Type	Mnemonic	Instruction	Type	Mnemonic	Instruction
Arithmetic Register	ADD	Add	Logical Immediate	ANDI	Bitwise AND Immediate
	ADDS	Add and set flags		ANDIS	Bitwise AND and set flags Immediate
	SUB	Subtract		ORRI	Bitwise inclusive OR Immediate
	SUBS	Subtract and set flags		EORI	Bitwise exclusive OR Immediate
	CMP	Compare		*TSTI*	Test bits Immediate
	CMN	Compare negative	Shift Register Shift Immed	LSL	Logical shift left Immediate
	NEG	Negate		LSR	Logical shift right Immediate
	NEGS	Negate and set flags		ASR	Arithmetic shift right Immediate
Arithmetic Immediate	ADDI	Add Immediate		ROR	Rotate right Immediate
	ADDIS	Add and set flags Immediate		LSRV	Logical shift right register
	SUBI	Subtract Immediate		LSLV	Logical shift left register
	SUBIS	Subtract and set flags Immediate		ASRV	Arithmetic shift right register
	CMPI	Compare Immediate		RORV	Rotate right register
	CMNI	Compare negative Immediate	Move Wide Immediate	MOVZ	Move wide with zero
Arithmetic Extended	ADD	Add Extended Register		MOVK	Move wide with keep
	ADDS	Add and set flags Extended		MOVN	Move wide with NOT
	SUB	Subtract Extended Register		*MOV*	Move register
	SUBS	Subtract and set flags Extended	Bit Field Insert & Extract	BFM	Bitfield move
	CMP	Compare Extended Register		SBFM	Signed bitfield move
	CMN	Compare negative Extended		UBFM	Unsigned bitfield move (32-bit)
Arithmetic with Carry	ADC	Add with carry		BFI	Bitfield insert
	ADCS	Add with carry and set flags		BFXIL	Bitfield extract and insert low
	SBC	Subtract with carry		SBFIZ	Signed bitfield insert in zero
	SBCS	Subtract with carry and set flags		SBFX	Signed bitfield extract
	NGC	Negate with carry		UBFIZ	Unsigned bitfield insert in zero
	NGCS	Negate with carry and set flags		UBFX	Unsigned bitfield extract
Logical Register	AND	Bitwise AND		EXTR	Extract register from pair
	ANDS	Bitwise AND and set flags	Sign Extend	*SXTB*	Sign-extend byte
	ORR	Bitwise inclusive OR		*SXTH*	Sign-extend halfword
	EOR	Bitwise exclusive OR		*SXTW*	Sign-extend word
	BIC	Bitwise bit clear		*UXTB*	Unsigned extend byte
	BICS	Bitwise bit clear and set flags		*UXTH*	Unsigned extend halfword
	ORN	Bitwise inclusive OR NOT	Bit Operation	CLS	Count leading sign bits
	EON	Bitwise exclusive OR NOT		CLZ	Count leading zero bits
	MVN	Bitwise NOT		RBIT	Reverse bit order
	TST	Test bits		REV	Reverse bytes in register
				REV16	Reverse bytes in halfwords
				REV32	Reverses bytes in words

FIGURE 2.41 The list of assembly language instructions for the integer operations in the full ARMv8 instruction set.
Bold means the instruction is also in LEGv8, italic means it is a pseudoinstruction, and bold italic means it is a pseudoinstruction that is also in LEGv8.

It uses those instructions to create five more pseudoinstructions that sign extend or zero extend narrow operands to the full width of a register. Finally, there are another six "bit twiddling" instructions that count leadings zeros or ones and reverse order of bits and bytes. The total is now 74 integer operations in the full ARMv8 integer operation assembly language.

Full ARMv8 Integer Data Transfer Instructions

We did not show all of the addressing modes available for the 10 integer data transfer instructions in LEGv8. The addressing mode we described earlier is called *unscaled signed immediate offset* in ARMv8 lingo. Here are five more:

1. Base plus a scaled 12-bit unsigned immediate offset.

2. Base plus a 64-bit register offset, optionally scaled.

3. Base plus a 32-bit extended register offset, optionally scaled.

4. Pre-indexed by an unscaled 9-bit signed immediate offset.

5. Post-indexed by an unscaled 9-bit signed immediate offset.

The scaling options of the first three addressing modes multiply or scale the address in the immediate field or in the register by the size of the data being transferred in bytes. Since you would normally want that address to be a multiple of the data size, these addressing modes are built into the hardware so the programmer doesn't have to convert an index in a register into a byte address. They also increase the range of what the immediate field can represent within an instruction. Thus, if X11 contains $100,000_{ten}$

```
    LDR X10, [X11, #16] // scaled addressing mode
```

will load the double word (8 bytes) at address $100,128_{ten}$ ($100,000 + 8*16$) into register X10.

The address of the second addressing mode is simply the sum of two registers, with the option of shifting the second operand by 1, 2, or 3 bits, which multiplies the value by 2, 4, or 8. Thus, if X11 contains $100,000_{ten}$ and X12 contains $1,000_{ten}$

```
    LDR X10, [X11, X12 LSL #3] // base + register, scaled
```

will load the double word (8 bytes) at address $108,000_{ten}$ ($100,000 + 2^{3}*1000$) into register X10. The third addressing mode simply uses a 32-bit register (e.g., W12) instead of a 64-bit register (e.g., X12), with the same option of scaling.

The last two addressing modes *change the base register* as part of the address calculation. The use case is when marching through a sequential array, the address increment can happen as part of the addressing mode rather than as part of a separate add or subtract instruction. Thus, if X11 contains $100,000_{ten}$

```
    LDR X10, [X11,#16]! // pre-indexed addressing mode
```

will load the double word (8 bytes) at address $100,0\underline{16}_{ten}$ into register X10 and change X11 to $100,016_{ten}$. On the other hand,

```
LDR X10, [X11],#16 // post-indexed addressing mode
```

will load the double word (8 bytes) at address 100,0̲0̲0̲ $_{ten}$ into register X10 and change X11 to 100,016 $_{ten}$. The pre- and post- name indicates whether the address arithmetic occurs before the operand is accessed or afterwards.

The main reason there are so many more machine data transfer instructions (125) than assembly data transfer instructions (49) for integers is that the addressing modes require separate opcodes in machine language, but as we see above, they are just a different addressing notation for the same mnemonics in assembly language.

Figure 2.42 lists the full set of integer data transfer instructions for ARMv8. To get from 10 integer data-transfer instructions to the ARMv8 total of 49, we first add the LDA pseudoinstruction, to get us to 11. Next, we note that there are signed data transfer versions of the *load byte* (LDURSB) and *load halfword* (LDURSH), which brings us to 13. In addition to the doubleword load exclusive and store exclusive instructions, ARMv8 includes exclusive data transfers of bytes, halfwords, and paired double words, yielding another six instructions, or a total of 19.

Type	Mnemonic	Instruction	Type	Mnemonic	Instruction
Unscaled	LDUR	Load register (unscaled offset)	Exclusive	LDXR	Load Exclusive register
	LDURB	Load byte (unscaled offset)		LDXRB	Load Exclusive byte
	LDURSB	Load signed byte (unscaled offset)		LDXRH	Load Exclusive halfword
	LDURH	Load halfword (unscaled offset)		LDXP	Load Exclusive Pair
	LDURSH	Load signed halfword (unscaled offset)		STXR	Store Exclusive register
	LDURSW	Load signed word (unscaled offset)		STXRB	Store Exclusive byte
	STUR	Store register (unscaled offset)		STXRH	Store Exclusive halfword
	STURB	Store byte (unscaled offset)		STXP	Store Exclusive Pair
	STURH	Store halfword (unscaled offset)	Exclusive Aquire/Release	LDAXR	Load-aquire Exclusive register
	STURW	Store word (unscaled offset)		LDAXRB	Load-aquire Exclusive byte
	LDA	Load address		LDAXRH	Load-aquire Exclusive halfword
Scaled, Extended, Pre-& Post-Indexed	LDR	Load register		LDAXP	Load-aquire Exclusive Pair
	LDRB	Load byte		STLXR	Store-release Exclusive register
	LDRSB	Load signed byte		STLXRB	Store-release Exclusive byte
	LDRH	Load halfword		STLXRH	Store-release Exclusive halfword
	LDRSH	Load signed halfword		STLXP	Store-release Exclusive Pair
	LDRSW	Load signed word	Pair	LDP	Load Pair
	STR	Store register		LDPSW	Load Pair signed words
	STRB	Store byte		STP	Store Pair
	STRH	Store halfword	PC	ADRP	Compute address of 4KB page at a PC-relative offset
				ADR	Compute address of label at a PC-relative offset

FIGURE 2.42 The list of assembly language instructions for the integer data transfer operations in the full ARMv8 instruction set. Bold means the instruction is also in LEGv8, italic means it is a pseudoinstruction, and bold italic means it is a pseudoinstruction that is also in LEGv8.

The data transfers with the other addressing modes use a different mnemonic (LDR, STR, ...), which leads to another nine instructions to get the various sizes both signed and unsigned. To accelerate data transfers, ARMv8 includes three load pair and store pair instructions, which transfer two doublewords at a time. To create PC-relative addresses that can be passed as arguments, ARMv8 has two instructions that add a 21-bit immediate in the current instruction to the current value of the PC and then load it (or the value shifted left by 12 bits) into a register. These 14 instructions bring us to 33.

The final set of data transfer instructions for integers perform exclusive access to memory in multiprocessor environments. We saw two examples earlier in LDXR and STXR. In addition to providing exclusive access to doublewords, there are versions for bytes, halfwords, and paired registers. ARMv8 also offers load acquire and store release versions, which implement the release-consistency memory model to support programming language standards for shared memory semantics. This set adds 16 exclusive data transfer instructions in the second column of Figure 2.42, bringing us to the final total of 49 data transfer instructions.

Full ARMv8 Branch Instructions

Figure 2.43 lists all the branch instructions for ARMv8. To get from six branch instructions in LEGv8 to the ARMv8 total of 23, we start with two more conditional branch instructions that test to see if a bit within a register is zero or one. There are also two more unconditional branches. The first is a variation of branch and link that uses a register for the *branch address* (BALR). The second is return from *subroutine* (RET), which sounds a lot like *branch register* (BR); the reason they sound similar is that they have the identical semantics. The reason ARMv8 has different opcodes for the same operation is so that hardware branch predictors (Chapter 4) can know whether it is really return from a *subroutine* (RET), which is easy to predict, or

Type	Mnemonic	Instruction	Type	Mnemonic	Instruction
Conditional Branch	**B.cond**	Branch conditionally	Conditional Select	CSEL	Conditional select
	CBNZ	Compare and branch if nonzero		CSINC	Conditional select increment
	CBZ	Compare and branch if zero		CSINV	Conditional select inversion
	TBNZ	Test bit and branch if nonzero		CSNEG	Conditional select negation
	TBZ	Test bit and branch if zero		*CSET*	Conditional set
Unconditional Branch	**B**	Branch unconditionally		*CSETM*	Conditional set mask
	BL	Branch with link		*CINC*	Conditional increment
	BLR	Branch with link to register		*CINV*	Conditional invert
	BR	Branch to register		*CNEG*	Conditional negate
	RET	Return from subroutine	Conditional Compare	CCMP	Conditional compare register
				CCMPI	Conditional compare immediate
				CCMN	Conditional compare negative register
				CCMNI	Conditional compare negative immediate

FIGURE 2.43 The list of assembly language instructions for the branches of the ARMv8 instruction set. Bold means the instruction is also in LEGv8, italic means it is a pseudoinstruction, and bold italic means it is a pseudoinstruction that is also in LEGv8.

being used in a *branch table* (BR), which is much harder to predict and should use a distinct mechanism. These additions get us to 10 branches.

The next nine instructions store a value into a register based on the condition codes. If the condition is true, the destination register gets the first register. If not, it gets the second register. The idea behind condition select instructions is to replace conditional branches, which can cause problems in pipelined execution if they can't be predicted (see Chapter 4). After adding the nine condition select instructions, we're up to 19.

The final four instructions are similar to the conditional select instructions, except the destination for these instructions is the condition codes. That is, these instructions conditionally set the condition codes based on the current condition codes, which is a bit confusing. If the selected condition is true, then the condition codes are set by the result of an arithmetic comparison of two operands, or to an immediate value if the condition is false. One version of conditional compare instruction checks two registers, and the other compares a register to a small 5-bit unsigned value. These last four instructions raise the total to 23 branch instructions in ARMv8.

2.20 Fallacies and Pitfalls

Fallacy: More powerful instructions mean higher performance.

Part of the power of the Intel x86 is the prefixes that can modify the execution of the following instruction. One prefix can repeat the subsequent instruction until a counter steps down to 0. Thus, to move data in memory, it would seem that the natural instruction sequence is to use move with the repeat prefix to perform 32-bit memory-to-memory moves.

An alternative method, which uses the standard instructions found in all computers, is to load the data into the registers and then store the registers back to memory. This second version of this program, with the code replicated to reduce loop overhead, copies at about 1.5 times as fast. A third version, which uses the larger floating-point registers instead of the integer registers of the x86, copies at about 2.0 times as fast as the complex move instruction.

Fallacy: Write in assembly language to obtain the highest performance.

At one time compilers for programming languages produced naïve instruction sequences; the increasing sophistication of compilers means the gap between compiled code and code produced by hand is closing fast. In fact, to compete with current compilers, the assembly language programmer needs to understand the concepts in Chapters 4 and 5 thoroughly (processor pipelining and memory hierarchy).

This battle between compilers and assembly language coders is another situation in which humans are losing ground. For example, C offers the programmer a chance to give a hint to the compiler about which variables to keep in registers versus spilled to memory. When compilers were poor at register allocation, such hints were vital to performance. In fact, some old C textbooks spent a fair amount of time giving examples that effectively use register hints. Today's C compilers generally ignore these hints, because the compiler does a better job at allocation than the programmer does.

Even *if* writing by hand resulted in faster code, the dangers of writing in assembly language are the protracted time spent coding and debugging, the loss in portability, and the difficulty of maintaining such code. One of the few widely accepted axioms of software engineering is that coding takes longer if you write more lines, and it clearly takes many more lines to write a program in assembly language than in C or Java. Moreover, once it is coded, the next danger is that it will become a popular program. Such programs always live longer than expected, meaning that someone will have to update the code over several years and make it work with new releases of operating systems and recent computers. Writing in higher-level language instead of assembly language not only allows future compilers to tailor the code to forthcoming machines; it also makes the software easier to maintain and allows the program to run on more brands of computers.

Fallacy: The importance of commercial binary compatibility means successful instruction sets don't change.

While backwards binary compatibility is sacrosanct, Figure 2.44 shows that the x86 architecture has grown dramatically. The average is more than one instruction per month over its 35-year lifetime!

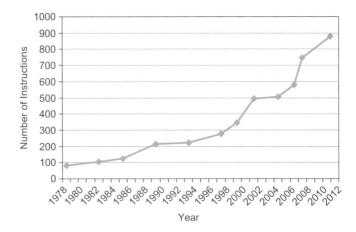

FIGURE 2.44 Growth of x86 instruction set over time. While there is clear technical value to some of these extensions, this rapid change also increases the difficulty for other companies to try to build compatible processors.

Pitfall: Forgetting that sequential word or doubleword addresses in machines with byte addressing do not differ by one.

Many an assembly language programmer has toiled over errors made by assuming that the address of the next word or doubleword can be found by incrementing the address in a register by one instead of by the word or doubleword size in bytes. Forewarned is forearmed!

Pitfall: Using a pointer to an automatic variable outside its defining procedure.

A common mistake in dealing with pointers is to pass a result from a procedure that includes a pointer to an array that is local to that procedure. Following the stack discipline in Figure 2.13, the memory that contains the local array will be reused as soon as the procedure returns. Pointers to automatic variables can lead to chaos.

2.21 Concluding Remarks

The two principles of the *stored-program* computer are the use of instructions that are indistinguishable from numbers and the use of alterable memory for programs. These principles allow a single machine to aid cancer researchers, financial advisers, and novelists in their specialties. The selection of a set of instructions that the machine can understand demands a delicate balance among the number of instructions needed to execute a program, the number of clock cycles needed by an instruction, and the speed of the clock. As illustrated in this chapter, three design principles guide the authors of instruction sets in making that tricky tradeoff:

Less is more.

Robert Browning,
Andrea del Sarto, 1855

1. *Simplicity favors regularity.* Regularity motivates many features of the ARMv8 instruction set: keeping all instructions a single size, always requiring three register operands in arithmetic instructions, and keeping the register fields in the same place in most instruction formats.

2. *Smaller is faster.* The desire for speed is the reason that ARMv8 has 32 registers rather than many more.

3. *Good design demands good compromises.* One ARMv8 example was the compromise between providing for larger addresses and constants in instructions and keeping all instructions the same length.

We also saw the great idea from Chapter 1 of making the **common cast fast** applied to instruction sets as well as computer architecture. Examples of making the common ARMv8 case fast include PC-relative addressing for conditional branches and immediate addressing for larger constant operands.

COMMON CASE FAST

Above this machine level is assembly language, a language that humans can read. The assembler translates it into the binary numbers that machines can understand,

and it even "extends" the instruction set by creating symbolic instructions that aren't in the hardware. For instance, constants or addresses that are too big are broken into properly sized pieces, common variations of instructions are given their own name, and so on. Figure 2.45 lists the LEGv8 instructions we have covered so far,

LEGv8 instructions	Name	Format	Pseudo LEGv8	Name	Format
add	ADD	R	move	MOV	R
subtract	SUB	R	compare	CMP	R
add immediate	ADDI	I	compare immediate	CMPI	I
subtract immediate	SUBI	I	load address	LDA	M
add and set flags	ADDS	R			
subtract and set flags	SUBS	R			
add immediate and set flags	ADDIS	I			
subtract immediate and set flags	SUBIS	I			
load register	LDUR	D			
store register	STUR	D			
load signed word	LDURSW	D			
store word	STURW	D			
load half	LDURH	D			
store half	STURH	D			
load byte	LDURB	D			
store byte	STURB	D			
load exclusive register	LDXR	D			
store exclusive register	STXR	D			
move wide with zero	MOVZ	IM			
move wide with keep	MOVK	IM			
and	AND	R			
inclusive or	ORR	R			
exclusive or	EOR	R			
and immediate	ANDI	I			
inclusive or immediate	ORRI	I			
exclusive or immediate	EORI	I			
logical shift left	LSL	R			
logical shift right	LSR	R			
compare and branch on equal 0	CBZ	CB			
compare and branch on not equal 0	CBNZ	CB			
branch conditionally	B.cond	CB			
branch	B	B			
branch to register	BR	R			
branch with link	BL	B			

FIGURE 2.45 The LEGv8 instruction set covered so far, with the real LEGv8 instructions on the left and the pseudoinstructions on the right. Section 2.19 describes the full ARMv8 architecture. Figure 2.1 shows more details of the LEGv8 architecture revealed in this chapter. The information given here is also found in Columns 1 and 2 of the LEGv8 Reference Data Card at the front of the book.

both real and pseudoinstructions. Hiding details from the higher level is another example of the great idea of **abstraction**.

Each category of ARMv8 instructions is associated with constructs that appear in programming languages:

ABSTRACTION

■ Arithmetic instructions correspond to the operations found in assignment statements.

■ Transfer instructions are most likely to occur when dealing with data structures like arrays or structures.

■ Conditional branches are used in *if* statements and in loops.

■ Unconditional branches are used in procedure calls and returns and for *case/switch* statements.

These instructions are not born equal; the popularity of the few dominates the many. For example, Figure 2.46 shows the popularity of each class of instructions for SPEC CPU2006. The varying popularity of instructions plays an important role in the chapters about datapath, control, and pipelining.

Instruction class	LEGv8 examples	HLL correspondence	Frequency	
			Integer	Ft. pt.
Arithmetic	ADD, SUB, ADDI, SUBI	Operations in assignment statements	16%	48%
Data transfer	LDUR, STUR, LDURSW, STURW, LDURH, STURH, LDURB, STURB, MOVZ, MOVK	References to data structures, such as arrays	35%	36%
Logical	AND, ORR, EOR, ANDI, ORRI, EORI, LSL, LSR	Operations in assignment statements	12%	4%
Conditional branch	CBZ, CBNZ, B.cond, CMP, CMPI	*If* statements and loops	34%	8%
Jump	B, BR, BL	Procedure calls, returns, and *case/switch* statements	2%	0%

FIGURE 2.46 LEGv8 instruction classes, examples, correspondence to high-level program language constructs, and percentage of LEGv8 instructions executed by category for the average integer and floating point SPEC CPU2006 benchmarks. Figure 3.27 in Chapter 3 shows average percentage of the individual LEGv8 instructions executed.

After we explain computer arithmetic in Chapter 3, we reveal more of the ARMv8 instruction set architecture.

Historical Perspective and Further Reading

This section surveys the history of *instruction set architectures* (ISAs) over time, and we give a short history of programming languages and compilers. ISAs include accumulator architectures, general-purpose register architectures, stack architectures, and a brief history of ARMv7 and the x86. We also review the controversial subjects of high-level-language computer architectures and reduced instruction set computer architectures. The history of programming languages includes Fortran, Lisp, Algol, C, Cobol, Pascal, Simula, Smalltalk, C++, and Java, and the history of compilers includes the key milestones and the pioneers who achieved them. The rest of Section 2.22 is found online.

2.23 Exercises

(Readers may wish to download and install a free Community Edition of the ARM DS-5 professional software suite which contains an ARMv8-A (64-bit) architecture simulator before attempting these exercises. More details are available on the companion Web site for this book.)

The main difference between LEGv8 and ARMv8, is that LEGv8 is a subset of ARMv8. Thus, compilers may generate ARMv8 instructions that are not found in LEGv8. That is one reason we list the rest of the ARMv8 instructions in Sections 2.19, 3.8, and 5.14. If you were to look at the output of a compiler, the most obvious difference would be the likely use of scaled loads and stores (LDR and STR) instead of the unscaled versions that we feature in LEGv8 (LDUR and STUR).

We also made some changes to ARMv8 for pedagogic reasons in LEGv8, which we noted in the *Elaboration* sections in the text, that will not work with the ARMv8 assembler and simulators. Here are the four differences:

- Immediate versions are not separate assembly language instructions in ARMv8. While LEGv8 has ADDI, SUBI, ADDIS, SUBIS, ANDI, ORRI, and EORI, ARMv8 does not. You simply write the non-immediate version and use an immediate operand. For example, the ARMv8 assembler accepts

  ADD X0, X1, #4

  and produces the immediate version of the opcode for ADD, which LEGv8 associates with ADDI.

- Similarly, single- and double-precision floating-point instructions do not have separate assembly language instructions. Thus, ARMv8 does not have FADDS, FADDD, FSUBS, FSUBD, FMULS, FMULD, FDIVS, FDIVD, FCMPS, FCMPD, LDURS, LDURD, STURS, and STURD, as does LEGv8. Instead, you just write the floating-point operation, and the assembler produces the right opcode depending on which register name you use: S for single and D for double. For example, the ARMv8 assembler accepts

```
FADD D0, D1, D2
```

and produces the double-precision version of the opcode for FADD, which LEGv8 associates with FADDD.

■ In the full ARMv8 instruction set, register 31 is XZR in most instructions but the *stack pointer* (SP) in others. We think it is confusing, so register 31 is always XZR and SP is always register 28 in LEGv8. If you stick to using XZR and SP for register names when using the assembler and simulator, it should not be a problem, but this subtle distinction might show up in visualizations of the state of the processor, for example.

■ The immediate fields for ANDI, ORRI, and EORI of the full ARMv8 instruction set are not simple 12-bit immediates that we assume in LEGv8. ARMv8 has an algorithm for encoding immediate values as repeating patterns. This means that some small constants (e.g., 1, 2, 3, 4, and 6) are valid, while others (e.g., 0, 5) are not. Thus, the assembler may insert more instructions than you might expect to create simple constants, or fewer if you pick a lucky large constant. The official definition is the bit pattern can be viewed "as a vector of identical elements of size e = 2, 4, 8, 16, 32, or 64 bits. Each element contains the same sub-pattern, that is a single run of 1 to (e − 1) nonzero bits from bit 0 followed by zero bits, then rotated by 0 to (e − 1) bits." Don't try to figure this out yourself; just leave it to the assembler!

2.1 [5] <§2.2> For the following C statement, write the corresponding LEGv8 assembly code. Assume that the C variables f, g, and h, have already been placed in registers X0, X1, and X2 respectively. Use a minimal number of LEGv8 assembly instructions.

```
f = g + (h − 5);
```

2.2 [5] <§2.2> Write a single C statement that corresponds to the two LEGv8 assembly instructions below.

```
ADD f, g, h
ADD f, i, f
```

2.3 [5] <§§2.2, 2.3> For the following C statement, write the corresponding LEGv8 assembly code. Assume that the variables f, g, h, i, and j are assigned to registers X0, X1, X2, X3, and X4, respectively. Assume that the base address of the arrays A and B are in registers X6 and X7, respectively.

```
B[8] = A[i−j];
```

2.4 [10] <§§2.2, 2.3> For the LEGv8 assembly instructions below, what is the corresponding C statement? Assume that the variables f, g, h, i, and j are assigned to registers X0, X1, X2, X3, and X4, respectively. Assume that the base address of the arrays A and B are in registers X6 and X7, respectively.

```
LSL   X9,  X0,  #3      // X9 = f*8
ADD   X9,  X6,  X9      // X9 = &A[f]
LSL   X10, X1,  #3      // X10 = g*8
ADD   X10, X7,  X10     // X10 = &B[g]
LDUR  X0,  [X9, #0]     // f = A[f]

ADDI  X11, X9,  #8
LDUR  X9,  [X11, #0]
ADD   X9,  X9,  X0
STUR  X9,  [X10, #0]
```

2.5 [5] <§§2.2, 2.3, 2.6> For the LEGv8 assembly instructions in Exercise 2.4, rewrite the assembly code to minimize the number of LEGv8 instructions needed to carry out the same function.

2.6 [5] <§2.3> Show how the value 0xabcdef12 would be arranged in memory of a little-endian and a big-endian machine. Assume the data are stored starting at address 0 and that the word size is 4 bytes.

2.7 [5] <§2.4> Translate 0xabcdef12 into decimal.

2.8 [5] <§§2.2, 2.3> Translate the following C code to LEGv8. Assume that the variables f, g, h, i, and j are assigned to registers X0, X1, X2, X3, and X4, respectively. Assume that the base address of the arrays A and B are in registers X6 and X7, respectively. Assume that the elements of the arrays A and B are 8-byte words:

```
B[8] = A[i] + A[j];
```

2.9 [10] <§§2.2, 2.3> Translate the following LEGv8 code to C. Assume that the variables f, g, h, i, and j are assigned to registers X0, X1, X2, X3, and X4, respectively. Assume that the base address of the arrays A and B are in registers X6 and X7, respectively.

```
ADDI  X9,  X6,  #8
ADD   X10, X6,  XZR
STUR  X10, [X9, #0]
LDUR  X9,  [X9, #0]
ADD   X0,  X9,  X10
```

2.10 [20] <§§2.2, 2.5> For each LEGv8 instruction in Exercise 2.9, show the value of the opcode (Op), source register (Rn), and target register (Rd or Rt) fields. For the I-type instructions, show the value of the immediate field, and for the R-type instructions, show the value of the second source register (Rm).

2.11 Assume that registers X0 and X1 hold the values 0×8000000000000000 and 0×D000000000000000, respectively.

2.11.1 [5] <§2.4> What is the value of X9 for the following assembly code?

```
ADD X9, X0, X1
```

2.11.2 [5] <§2.4> Is the result in X9 the desired result, or has there been overflow?

2.11.3 [5] <§2.4> For the contents of registers X0 and X1 as specified above, what is the value of X9 for the following assembly code?

```
SUB X9, X0, X1
```

2.11.4 [5] <§2.4> Is the result in X9 the desired result, or has there been overflow?

2.11.5 [5] <§2.4> For the contents of registers X0 and X1 as specified above, what is the value of X9 for the following assembly code?

```
ADD X9, X0, X1
ADD X9, X9, X0
```

2.11.6 [5] <§2.4> Is the result in X9 the desired result, or has there been overflow?

2.12 Assume that X0 holds the value 128_{ten}.

2.12.1 [5] <§2.4> For the instruction ADD X9,X0,X1, what is the range(s) of values for X1 that would result in overflow?

2.12.2 [5] <§2.4> For the instruction SUB X9,X0,X1, what is the range(s) of values for X1 that would result in overflow?

2.12.3 [5] <§2.4> For the instruction SUB X9,X1,X0, what is the range(s) of values for X1 that would result in overflow?

2.13 [5] <§§2.2, 2.5> Provide the instruction type and assembly language instruction for the following binary value:

```
1000 1011 0000 0000 0000 0000 0000 0000₂
```

Hint: Figure 2.20 may be helpful.

2.14 [5] <§§2.2, 2.5> Provide the instruction type and hexadecimal representation of the following instruction:

```
STUR X9, [X10,#32]
```

2.15 [5] <§2.5> Provide the instruction type, assembly language instruction, and binary representation of instruction described by the following LEGv8 fields:

```
op=0x658, Rm=13, Rn=15, Rd=17, shamt=0
```

2.16 [5] <§2.5> Provide the instruction type, assembly language instruction, and binary representation of instruction described by the following LEGv8 fields:

```
op=0×7c2, Rn=12, Rt=3, const=0×4
```

2.17 Assume that we would like to expand the LEGv8 register file to 128 registers and expand the instruction set to contain four times as many instructions.

2.17.1 [5] <§2.5> How would this affect the size of each of the bit fields in the R-type instructions?

2.17.2 [5] <§2.5> How would this affect the size of each of the bit fields in the I-type instructions?

2.17.3 [5] <§§2.5, 2.8, 2.10> How could each of the two proposed changes decrease the size of a LEGv8 assembly program? On the other hand, how could the proposed change increase the size of an LEGv8 assembly program?

2.18 Assume the following register contents:

```
X10 = 0x00000000AAAAAAAA, X11 = 0x1234567812345678
```

2.18.1 [5] <§2.6> For the register values shown above, what is the value of X12 for the following sequence of instructions?

```
LSL X12, X10, #4
ORR X12, X12, X11
```

2.18.2 [5] <§2.6> For the register values shown above, what is the value of X12 for the following sequence of instructions?

```
LSL X12, X11, #4
```

2.18.3 [5] <§2.6> For the register values shown above, what is the value of X12 for the following sequence of instructions?

```
LSR  X12, X10, #3
ANDI X12, X12, 0xFEF
```

2.19 [10] <§2.6> Find the shortest sequence of LEGv8 instructions that extracts bits 16 down to 11 from register X10 and uses the value of this field to replace bits 31 down to 26 in register X11 without changing the other bits of registers X10 or X11. (Be sure to test your code using X10 = 0 and X11 = 0xffffffffffffffff. Doing so may reveal a common oversight.)

2.20 [5] <§2.6> Provide a minimal set of LEGv8 instructions that may be used to implement the following pseudoinstruction:

```
NOT X10, X11        // bit-wise invert
```

2.21 [5] <§2.6> For the following C statement, write a minimal sequence of LEGv8 assembly instructions that performs the identical operation. Assume X11 = A, and X13 is the base address of C.

```
A = C[0] << 4;
```

2.22 [5] <§2.7> Assume X0 holds the value 0×0000000000101000. What is the value of X1 after the following instructions?

```
        CMP X0, #0
        B.GE ELSE
        B DONE
ELSE:   ORRI X1, XZR, #2
DONE:
```

2.23 Suppose the *program counter* (PC) is set to 0×2000 0000.

2.23.1 [5] <§2.10> What range of addresses can be reached using the LEGv8 *branch* (B) instruction? (In other words, what is the set of possible values for the PC after the branch instruction executes?).

2.23.2 [5] <§2.10> What range of addresses can be reached using the LEGv8 *conditional branch-on-equal* (CBZ) instruction? (In other words, what is the set of possible values for the PC after the branch instruction executes?)

2.24 Consider a proposed new instruction named RPT. This instruction combines a loop's condition check and counter decrement into a single instruction. For example RPT X12, loop would do the following:

```
if (X12 >0) {
        X12 = X12 -1;
        goto loop
    }
```

2.24.1 [5] <§2.7, 2.10> If this instruction were to be added to the ARMv8 instruction set, what is the most appropriate instruction format?

2.24.2 [5] <§2.7> What is the shortest sequence of LEGv8 instructions that performs the same operation?

2.25 Consider the following LEGv8 loop:

```
LOOP:   SUBIS X1, X1, #0
        B.LE DONE
        SUBI X1, X1, #1
        ADDI X0, X0, #2
        B LOOP
DONE:
```

2.25.1 [5] <§2.7> Assume that the register X1 is initialized to the value 10. What is the final value in register X0 assuming the X0 is initially zero?

2.25.2 [5] <§2.7> For the loop above, write the equivalent C code. Assume that the registers X0, and X1 are integers acc and i respectively.

2.25.3 [5] <§2.7> For the loop written in LEGv8 assembly above, assume that the register X1 is initialized to the value N. How many LEGv8 instructions are executed?

2.25.4 [5] <§2.7> For the loop written in LEGv8 assembly above, replace the instruction "B.LE DONE" instruction with "B.MI DONE". What is the final value in register X0 assuming the X0 is initially zero?

2.25.5 [5] <§2.7> For the loop written in LEGv8 assembly above, replace the instruction "B.LE DONE" instruction with "B.MI DONE" and write equivalent C code.

2.25.6 [5] <§2.7> What is the purpose of the SUBIS instruction in the assembly code above?

2.25.7 [5] <§2.7> Show how you can reduce the number of instructions by combining the SUBIS and SUBI instructions. (Hint: Add one instruction outside the loop.)

2.26 [10] <§2.7> Translate the following C code to LEGv8 assembly code. Use a minimum number of instructions. Assume that the values of a, b, i, and j are in registers X0, X1, X10, and X11, respectively. Also, assume that register X2 holds the base address of the array D.

```
for(i=0; i<a; i++)
    for(j=0; j<b; j++)
        D[4*j] = i + j;
```

2.27 [5] <§2.7> How many LEGv8 instructions does it take to implement the C code from Exercise 2.26? If the variables a and b are initialized to 10 and 1 and all elements of D are initially 0, what is the total number of LEGv8 instructions executed to complete the loop?

2.28 [5] <§2.7> Translate the following loop into C. Assume that the C-level integer i is held in register X10, X0 holds the C-level integer called result, and X1 holds the base address of the integer MemArray.

```
        ORR  X10, XZR, XZR
LOOP:   LDUR X11, [X1, #0]
        ADD  X0, X0, X11
        ADDI X1, X1, #8
        ADDI X10, X10, #1
        CMPI X10, 100
        B.LT LOOP
```

2.29 [10] <§2.7> Rewrite the loop from Exercise 2.28 to reduce the number of LEGv8 instructions executed. Hint: Notice that variable i is used only for loop control.

2.30 [30] <§2.8> Implement the following C code in LEGv8 assembly. Hint: Remember that the stack pointer must remain aligned on a multiple of 16.

```
int fib(int n){
    if (n==0)
        return 0;
    else if (n == 1)
        return 1;
    else
        return fib(n-1) + fib(n-2);
}
```

2.31 [20] <§2.8> For each function call, show the contents of the stack after the function call is made. Assume the stack pointer is originally at address 0x7ffffffc, and follow the register conventions as specified in Figure 2.12.

2.32 [20] <§2.8> Translate function f into LEGv8 assembly language. If you need to use registers X10 through X27, use the lower-numbered registers first. Assume the function declaration for g is "int g(int a, int b)". The code for function f is as follows:

```
int f(int a, int b, int c, int d){
  return g(g(a,b),c+d);
}
```

2.33 [5] <§2.8> Can we use the tail-call optimization in this function? If no, explain why not. If yes, what is the difference in the number of executed instructions in f with and without the optimization?

2.34 [5] <§2.8> Right before your function f from Exercise 2.32 returns, what do we know about contents of registers X5, X29, X30, and SP? Keep in mind that we know what the entire function f looks like, but for function g we only know its declaration.

2.35 [30] <§2.9> Write a program in LEGv8 assembly to convert an ASCII string containing a positive or negative integer decimal string to an integer. Your program should expect register X0 to hold the address of a null-terminated string containing an optional '+' or '−' followed by some combination of the digits 0 through 9. Your program should compute the integer value equivalent to this string of digits, then place the number in register X0. If a non-digit character appears anywhere in the string, your program should stop with the value −1 in register X0. For example, if register X0 points to a sequence of three bytes 50_{ten}, 52_{ten}, 0_{ten} (the null-terminated string "24"), then when the program stops, register X0 should contain the value 24_{ten}. The ARMv8 MUL instruction takes two registers as input. There is no "MULI" instruction. Thus, just store the constant 10 in a register.

2.36 Consider the following code:

```
LDURB X10, [X11, #0]
STUR  X10, [X11, #8]
```

Assume that the register X11 contains the address 0×10000000 and the data at address is 0×1122334455667788.

2.36.1 [5] <§2.3, 2.9> What value is stored in 0×10000008 on a big-endian machine?

2.36.2 [5] <§2.3, 2.9> What value is stored in 0×10000008 on a little-endian machine?

2.37 [5] <§2.10> Write the LEGv8 assembly code that creates the 64-bit constant $0\times1122334455667788_{two}$ and stores that value to register X0.

2.38 [10] <§2.11> Write the LEGv8 assembly code to implement the following C code:

```
lock(lk);
shvar=max(shvar,x);
unlock(lk);
```

Assume that the address of the lk variable is in X0, the address of the shvar variable is in X1, and the value of variable x is in X2. Your critical section should not contain any function calls. Use LDXR/STXR instructions to implement the lock() operation, the unlock() operation is simply an ordinary store instruction.

2.39 [10] <§2.11> Repeat Exercise 2.38, but this time use LDXR/STXR to perform an atomic update of the shvar variable directly, without using lock() and unlock(). Note that in this exercise there is no variable lk.

2.40 [5] <§2.11> Using your code from Exercise 2.38 as an example, explain what happens when two processors begin to execute this critical section at the same time, assuming that each processor executes exactly one instruction per cycle.

2.41 Assume for a given processor the CPI of arithmetic instructions is 1, the CPI of load/store instructions is 10, and the CPI of branch instructions is 3. Assume a program has the following instruction breakdowns: 500 million arithmetic instructions, 300 million load/store instructions, 100 million branch instructions.

2.41.1 [5] <§§1.6, 2.13> Suppose that new, more powerful arithmetic instructions are added to the instruction set. On average, through the use of these more powerful arithmetic instructions, we can reduce the number of arithmetic instructions needed to execute a program by 25%, while increasing the clock cycle time by only 10%. Is this a good design choice? Why?

2.41.2 [5] <§§1.6, 2.13> Suppose that we find a way to double the performance of arithmetic instructions. What is the overall speedup of our machine? What if we find a way to improve the performance of arithmetic instructions by 10 times?

2.42 Assume that for a given program 70% of the executed instructions are arithmetic, 10% are load/store, and 20% are branch.

2.42.1 [5] <§§1.6, 2.13> Given this instruction mix and the assumption that an arithmetic instruction requires two cycles, a load/store instruction takes six cycles, and a branch instruction takes three cycles, find the average CPI.

2.42.2 [5] <§§1.6, 2.13> For a 25% improvement in performance, how many cycles, on average, may an arithmetic instruction take if load/store and branch instructions are not improved at all?

2.42.3 [5] <§§1.6, 2.13> For a 50% improvement in performance, how many cycles, on average, may an arithmetic instruction take if load/store and branch instructions are not improved at all?

2.43 [10] <§2.19> Use the full ARMv8 instruction set (in particular the ability to scale a register offset when loading data) to further reduce the number of assembly instructions needed to carry out the function given in Exercise 2.4.

2.44 [10] <§2.19> Use the full ARMv8 instruction set (in particular the ability to scale a register offset when loading data) to further reduce the number of assembly instructions needed to implement the C code given in Exercise 2.8.

2.45 [10] <§2.19> Use the full ARMv8 instruction set to further reduce the number of instructions needed to complete Exercise 2.19.

2.46 [10] <§2.19> Provide a minimal set of ARMv8 instructions that may be used to implement the following pseudoinstruction:

```
NOT X10, X11        // bit-wise invert
```

(Note that some LEGv8 answers to Exercise 2.20 may not be valid ARMv8.)

2.47 [10] <§2.19> Use the pre-indexed addressing mode to minimize the number of ARMv8 assembly instructions needed to implement the code given in Exercise 2.28.

2.48 [5] <§2.19> Use the post-indexed addressing mode to minimize the number of ARMv8 assembly instructions needed to implement the code given in Exercise 2.28.

2.49 [5] <§2.19> Implement an ARMv8 assembly function to find the maximum value in an integer array. Optimize your function to require as few instructions as possible. Hint: using the CSEL instruction, it is possible to reduce the body of the loop to five instructions.

Answers to Check Yourself

§2.2, page 67: ARMv8, C, Java.

§2.3, page 73: 2) Very slow.

§2.4, page 81: 2) -8_{ten}

§2.5, page 89: 3) SUB X11, X10, X9

§2.6, page 92: Both. AND with a mask pattern of 1s will leaves 0s everywhere but the desired field. Shifting left by the correct amount removes the bits from the left of the field. Shifting right by the appropriate amount puts the field into the right-most bits of the doubleword, with 0s in the rest of the doubleword. Note that AND leaves the field where it was originally, and the shift pair moves the field into the rightmost part of the doubleword.

§2.7, page 99: I. All are true. II. 1).

§2.8, page 110: Both are true.

§2.9, page 115: I. 1) and 2) II. 3).

§2.10, page 124: I. 4) ± 1024 K. II. 4) $\pm$ 128 M.

§2.11, page 128: Both are true.

§2.12, page 137: 4) Machine independence.

Arithmetic for Computers

Numerical precision
is the very soul of
science.

Sir D'arcy Wentworth Thompson,
On Growth and Form, 1917

The Five Classic Components of a Computer

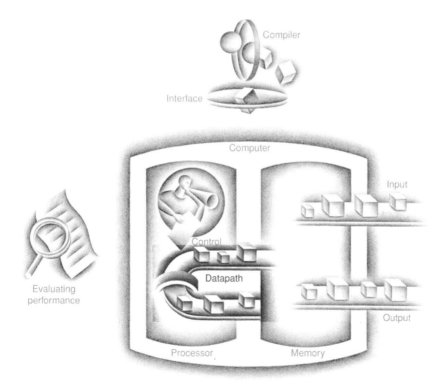

3.1   Introduction

Computer words are composed of bits; thus, words can be represented as binary numbers. Chapter 2 shows that integers can be represented either in decimal or binary form, but what about the other numbers that commonly occur? For example:

- What about fractions and other real numbers?

- What happens if an operation creates a number bigger than can be represented?

- And underlying these questions is a mystery: How does hardware really multiply or divide numbers?

The goal of this chapter is to unravel these mysteries—including representation of real numbers, arithmetic algorithms, hardware that follows these algorithms—and the implications of all this for instruction sets. These insights may explain quirks that you have already encountered with computers. Moreover, we show how to use this knowledge to make arithmetic-intensive programs go much faster.

Subtraction: Addition's Tricky Pal

No. 10, Top Ten Courses for Athletes at a Football Factory, David Letterman et al., *Book of Top Ten Lists*, 1990

3.2   Addition and Subtraction

Addition is just what you would expect in computers. Digits are added bit by bit from right to left, with carries passed to the next digit to the left, just as you would do by hand. Subtraction uses addition: the appropriate operand is simply negated before being added.

Binary Addition and Subtraction

Let's try adding 6_{ten} to 7_{ten} in binary and then subtracting 6_{ten} from 7_{ten} in binary.

$$00000000\ 00000000\ 00000000\ 00000000\ 00000000\ 00000000\ 00000000\ 00000111_{two} = 7_{ten}$$
$$+\ 00000000\ 00000000\ 00000000\ 00000000\ 00000000\ 00000000\ 00000000\ 00000110_{two} = 6_{ten}$$

$$=\ 00000000\ 00000000\ 00000000\ 00000000\ 00000000\ 00000000\ 00000000\ 00001101_{two} = 13_{ten}$$

The 4 bits to the right have all the action; Figure 3.1 shows the sums and carries. Parentheses identify the carries, with the arrows illustrating how they are passed.

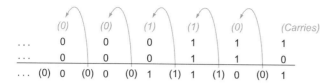

FIGURE 3.1 Binary addition, showing carries from right to left. The rightmost bit adds 1 to 0, resulting in the sum of this bit being 1 and the carry out from this bit being 0. Hence, the operation for the second digit to the right is $0 + 1 + 1$. This generates a 0 for this sum bit and a carry out of 1. The third digit is the sum of $1 + 1 + 1$, resulting in a carry out of 1 and a sum bit of 1. The fourth bit is $1 + 0 + 0$, yielding a 1 sum and no carry.

Subtracting 6_{ten} from 7_{ten} can be done directly:

ANSWER

$$00000000\ 00000000\ 00000000\ 00000000\ 00000000\ 00000000\ 00000000\ 00000111_{two} = 7_{ten}$$
$$+\ 00000000\ 00000000\ 00000000\ 00000000\ 00000000\ 00000000\ 00000000\ 00000110_{two} = 6_{ten}$$
$$=\ 00000000\ 00000000\ 00000000\ 00000000\ 00000000\ 00000000\ 00000000\ 00000001_{two} = 1_{ten}$$

or via addition using the two's complement representation of -6:

$$00000000\ 00000000\ 00000000\ 00000000\ 00000000\ 00000000\ 00000000\ 00000111_{two} = 7_{ten}$$
$$+\ 11111111\ 11111111\ 11111111\ 11111111\ 11111111\ 11111111\ 11111111\ 11111010_{two} = -6_{ten}$$
$$=\ 00000000\ 00000000\ 00000000\ 00000000\ 00000000\ 00000000\ 00000000\ 00000001_{two} = 1_{ten}$$

Recall that overflow occurs when the result from an operation cannot be represented with the available hardware, in this case a 64-bit word. When can overflow occur in addition? When adding operands with different signs, overflow cannot occur. The reason is the sum must be no larger than one of the operands. For example, $-10 + 4 = -6$. Since the operands fit in 64 bits and the sum is no larger than an operand, the sum must fit in 64 bits as well. Therefore, no overflow can occur when adding positive and negative operands.

There are similar restrictions to the occurrence of overflow during subtract, but it's just the opposite principle: when the signs of the operands are the *same*, overflow cannot occur. To see this, remember that $c - a = c + (-a)$ because we subtract by negating the second operand and then add. Therefore, when we subtract operands of the same sign we end up *adding* operands of *different* signs. From the prior paragraph, we know that overflow cannot occur in this case either.

Knowing when an overflow cannot occur in addition and subtraction is all well and good, but how do we detect it when it *does* occur? Clearly, adding or subtracting two 64-bit numbers can yield a result that needs 65 bits to be fully expressed.

Operation	Operand A	Operand B	Result indicating overflow
A + B	≥ 0	≥ 0	< 0
A + B	< 0	< 0	≥ 0
A − B	≥ 0	< 0	< 0
A − B	< 0	≥ 0	≥ 0

FIGURE 3.2   Overflow conditions for addition and subtraction.

The lack of a 65th bit means that when an overflow occurs, the sign bit is set with the *value* of the result instead of the proper sign of the result. Since we need just one extra bit, only the sign bit can be wrong. Hence, overflow occurs when adding two positive numbers and the sum is negative, or vice versa. This spurious sum means a carry out occurred into the sign bit.

Overflow occurs in subtraction when we subtract a negative number from a positive number and get a negative result, or when we subtract a positive number from a negative number and get a positive result. Such a ridiculous result means a borrow occurred from the sign bit. Figure 3.2 shows the combination of operations, operands, and results that indicate an overflow.

We have just seen how to detect overflow for two's complement numbers in a computer. What about overflow with unsigned integers? Unsigned integers are commonly used for memory addresses where overflows are ignored.

Arithmetic Logic Unit (ALU) Hardware that performs addition, subtraction, and usually logical operations such as AND and OR.

Given that overflow is one of the four condition codes, it is easy for the compiler to check for overflow for add and subtract instructions if it needs to. The programming language C ignores overflows, but the programming language Fortran needs them discovered.

Appendix A describes the hardware that performs addition and subtraction, which is called an Arithmetic Logic Unit or ALU.

Hardware/Software Interface

The computer designer must decide how to handle arithmetic overflows. Although some languages like C and Java ignore integer overflow, languages like Ada and Fortran require that the program be notified. The programmer or the programming environment must then decide what to do when an overflow occurs.

Summary

A major point of this section is that, independent of the representation, the finite word size of computers means that arithmetic operations can create results that are too large to fit in this fixed word size. It's easy to detect overflow in unsigned numbers, although these are almost always ignored because programs don't want to detect overflow for address arithmetic, the most common use of natural numbers. Two's complement presents a greater challenge, yet some software systems require recognizing overflow, so today all computers have a way to detect it.

Some programming languages allow two's complement integer arithmetic on variables declared byte and half, whereas the LEGv8 core only has integer arithmetic operations on full words. As we recall from Chapter 2, LEGv8 core does have data transfer operations for bytes and halfwords. What LEGv8 instructions should be generated for byte and halfword arithmetic operations?

Check Yourself

1. Load with LDURB, LDURH; arithmetic with ADD, SUB, MUL, DIV, using AND to mask result to 8 or 16 bits after each operation; then store using STURB, STURH.

2. Load with LDURB, LDURH; arithmetic with ADD, SUB, MUL, DIV; then store using STURB, STURH.

Elaboration: One feature not generally found in general-purpose microprocessors is saturating operations. *Saturation* means that when a calculation overflows, the result is set to the largest positive number or the most negative number, rather than a modulo calculation as in two's complement arithmetic. Saturation is likely what you want for media operations. For example, the volume knob on a radio set would be frustrating if, as you turned it, the volume would get continuously louder for a while and then immediately very soft. A knob with saturation would stop at the highest volume no matter how far you turned it. Multimedia extensions to standard instruction sets often offer saturating arithmetic.

Elaboration: Determining the carry into the high-order bits earlier increases the speed of addition. There are a variety of schemes to anticipate the carry so that the worst-case scenario is a function of the $\log_2$ of the number of bits in the adder. These anticipatory signals are faster because they go through fewer gates in sequence, but it takes many more gates to anticipate the proper carry. The most popular is *carry lookahead*, which Section A.6 in Appendix A describes.

3.3   Multiplication

Now that we have completed the explanation of addition and subtraction, we are ready to build the more vexing operation of multiplication.

First, let's review the multiplication of decimal numbers in longhand to remind ourselves of the steps of multiplication and the names of the operands. For reasons that will become clear shortly, we limit this decimal example to using only the digits 0 and 1. Multiplying 1000_{ten} by 1001_{ten}:

Multiplication is vexation, Division is as bad; The rule of three doth puzzle me, And practice drives me mad.

Anonymous, Elizabethan manuscript, 1570

Multiplicand		1000_{ten}
Multiplier	X	1001_{ten}
		1000
		0000
		0000
		1000
Product		1001000_{ten}

The first operand is called the *multiplicand* and the second the *multiplier*. The final result is called the *product*. As you may recall, the algorithm learned in grammar school is to take the digits of the multiplier one at a time from right to left, multiplying the multiplicand by the single digit of the multiplier, and shifting the intermediate product one digit to the left of the earlier intermediate products.

The first observation is that the number of digits in the product is considerably larger than the number in either the multiplicand or the multiplier. In fact, if we ignore the sign bits, the length of the multiplication of an n-bit multiplicand and an m-bit multiplier is a product that is $n + m$ bits long. That is, $n + m$ bits are required to represent all possible products. Hence, like add, multiply must cope with overflow because we frequently want a 64-bit product as the result of multiplying two 64-bit numbers.

In this example, we restricted the decimal digits to 0 and 1. With only two choices, each step of the multiplication is simple:

1. Just place a copy of the multiplicand (1 × multiplicand) in the proper place if the multiplier digit is a 1, or

2. Place 0 (0 × multiplicand) in the proper place if the digit is 0.

Although the decimal example above happens to use only 0 and 1, multiplication of binary numbers must always use 0 and 1, and thus always offers only these two choices.

Now that we have reviewed the basics of multiplication, the traditional next step is to provide the highly optimized multiply hardware. We break with tradition in the belief that you will gain a better understanding by seeing the evolution of the multiply hardware and algorithm through multiple generations. For now, let's assume that we are multiplying only positive numbers.

Sequential Version of the Multiplication Algorithm and Hardware

This design mimics the algorithm we learned in grammar school; Figure 3.3 shows the hardware. We have drawn the hardware so that data flow from top to bottom to resemble more closely the paper-and-pencil method.

Let's assume that the multiplier is in the 64-bit Multiplier register and that the 128-bit Product register is initialized to 0. From the paper-and-pencil example above, it's clear that we will need to move the multiplicand left one digit each step, as it may be added to the intermediate products. Over 64 steps, a 64-bit multiplicand would move 64 bits to the left. Hence, we need a 128-bit Multiplicand register, initialized with the 64-bit multiplicand in the right half and zero in the left half. This register is then shifted left 1 bit each step to align the multiplicand with the sum being accumulated in the 128-bit Product register.

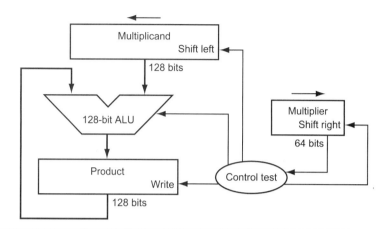

FIGURE 3.3 First version of the multiplication hardware. The Multiplicand register, ALU, and Product register are all 128 bits wide, with only the Multiplier register containing 64 bits. (Appendix A describes ALUs.) The 64-bit multiplicand starts in the right half of the Multiplicand register and is shifted left 1 bit on each step. The multiplier is shifted in the opposite direction at each step. The algorithm starts with the product initialized to 0. Control decides when to shift the Multiplicand and Multiplier registers and when to write new values into the Product register.

Figure 3.4 shows the three basic steps needed for each bit. The least significant bit of the multiplier (Multiplier0) determines whether the multiplicand is added to the Product register. The left shift in step 2 has the effect of moving the intermediate operands to the left, just as when multiplying with paper and pencil. The shift right in step 3 gives us the next bit of the multiplier to examine in the following iteration. These three steps are repeated 64 times to obtain the product. If each step took a clock cycle, this algorithm would require almost 200 clock cycles to multiply two 64-bit numbers. The relative importance of arithmetic operations like multiply varies with the program, but addition and subtraction may be anywhere from 5 to 100 times more popular than multiply. Accordingly, in many applications, multiply can take several clock cycles without significantly affecting performance. However, Amdahl's Law (see Section 1.10) reminds us that even a moderate frequency for a slow operation can limit performance.

This algorithm and hardware are easily refined to take one clock cycle per step. The speed up comes from performing the operations in parallel: the multiplier and multiplicand are shifted while the multiplicand is added to the product if the multiplier bit is a 1. The hardware just has to ensure that it tests the right bit of the multiplier and gets the preshifted version of the multiplicand. The hardware is usually further optimized to halve the width of the adder and registers by noticing where there are unused portions of registers and adders. Figure 3.5 shows the revised hardware.

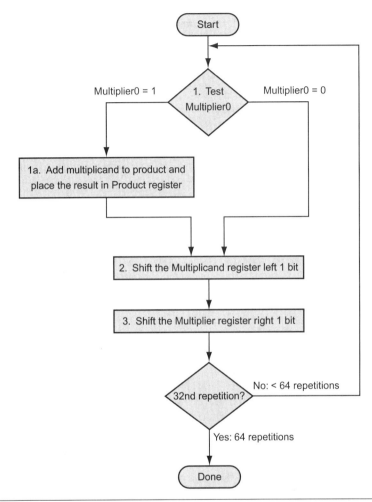

FIGURE 3.4 The first multiplication algorithm, using the hardware shown in Figure 3.3. If the least significant bit of the multiplier is 1, add the multiplicand to the product. If not, go to the next step. Shift the multiplicand left and the multiplier right in the next two steps. These three steps are repeated 64 times.

Hardware/ Software Interface

Replacing arithmetic by shifts can also occur when multiplying by constants. Some compilers replace multiplies by short constants with a series of shifts and adds. Because one bit to the left represents a number twice as large in base 2, shifting the bits left has the same effect as multiplying by a power of 2. As mentioned in Chapter 2, almost every compiler will perform the strength reduction optimization of substituting a left shift for a multiply by a power of 2.

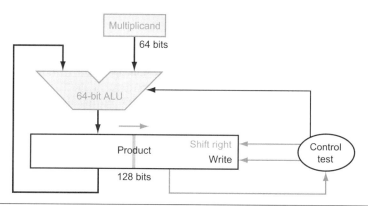

FIGURE 3.5 Refined version of the multiplication hardware. Compare with the first version in Figure 3.3. The Multiplicand register, ALU, and Multiplier register are all 64 bits wide, with only the Product register left at 128 bits. Now the product is shifted right. The separate Multiplier register also disappeared. The multiplier is placed instead in the right half of the Product register. These changes are highlighted in color. (The Product register should really be 129 bits to hold the carry out of the adder, but it's shown here as 128 bits to highlight the evolution from Figure 3.3.)

Iteration	Step	Multiplier	Multiplicand	Product
0	Initial values	0011	0000 0010	0000 0000
1	1a: 1 ⟹ Prod = Prod + Mcand	0011	0000 0010	0000 0010
	2: Shift left Multiplicand	0011	0000 0100	0000 0010
	3: Shift right Multiplier	0001	0000 0100	0000 0010
2	1a: 1 ⟹ Prod = Prod + Mcand	0001	0000 0100	0000 0110
	2: Shift left Multiplicand	0001	0000 1000	0000 0110
	3: Shift right Multiplier	0000	0000 1000	0000 0110
3	1: 0 ⟹ No operation	0000	0000 1000	0000 0110
	2: Shift left Multiplicand	0000	0001 0000	0000 0110
	3: Shift right Multiplier	0000	0001 0000	0000 0110
4	1: 0 ⟹ No operation	0000	0001 0000	0000 0110
	2: Shift left Multiplicand	0000	0010 0000	0000 0110
	3: Shift right Multiplier	0000	0010 0000	0000 0110

FIGURE 3.6 Multiply example using algorithm in Figure 3.4. The bit examined to determine the next step is circled in color.

A Multiply Algorithm

Using 4-bit numbers to save space, multiply $2_{ten} \times 3_{ten}$, or $0010_{two} \times 0011_{two}$.

EXAMPLE

Figure 3.6 shows the value of each register for each of the steps labeled according to Figure 3.4, with the final value of $0000\ 0110_{two}$ or 6_{ten}. Color is used to indicate the register values that change on that step, and the bit circled is the one examined to determine the operation of the next step.

ANSWER

Signed Multiplication

So far, we have dealt with positive numbers. The easiest way to understand how to deal with signed numbers is to first convert the multiplier and multiplicand to positive numbers and then remember their original signs. The algorithms should next be run for 31 iterations, leaving the signs out of the calculation. As we learned in grammar school, we need negate the product only if the original signs disagree.

It turns out that the last algorithm will work for signed numbers, if we remember that we are dealing with numbers that have infinite digits, and we are only representing them with 64 bits. Hence, the shifting steps would need to extend the sign of the product for signed numbers. When the algorithm completes, the lower doubleword would have the 64-bit product.

Faster Multiplication

MOORE'S LAW

Moore's Law has provided so much more in resources that hardware designers can now build much faster multiplication hardware. Whether the multiplicand is to be added or not is known at the beginning of the multiplication by looking at each of the 64 multiplier bits. Faster multiplications are possible by essentially providing one 64-bit adder for each bit of the multiplier: one input is the multiplicand ANDed with a multiplier bit, and the other is the output of a prior adder.

A straightforward approach would be to connect the outputs of adders on the right to the inputs of adders on the left, making a stack of adders 64 high. An alternative way to organize these 64 additions is in a parallel tree, as Figure 3.7 shows. Instead of waiting for 64 add times, we wait just the $\log_2$ (64) or six 64-bit add times.

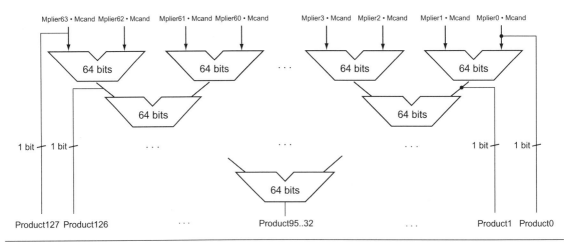

FIGURE 3.7   Fast multiplication hardware. Rather than use a single 64-bit adder 63 times, this hardware "unrolls the loop" to use 63 adders and then organizes them to minimize delay.

In fact, multiply can go even faster than six add times because of the use of *carry save adders* (see Section A.6 in Appendix A), and because it is easy to **pipeline** such a design to be able to support many multiplies simultaneously (see Chapter 4).

Multiply in LEGv8

To produce a properly signed or unsigned 128-bit product, LEGv8 has three instructions: *multiply* (MUL), *signed multiply high* (SMULH) and *unsigned multiply high* (UMULH). To get the integer 64-bit product, the programmer use MUL. To get the upper 64 bits of the 128-bit product, the programmer uses either SMULH or UMULH, depending on the types of multiplier and multiplicand.

PIPELINING

Summary

Multiplication hardware simply shifts and adds, as derived from the paper-and-pencil method learned in grammar school. Compilers even use shift instructions for multiplications by powers of 2. With much more hardware we can do the adds in **parallel**, and do them much faster.

PARALLELISM

LEGv8 multiply instructions do not set the overflow condition code, so it is up to the software to check to see if the product is too big to fit in 64 bits. There is no overflow if the upper 64 bits is 0 for UMULH or the replicated sign of the lower 64 bits for SMULH.

Hardware/ Software Interface

3.4 Division

The reciprocal operation of multiply is divide, an operation that is even less frequent and even quirkier. It even offers the opportunity to perform a mathematically invalid operation: dividing by 0.

Let's start with an example of long division using decimal numbers to recall the names of the operands and the division algorithm from grammar school. For reasons similar to those in the previous section, we limit the decimal digits to just 0 or 1. The example is dividing $1,001,010_{ten}$ by 1000_{ten}:

Divide et impera.

Latin for "Divide and rule," ancient political maxim cited by Machiavelli, 1532

```
                    1001_ten    Quotient
Divisor 1000_ten  |1001010_ten  Dividend
                  −1000
                      10
                     101
                    1010
                   −1000
                      10_ten     Remainder
```

dividend A number being divided.

divisor A number that the dividend is divided by.

quotient The primary result of a division; a number that when multiplied by the divisor and added to the remainder produces the dividend.

remainder The secondary result of a division; a number that when added to the product of the quotient and the divisor produces the dividend.

Divide's two operands, called the dividend and divisor, and the result, called the quotient, are accompanied by a second result, called the remainder. Here is another way to express the relationship between the components:

$$\text{Dividend} = \text{Quotient} \times \text{Divisor} + \text{Remainder}$$

where the remainder is smaller than the divisor. Infrequently, programs use the divide instruction just to get the remainder, ignoring the quotient.

The basic division algorithm from grammar school tries to see how big a number can be subtracted, creating a digit of the quotient on each attempt. Our carefully selected decimal example uses just the numbers 0 and 1, so it's easy to figure out how many times the divisor goes into the portion of the dividend: it's either 0 times or 1 time. Binary numbers contain only 0 or 1, so binary division is restricted to these two choices, thereby simplifying binary division.

Let's assume that both the dividend and the divisor are positive and hence the quotient and the remainder are nonnegative. The division operands and both results are 64-bit values, and we will ignore the sign for now.

A Division Algorithm and Hardware

Figure 3.8 shows hardware to mimic our grammar school algorithm. We start with the 64-bit Quotient register set to 0. Each iteration of the algorithm needs to move the divisor to the right one digit, so we start with the divisor placed in the left half of the 128-bit Divisor register and shift it right 1 bit each step to align it with the dividend. The Remainder register is initialized with the dividend.

Figure 3.9 shows three steps of the first division algorithm. Unlike a human, the computer isn't smart enough to know in advance whether the divisor is smaller

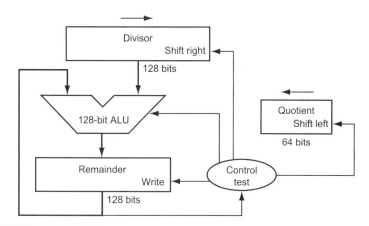

FIGURE 3.8 First version of the division hardware. The Divisor register, ALU, and Remainder register are all 128 bits wide, with only the Quotient register being 62 bits. The 64-bit divisor starts in the left half of the Divisor register and is shifted right 1 bit each iteration. The remainder is initialized with the dividend. Control decides when to shift the Divisor and Quotient registers and when to write the new value into the Remainder register.

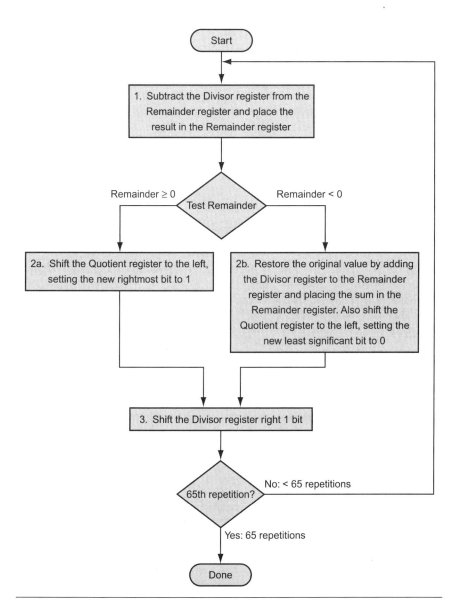

FIGURE 3.9   A division algorithm, using the hardware in Figure 3.8. If the remainder is positive, the divisor did go into the dividend, so step 2a generates a 1 in the quotient. A negative remainder after step 1 means that the divisor did not go into the dividend, so step 2b generates a 0 in the quotient and adds the divisor to the remainder, thereby reversing the subtraction of step 1. The final shift, in step 3, aligns the divisor properly, relative to the dividend for the next iteration. These steps are repeated 65 times.

than the dividend. It must first subtract the divisor in step 1; remember that this is how we performed comparison. If the result is positive, the divisor was smaller or equal to the dividend, so we generate a 1 in the quotient (step 2a). If the result is negative, the next step is to restore the original value by adding the divisor back to the remainder and generate a 0 in the quotient (step 2b). The divisor is shifted right, and then we iterate again. The remainder and quotient will be found in their namesake registers after the iterations complete.

EXAMPLE

A Divide Algorithm

Using a 4-bit version of the algorithm to save pages, let's try dividing 7_{ten} by 2_{ten}, or 0000 0111$_{two}$ by 0010$_{two}$.

ANSWER

Figure 3.10 shows the value of each register for each of the steps, with the quotient being 3_{ten} and the remainder 1_{ten}. Notice that the test in step 2 of whether the remainder is positive or negative simply checks whether the sign bit of the Remainder register is a 0 or 1. The surprising requirement of this algorithm is that it takes $n + 1$ steps to get the proper quotient and remainder.

This algorithm and hardware can be refined to be faster and cheaper. The speed-up comes from shifting the operands and the quotient simultaneously with the

Iteration	Step	Quotient	Divisor	Remainder
0	Initial values	0000	0010 0000	0000 0111
1	1: Rem = Rem – Div	0000	0010 0000	⓪110 0111
	2b: Rem < 0 ⟹ +Div, LSL Q, Q0 = 0	0000	0010 0000	0000 0111
	3: Shift Div right	0000	0001 0000	0000 0111
2	1: Rem = Rem – Div	0000	0001 0000	⓪111 0111
	2b: Rem < 0 ⟹ +Div, LSL Q, Q0 = 0	0000	0001 0000	0000 0111
	3: Shift Div right	0000	0000 1000	0000 0111
3	1: Rem = Rem – Div	0000	0000 1000	⓪111 1111
	2b: Rem < 0 ⟹ +Div, LSL Q, Q0 = 0	0000	0000 1000	0000 0111
	3: Shift Div right	0000	0000 0100	0000 0111
4	1: Rem = Rem – Div	0000	0000 0100	⓪000 0011
	2a: Rem ≥ 0 ⟹ LSL Q, Q0 = 1	0001	0000 0100	0000 0011
	3: Shift Div right	0001	0000 0010	0000 0011
5	1: Rem = Rem – Div	0001	0000 0010	⓪000 0001
	2a: Rem ≥ 0 ⟹ LSL Q, Q0 = 1	0011	0000 0010	0000 0001
	3: Shift Div right	0011	0000 0001	0000 0001

FIGURE 3.10 Division example using the algorithm in Figure 3.9. The bit examined to determine the next step is circled in color.

subtraction. This refinement halves the width of the adder and registers by noticing where there are unused portions of registers and adders. Figure 3.11 shows the revised hardware.

Signed Division

So far, we have ignored signed numbers in division. The simplest solution is to remember the signs of the divisor and dividend and then negate the quotient if the signs disagree.

Elaboration: The one complication of signed division is that we must also set the sign of the remainder. Remember that the following equation must always hold:

$$\text{Dividend} = \text{Quotient} \times \text{Divisor} + \text{Remainder}$$

To understand how to set the sign of the remainder, let's look at the example of dividing all the combinations of $\pm 7_{ten}$ by $\pm 2_{ten}$. The first case is easy:

$$+7 \div +2: \text{Quotient} = +3, + \text{Remainder} = +1$$

Checking the results:

$$+7 = 3 \times 2 + (+1) = 6 + 1$$

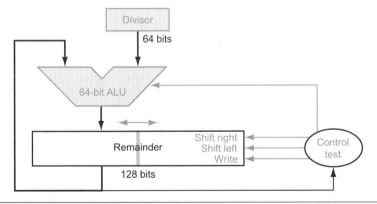

FIGURE 3.11   An improved version of the division hardware. The Divisor register, ALU, and Quotient register are all 64 bits wide, with only the Remainder register left at 128 bits. Compared to Figure 3.8, the ALU and Divisor registers are halved and the remainder is shifted left. This version also combines the Quotient register with the right half of the Remainder register. (As in Figure 3.5, the Remainder register should really be 129 bits to make sure the carry out of the adder is not lost.)

If we change the sign of the dividend, the quotient must change as well:

$$-7 \div +2: \text{Quotient} = -3$$

Rewriting our basic formula to calculate the remainder:

$$\text{Remainder} = (\text{Dividend} - \text{Quotient} \times \text{Divisor}) = -7 - (-3x + 2)$$
$$= -7 - (-6) = -1$$

So,

$$-7 \div +2: \text{Quotient} = -3, \text{Remainder} = -1$$

Checking the results again:

$$-7 = -3 \times 2 + (-1) = -6 - 1$$

The reason the answer isn't a quotient of −4 and a remainder of +1, which would also fit this formula, is that the absolute value of the quotient would then change depending on the sign of the dividend and the divisor! Clearly, if

$$-(x \div y) \neq (-x) \div y$$

programming would be an even greater challenge. This anomalous behavior is avoided by following the rule that the dividend and remainder must have identical signs, no matter what the signs of the divisor and quotient.

We calculate the other combinations by following the same rule:

$$+7 \div -2: \text{Quotient} = -3, \text{Remainder} = +1$$
$$-7 \div -2: \text{Quotient} = +3, \text{Remainder} = -1$$

Thus, the correctly signed division algorithm negates the quotient if the signs of the operands are opposite and makes the sign of the nonzero remainder match the dividend.

Faster Division

MOORE'S LAW

Moore's Law applies to division hardware as well as multiplication, so we would like to be able to speed up division by throwing hardware at it. We used many adders to speed up multiply, but we cannot do the same trick for divide. The reason is that we need to know the sign of the difference before we can perform the next step of the algorithm, whereas with multiply we could calculate the 64 partial products immediately.

There are techniques to produce more than one bit of the quotient per step. The *SRT division* technique tries to **predict** several quotient bits per step, using a table lookup based on the upper bits of the dividend and remainder. It relies on subsequent steps to correct wrong predictions. A typical value today is 4 bits. The key is guessing the value to subtract. With binary division, there is only a single choice. These algorithms use 6 bits from the remainder and 4 bits from the divisor to index a table that determines the guess for each step.

The accuracy of this fast method depends on having proper values in the lookup table. The *Fallacy* on page 244 in Section 3.9 shows what can happen if the table is incorrect.

P R E D I C T I O N

Divide in LEGv8

You may have already observed that the same sequential hardware can be used for both multiply and divide in Figures 3.5 and 3.11. The only requirement is a 128-bit register that can shift left or right and a 64-bit ALU that adds or subtracts.

To handle both signed integers and unsigned integers, LEGv8 has two instructions: *signed divide* (SDIV) and *divide unsigned* (UDIV).

Summary

The common hardware support for multiply and divide allows LEGv8 to provide a single pair of 64-bit registers that are used both for multiply and divide. We accelerate division by predicting multiple quotient bits and then correcting mispredictions later. Figure 3.12 summarizes the enhancements to the LEGv8 architecture for the last two sections.

LEGv8 divide instructions ignore overflow, so software must determine whether the quotient is too large. In addition to overflow, division can also result in an improper calculation: division by 0. Some computers distinguish these two anomalous events. LEGv8 software must check the divisor to discover division by 0 as well as overflow.

Hardware/ Software Interface

Elaboration: An even faster algorithm does not immediately add the divisor back if the remainder is negative. It simply *adds* the dividend to the shifted remainder in the following step, since $(r + d) \times 2 - d = r - 2 + d \times 2 - d = r \times 2 + d$. This *nonrestoring* division algorithm, which takes one clock cycle per step, is explored further in the exercises; the algorithm above is called *restoring* division. A third algorithm that doesn't save the result of the subtract if it's negative is called a *nonperforming* division algorithm. It averages one-third fewer arithmetic operations.

LEGv8 assembly language

Category	Instruction	Example	Meaning	Comments
Arithmetic	add	ADD X1, X2, X3	X1 = X2 + X3	Three register operands
	subtract	SUB X1, X2, X3	X1 = X2 - X3	Three register operands
	add immediate	ADDI X1, X2, 20	X1 = X2 + 20	Used to add constants
	subtract immediate	SUBI X1, X2, 20	X1 = X2 - 20	Used to subtract constants
	add and set flags	ADDS X1, X2, X3	X1 = X2 + X3	Add, set condition codes
	subtract and set flags	SUBS X1, X2, X3	X1 = X2 - X3	Subtract, set condition codes
	add immediate and set flags	ADDIS X1, X2, 20	X1 = X2 + 20	Add constant, set condition codes
	subtract immediate and set flags	SUBIS X1, X2, 20	X1 = X2 - 20	Subtract constant, set condition codes
	multiply	MUL X1, X2, X3	X1 = X2 × X3	Lower 64-bits of 128-bit product
	signed multiply high	SMULH X1, X2, X3	X1 = X2 × X3	Upper 64-bits of 128-bit signed product
	unsigned multiply high	UMULH X1, X2, X3	X1 = X2 × X3	Upper 64-bits of 128-bit unsigned product
	signed divide	SDIV X1, X2, X3	X1 = X2 / X3	Divide, treating operands as signed
	unsigned divide	UDIV X1, X2, X3	X1 = X2 / X3	Divide, treating operands as unsigned
Data transfer	load register	LDUR X1, [X2,40]	X1 = Memory[X2 + 40]	Doubleword from memory to register
	store register	STUR X1, [X2,40]	Memory[X2 + 40] = X1	Doubleword from register to memory
	load signed word	LDURSW X1, [X2,40]	X1 = Memory[X2 + 40]	Word from memory to register
	store word	STURW X1, [X2,40]	Memory[X2 + 40] = X1	Word from register to memory
	load half	LDURH X1, [X2,40]	X1 = Memory[X2 + 40]	Halfword memory to register
	store half	STURH X1, [X2,40]	Memory[X2 + 40] = X1	Halfword register to memory
	load byte	LDURB X1, [X2,40]	X1 = Memory[X2 + 40]	Byte from memory to register
	store byte	STURB X1, [X2,40]	Memory[X2 + 40] = X1	Byte from register to memory
	load exclusive register	LDXR X1, [X2,0]	X1 = Memory[X2]	Load; 1st half of atomic swap
	store exclusive register	STXR X1, X3, [X2]	Memory[X2]=X1;X3=0 or 1	Store; 2nd half of atomic swap
	move wide with zero	MOVZ X1,20	X1 = 20 or 20 * 2^{16} or 20 * 2^{32} or 20 * 2^{48}	Loads 16-bit constant, rest zeros
	move wide with keep	MOVK X1,20	X1 = 20 or 20 * 2^{16} or 20 * 2^{32} or 20 * 2^{48}	Loads 16-bit constant, rest unchanged
Logical	and	AND X1, X2, X3	X1 = X2 & X3	Three reg. operands; bit-by-bit AND
	inclusive or	ORR X1, X2, X3	X1 = X2 | X3	Three reg. operands; bit-by-bit OR
	exclusive or	EOR X1, X2, X3	X1 = X2 ^ X3	Three reg. operands; bit-by-bit XOR
	and immediate	ANDI X1, X2, 20	X1 = X2 & 20	Bit-by-bit AND reg with constant
	inclusive or immediate	ORRI X1, X2, 20	X1 = X2 | 20	Bit-by-bit OR reg with constant
	exclusive or immediate	EORI X1, X2, 20	X1 = X2 ^ 20	Bit-by-bit XOR reg with constant
	logical shift left	LSL X1, X2, 10	X1 = X2 << 10	Shift left by constant
	logical shift right	LSR X1, X2, 10	X1 = X2 >> 10	Shift right by constant
Conditional branch	compare and branch on equal 0	CBZ X1, 25	if (X1 == 0) go to PC + 4 + 100	Equal 0 test; PC-relative branch
	compare and branch on not equal 0	CBNZ X1, 25	if (X1!= 0) go to PC + 4 + 100	Not equal 0 test; PC-relative
	branch conditionally	B.cond 25	if (condition true) go to PC + 4 + 100	Test condition codes; if true, branch
Unconditional jump	branch	B 2500	go to PC + 4 + 10000	Branch to target address; PC-relative
	branch to register	BR X30	go to X30	For switch, procedure return
	branch with link	BL 2500	X30 = PC + 4; PC + 4 + 10000	For procedure call PC-relative

FIGURE 3.12 LEGv8 core architecture. LEGv8 machine language is listed in the LEGv8 Reference Data Card at the front of this book.

3.5 Floating Point

Going beyond signed and unsigned integers, programming languages support numbers with fractions, which are called *reals* in mathematics. Here are some examples of reals:

$3.14159265\ldots_{ten}$ (pi)

$2.71828\ldots_{ten}$ *(e)*

0.000000001_{ten} or $1.0_{ten} \times 10^{-9}$ (seconds in a nanosecond)

$3{,}155{,}760{,}000_{ten}$ or $3.15576_{ten} \times 10^{9}$ (seconds in a typical century)

Notice that in the last case, the number didn't represent a small fraction, but it was bigger than we could represent with a 32-bit signed integer. The alternative notation for the last two numbers is called scientific notation, which has a single digit to the left of the decimal point. A number in scientific notation that has no leading 0s is called a normalized number, which is the usual way to write it. For example, $1.0_{ten} \times 10^{-9}$ is in normalized scientific notation, but $0.1_{ten} \times 10^{-8}$ and $10.0_{ten} \times 10^{-10}$ are not.

Just as we can show decimal numbers in scientific notation, we can also show binary numbers in scientific notation:

$$1.0_{two} \times 2^{-1}$$

To keep a binary number in the normalized form, we need a base that we can increase or decrease by exactly the number of bits the number must be shifted to have one nonzero digit to the left of the decimal point. Only a base of 2 fulfills our need. Since the base is not 10, we also need a new name for decimal point; *binary point* will do fine.

Computer arithmetic that supports such numbers is called floating point because it represents numbers in which the binary point is not fixed, as it is for integers. The programming language C uses the name *float* for such numbers. Just as in scientific notation, numbers are represented as a single nonzero digit to the left of the binary point. In binary, the form is

$$1.xxxxxxxxx_{two} \times 2^{yyyy}$$

(Although the computer represents the exponent in base 2 as well as the rest of the number, to simplify the notation we show the exponent in decimal.)

A standard scientific notation for reals in the normalized form offers three advantages. It simplifies exchange of data that includes floating-point numbers; it simplifies the floating-point arithmetic algorithms to know that numbers will

Speed gets you nowhere if you're headed the wrong way.

American proverb

scientific notation A notation that renders numbers with a single digit to the left of the decimal point.

normalized A number in floating-point notation that has no leading 0s.

floating point Computer arithmetic that represents numbers in which the binary point is not fixed.

always be in this form; and it increases the accuracy of the numbers that can be stored in a word, since real digits to the right of the binary point replace the unnecessary leading 0s.

Floating-Point Representation

fraction The value, generally between 0 and 1, placed in the fraction field. The fraction is also called the *mantissa*.

exponent In the numerical representation system of floating-point arithmetic, the value that is placed in the exponent field.

A designer of a floating-point representation must find a compromise between the size of the fraction and the size of the exponent, because a fixed word size means you must take a bit from one to add a bit to the other. This tradeoff is between precision and range: increasing the size of the fraction enhances the precision of the fraction, while increasing the size of the exponent increases the range of numbers that can be represented. As our design guideline from Chapter 2 reminds us, good design demands good compromise.

Floating-point numbers are usually a multiple of the size of a word. The representation of an LEGv8 floating-point number is shown below, where s is the sign of the floating-point number (1 meaning negative), *exponent* is the value of the 8-bit exponent field (including the sign of the exponent), and *fraction* is the 23-bit number. As we recall from Chapter 2, this representation is *sign and magnitude*, since the sign is a separate bit from the rest of the number.

31	30	29	28	27	26	25	24	23	22	21	20	19	18	17	16	15	14	13	12	11	10	9	8	7	6	5	4	3	2	1	0
s		exponent							fraction																						
1 bit		8 bits							23 bits																						

In general, floating-point numbers are of the form

$$(-1)^S \times F \times 2^E$$

F involves the value in the fraction field and E involves the value in the exponent field; the exact relationship to these fields will be spelled out soon. (We will shortly see that LEGv8 does something slightly more sophisticated.)

These chosen sizes of exponent and fraction give LEGv8 computer arithmetic an extraordinary range. Fractions almost as small as $2.0_{ten} \times 10^{-38}$ and numbers almost as large as $2.0_{ten} \times 10^{38}$ can be represented in a computer. Alas, extraordinary differs from infinite, so it is still possible for numbers to be too large. Thus, overflow interrupts can occur in floating-point arithmetic as well as in integer arithmetic. Notice that overflow here means that the exponent is too large to be represented in the exponent field.

overflow (floating-point) A situation in which a positive exponent becomes too large to fit in the exponent field.

underflow (floating-point) A situation in which a negative exponent becomes too large to fit in the exponent field.

Floating point offers a new kind of exceptional event as well. Just as programmers will want to know when they have calculated a number that is too large to be represented, they will want to know if the nonzero fraction they are calculating has become so small that it cannot be represented; either event could result in a program giving incorrect answers. To distinguish it from overflow, we call this event underflow. This situation occurs when the negative exponent is too large to fit in the exponent field.

One way to reduce the chances of underflow or overflow is to offer another format that has a larger exponent. In C, this number is called *double*, and operations on doubles are called double precision floating-point arithmetic; single precision floating point is the name of the earlier format.

double precision A floating-point value represented in a 64-bit doubleword.

The representation of a double precision floating-point number takes one LEGv8 doubleword, as shown below, where *s* is still the sign of the number, *exponent* is the value of the 11-bit exponent field, and *fraction* is the 52-bit number in the fraction field.

single precision A floating-point value represented in a 32-bit word.

63	62	61	60	59	58	57	56	55	54	53	52	51	50	49	48	47	46	45	44	43	42	41	40	39	38	37	36	35	34	33	32
s	exponent												fraction																		

| 1 bit | 11 bits | | | | | | | | | | | | 20 bits | | | | | | | | | | | | | | | | | | |

31	30	29	28	27	26	25	24	23	22	21	20	19	18	17	16	15	14	13	12	11	10	9	8	7	6	5	4	3	2	1	0
fraction																															

32 bits

LEGv8 double precision allows numbers almost as small as $2.0_{ten} \times 10^{-308}$ and almost as large as $2.0_{ten} \times 10^{308}$. Although double precision does increase the exponent range, its primary advantage is its greater precision because of the much larger fraction.

Exceptions and Interrupts

What should happen on an overflow or underflow to let the user know that a problem occurred? LEGv8 can raise an exception, also called an interrupt on many computers. An exception or interrupt is essentially an unscheduled procedure call. The address of the instruction that overflowed is saved in a register, and the computer jumps to a predefined address to invoke the appropriate routine for that exception. The interrupted address is saved so that in some situations the program can continue after corrective code is executed. (Section 4.9 covers exceptions in more detail; Chapter 5 describes other situations where exceptions and interrupts occur.)

exception Also called **interrupt**. An unscheduled event that disrupts program execution; used to detect overflow.

interrupt An exception that comes from outside of the processor. (Some architectures use the term *interrupt* for all exceptions.)

IEEE 754 Floating-Point Standard

These formats go beyond LEGv8. They are part of the *IEEE 754 floating-point standard*, found in virtually every computer invented since 1980. This standard has greatly improved both the ease of porting floating-point programs and the quality of computer arithmetic.

To pack even more bits into the number, IEEE 754 makes the leading 1 bit of normalized binary numbers implicit. Hence, the number is actually 24 bits long in single precision (implied 1 and a 23-bit fraction), and 53 bits long in double precision (1 + 52). To be precise, we use the term *significand* to represent the 24- or 53-bit number that is 1 plus the fraction, and *fraction* when we mean the 23- or

52-bit number. Since 0 has no leading 1, it is given the reserved exponent value 0 so that the hardware won't attach a leading 1 to it.

Thus $00 \ldots 00_{two}$ represents 0; the representation of the rest of the numbers uses the form from before with the hidden 1 added:

$$(-1)^S \times (1 + \text{Fraction}) \times 2^E$$

where the bits of the fraction represent a number between 0 and 1 and E specifies the value in the exponent field, to be given in detail shortly. If we number the bits of the fraction from *left to right* s1, s2, s3, …, then the value is

$$(-1)^S \times (1 + (s1 \times 2^{-1}) + (s2 \times 2^{-2}) + (s3 \times 2^{-3}) + (s4 \times 2^{-4}) + \ldots) \times 2^E$$

Figure 3.13 shows the encodings of IEEE 754 floating-point numbers. Other features of IEEE 754 are special symbols to represent unusual events. For example, instead of interrupting on a divide by 0, software can set the result to a bit pattern representing $+\infty$ or $-\infty$; the largest exponent is reserved for these special symbols. When the programmer prints the results, the program will output an infinity symbol. (For the mathematically trained, the purpose of infinity is to form topological closure of the reals.)

IEEE 754 even has a symbol for the result of invalid operations, such as 0/0 or subtracting infinity from infinity. This symbol is *NaN*, for *Not a Number*. The purpose of NaNs is to allow programmers to postpone some tests and decisions to a later time in the program when they are convenient.

The designers of IEEE 754 also wanted a floating-point representation that could be easily processed by integer comparisons, especially for sorting. This desire is why the sign is in the most significant bit, allowing a quick test of less than, greater than, or equal to 0. (It's a little more complicated than a simple integer sort, since this notation is essentially sign and magnitude rather than two's complement.)

Single precision		Double precision		Object represented
Exponent	Fraction	Exponent	Fraction	
0	0	0	0	0
0	Nonzero	0	Nonzero	± denormalized number
1–254	Anything	1–2046	Anything	± floating-point number
255	0	2047	0	± infinity
255	Nonzero	2047	Nonzero	NaN (Not a Number)

FIGURE 3.13 EEE 754 encoding of floating-point numbers. A separate sign bit determines the sign. Denormalized numbers are described in the *Elaboration* on page 230. This information is also found in Column 4 of the LEGv8 Reference Data Card at the front of this book.

Placing the exponent before the significand also simplifies the sorting of floating-point numbers using integer comparison instructions, since numbers with bigger exponents look larger than numbers with smaller exponents, as long as both exponents have the same sign.

Negative exponents pose a challenge to simplified sorting. If we use two's complement or any other notation in which negative exponents have a 1 in the most significant bit of the exponent field, a negative exponent will look like a big number. For example, $1.0_{two} \times 2^{-1}$ would be represented in a single precision as

31	30	29	28	27	26	25	24	23	22	21	20	19	18	17	16	15	14	13	12	11	10	9	8	7	6	5	4	3	2	1	0
1	1	1	1	1	1	1	1	0	0	0	0	0	0	0	0	0	0	0	0	0	0	0	0	0	0	0	0	0	0	0	0

(Remember that the leading 1 is implicit in the significand.) The value $1.0_{two} \times 2^{+1}$ would look like the smaller binary number

31	30	29	28	27	26	25	24	23	22	21	20	19	18	17	16	15	14	13	12	11	10	9	8	7	6	5	4	3	2	1	0
0	0	0	0	0	0	0	0	1	0	0	0	0	0	0	0	0	0	0	0	0	0	0	0	0	0	0	0	0	0	0	0

The desirable notation must therefore represent the most negative exponent as $00 \ldots 00_{two}$ and the most positive as $11 \ldots 11_{two}$. This convention is called *biased notation*, with the bias being the number subtracted from the normal, unsigned representation to determine the real value.

IEEE 754 uses a bias of 127 for single precision, so an exponent of -1 is represented by the bit pattern of the value $-1 + 127_{ten}$, or $126_{ten} = 0111\ 1110_{two}$, and $+1$ is represented by $1 + 127$, or $128_{ten} = 1000\ 0000_{two}$. The exponent bias for double precision is 1023. Biased exponent means that the value represented by a floating-point number is really

$$(-1)^S \times (1 + \text{Fraction}) \times 2^{(\text{Exponent} - \text{Bias})}$$

The range of single precision numbers is then from as small as

$$\pm 1.00000000000000000000000_{two} \times 2^{-126}$$

to as large as

$$\pm 1.11111111111111111111111_{two} \times 2^{+127}.$$

Let's demonstrate.

EXAMPLE

Show the IEEE 754 binary representation of the number -0.75_{ten} in single and double precision.

ANSWER

The number -0.75_{ten} is also

$$-3/4_{ten} \text{ or } -3/2^2_{ten}$$

It is also represented by the binary fraction

$$-11_{two}/2^2_{ten} \text{ or } -0.11_{two}$$

In scientific notation, the value is

$$-0.11_{two} \times 2^0$$

and in normalized scientific notation, it is

$$-1.1_{two} \times 2^{-1}$$

The general representation for a single precision number is

$$(-1)^S \times (1 + \text{Fraction}) \times 2^{(\text{Exponent}-127)}$$

Subtracting the bias 127 from the exponent of $-1.1_{two} \times 2^{-1}$ yields

$$(-1)^1 \times (1 + .1000\,0000\,0000\,0000\,0000\,000_{two}) \times 2^{(126-127)}$$

The single precision binary representation of -0.75_{ten} is then

31	30	29	28	27	26	25	24	23	22	21	20	19	18	17	16	15	14	13	12	11	10	9	8	7	6	5	4	3	2	1	0
1	0	1	1	1	1	1	1	0	1	0	0	0	0	0	0	0	0	0	0	0	0	0	0	0	0	0	0	0	0	0	0

1 bit   8 bits   23 bits

The double precision representation is

$$(-1)^1 \times (1 + .1000\,0000\,0000\,0000\,0000\,0000\,0000\,0000\,0000\,0000\,0000\,0000\,0000_{two}) \times 2^{(1022-1023)}$$

31	30	29	28	27	26	25	24	23	22	21	20	19	18	17	16	15	14	13	12	11	10	9	8	7	6	5	4	3	2	1	0
1	0	1	1	1	1	1	1	1	1	1	0	1	0	0	0	0	0	0	0	0	0	0	0	0	0	0	0	0	0	0	0

1 bit   11 bits   20 bits

0	0	0	0	0	0	0	0	0	0	0	0	0	0	0	0	0	0	0	0	0	0	0	0	0	0	0	0	0	0	0	0

32 bits

Now let's try going the other direction.

Converting Binary to Decimal Floating Point

What decimal number does this single precision float represent?

31	30	29	28	27	26	25	24	23	22	21	20	19	18	17	16	15	14	13	12	11	10	9	8	7	6	5	4	3	2	1	0
1	1	0	0	0	0	0	0	1	0	1	0	0	0	0	0	0	0	0	0	0	0	0	0	0	0	0	0	0	0	0	0

The sign bit is 1, the exponent field contains 129, and the fraction field contains $1 \times 2^{-2} = 1/4$, or 0.25. Using the basic equation,

$$
\begin{aligned}
(-1)^S \times (1 + \text{Fraction}) \times 2^{(\text{Exponent} - \text{Bias})} &= (-1)^1 \times (1 + 0.25) \times 2^{(129-127)} \\
&= -1 \times 1.25 \times 2^2 \\
&= -1.25 \times 4 \\
&= -5.0
\end{aligned}
$$

In the next few subsections, we will give the algorithms for floating-point addition and multiplication. At their core, they use the corresponding integer operations on the significands, but extra bookkeeping is necessary to handle the exponents and normalize the result. We first give an intuitive derivation of the algorithms in decimal and then give a more detailed, binary version in the figures.

Elaboration: Following IEEE guidelines, the IEEE 754 committee was reformed 20 years after the standard to see what changes, if any, should be made. The revised standard IEEE 754-2008 includes nearly all the IEEE 754-1985 and adds a 16-bit format ("half precision") and a 128-bit format ("quadruple precision"). No hardware has yet been built that supports quadruple precision, but it will surely come. The revised standard also adds decimal floating point arithmetic, which IBM mainframes have implemented.

Elaboration: In an attempt to increase range without removing bits from the significand, some computers before the IEEE 754 standard used a base other than 2. For example, the IBM 360 and 370 mainframe computers use base 16. Since changing the IBM exponent by one means shifting the significand by 4 bits, "normalized" base 16 numbers can have up to 3 leading bits of 0s! Hence, hexadecimal digits mean that up to 3 bits must be dropped from the significand, which leads to surprising problems in the accuracy of floating-point arithmetic. IBM mainframes now support IEEE 754 as well as the old hex format.

Floating-Point Addition

Let's add numbers in scientific notation by hand to illustrate the problems in floating-point addition: $9.999_{ten} \times 10^1 + 1.610_{ten} \times 10^{-1}$. Assume that we can store only four decimal digits of the significand and two decimal digits of the exponent.

Step 1. To be able to add these numbers properly, we must align the decimal point of the number that has the smaller exponent. Hence, we need a form of the smaller number, $1.610_{ten} \times 10^{-1}$, that matches the larger exponent. We obtain this by observing that there are multiple representations of an unnormalized floating-point number in scientific notation:

$$1.610_{ten} \times 10^{-1} = 0.1610_{ten} \times 10^0 = 0.01610_{ten} \times 10^1$$

The number on the right is the version we desire, since its exponent matches the exponent of the larger number, $9.999_{ten} \times 10^1$. Thus, the first step shifts the significand of the smaller number to the right until its corrected exponent matches that of the larger number. But we can represent only four decimal digits so, after shifting, the number is really

$$0.016 \times 10^1$$

Step 2. Next comes the addition of the significands:

$$
\begin{array}{r}
9.999_{ten} \\
+ \quad 0.016_{ten} \\
\hline
10.015_{ten}
\end{array}
$$

The sum is $10.015_{ten} \times 10^1$.

Step 3. This sum is not in normalized scientific notation, so we need to adjust it:

$$10.015_{ten} \times 10^1 = 1.0015_{ten} \times 10^2$$

Thus, after the addition we may have to shift the sum to put it into normalized form, adjusting the exponent appropriately. This example shows shifting to the right, but if one number were positive and the other were negative, it would be possible for the sum to have many leading 0s, requiring left shifts. Whenever the exponent is increased or decreased, we must check for overflow or underflow—that is, we must make sure that the exponent still fits in its field.

Step 4. Since we assumed that the significand could be only four digits long (excluding the sign), we must round the number. In our grammar school algorithm, the rules truncate the number if the digit to the right of the desired point is between 0 and 4 and add 1 to the digit if the number to the right is between 5 and 9. The number

$$1.0015_{ten} \times 10^2$$

is rounded to four digits in the significand to

$$1.002_{ten} \times 10^2$$

since the fourth digit to the right of the decimal point was between 5 and 9. Notice that if we have bad luck on rounding, such as adding 1 to a string of 9s, the sum may no longer be normalized and we would need to perform step 3 again.

Figure 3.14 shows the algorithm for binary floating-point addition that follows this decimal example. Steps 1 and 2 are similar to the example just discussed: adjust the significand of the number with the smaller exponent and then add the two significands. Step 3 normalizes the results, forcing a check for overflow or underflow. The test for overflow and underflow in step 3 depends on the precision of the operands. Recall that the pattern of all 0 bits in the exponent is reserved and used for the floating-point representation of zero. Moreover, the pattern of all 1 bits in the exponent is reserved for indicating values and situations outside the scope of normal floating-point numbers (see the *Elaboration* on page 230). For the example below, remember that for single precision, the maximum exponent is 127, and the minimum exponent is −126.

Binary Floating-Point Addition

Try adding the numbers 0.5_{ten} and -0.4375_{ten} in binary using the algorithm in Figure 3.14.

EXAMPLE

Let's first look at the binary version of the two numbers in normalized scientific notation, assuming that we keep 4 bits of precision:

ANSWER

$$
\begin{aligned}
0.5_{ten} \quad &= 1/2_{ten} \quad &&= 1/2^1_{ten} \\
&= 0.1_{two} \quad &&= 0.1_{two} \times 2^0 \quad &&&= 1.000_{two} \times 2^{-1} \\
-0.4375_{ten} \quad &= -7/16_{ten} \quad &&= -7/2^4_{ten} \\
&= -0.0111_{two} \quad &&= -0.0111_{two} \times 2^0 \quad &&&= -1.110_{two} \times 2^{-2}
\end{aligned}
$$

Now we follow the algorithm:

Step 1. The significand of the number with the lesser exponent ($-1.11_{two} \times 2^{-2}$) is shifted right until its exponent matches the larger number:

$$-1.110_{two} \times 2^{-2} = -0.111_{two} \times 2^{-1}$$

Step 2. Add the significands:

$$1.000_{two} \times 2^{-1} + (-0.111_{two} \times 2^{-1}) = 0.001_{two} \times 2^{-1}$$

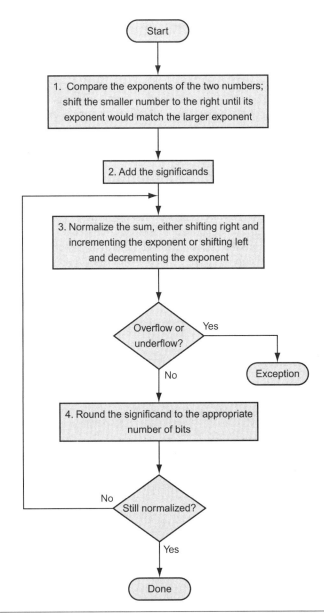

FIGURE 3.14 Floating-point addition. The normal path is to execute steps 3 and 4 once, but if rounding causes the sum to be unnormalized, we must repeat step 3.

Step 3. Normalize the sum, checking for overflow or underflow:

$$0.001_{two} \times 2^{-1} = 0.010_{two} \times 2^{-2} = 0.100_{two} \times 2^{-3}$$
$$= 1.000_{two} \times 2^{-4}$$

Since $127 \geq -4 \geq -126$, there is no overflow or underflow. (The biased exponent would be $-4 + 127$, or 123, which is between 1 and 254, the smallest and largest unreserved biased exponents.)

Step 4. Round the sum:

$$1.000_{two} \times 2^{-4}$$

The sum already fits exactly in 4 bits, so there is no change to the bits due to rounding.

This sum is then

$$1.000_{two} \times 2^{-4} \quad = 0.0001000_{two} \quad = 0.0001_{two}$$
$$= 1/2_{ten}^{4} \quad = 1/16_{ten} \quad = 0.0625_{ten}$$

This sum is what we would expect from adding 0.5_{ten} to -0.4375_{ten}.

Many computers dedicate hardware to run floating-point operations as fast as possible. Figure 3.15 sketches the basic organization of hardware for floating-point addition.

Floating-Point Multiplication

Now that we have explained floating-point addition, let's try floating-point multiplication. We start by multiplying decimal numbers in scientific notation by hand: $1.110_{ten} \times 10^{10} \times 9.200_{ten} \times 10^{-5}$. Assume that we can store only four digits of the significand and two digits of the exponent.

Step 1. Unlike addition, we calculate the exponent of the product by simply adding the exponents of the operands together:

$$New\,exponent = 10 + (-5) = 5$$

Let's do this with the biased exponents as well to make sure we obtain the same result: $10 + 127 = 137$, and $-5 + 127 = 122$, so

$$New\,exponent = 137 + 122 = 259$$

This result is too large for the 8-bit exponent field, so something is amiss! The problem is with the bias because we are adding the biases as well as the exponents:

$$New\,exponent = (10 + 127) + (-5 + 127) = (5 + 2 \times 127) = 259$$

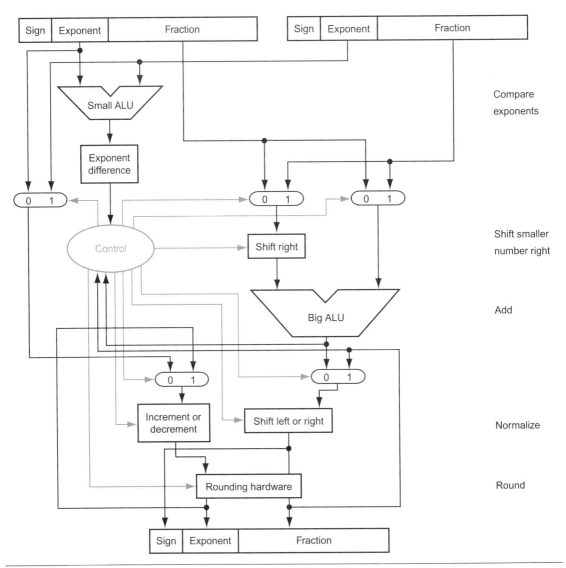

FIGURE 3.15 **Block diagram of an arithmetic unit dedicated to floating-point addition.** The steps of Figure 3.14 correspond to each block, from top to bottom. First, the exponent of one operand is subtracted from the other using the small ALU to determine which is larger and by how much. This difference controls the three multiplexors; from left to right, they select the larger exponent, the significand of the smaller number, and the significand of the larger number. The smaller significand is shifted right, and then the significands are added together using the big ALU. The normalization step then shifts the sum left or right and increments or decrements the exponent. Rounding then creates the final result, which may require normalizing again to produce the actual final result.

Accordingly, to get the correct biased sum when we add biased numbers, we must subtract the bias from the sum:

$$\text{New exponent} = 137 + 122 - 127 = 259 - 127 = 132 = (5 + 127)$$

and 5 is indeed the exponent we calculated initially.

Step 2. Next comes the multiplication of the significands:

$$
\begin{array}{r}
1.110_{ten} \\
\times \quad 9.200_{ten} \\
\hline
0000 \\
0000 \\
2220 \\
9990 \\
\hline
10212000_{ten}
\end{array}
$$

There are three digits to the right of the decimal point for each operand, so the decimal point is placed six digits from the right in the product significand:

$$10.212000_{ten}$$

If we can keep only three digits to the right of the decimal point, the product is 10.212×10^5.

Step 3. This product is unnormalized, so we need to normalize it:

$$10.212_{ten} \times 10^5 = 1.0212_{ten} \times 10^6$$

Thus, after the multiplication, the product can be shifted right one digit to put it in normalized form, adding 1 to the exponent. At this point, we can check for overflow and underflow. Underflow may occur if both operands are small—that is, if both have large negative exponents.

Step 4. We assumed that the significand is only four digits long (excluding the sign), so we must round the number. The number

$$1.0212_{ten} \times 10^6$$

is rounded to four digits in the significand to

$$1.021_{ten} \times 10^6$$

Step 5. The sign of the product depends on the signs of the original operands. If they are both the same, the sign is positive; otherwise, it's negative. Hence, the product is

$$+1.021_{ten} \times 10^6$$

The sign of the sum in the addition algorithm was determined by addition of the significands, but in multiplication, the signs of the operands determine the sign of the product.

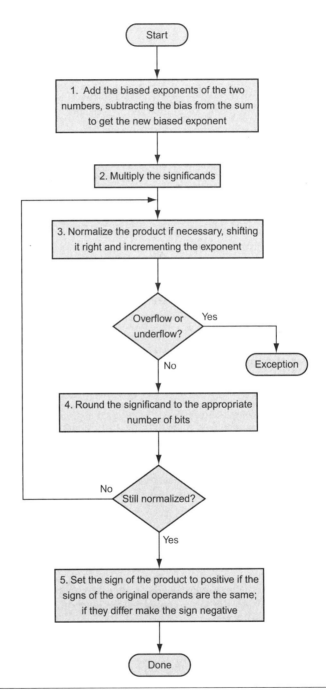

FIGURE 3.16   Floating-point multiplication. The normal path is to execute steps 3 and 4 once, but if rounding causes the sum to be unnormalized, we must repeat step 3.

Once again, as Figure 3.16 shows, multiplication of binary floating-point numbers is quite similar to the steps we have just completed. We start with calculating the new exponent of the product by adding the biased exponents, being sure to subtract one bias to get the proper result. Next is multiplication of significands, followed by an optional normalization step. The size of the exponent is checked for overflow or underflow, and then the product is rounded. If rounding leads to further normalization, we once again check for exponent size. Finally, set the sign bit to 1 if the signs of the operands were different (negative product) or to 0 if they were the same (positive product).

Binary Floating-Point Multiplication

EXAMPLE

Let's try multiplying the numbers 0.5_{ten} and -0.4375_{ten}, using the steps in Figure 3.16.

ANSWER

In binary, the task is multiplying $1.000_{two} \times 2^{-1}$ by $-1.110_{two} \times 2^{-2}$.

Step 1. Adding the exponents without bias:

$$-1 + (-2) = -3$$

or, using the biased representation:

$$(-1 + 127) + (-2 + 127) - 127 = (-1 - 2) + (127 + 127 - 127)$$
$$= -3 + 127 = 124$$

Step 2. Multiplying the significands:

$$
\begin{array}{r}
1.000_{two} \\
\times \quad 1.110_{two} \\
\hline
0000 \\
1000 \\
1000 \\
1000 \\
\hline
1110000_{two}
\end{array}
$$

The product is $1.110000_{two} \times 2^{-3}$, but we need to keep it to 4 bits, so it is $1.110_{two} \times 2^{-3}$.

Step 3. Now we check the product to make sure it is normalized, and then check the exponent for overflow or underflow. The product is already normalized and, since $127 \geq -3 \geq -126$, there is no overflow or underflow. (Using the biased representation, $254 \geq 124 \geq 1$, so the exponent fits.)

Step 4. Rounding the product makes no change:

$$1.110_{two} \times 2^{-3}$$

Step 5. Since the signs of the original operands differ, make the sign of the product negative. Hence, the product is

$$-1.110_{two} \times 2^{-3}$$

Converting to decimal to check our results:

$$-1.110_{two} \times 2^{-3} = -0.001110_{two} = -0.00111_{two}$$
$$= -7/2^5_{ten} = -7/32_{ten} = -0.21875_{ten}$$

The product of 0.5_{ten} and -0.4375_{ten} is indeed -0.21875_{ten}.

Floating-Point Instructions in LEGv8

LEGv8 supports the IEEE 754 single-precision and double-precision formats with these instructions:

- Floating-point *addition, single* (FADDS) and *addition, double* (FADDD)

- Floating-point *subtraction, single* (FSUBS) and *subtraction, double* (FSUBD)

- Floating-point *multiplication, single* (FMULS) and *multiplication, double* (FMULD)

- Floating-point *division, single* (FDIVS) and *division, double* (FDIVD)

- Floating-point *comparison, single* (FCMPS) and *comparison, double* (FCMPD), with the condition codes given slightly different interpretations

Programmers use B.cond to branch based on floating-point comparisons.

The LEGv8 designers decided to add separate floating-point registers. They are called S0, S1, S2, ... for single precision and D0, D1, D2, ... for double precision. Hence, they included separate loads and stores for floating-point registers: LDURS and STURS. The base registers for floating-point data transfers which are used for addresses remain integer registers. The LEGv8 code to load two single precision numbers from memory, add them, and then store the sum might look like this:

```
LDURS   S4, [X28,c]  // Load 32-bit F.P. number into S4
LDURS   S6, [X28,a]  // Load 32-bit F.P. number into S6
FADDS   S2, S4, S6   // S2 = S4 + S6 single precision
STURS   S2, [X28,b]  // Store 32-bit F.P. number from S2
```

A single precision register is just the lower half of a double-precision register.

Figure 3.17 summarizes the floating-point portion of the LEGv8 architecture revealed in this chapter, with the new pieces to support floating point shown in color.

LEGv8 floating-point operands

Name	Example	Comments
32 floating-point registers	S0, S1, ... S31 or D0, D1, ... D31	LEGv8 floating-point single precession registers (S0, S1, ..., S31) are just the lower half of the double precision registers (D0, D1, ..., D31)
2^{61} memory doublewords	Memory[0], Memory[8], . . . , Memory[4,611,686,018,427,387,900]	Accessed only by data transfer instructions. LEGv8 uses byte addresses, so sequential doubleword addresses differ by 8. Memory holds data structures, such as arrays, and spilled registers, such as those saved on procedure calls.

LEGv8 floating-point assembly language

Category	Instruction	Example	Meaning	Comments
Arithmetic	FP add single	FADDS S2, S4, S6	S2 = S4 + S6	FP add (single precision)
	FP subtract single	FSUBS S2, S4, S6	S2 = S4 − S6	FP sub (single precision)
	FP multiply single	FMULS S2, S4, S6	S2 = S4 × S6	FP multiply (single precision)
	FP divide single	FDIVS S2, S4, S6	S2 = S4 / S6	FP divide (single precision)
	FP add double	FADDD D2, D4, D6	D2 = D4 + D6	FP add (double precision)
	FP subtract double	FSUBD D2, D4, D6	D2 = D4 − D6	FP sub (double precision)
	FP multiply double	FMULD D2, D4, D6	D2 = D4 × D6	FP multiply (double precision)
	FP divide double	FDIVD D2, D4, D6	D2 = D4 / D6	FP divide (double precision)
Conditional branch	FP compare single	FCMPS S4, S6	Test S4 vs. S6	FP compare single precision
	FP compare double	FCMPD D4, D6	Test D4 vs. D6	FP compare double precision
Data transfer	Load single FP	LDURS S1, [X23,100]	S1 = Memory[X23 + 100]	32-bit data to FP register
	Load double FP	LDURD D1, [X23,100]	D1 = Memory[X23 + 100]	64-bit data to FP register
	Store single FP	STURS S1, [X23,100]	Memory[X23 + 100] = S1	32-bit data to memory
	Store double FP	STURD D1, [X23,100]	Memory[X23 + 100] = D1	64-bit data to memory

LEGv8 floating-point machine language

Name	Format			Example			Comments
FADDS	R	241	6	10	4	2	FADDS S2, S4, S6
FSUBS	R	241	6	14	4	2	FSUBS S2, S4, S6
FMULS	R	241	6	2	4	2	FMULS S2, S4, S6
FDIVS	R	241	6	6	4	2	FDIVS S2, S4, S6
FADDD	R	243	6	10	4	2	FADDD D2, D4, D6
FSUBD	R	243	6	14	4	2	FSUBD D2, D4, D6
FMULD	R	243	6	2	4	2	FMULD D2, D4, D6
FDIVD	R	243	6	6	4	2	FDIVD D2, D4, D6
FCMPS	R	241	6	8	4	0	FCMPS S4, S6
FCMPD	R	243	6	8	4	0	FCMPD D4, D6
LDURS	D	1506	100	0	4	2	LDURS S2, [X23,100]
LDURD	D	2018	100	0	4	2	LDURD S2, [X23,100]
STURS	D	1504	100	0	4	2	STURS D2, [X23,100]
STURD	D	2016	100	0	4	2	STURD D2, [X23,100]
Field size		11 bits	5 or 9 bits	6 or 2 bits	5 bits	5 bits	All LEGv8 instructions 32 bits

FIGURE 3.17 LEGv8 floating-point architecture revealed thus far. This information is also found in column 2 of the LEGv8 Reference Data Card at the front of this book.

Elaboration: The full ARMv8 instruction set does not use the mnemonics FADDS or FADDD. It just uses FADD, and lets the assembler pick the proper opcode depending if the registers used are S registers and D registers. We worry that it might be confusing to use the same mnemonic for both opcodes, so for teaching purposes LEGv8 distinguishes the two cases with different mnemonics. LEGv8 follows the same decision for the rest of the floating-point arithmetic and data transfer operations.

Hardware/ Software Interface

One issue that architects face in supporting floating-point arithmetic is whether to select the same registers used by the integer instructions or to add a special set for floating point. Because programs normally perform integer operations and floating-point operations on different data, separating the registers will only slightly increase the number of instructions needed to execute a program. The major impact is to create a distinct set of data transfer instructions to move data between floating-point registers and memory.

The benefits of separate floating-point registers are having twice as many registers without using up more bits in the instruction format, having twice the register bandwidth by having separate integer and floating-point register sets, and being able to customize registers to floating point; for example, some computers convert all sized operands in registers into a single internal format.

EXAMPLE

Compiling a Floating-Point C Program into LEGv8 Assembly Code

Let's convert a temperature in Fahrenheit to Celsius:

```
float f2c (float fahr)
      {
            return ((5.0/9.0) *(fahr - 32.0));
      }
```

Assume that the floating-point argument fahr is passed in S12 and the result should go in S0. What is the LEGv8 assembly code?

ANSWER

We assume that the compiler places the three floating-point constants in memory within easy reach of register X27. The first two instructions load the constants 5.0 and 9.0 into floating-point registers:

```
f2c:
    LDURS S16, [X27,const5] // S16 = 5.0 (5.0 in memory)
    LDURS S18, [X27,const9] // S18 = 9.0 (9.0 in memory)
```

They are then divided to get the fraction 5.0/9.0:

```
FDIVS S16, S16, S18 // S16 = 5.0 / 9.0
```

(Many compilers would divide 5.0 by 9.0 at compile time and save the single constant 5.0/9.0 in memory, thereby avoiding the divide at runtime.) Next, we load the constant 32.0 and then subtract it from fahr (S12):

```
LDURS S18, [X27,const32] // S18 = 32.0
FSUBS S18, S12, S18       // S18 = fahr - 32.0
```

Finally, we multiply the two intermediate results, placing the product in S0 as the return result, and then return

```
FMULS S0, S16, S18 // S0 = (5/9)*(fahr - 32.0)
BR LR              // return
```

Now let's perform floating-point operations on matrices, code commonly found in scientific programs.

Compiling Floating-Point C Procedure with Two-Dimensional Matrices into LEGv8

EXAMPLE

Most floating-point calculations are performed in double precision. Let's perform matrix multiply of C = C + A * B. It is commonly called *DGEMM*, for Double precision, General Matrix Multiply. We'll see versions of DGEMM again in Section 3.9 and subsequently in Chapters 4, 5, and 6. Let's assume C, A, and B are all square matrices with 32 elements in each dimension.

```
void mm (double c[][], double a[][],  double b[][])
{
        size_t i, j, k;
        for (i = 0; i < 32; i = i + 1)
        for (j = 0; j < 32; j = j + 1)
        for (k = 0; k < 32; k = k + 1)
          c[i][j] = c[i][j] + a[i][k] *b[k][j];
}
```

The array starting addresses are parameters, so they are in X0, X1, and X2. Assume that the integer variables are in X19, X20, and X21, respectively. What is the LEGv8 assembly code for the body of the procedure?

Note that c[i][j] is used in the innermost loop above. Since the loop index is k, the index does not affect c[i][j], so we can avoid loading and storing

ANSWER

c[i][j] each iteration. Instead, the compiler loads c[i][j] into a register outside the loop, accumulates the sum of the products of a[i][k] and b[k][j] in that same register, and then stores the sum into c[i][j] upon termination of the innermost loop. We keep the code simpler by using the assembly language pseudoinstruction LDI, which loads a constant into a register.

The body of the procedure starts with saving the loop termination value of 32 in a temporary register and then initializing the three *for* loop variables:

```
mm: . . .
        LDI    X10, 32    // X10 = 32 (row size/loop end)
        LDI    X19, 0     // i = 0; initialize 1st for loop
L1:     LDI    X20, 0     // j = 0; restart 2nd for loop
L2:     LDI    X21, 0     // k = 0; restart 3rd for loop
```

To calculate the address of c[i][j], we need to know how a 32 × 32, two-dimensional array is stored in memory. As you might expect, its layout is the same as if there were 32 single-dimensional arrays, each with 32 elements. So the first step is to skip over the i "single-dimensional arrays," or rows, to get the one we want. Thus, we multiply the index in the first dimension by the size of the row, 32. Since 32 is a power of 2, we can use a shift instead:

```
LSL  X11, X19, 5    // X11 = i * 2⁵(size of row of c)
```

Now we add the second index to select the jth element of the desired row:

```
ADD  X11, X11, X20    // X11 = i * size(row) + j
```

To turn this sum into a byte index, we multiply it by the size of a matrix element in bytes. Since each element is 8 bytes for double precision, we can instead shift left by three:

```
LSL  X11, X11, 3    // X11 = byte offset of [i][j]
```

Next we add this sum to the base address of c, giving the address of c[i][j], and then load the double precision number c[i][j] into D4:

```
ADD    X11, X0, X11    // X11 = byte address of c[i][j]
LDURD  D4, [X11,#0]    // D4 = 8 bytes of c[i][j]
```

The following five instructions are virtually identical to the last five: calculate the address and then load the double precision number b[k][j].

```
L3: LSL    X9, X21, 5    // X9 = k * 2⁵(size of row of b)
    ADD    X9, X9, X20   // X9 = k * size(row) + j
    LSL    X9, X9, 3     // X9 = byte offset of [k][j]
    ADD    X9, X2, X9    // X9 = byte address of b[k][j]
    LDURD  D16, [X9,#0]  // D16 = 8 bytes of b[k][j]
```

Similarly, the next five instructions are like the last five: calculate the address and then load the double precision number a[i][k].

```
LSL   X9, X19, 5    // X9 = i * 2⁵(size of row of a)
ADD   X9, X9, X21   // X9 = i * size(row) + k
LSL   X9, X9, 3     // X9 = byte offset of [i][k]
ADD   X9, X1, X9    // X9 = byte address of a[i][k]
LDURD D18, [X9,#0]  // D18 = 8 bytes of a[i][k]
```

Now that we have loaded all the data, we are finally ready to do some floating-point operations! We multiply elements of a and b located in registers D18 and D16, and then accumulate the sum in D4.

```
FMULD D16, D18, D16 // D16 = a[i][k] * b[k][j]
FADDD D4, D4, D16   // f4 = c[i][j] + a[i][k] * b[k][j]
```

The final block increments the index k and loops back if the index is not 32. If it is 32, and thus the end of the innermost loop, we need to store the sum accumulated in D4 into c[i][j].

```
ADDI  X21, X21, 1   // $k = k + 1
CMP   X21, X10      // test k vs. 32
B.LT  L3            // if (k < 32) go to L3
STURD D4, [X11,0]   // = D4
```

Similarly, these final six instructions increment the index variable of the middle and outermost loops, looping back if the index is not 32 and exiting if the index is 32.

```
ADDI  X20, X20, #1  // $j = j + 1
CMP   X20, X10      // test j vs. 32
B.LT  L2            // if (j < 32) go to L2
ADDI  X19, X19, #1  // $i = i + 1
CMP   X19, X10      // test i vs. 32
B.LT  L1            // if (i < 32) go to L1
. . .
```

Looking ahead, Figure 3.23 below shows the x86 assembly language code for a slightly different version of DGEMM in Figure 3.22.

Elaboration: C and many other programming languages use the array layout discussed in the example, called *row-major order*. Fortran instead uses *column-major order*, whereby the array is stored column by column.

Elaboration: Another reason for separate integers and floating-point registers is that microprocessors in the 1980s didn't have enough transistors to put the floating-point unit on the same chip as the integer unit. Hence, the floating-point unit, including the floating-point registers, was optionally available as a second chip. Such optional accelerator chips are called *coprocessor chips*. Since the early 1990s, microprocessors have integrated floating point (and just about everything else) on chip, and thus the term *coprocessor chip* joins *accumulator* and *core memory* as quaint terms that date the speaker.

Elaboration: As mentioned in Section 3.4, accelerating division is more challenging than multiplication. In addition to SRT, another technique to leverage a fast multiplier is *Newton's iteration*, where division is recast as finding the zero of a function to produce the reciprocal $1/c$, which is then multiplied by the other operand. Iteration techniques *cannot* be rounded properly without calculating many extra bits. A TI chip solved this problem by calculating an extra-precise reciprocal.

Elaboration: Java embraces IEEE 754 by name in its definition of Java floating-point data types and operations. Thus, the code in the first example could have well been generated for a class method that converted Fahrenheit to Celsius.

The second example above uses multiple dimensional arrays, which are not explicitly supported in Java. Java allows arrays of arrays, but each array may have its own length, unlike multiple dimensional arrays in C. Like the examples in Chapter 2, a Java version of this second example would require a good deal of checking code for array bounds, including a new length calculation at the end of row accesses. It would also need to check that the object reference is not null.

Accurate Arithmetic

guard The first of two extra bits kept on the right during intermediate calculations of floating-point numbers; used to improve rounding accuracy.

round Method to make the intermediate floating-point result fit the floating-point format; the goal is typically to find the nearest number that can be represented in the format. It is also the name of the second of two extra bits kept on the right during intermediate floating-point calculations, which improves rounding accuracy.

Unlike integers, which can represent exactly every number between the smallest and largest number, floating-point numbers are normally approximations for a number they can't really represent. The reason is that an infinite variety of real numbers exists between, say, 0 and 1, but no more than 2^{53} can be represented exactly in double precision floating point. The best we can do is getting the floating-point representation close to the actual number. Thus, IEEE 754 offers several modes of rounding to let the programmer pick the desired approximation.

Rounding sounds simple enough, but to round accurately requires the hardware to include extra bits in the calculation. In the preceding examples, we were vague on the number of bits that an intermediate representation can occupy, but clearly, if every intermediate result had to be truncated to the exact number of digits, there would be no opportunity to round. IEEE 754, therefore, always keeps two extra bits on the right during intervening additions, called guard and round, respectively. Let's do a decimal example to illustrate their value.

Rounding with Guard Digits

Add $2.56_{ten} \times 10^0$ to $2.34_{ten} \times 10^2$, assuming that we have three significant decimal digits. Round to the nearest decimal number with three significant decimal digits, first with guard and round digits, and then without them.

First we must shift the smaller number to the right to align the exponents, so $2.56_{ten} \times 10^0$ becomes $0.0256_{ten} \times 10^2$. Since we have guard and round digits, we are able to represent the two least significant digits when we align exponents. The guard digit holds 5 and the round digit holds 6. The sum is

$$
\begin{array}{r}
2.3400_{ten} \\
+0.0256_{ten} \\
\hline
2.3656_{ten}
\end{array}
$$

Thus the sum is $2.3656_{ten} \times 10^2$. Since we have two digits to round, we want values 0 to 49 to round down and 51 to 99 to round up, with 50 being the tiebreaker. Rounding the sum up with three significant digits yields $2.37_{ten} \times 10^2$.

Doing this *without* guard and round digits drops two digits from the calculation. The new sum is then

$$
\begin{array}{r}
2.34_{ten} \\
+0.02_{ten} \\
\hline
2.36_{ten}
\end{array}
$$

The answer is $2.36_{ten} \times 10^2$, off by 1 in the last digit from the sum above.

Since the worst case for rounding would be when the actual number is halfway between two floating-point representations, accuracy in floating point is normally measured in terms of the number of bits in error in the least significant bits of the significand; the measure is called the number of **units in the last place**, or **ulp**. If a number were off by 2 in the least significant bits, it would be called off by 2 ulps. Provided there are no overflow, underflow, or invalid operation exceptions, IEEE 754 guarantees that the computer uses the number that is within one-half ulp.

units in the last place (ulp) The number of bits in error in the least significant bits of the significand between the actual number and the number that can be represented.

Elaboration: Although the example above really needed just one extra digit, multiply can require two. A binary product may have one leading 0 bit; hence, the normalizing step must shift the product one bit left. This shifts the guard digit into the least significant bit of the product, leaving the round bit to help accurately round the product.

IEEE 754 has four rounding modes: always round up (toward $+\infty$), always round down (toward $-\infty$), truncate, and round to nearest even. The final mode determines what to do if the number is exactly halfway in between. The U.S. *Internal Revenue Service* (IRS) always rounds 0.50 dollars up, possibly to the benefit of the IRS. A more equitable way would be to round up this case half the time and round down the other half. IEEE 754

sticky bit A bit used in rounding in addition to guard and round that is set whenever there are nonzero bits to the right of the round bit.

says that if the least significant bit retained in a halfway case would be odd, add one; if it's even, truncate. This method always creates a 0 in the least significant bit in the tie-breaking case, giving the rounding mode its name. This mode is the most commonly used, and the only one that Java supports.

The goal of the extra rounding bits is to allow the computer to get the same results as if the intermediate results were calculated to infinite precision and then rounded. To support this goal and round to the nearest even, the standard has a third bit in addition to guard and round; it is set whenever there are nonzero bits to the right of the round bit. This **sticky bit** allows the computer to see the difference between $0.50 \ldots 00_{ten}$ and $0.50 \ldots 01_{ten}$ when rounding.

The sticky bit may be set, for example, during addition, when the smaller number is shifted to the right. Suppose we added $5.01_{ten} \times 10^{-1}$ to $2.34_{ten} \times 10^2$ in the example above. Even with guard and round, we would be adding 0.0050 to 2.34, with a sum of 2.3450. The sticky bit would be set, since there are nonzero bits to the right. Without the sticky bit to remember whether any 1s were shifted off, we would assume the number is equal to $2.345000 \ldots 00$ and round to the nearest even of 2.34. With the sticky bit to remember that the number is larger than $2.345000 \ldots 00$, we round instead to 2.35.

fused multiply add A floating-point instruction that performs both a multiply and an add, but rounds only once after the add.

Elaboration: MIPS-64, PowerPC, SPARC64, AMD SSE5, and Intel AVX architectures provide a single instruction that does a multiply and add on three registers: $a = a + (b \times c)$. Obviously, this instruction allows potentially higher floating-point performance for this common operation. Equally important is that instead of performing two roundings—after the multiply and then after the add—which would happen with separate instructions, the multiply add instruction can perform a single rounding after the add. A single rounding step increases the precision of multiply add. Such operations with a single rounding are called **fused multiply add**. It was added to the revised IEEE 754-2008 standard (see 📖 Section 3.12). Thus, the full ARMv8 instruction offers fused multiply-adds as well (see Section 3.8).

Summary

The *Big Picture* that follows reinforces the stored-program concept from Chapter 2; the meaning of the information cannot be determined just by looking at the bits, for the same bits can represent a variety of objects. This section shows that computer arithmetic is finite and thus can disagree with natural arithmetic. For example, the IEEE 754 standard floating-point representation

$$(-1)^5 \times (1 + \text{Fraction}) \times 2^{(\text{Exponent} - \text{Bias})}$$

is almost always an approximation of the real number. Computer systems must take care to minimize this gap between computer arithmetic and arithmetic in the real world, and programmers at times need to be aware of the implications of this approximation.

The **BIG** Picture

Bit patterns have no inherent meaning. They may represent signed integers, unsigned integers, floating-point numbers, instructions, character strings, and so on. What is represented depends on the instruction that operates on the bits in the word.

The major difference between computer numbers and numbers in the real world is that computer numbers have limited size and hence limited precision; it's possible to calculate a number too big or too small to be represented in a computer word. Programmers must remember these limits and write programs accordingly.

C type	Java type	Data transfers	Operations
long long int	int	LDUR, STUR, MOVZ, MOVK	ADD, SUB, ADDI, SUBI, ADDS, SUBS, ADDIS, SUBIS, MUL, SMULH, SDIV, AND, ANDI, ORR, ORRI, EOR, EORI
unsigned long long int	—	LDUR, STUR, MOVZ, MOVK	ADD, SUB, ADDI, SUBI, ADDS, SUBS, ADDIS, SUBIS, MUL, UMULH, UDIV, AND, ANDI, ORR, ORRI, EOR, EORI
char	—	LDURB, STURB, MOVZ, MOVK	ADD, SUB, ADDI, SUBI, ADDS, SUBS, ADDIS, SUBIS, MUL, SMULH, SDIV, AND, ANDI, ORR, ORRI, EOR, EORI
—	char	LDURH, STURH, MOVZ, MOVK	ADD, SUB, ADDI, SUBI, ADDS, SUBS, ADDIS, SUBIS, MUL, UMULH, UDIV, AND, ANDI, ORR, ORRI, EOR, EORI
float	float	LDURS, STURS	FADDS, FSUBS, FMULS, FDIVS, FCMPS
double	double	LDURD, STURD	FADDD, FDUBD, FMULD, FDIVD, FCMPD

In the last chapter, we presented the storage classes of the programming language C (see the *Hardware/Software Interface* section in Section 2.7). The table above shows some of the C and Java data types, the data transfer instructions, and instructions that operate on those types that appear in Chapter 2 and this chapter. Note that Java omits unsigned integers.

Hardware/ Software Interface

The revised IEEE 754-2008 standard added a 16-bit floating-point format with five exponent bits. What do you think is the likely range of numbers it could represent?

Check Yourself

1. $1.0000\ 00 \times 2^0$ to $1.1111\ 1111\ 11 \times 2^{31}$, 0

2. $\pm 1.0000\ 0000\ 0 \times 2^{-14}$ to $\pm\ 1.1111\ 1111\ 1 \times 2^{15}$, $\pm\ 0$, $\pm\ \infty$, NaN

3. $\pm 1.0000\ 0000\ 00 \times 2^{-14}$ to $\pm\ 1.1111\ 1111\ 11 \times 2^{15}$, $\pm\ 0$, $\pm\ \infty$, NaN

4. $\pm 1.0000\ 0000\ 00 \times 2^{-15}$ to $\pm\ 1.1111\ 1111\ 11 \times 2^{14}$, $\pm\ 0$, $\pm\ \infty$, NaN

Elaboration: To accommodate comparisons that may include NaNs, the standard includes *ordered* and *unordered* as options for compares. Hence, the full ARMv8 instruction set has many flavors of compares to support NaNs. (Java does not support unordered compares.)

In an attempt to squeeze every bit of precision from a floating-point operation, the standard allows some numbers to be represented in unnormalized form. Rather than having a gap between 0 and the smallest normalized number, IEEE allows *denormalized numbers* (also known as *denorms* or *subnormals*). They have the same exponent as zero but a nonzero fraction. They allow a number to degrade in significance until it becomes 0, called *gradual underflow*. For example, the smallest positive single precision normalized number is

$$1.0000\,0000\,0000\,0000\,0000\,000_{two} \times 2^{-126}$$

but the smallest single precision denormalized number is

$$0.0000\,0000\,0000\,0000\,0000\,001_{two} \times 2^{-126},\ or\ 1.0_{two} \times 2^{-149}$$

For double precision, the denorm gap goes from 1.0×2^{-1022} to 1.0×2^{-1074}.

The possibility of an occasional unnormalized operand has given headaches to floating-point designers who are trying to build fast floating-point units. Hence, many computers cause an exception if an operand is denormalized, letting software complete the operation. Although software implementations are perfectly valid, their lower performance has lessened the popularity of denorms in portable floating-point software. Moreover, if programmers do not expect denorms, their programs may surprise them.

3.6   Parallelism and Computer Arithmetic: Subword Parallelism

Since every microprocessor in a phone, tablet, or laptop by definition has its own graphical display, as transistor budgets increased it was inevitable that support would be added for graphics operations.

Many graphics systems originally used 8 bits to represent each of the three primary colors plus 8 bits for a location of a pixel. The addition of speakers and microphones for teleconferencing and video games suggested support of sound as well. Audio samples need more than 8 bits of precision, but 16 bits are sufficient.

Every microprocessor has special support so that bytes and halfwords take up less space when stored in memory (see Section 2.9), but due to the infrequency of arithmetic operations on these data sizes in typical integer programs, there was little support beyond data transfers. Architects recognized that many graphics and audio applications would perform the same operation on vectors of these data. By partitioning the carry chains within a 128-bit adder, a processor could use **parallelism** to perform simultaneous operations on short vectors of sixteen 8-bit operands, eight 16-bit operands, four 32-bit operands, or two 64-bit operands.

PARALLELISM

The cost of such partitioned adders was small yet the speedups could be large.

Given that the parallelism occurs within a wide word, the extensions are classified as *subword parallelism*. It is also classified under the more general name of *data level parallelism*. They are known as well as vector or SIMD, for single instruction, multiple data (see Section 6.6). The rising popularity of multimedia applications led to arithmetic instructions that support narrower operations that can easily compute in parallel.

For example, ARMv8 added 32 128-bit registers (V0, V1, ..., V31) and more than 500 machine-language instructions to support subword parallelism. It supports all the subword data types you can imagine:

- 8-bit, 16-bit, 32-bit, 64-bit, and 128-bit signed and unsigned integers

- 32-bit and 64-bit floating point numbers

Figure 3.18 gives a quick summary of some basic ARMv8 SIMD instructions, while Section 3.8 surveys the full SIMD architecture.

Type	Description	Name	Size (bits)					FP Precision	
			8	16	32	64	128	SP	DP
Add/ Subtract	Integer add	ADD	✓	✓	✓	✓	✓		
	FP add	FADD						✓	✓
	Integer subtract	SUB	✓	✓	✓	✓	✓		
	FP subtract	FSUB						✓	✓
Multiply	Unsigned integer multiply	UMUL	✓	✓	✓	✓	✓		
	Signed integer multiply	SMUL	✓	✓	✓	✓	✓		
	FP multiply	FMUL						✓	✓
Compare	Integer compare equal	CMEQ	✓	✓	✓	✓	✓		
	FP compare equal	FCMEQ						✓	✓
Min/Max	Unsigned integer minmum	UMIN	✓	✓	✓	✓	✓		
	Signed integer minmum	SMIN	✓	✓	✓	✓	✓		
	FP minmum	FMIN						✓	✓
	Unsigned integer maximum	UMAX	✓	✓	✓	✓	✓		
	Signed integer maximum	SMAX	✓	✓	✓	✓	✓		
	FP maximum	FMAX						✓	✓
Shift	Integer shift left	SHL	✓	✓	✓	✓	✓		
	Unsigned integer shift right	USHR	✓	✓	✓	✓	✓		
	Signed integer shift right	SSHR	✓	✓	✓	✓	✓		
Logical	Bitwise AND	AND	✓	✓	✓	✓	✓		
	Bitwise OR	ORR	✓	✓	✓	✓	✓		
	Bitwise exclusive OR	EOR	✓	✓	✓	✓	✓		
Data Transfer	Load register	LDR	✓	✓	✓	✓	✓	✓	✓
	Store register	STR	✓	✓	✓	✓	✓	✓	✓

FIGURE 3.18 Example of ARMv8 SIMD instructions for subword parallelism. Figure 3.21 shows the full set of SIMD instructions.

Rather than have a different assembly mnemonic for each data width, the ARMv8 assembler uses different suffixes for the SIMD registers to represent different widths, which allows it to pick the proper machine language opcode. The suffixes are B (*byte*) for 8-bit operands, H (*half*) for 16-bit operands, S (*single*) for 32-bit operands, D (*double*) for 64-bit operands, and Q (*quad*) for 128-bit operands. The programmer also specifies the number of subword operations for that data width with a number before the register name. Thus, to perform 16 parallel integer adds between the 8-bit elements in registers V2 and V3 and place the sixteen 8-bit sums in the elements of V1, we write:

```
ADD  V1.16B, V2.16B, V3.16B // 16 8-bit integer adds
```

To perform four parallel 32-bit floating-point additions, we write:

```
FADD  V1.4S, V2.4S, V3.4S  // 4 32-bit floating-point adds
```

3.7 Real Stuff: Streaming SIMD Extensions and Advanced Vector Extensions in x86

The original MMX (*MultiMedia eXtension*) and SSE (*Streaming SIMD Extension*) instructions for the x86 included similar operations to those found in ARMv8. Chapter 2 notes that in 2001 Intel added 144 instructions to its architecture as part of SSE2, including double precision floating-point registers and operations. It included eight 64-bit registers that can be used for floating-point operands. AMD expanded the number to 16 registers, called XMM, as part of AMD64, which Intel relabeled EM64T for its use. Figure 3.19 summarizes the SSE and SSE2 instructions.

In addition to holding a single precision or double precision number in a register, Intel allows multiple floating-point operands to be packed into a single 128-bit SSE2 register: four single precision or two double precision. Thus, the 16 floating-point registers for SSE2 are actually 128 bits wide. If the operands can be arranged in memory as 128-bit aligned data, then 128-bit data transfers can load and store multiple operands per instruction. This packed floating-point format is supported by arithmetic operations that can compute simultaneously on four singles (PS) or two doubles (PD).

In 2011, Intel doubled the width of the registers again, now called YMM, with *Advanced Vector Extensions* (AVX). Thus, a single operation can now specify eight 32-bit floating-point operations or four 64-bit floating-point operations. The legacy SSE and SSE2 instructions now operate on the lower 128 bits of the YMM registers. Thus, to go from 128-bit and 256-bit operations, you prepend the letter "v" (for vector) in front of the SSE2 assembly language operations and then use the

Data transfer	Arithmetic		Compare
MOV[AU]{SS|PS|SD|PD} xmm, {mem|xmm}	ADD{SS|PS|SD|PD} xmm,{mem|xmm}		CMP{SS|PS|SD|PD}
	SUB{SS|PS|SD|PD} xmm,{mem|xmm}		
MOV[HL]{PS|PD} xmm, {mem|xmm}	MUL{SS|PS|SD|PD} xmm,{mem|xmm}		
	DIV{SS|PS|SD|PD} xmm,{mem|xmm}		
	SQRT{SS|PS|SD|PD} {mem|xmm}		
	MAX{SS|PS|SD|PD} {mem|xmm}		
	MIN{SS|PS|SD|PD} {mem|xmm}		

FIGURE 3.19   The SSE/SSE2 floating-point instructions of the x86. xmm means one operand is a 128-bit SSE2 register, and {mem|xmm} means the other operand is either in memory or it is an SSE2 register. The table uses regular expressions to show the variations of instructions. Thus, MOV[AU]{SS|PS|SD|PD} represents the eight instructions MOVASS, MOVAPS, MOVASD, MOVAPD, MOVUSS, MOVUPS, MOVUSD, and MOVUPD. We use square brackets [ ] to show single-letter alternatives: A means the 128-bit operand is aligned in memory; U means the 128-bit operand is unaligned in memory; H means move the high half of the 128-bit operand; and L means move the low half of the 128-bit operand. We use the curly brackets { } with a vertical bar | to show multiple letter variations of the basic operations: SS stands for *Scalar Single* precision floating point, or one 32-bit operand in a 128-bit register; PS stands for *Packed Single* precision floating point, or four 32-bit operands in a 128-bit register; SD stands for Scalar Double precision floating point, or one 64-bit operand in a 128-bit register; PD stands for *Packed Double* precision floating point, or two 64-bit operands in a 128-bit register.

YMM register names instead of the XMM register name. For example, the SSE2 instruction to perform two 64-bit floating-point additions

```
addpd  %xmm0, %xmm4
```

becomes

```
vaddpd  %ymm0, %ymm4
```

which now produces four 64-bit floating-point additions. Intel has announced plans to widen the AVX registers to first 512 bits and later 1024 bits in later editions of the x86 architecture.

Elaboration: AVX also added three address instructions to x86. For example, vaddpd can now specify

```
vaddpd  %ymm0, %ymm1, %ymm4    // %ymm4 = %ymm0 + %ymm1
```

instead of the standard, two address version

```
addpd  %xmm0, %xmm4  // %xmm4 = %xmm4 + %xmm0
```

(Unlike LEGv8, the destination is on the right in x86.) Three addresses can reduce the number of registers and instructions needed for a computation.

Real Stuff: The Rest of the ARMv8 Arithmetic Instructions

Figure 2.40 lists 63 assembly-language instructions for integer multiply, integer divide, and floating-point operations in the full ARMv8 instruction set and an impressive 245 SIMD assembly language instructions.

It also shows 18 floating-point data transfer instructions in machine language but none in assembly language. Figure 3.17, however, lists four assembly language instructions for floating-point data transfer. This apparent contradiction occurs because ARMv8 assemblers can use the names of the registers along with the generic data transfer instruction names to generate the proper opcode. For example, the ARMv8 assembler turns three instructions:

```
LDUR S1, [X23,#100]
LDUR D1, [X23,#100]
LDUR X1, [X23,#100]
```

into machine language versions of the following LEGv8 instructions:

```
LDURS S1, [X23,#100]
LDURD D1, [X23,#100]
LDUR  X1, [X23,#100]
```

As mentioned in an earlier elaboration, we use distinct assembly language names for different floating-point machine language instructions in this book in the belief that there will be less confusion about how the hardware works if we keep the relationship between the two levels one-to-one, but the register names alone are sufficient for an assembler to keep things straight.

Full ARMv8 Integer and Floating-point Arithmetic Instructions

Figure 3.20 shows all 63 assembly-language integer arithmetic and floating-point instructions in the full ARMv8 instruction set, with the 15 ARMv8 arithmetic core instructions highlighted in bold. Pseudoinstructions are italicized.

Like many other ARMv8 instruction categories, there is a version of the integer multiply instruction that supplies the negative of the result (MNEG). There also "long" versions of the four multiply instructions, where the operands are 32 bits (long) instead of 64 bits (long long). ARMv8 also has six instructions that do both an integer multiply *and* an add or subtract for either three long (32-bit) or three long long (64-bit) operands. In fact, six of the integer multiply instructions are just pseudoinstructions of the mutiply-add instructions with one of the three operands being the zero register (XZR). These 11 new ARMv8 integer instructions join the five multiply and divide instructions in the ARMv8 arithmetic core account for 16 instructions.

Type	Mnemonic	Instruction	Type	Mnemonic	Instruction
Integer Multiply & Divide	**MUL**	Multiply	Integer Mul-Add	MADD	Multiply-add
	SMULH	Signed multiply high		MSUB	Multiply-subtract
	UMULH	Unsigned multiply high		SMADDL	Signed multiply-add long
	SDIV	Signed divide		SMSUBL	Signed multiply-subtract long
	UDIV	Unsigned divide		UMADDL	Unsigned multiply-add long
	SMULL	Signed multiply long		UMSUBL	Unsigned multiply-subtract long
	UMULL	Unsigned multiply long	FP Mul-Add	FMADD	Floating-point fused multiply-add
	MNEG	Multiply-negate		FMSUB	Floating-point fused multiply-subtract
	UMNEGL	Unsigned multiply-negate long		FNMADD	Floating-point negated fused multiply-add
	SMNEGL	Signed multiply-negate long		FNMSUB	Floating-point negated fused multiply-subtract
FP two source operands	**FADDS**	Floating-point add single	FP move	FMOV	Floating-point move to/from integer or FP register
	FSUBS	Floating-point subtract single		FMOVI	Floating-point move immediate
	FMULS	Floating-point multiply single	FP sel	FCSEL	Floating-point conditional select
	FDIVS	Floating-point divide single			
	FADDD	Floating-point add double	FP round	FRINTA	Floating-point round to nearest with ties to odd
	FSUBD	Floating-point subtract double		FRINTI	Floating-point round using current rounding mode
	FNMUL	Floating-point scalar multiply-negate		FRINTM	Floating-point round toward -infinity
	FMULD	Floating-point multiply double		FRINTN	Floating-point round to nearest with ties to even
	FDIVD	Floating-point divide double		FRINTP	Floating-point round toward +infinity
	FCMPS	Floating-point compare single (quiet)		FRINTX	Floating-point exact using current rounding mode
	FCMPD	Floating-point compare double (quiet)		FRINTZ	Floating-point round toward 0
	FCMPE	Floating-point signaling compare	FP convert	FCVTAS	FP convert to signed integer, rounding to nearest odd
	FCCMP	Floating-point conditional quiet compare		FCVTAU	FP convert to unsigned integer, rounding to nearest odd
	FCCMPE	Floating-point conditional signaling compare		FCVTMS	FP convert to signed integer, rounding toward -infinity
FP one operand	FABS	Floating-point scalar absolute value		FCVTMU	FP convert to unsigned integer, rounding toward -infinity
	FNEG	Floating-point scalar negate		FCVTNS	FP convert to signed integer, rounding to nearest even
	FSQRT	Floating-point scalar square root		FCVTNU	FP convert to unsigned integer, rounding to nearest even
FP Min/Max	FMAX	Floating-point scalar maximum		FCVTPS	FP convert to signed integer, rounding toward +infinity
	FMIN	Floating-point scalar minimum		FCVTPU	FP convert to unsigned integer, rounding toward +infinity
	FMAXNM	Floating-point scalar maximum number (NaN = –Inf)		FCVTZS	FP convert to signed integer, rounding toward 0
				FCVTZU	FP convert to unsigned integer, rounding toward 0
	FMINNM	Floating-point scalar minimum number (NaN = +Inf)		SCVTF	Signed integer convert to FP, current rounding mode
				UCVTF	Unsigned integer convert to FP, current rounding mode

FIGURE 3.20 Full ARMv8 assembly-language instructions for integer and floating-point arithmetic. Instruction in bold font is in the LEGv8 core. Italicized instructions are pseudoinstructions.

Like integer multiply, there is a version of floating-point multiply that produces a negative product (FNMUL). Like integer conditional compare instruction (CCMP), there is a conditional version of compare that (confusingly) does a comparison only if the initial condition is true (FCCMP). To allow programmers to check to see if an operand is a *Not-A-Number* (NaN), there are two versions of the comparison instructions: one that doesn't cause an exception whenever one of the operands is the value NaN (quiet compare) as well as one that does (signaling compare) in case the programmer wants an exception whenever a NaN is found. Again like integer multiply, there are four floating-point instructions that do multiply followed by an add or a subtract. ARMv8 has three more instructions for floating-point operations with a single operand: absolute value, negate, and square root. These 11 instructions join the 10 other existing instructions from the ARMv8 arithmetic core to bring our running total in 37.

Minimum and maximum floating-point operations are again a bit more complicated due to NaNs. There are two instructions that trap if an operand is a NaN and two that treat NaN as an extreme number: minus infinity for maximum or plus infinity for minimum. Not only does ARMv8 have one floating-point move instruction that can copy a value between floating-point registers or between integer registers and floating-point registers, it has one that can load a floating-point constant into a register. And once again like integer conditional select (CSEL), there is a floating-point version (FCSEL). These seven instructions get us to 44 arithmetic assembly-language instructions.

The final two categories are for rounding floating-point numbers and converting between integers and floating-point numbers. To round according to the many modes of IEEE 754, ARMv8 has seven instructions. To cover all combinations of conversions of signed and integers for the different rounding modes requires 12 more instructions. These last two categories bring us to 63 arithmetic assembly-language instructions, which matches Figure 2.40.

Full ARMv8 SIMD Instructions

Figure 3.21 shows all 245 assembly-language SIMD instructions in the full ARMv8 instruction set. To fit these 245 instructions into a single table, we use regular expressions to represent several instructions within a single table entry. The figure caption reviews the three regular expression operators we use.

Many SIMD instructions offer three versions:

- *Wide* means the width of the elements in the destination register and the first source registers is twice the width of the elements in the *second* source register.

- *Long* means the width of the elements in the destination register is twice the width of the elements in *all* source registers.

- *Narrow* means the width of the elements in the destination register is *half* the width of the elements in all source registers.

These three options are naturally enough indicated by the three suffixes W, L, and N. Like the non-SIMD instructions, *pair* means do the operation to a pair of SIMD registers. These instructions use the suffix P.

When dealing with narrow operations, the instruction could operate on either the lower half or the upper half of the 128-bit SIMD register containing the narrower elements. The default is use the lower half, with the suffix 2 indicating the operation is working on the upper half of the SIMD register.

The *prefixes* U, S, and F refer to the data types of unsigned integers, signed integers, and floating-point. When the elements are just plain bits, there is no prefix. There are also three instructions that use the type polynomial, and they use P as the prefix.

Most SIMD categories are self-explanatory, so we'll explain just the ones that may not be obvious. Instead of setting condition codes, SIMD compare vector sets the destination vector element to all 1s if the condition holds, or to 0 if not. While ARMv8

Type	Description	Name	Type	Description	Name
Add / Subtract	Vector add	F?ADD	Saturating Arithmetic	Integer saturating vector add	[US]QADD
	Integer vector add returning high, narrow	ADDHN2?		Integer saturating vector subtract	[US]QSUB
	Integer vector add long	[US]ADDL2?		Signed integer saturating vector accumulate of unsigned value	SUQADD
	Integer vector add wide	[US]ADDW2?		Unsigned integer saturating vector accumulate of unsigned value	USQADD
	Vector add pair	FADDP		Signed integer saturating vector absolute	SQABS
	Integer vector add long pair	[US]ADDLP		Signed integer saturating vector doubling multiply-add long	SQDMLAL2?
	Integer vector add and accumulate long pair	[US]ADALP		Signed integer saturating vector doubling multiply-subtract long	SQDMLSL2?
	Vector subtract	F?SUB		Signed integer saturating vector doubling multiply high half	SQDMULH
	Integer vector subtract returning high, narrow	SUBHN2?		Signed integer saturating vector doubling multiply long	SQDMULL2?
	Integer vector subtract long	[US]SUBL2?		Integer saturating vector narrow	[US]QXTN2?
	Integer vector subtract wide	[US]SUBW2?		Signed integer saturating vector and unsigned narrow	SQXTUN2?
	Vector negate	F?NEG		Signed integer saturating vector negate	SQNEG
Multiply / Divide / Square Root	Vector multiply	[FP]MUL		Signed integer vector saturating rounding doubling multiply high half	SQRDMULH
	FP vector multiply extended (0xINF→2)	FMULX	Multiply-Add	Vector chained multiply-add	MLA
	Vector multiply long	[USP]MULL2?		Vector fused multiply-add	FMLA
	Vector FP divide	FDIV		Integer vector multiply-add long	[US]MLAL2?
	FP vector square root	FSQRT		Vector chained multiply-subtract	MLS
	FP reciprocal square root	FRSQRTS		Vector fused multiply-subtract	FMLS
	Vector reciprocal square root estimate	[UF]RSQRTE		Integer vector multiply-subtract long	[US]MLSL2?
	Vector reciprocal estimate	[UF]RECPE	Reduction	Integer sum elements in vector	ADDV
	FP vector reciprocal step	FRECPS		Integer sum elements in vector long	[US]ADDLV
	FP vector reciprocal exponent	FRECPX		Maximum element in vector	[USF]MAXV
Compare	FP vector absolute compare greater than or equal	FACGE		FP maxNum element in vector	FMAXNMV
	FP vector absolute compare greater than	FACGT		Minimum element in vector	[USF]MINV
	FP vector absolute compare less than or equal	FACLE		FP minNum element in vector	FMINNMV
	FP vector absolute compare less than	FACLT	Saturating Shifts	Integer saturating vector rounding shift left	[US]QRSHL
	Vector compare equal	F?CMEQ		Integer saturating vector shift right rounded narrow	[US]QRSHRN
	Vector compare greater than or equal	{CMHS|F?CMGE}		Signed integer saturating vector shift right rounded unsigned narrow	SQRSHRUN
	Vector compare greater than	{CMHI|F?CMGT}		Integer saturating vector shift left	[US]QSHL
	Vector compare less than or equal	{CMLS|F?CMLE}		Signed integer saturating vector shift left unsigned	SQSHLU
	Vector compare less than	{CMLO|F?CMLT}		Integer saturating vector shift right narrow	[US]QSHRN
	Vector test bits	CMTST		Signed integer saturating vector shift right unsigned narrow	SQSHRUN
Logical	Bitwise vector AND	AND	Min / Max	Vector maximum	[USF]MAX
	Bitwise vector bit clear	BIC		FP vector maxNum	FMAXNM
	Bitwise vector OR	ORR		Vector minimum	[USF]MIN
	Bitwise vector OR NOT	ORN		FP vector minNum	FMINNM
	Bitwise vector exclusive OR	EOR		Vector max pair	[USF]MAXP
	Bitwise vector NOT	MVN		FP vector maxNum pair	FMAXNMP
Shifts	Integer vector shift left	SHL		Vector min pair	[USF]MINP
	Integer vector shift left long	[US]SHLL		FP vector minNum pair	FMINNMP
	Integer vector shift right	[US]SHR	Convert	Vector FP convert to {signed | unsigned} integer (round to {0|x})	FCVT[Zx][SU]
	Integer vector shift right narrow	SHRN2?		Vector integer convert to FP	[US]CVTF
	Integer vector shift left and insert	SLI		Vector convert FP precision	FCVT[NL]
	Integer vector shift right and accumulate	[US]SRA		Vector convert double to single-precision (rounding to odd)	FCVTXN
	Integer vector shift right and insert	SRI	Round	Vector FP round to integral FP value (towards x)	FRINTx
	Integer rounding vector shift left	[US]RSHL		Integer rounding vector shift right and accumulate	[US]RSRA
	Integer rounding vector shift right	[US]RSHR		Integer rounding vector subtract returning high, narrow	RSUBHN2
	Integer rounding vector shift right narrow	RSHRN2?		Integer rounding vector halving add	[US]RHADD
Absolute value / difference	Integer vector absolute difference and accumulate	[US]ABA		Integer vector rounding add returning high, narrow	RADDHN
	Integer vector absolute difference and accumulate long	[US]ABAL2?	Insert / Extract / Reverse / Duplicate	Bitwise vector select	BSL
	Vector absolute difference	[USF]ABD		Bitwise vector extract	EXT
	Integer vector absolute difference long	[USF]ABDL2?		Bitwise vector insert if {true | false}	BI[TF]
	Vector absolute value	F?ABS		Vector reverse bits in bytes	RBIT
Data Transfer	Vector load {pair | register}	LD[PR]		Vector reverse elements	REV{16|32|64}
	Vector store {pair | register}	ST[PR]		Duplicate single vector element to all elements	DUP
	Vector move	[USF]MOV		Insert single element in another element	INS
	Vector structure/element load	LD[1234]		Vector element transpose	TRN[12]
	Vector replicated element load	LD[1234]R		Vector element zip	ZIP2?
	Vector structure store	ST[1234]		Vector element unzip	UZP[12]
Count Bits	Integer vector count leading {sign | zero} bits	CL[SZ]	Vector Length	Integer vector lengthen	[US]XTL2?
	Vector count non-zero bits	CNT		Integer vector narrow	XTN
Table Lookup	Vector table lookup	TBL		Integer vector halving add	[US]HADD
	Vector table extension	TBX		Integer vector halving subtract	[US]HSUB

FIGURE 3.21 Full ARMv8 assembly language for SIMD instructions, which uses the term vector to distinguish them from single operands (scalar). To fit all 245 assembly language instructions into this small space, we use regular expressions to show the valid combinations. Question mark means 0 or 1 copies of the letter before it, so F?ADD represents the two instructions ADD and FADD. Square brackets mean there is a version for each letter in the brackets, so [US]QADD represents the two instructions UQADD and SQADD. Finally, curly brackets and a vertical line to separate the options show how to form multiple letter versions of the instructions. For example, REV{16|32|64} stands for the three instructions REV16, REV32, and REV64.

has only half of the comparisons (HS, GE, HI, GT), the programmer gets the others (LS, LE, LO, LT) by reversing the operands and using the complementary comparison. That is, A < B is the same as B ≥ A. Reductions do the operations across the elements *within* a *single* SIMD register rather than *between* elements of *different* SIMD registers as is the case for the rest of the SIMD instructions. As the category name suggests, these instructions perform the classic reduction operations of sum, minimum, and maximum. Finally, the table lookup instructions use one to four SIMD registers to act as a table. The elements of the other source register hold the indexes of the table and then the results of the parallel table lookups are stored in the elements of the destination register.

With the information above describing the terms wide, long, narrow, pair, and so on, the descriptions of the instructions within the categories are usually understandable. Here are a few that might not be:

- The elaboration in Section 3.5 explains how fused multiply-add only does one rounding instead of two as you might expect from doing two operations in one instruction. As is often the case, rather than choosing between them, ARMv8 offers both *fused* mutiply-adds (one rounding) and *chained* multiply-adds (two roundings).

- Integer vector shift left and insert (SLI) provides a way to combine bits from two vectors.

- Integer vector shift right and accumulate instructions (SSRA, USRA) are useful when intermediate calculations are made at a higher precision before the result is added to a lower precision accumulator.

- Vector structure/element load instructions (LD1, LD2, LD3, LD4) loads structures of one, two, three, or four elements into SIMD registers.

3.9   Going Faster: Subword Parallelism and Matrix Multiply

To demonstrate the performance impact of subword parallelism, we'll run the same code on the Intel Core i7 first without AVX and then with it. Figure 3.22 shows an unoptimized version of a matrix-matrix multiply written in C. As we saw in Section 3.5, this program is commonly called *DGEMM*, which stands for Double precision GEneral Matrix Multiply. Starting with this edition, we have added a new section entitled "Going Faster" to demonstrate the performance benefit of adapting software to the underlying hardware, in this case the Sandy Bridge version of the Intel Core i7 microprocessor. This new section in Chapters 3, 4, 5, and 6 will incrementally improve DGEMM performance using the ideas that each chapter introduces.

Figure 3.23 shows the x86 assembly language output for the inner loop of Figure 3.22. The five floating point-instructions start with a v like the AVX instructions, but note that they use the XMM registers instead of YMM, and they include sd

```
1.  void dgemm (size_t n, double* A, double* B, double* C)

2.  {

3.    for (size_t i = 0; i < n; ++i)

4.      for (size_t j = 0; j < n; ++j)

5.      {

6.        double cij = C[i+j*n]; /* cij = C[i][j] */

7.        for(size_t k = 0; k < n; k++ )

8.          cij += A[i+k*n] * B[k+j*n]; /*cij+=A[i][k]*B[k][j]*/

9.        C[i+j*n] = cij; /* C[i][j] = cij */

10.     }

11. }
```

FIGURE 3.22   Unoptimized C version of a double precision matrix multiply, widely known as DGEMM for Double-precision GEneral Matrix Multiply. Because we are passing the matrix dimension as the parameter n, this version of DGEMM uses single-dimensional versions of matrices C, A, and B and address arithmetic to get better performance instead of using the more intuitive two-dimensional arrays that we saw in Section 3.5. The comments remind us of this more intuitive notation.

```
1.  vmovsd (%r10),%xmm0              // Load 1 element of C into %xmm0

2.  mov    %rsi,%rcx                 // register %rcx = %rsi

3.  xor    %eax,%eax                 // register %eax = 0

4.  vmovsd (%rcx),%xmm1              // Load 1 element of B into %xmm1

5.  add    %r9,%rcx                  // register %rcx = %rcx + %r9

6.  vmulsd (%r8,%rax,8),%xmm1,%xmm1  // Multiply %xmm1,element of A

7.  add    $0x1,%rax                 // register %rax = %rax + 1

8.  cmp    %eax,%edi                 // compare %eax to %edi

9.  vaddsd %xmm1,%xmm0,%xmm0         // Add %xmm1, %xmm0

10. jg     30 <dgemm+0x30>           // jump if %eax > %edi

11. add    $0x1,%r11                 // register %r11 = %r11 + 1

12. vmovsd %xmm0,(%r10)              // Store %xmm0 into C element
```

FIGURE 3.23   The x86 assembly language for the body of the nested loops generated by compiling the unoptimized C code in Figure 3.22. Although it is dealing with just 64 bits of data, the compiler uses the AVX version of the instructions instead of SSE2 presumably so that it can use three address per instruction instead of two (see the *Elaboration* in Section 3.7).

```
1.  //include <x86intrin.h>
2.  void dgemm (size_t n, double* A, double* B, double* C)
3.  {
4.    for ( size_t i = 0; i < n; i+=4 )
5.      for ( size_t j = 0; j < n; j++ ) {
6.        __m256d c0 = _mm256_load_pd(C+i+j*n); /* c0 = C[i][j] */
7.        for( size_t k = 0; k < n; k++ )
8.          c0 = _mm256_add_pd(c0, /* c0 += A[i][k]*B[k][j] */
9.                  _mm256_mul_pd(_mm256_load_pd(A+i+k*n),
10.                      _mm256_broadcast_sd(B+k+j*n)));
11.        _mm256_store_pd(C+i+j*n, c0); /* C[i][j] = c0 */
12.      }
13. }
```

FIGURE 3.24   Optimized C version of DGEMM using C intrinsics to generate the AVX subword-parallel instructions for the x86. Figure 3.25 shows the assembly language produced by the compiler for the inner loop.

in the name, which stands for scalar double precision. We'll define the subword parallel instructions shortly.

While compiler writers may eventually be able to produce high-quality code routinely that uses the AVX instructions of the x86, for now we must "cheat" by using C intrinsics that more or less tell the compiler exactly how to produce good code. Figure 3.24 shows the enhanced version of Figure 3.22 for which the Gnu C compiler produces AVX code. Figure 3.25 shows annotated x86 code that is the output of compiling using gcc with the –O3 level of optimization.

The declaration on line 6 of Figure 3.24 uses the __m256d data type, which tells the compiler the variable will hold four double-precision floating-point values. The intrinsic _mm256_load_pd() also on line 6 uses AVX instructions to load four double-precision floating-point numbers in parallel (_pd) from the matrix C into c0. The address calculation C+i+j*n on line 6 represents element C[i+j*n]. Symmetrically, the final step on line 11 uses the intrinsic _mm256_store_pd() to store four double-precision floating-point numbers from c0 into the matrix C. As we're going through four elements each iteration, the outer *for* loop on line 4 increments i by 4 instead of by 1 as on line 3 of Figure 3.22.

Inside the loops, on line 9 we first load four elements of A again using _mm256_load_pd(). To multiply these elements by one element of B, on line 10 we first use the intrinsic _mm256_broadcast_sd(), which makes four identical copies of the scalar double precision number—in this case an element of B—in one of the YMM

```
1.   vmovapd (%r11),%ymm0              // Load 4 elements of C into %ymm0
2.   mov     %rbx,%rcx                 // register %rcx = %rbx
3.   xor     %eax,%eax                 // register %eax = 0
4.   vbroadcastsd (%rax,%r8,1),%ymm1   // Make 4 copies of B element
5.   add     $0x8,%rax                 // register %rax = %rax + 8
6.   vmulpd (%rcx),%ymm1,%ymm1         // Parallel mul %ymm1,4 A elements
7.   add     %r9,%rcx                  // register %rcx = %rcx + %r9
8.   cmp     %r10,%rax                 // compare %r10 to %rax
9.   vaddpd %ymm1,%ymm0,%ymm0          // Parallel add %ymm1, %ymm0
10.  jne     50 <dgemm+0x50>           // jump if not %r10 != %rax
11.  add     $0x1,%esi                 // register % esi = % esi + 1
12.  vmovapd %ymm0,(%r11)              // Store %ymm0 into 4 C elements
```

FIGURE 3.25   The x86 assembly language for the body of the nested loops generated by compiling the optimized C code in Figure 3.24. Note the similarities to Figure 3.23, with the primary difference being that the five floating-point operations are now using YMM registers and using the pd versions of the instructions for parallel double precision instead of the sd version for scalar double precision.

registers. We then use _mm256_mul_pd() on line 9 to multiply the four double-precision results in parallel. Finally, _mm256_add_pd() on line 8 adds the four products to the four sums in c0.

Figure 3.25 shows resulting x86 code for the body of the inner loops produced by the compiler. You can see the five AVX instructions—they all start with v and four of the five use pd for parallel double precision—that correspond to the C intrinsics mentioned above. The code is very similar to that in Figure 3.23 above: both use 12 instructions, the integer instructions are nearly identical (but different registers), and the floating-point instruction differences are generally just going from *scalar double* (sd) using XMM registers to *parallel double* (pd) with YMM registers. The one exception is line 4 of Figure 3.25. Every element of A must be multiplied by one element of B. One solution is to place four identical copies of the 64-bit B element side-by-side into the 256-bit YMM register, which is just what the instruction vbroadcastsd does.

For matrices of dimensions of 32 by 32, the unoptimized DGEMM in Figure 3.22 runs at 1.7 GigaFLOPS (FLoating point Operations Per Second) on one core of a 2.6 GHz Intel Core i7 (Sandy Bridge). The optimized code in Figure 3.24 performs at 6.4 GigaFLOPS. The AVX version is 3.85 times as fast, which is very close to the factor of 4.0 increase that you might hope for from performing four times as many operations at a time by using **subword parallelism**.

PARALLELISM

Elaboration: As mentioned in the *Elaboration* in Section 1.6, Intel offers Turbo mode that temporarily runs at a higher clock rate until the chip gets too hot. This Intel Core i7 (Sandy Bridge) can increase from 2.6 GHz to 3.3 GHz in Turbo mode. The results above are with Turbo mode turned off. If we turn it on, we improve all the results by the increase in the clock rate of $3.3/2.6 = 1.27$ to 2.1 GFLOPS for unoptimized DGEMM and 8.1 GFLOPS with AVX. Turbo mode works particularly well when using only a single core of an eight-core chip, as in this case, as it lets that single core use much more than its fair share of power since the other cores are idle.

3.10   Fallacies and Pitfalls

Thus mathematics may be defined as the subject in which we never know what we are talking about, nor whether what we are saying is true.

Bertrand Russell, *Recent Words on the Principles of Mathematics*, 1901

Arithmetic fallacies and pitfalls generally stem from the difference between the limited precision of computer arithmetic and the unlimited precision of natural arithmetic.

Fallacy: Just as a left shift instruction can replace an integer multiply by a power of 2, a right shift is the same as an integer division by a power of 2.

Recall that a binary number x, where xi means the ith bit, represents the number

$$\ldots + (x^3 \times 2^3) + (x^2 \times 2^2) + (x^1 \times 2^1) + (x^0 \times 2^0)$$

Shifting the bits of c right by n bits would seem to be the same as dividing by 2^n. And this *is* true for unsigned integers. The problem is with signed integers. For example, suppose we want to divide -5_{ten} by 4_{ten}; the quotient should be -1_{ten}. The two's complement representation of -5_{ten} is

11111111 11111111 11111111 11111111 11111111 11111111 11111111 11111011$_{two}$

According to this fallacy, shifting right by two should divide by 4_{ten} (2^2):

00111111 11111111 11111111 11111111 11111111 11111111 11111111 11111110$_{two}$

With a 0 in the sign bit, this result is clearly wrong. The value created by the shift right is actually $4,611,686,018,427,387,902_{ten}$ instead of -1_{ten}.

A solution would be to have an arithmetic right shift that extends the sign bit instead of shifting in 0s. A 2-bit arithmetic shift right of -5_{ten} produces

11111111 11111111 11111111 11111111 11111111 11111111 11111111 11111110$_{two}$

The result is -2_{ten} instead of -1_{ten}; close, but no cigar.

Pitfall: Floating-point addition is not associative.

Associativity holds for a sequence of two's complement integer additions, even if the computation overflows. Alas, because floating-point numbers are approximations of real numbers and because computer arithmetic has limited precision, it does not hold for floating-point numbers. Given the great range of numbers that can be represented in floating point, problems occur when adding two large numbers of opposite signs plus a small number. For example, let's see if $c + (a + b) = (c + a) + b$. Assume $c = -1.5_{\text{ten}} \times 10^{38}$, $a = 1.5_{\text{ten}} \times 10^{38}$, and $b = 1.0$, and that these are all single precision numbers.

$$c + (a + b) = -1.5_{\text{ten}} \times 10^{38} + (1.5_{\text{ten}} \times 10^{38} + 1.0)$$
$$= -1.5_{\text{ten}} \times 10^{38} + (1.5_{\text{ten}} \times 10^{38})$$
$$= 0.0$$
$$c + (a + b) = (-1.5_{\text{ten}} \times 10^{38} + 1.5_{\text{ten}} \times 10^{38}) + 1.0$$
$$= (0.0_{\text{ten}}) + 1.0$$
$$= 1.0$$

Since floating-point numbers have limited precision and result in approximations of real results, $1.5_{\text{ten}} \times 10^{38}$ is so much larger than 1.0_{ten} that $1.5_{\text{ten}} \times 10^{38} + 1.0$ is still $1.5_{\text{ten}} \times 10^{38}$. That is why the sum of c, a, and b is 0.0 or 1.0, depending on the order of the floating-point additions, so $c + (a + b) \neq (c + a) + b$. Therefore, floating-point addition is *not* associative.

Fallacy: Parallel execution strategies that work for integer data types also work for floating-point data types.

Programs have typically been written first to run sequentially before being rewritten to run concurrently, so a natural question is, "Do the two versions get the same answer?" If the answer is no, you presume there is a bug in the parallel version that you need to track down.

This approach assumes that computer arithmetic does not affect the results when going from sequential to parallel. That is, if you were to add a million numbers together, you would get the same results whether you used one processor or 1000 processors. This assumption holds for two's complement integers, since integer addition is associative. Alas, since floating-point addition is not associative, the assumption does not hold.

A more vexing version of this fallacy occurs on a parallel computer where the operating system scheduler may use a different number of processors depending on what other programs are running on a parallel computer. As the varying number of processors from each run would cause the floating-point sums to be calculated in different orders, getting slightly different answers each time despite running identical code with identical input may flummox unaware parallel programmers.

Given this quandary, programmers who write parallel code with floating-point numbers need to verify whether the results are credible, even if they don't give the exact same answer as the sequential code. The field that deals with such issues is called numerical analysis, which is the subject of textbooks in its own right. Such concerns are one reason for the popularity of numerical libraries such as LAPACK and SCALAPAK, which have been validated in both their sequential and parallel forms.

Fallacy: Only theoretical mathematicians care about floating-point accuracy.

Newspaper headlines of November 1994 prove this statement is a fallacy (see Figure 3.26). The following is the inside story behind the headlines.

The Pentium uses a standard floating-point divide algorithm that generates multiple quotient bits per step, using the most significant bits of divisor and dividend to guess the next 2 bits of the quotient. The guess is taken from a lookup table containing -2, -1, 0, $+1$, or $+2$. The guess is multiplied by the divisor and subtracted from the remainder to generate a new remainder. Like nonrestoring division, if a previous guess gets too large a remainder, the partial remainder is adjusted in a subsequent pass.

Evidently, there were five elements of the table from the 80486 that Intel engineers thought could never be accessed, and they optimized the PLA to return 0 instead of 2 in these situations on the Pentium. Intel was wrong: while the first 11 bits were always correct, errors would show up occasionally in bits 12 to 52, or the 4th to 15th decimal digits.

FIGURE 3.26   A sampling of newspaper and magazine articles from November 1994, including the *New York Times, San Jose Mercury News, San Francisco Chronicle, and Infoworld.* The Pentium floating-point divide bug even made the "Top 10 List" of the *David Letterman Late Show* on television. Intel eventually took a $300 million write-off to replace the buggy chips.

A math professor at Lynchburg College in Virginia, Thomas Nicely, discovered the bug in September 1994. After calling Intel technical support and getting no official reaction, he posted his discovery on the Internet. This post led to a story in a trade magazine, which in turn caused Intel to issue a press release. It called the bug a glitch that would affect only theoretical mathematicians, with the average spreadsheet user seeing an error every 27,000 years. IBM Research soon counterclaimed that the average spreadsheet user would see an error every 24 days. Intel soon threw in the towel by making the following announcement on December 21:

> *We at Intel wish to sincerely apologize for our handling of the recently publicized Pentium processor flaw. The Intel Inside symbol means that your computer has a microprocessor second to none in quality and performance. Thousands of Intel employees work very hard to ensure that this is true. But no microprocessor is ever perfect. What Intel continues to believe is technically an extremely minor problem has taken on a life of its own. Although Intel firmly stands behind the quality of the current version of the Pentium processor, we recognize that many users have concerns. We want to resolve these concerns. Intel will exchange the current version of the Pentium processor for an updated version, in which this floating-point divide flaw is corrected, for any owner who requests it, free of charge anytime during the life of their computer.*

Analysts estimate that this recall cost Intel $500 million, and Intel engineers did not get a Christmas bonus that year.

This story brings up a few points for everyone to ponder. How much cheaper would it have been to fix the bug in July 1994? What was the cost to repair the damage to Intel's reputation? And what is the corporate responsibility in disclosing bugs in a product so widely used and relied upon as a microprocessor?

3.11 Concluding Remarks

Over the decades, computer arithmetic has become largely standardized, greatly enhancing the portability of programs. Two's complement binary integer arithmetic is found in every computer sold today, and if it includes floating point support, it offers the IEEE 754 binary floating-point arithmetic.

Computer arithmetic is distinguished from paper-and-pencil arithmetic by the constraints of limited precision. This limit may result in invalid operations through calculating numbers larger or smaller than the predefined limits. Such anomalies, called "overflow" or "underflow," may result in exceptions or interrupts, emergency events similar to unplanned subroutine calls. Chapters 4 and 5 discuss exceptions in more detail.

PARALLELISM

Floating-point arithmetic has the added challenge of being an approximation of real numbers, and care needs to be taken to ensure that the computer number selected is the representation closest to the actual number. The challenges of imprecision and limited representation of floating point are part of the inspiration for the field of numerical analysis. The switch to **parallelism** will shine the searchlight on numerical analysis again, as solutions that were long considered safe on sequential computers must be reconsidered when trying to find the fastest algorithm for parallel computers that still achieves a correct result.

Data-level parallelism, specifically subword parallelism, offers a simple path to higher performance for programs that are intensive in arithmetic operations for either integer or floating-point data. We showed that we could speed up matrix multiply nearly fourfold by using instructions that could execute four floating-point operations at a time.

With the explanation of computer arithmetic in this chapter comes a description of much more of the LEGv8 instruction set. One point of confusion is the instructions covered in these chapters versus instructions executed by LEGv8 chips versus the instructions accepted by LEGv8 assemblers. Two figures try to make this clear.

Figure 3.27 lists the LEGv8 instructions covered in this chapter and Chapter 2. We call the set of instructions on the left-hand side of the figure the *LEGv8 core*. The instructions on the right we call the *LEGv8 arithmetic core*.

Figure 3.28 gives the popularity of the LEGv8 instructions for SPEC CPU2006 integer and floating-point benchmarks. All instructions are listed that were responsible for at least 0.2% of the instructions executed.

Note that although programmers and compiler writers may use the full ARMv8 instruction set to have a richer menu of options, LEGv8 core instructions dominate integer SPEC CPU2006 execution, and the integer core plus arithmetic core dominate SPEC CPU2006 floating point, as the table below shows.

Instruction subset	Integer	Fl. pt.
LEGv8 core	98%	31%
LEGv8 arithmetic core	2%	66%
Remaining ARMv8	0%	3%

For the rest of the book, we concentrate on the LEGv8 core instructions—the integer instruction set excluding multiply and divide—to make the explanation of computer design easier. As you can see, the LEGv8 core includes the most popular LEGv8 instructions; be assured that understanding a computer that runs the LEGv8 core will give you sufficient background to understand even more ambitious computers. No matter what the instruction set or its size—ARMv8, ARMv7, MIPS, x86—never forget that bit patterns have no inherent meaning. The same bit pattern may represent a signed integer, unsigned integer, floating-point number, string, instruction, and so on. In stored-program computers, it is the operation on the bit pattern that determines its meaning.

LEGv8 core instructions	Name	Format	LEGv8 arithmetic core	Name	Format
add	ADD	R	multiply	MUL	R
subtract	SUB	R	signed multiply high	SMULH	R
add immediate	ADDI	I	unsigned multiply high	UMULH	R
subtract immediate	SUBI	I	signed divide	SDIV	R
add and set flags	ADDS	R	unsigned divide	UDIV	R
subtract and set flags	SUBS	R	floating-point add single	FADDS	R
add immediate and set flags	ADDIS	I	floating-point subtract single	FSUBS	R
subtract immediate and set flags	SUBIS	I	floating-point multiply single	FMULS	R
load register	LDUR	D	floating-point divide single	FDIVS	R
store register	STUR	D	floating-point add double	FADDD	R
load signed word	LDURSW	D	floating-point subtract double	FSUBD	R
store word	STURW	D	floating-point multiply double	FMULD	R
load half	LDURH	D	floating-point divide double	FDIVD	R
store half	STURH	D	floating-point compare single	FCMPS	R
load byte	LDURB	D	floating-point compare double	FCMPD	R
store byte	STURB	D	load single floating-point	LDURS	D
load exclusive register	LDXR	D	load double floating-point	LDURD	D
store exclusive register	STXR	D	store single floating-point	STURS	D
move wide with zero	MOVZ	IM	store double floating-point	STURD	D
move wide with keep	MOVK	IM			
and	AND	R			
inclusive or	ORR	R			
exclusive or	EOR	R			
and immediate	ANDI	I			
inclusive or immediate	ORRI	I			
exclusive or immediate	EORI	I			
logical shift left	LSL	R			
logical shift right	LSR	R			
compare and branch on equal 0	CBZ	CB			
compare and branch on not equal 0	CBNZ	CB			
branch conditionally	B.cond	CB			
branch	B	B			
branch to register	BR	R			
branch with link	BL	B			

FIGURE 3.27 The LEGv8 instruction set. This book concentrates on the instructions in the left column. This information is also found in columns 1 and 2 of the LEGv8 Reference Data Card at the front of this book.

Core ARMv8	Name	Integer	Fl. pt.	Arithmetic core + ARMv8	Name	Integer	Fl. pt.
add	ADD	5.2%	3.5%	FP add double	FADDD	0.0%	10.6%
subtract	SUB	2.2%	0.6%	FP subtract double	FSUBD	0.0%	4.9%
add immediate	ADDI	6.0%	4.8%	FP multiply double	FMULD	0.0%	15.0%
subtract immediate	SUBI	3.0%	2.4%	FP divide double	FDIVD	0.0%	0.2%
add and set flags	ADDS	1.3%	0.3%	FP add single	FADDS	0.0%	1.5%
subtract and set flags	SUBS	12.0%	2.8%	FP subtract single	FSUBS	0.0%	1.8%
add immediate and set flags	ADDIS	0.4%	0.0%	FP multiply single	FMULS	0.0%	2.4%
subtract immediate and set flags	SUBIS	3.8%	0.4%	FP divide single	FDIVS	0.0%	0.2%
load register	LDUR	18.6%	5.8%	load word to FP double	LDURD	0.0%	17.5%
store register	STUR	7.6%	2.0%	store word to FP double	STURD	0.0%	4.9%
load half	LDURH	1.3%	0.0%	load word to FP single	LDURS	0.0%	4.2%
store half	STURH	0.1%	0.0%	store word to FP single	STURS	0.0%	1.1%
load byte	LDURB	3.7%	0.1%	floating-point compare double	FCMPD	0.0%	0.6%
store byte	STURB	0.6%	0.0%	multiply	MUL	0.0%	0.2%
move wide with zero	MOVZ	3.0%	0.5%	shift right arithmetic	ASR	0.5%	0.3%
move wide with keep	MOVK	0.3%	0.1%				
and	AND	0.2%	0.1%				
inclusive or	ORR	4.0%	1.2%				
exclusive or	EOR	0.4%	0.2%				
and immediate	ANDI	0.7%	0.2%				
inclusive or immediate	ORRI	1.0%	0.2%				
logical shift left	LSL	4.4%	1.9%				
logical shift right	LSR	1.1%	0.5%				
compare and branch on equal 0	CBZ	0.9%	2.1%				
compare and branch on not equal 0	CBNZ	0.9%	1.3%				
branch conditionally	B.cond	15.3%	0.6%				
branch to register	BR	1.1%	0.2%				
branch with link	BL	0.7%	0.2%				

FIGURE 3.28   The frequency of the LEGv8 instructions for SPEC CPU2006 integer and floating point. All instructions that accounted for at least 0.2% of the instructions are included in the table. Pseudoinstructions are converted into LEGv8 before execution, and hence do not appear here.

Gresham's Law ("Bad money drives out Good") for computers would say, "The Fast drives out the Slow even if the Fast is wrong."

W. Kahan, *1992*

3.12   Historical Perspective and Further Reading

This section surveys the history of the floating point going back to von Neumann, including the surprisingly controversial IEEE standards effort, plus the rationale for the 80-bit stack architecture for floating point in the x86. See the rest of ▦ Section 3.12 online.

3.13 Exercises

3.1 [5] <§3.2> What is 5ED4 − 07A4 when these values represent unsigned 16-bit hexadecimal numbers? The result should be written in hexadecimal. Show your work.

3.2 [5] <§3.2> What is 5ED4 − 07A4 when these values represent signed 16-bit hexadecimal numbers stored in sign-magnitude format? The result should be written in hexadecimal. Show your work.

3.3 [10] <§3.2> Convert 5ED4 into a binary number. What makes base 16 (hexadecimal) an attractive numbering system for representing values in computers?

3.4 [5] <§3.2> What is 4365 − 3412 when these values represent unsigned 12-bit octal numbers? The result should be written in octal. Show your work.

3.5 [5] <§3.2> What is 4365 − 3412 when these values represent signed 12-bit octal numbers stored in sign-magnitude format? The result should be written in octal. Show your work.

3.6 [5] <§3.2> Assume 185 and 122 are unsigned 8-bit decimal integers. Calculate 185−122. Is there overflow, underflow, or neither?

3.7 [5] <§3.2> Assume 185 and 122 are signed 8-bit decimal integers stored in sign-magnitude format. Calculate 185 + 122. Is there overflow, underflow, or neither?

3.8 [5] <§3.2> Assume 185 and 122 are signed 8-bit decimal integers stored in sign-magnitude format. Calculate 185 − 122. Is there overflow, underflow, or neither?

3.9 [10] <§3.2> Assume 151 and 214 are signed 8-bit decimal integers stored in two's complement format. Calculate 151 + 214 using saturating arithmetic. The result should be written in decimal. Show your work.

3.10 [10] <§3.2> Assume 151 and 214 are signed 8-bit decimal integers stored in two's complement format. Calculate 151 − 214 using saturating arithmetic. The result should be written in decimal. Show your work.

3.11 [10] <§3.2> Assume 151 and 214 are unsigned 8-bit integers. Calculate 151+ 214 using saturating arithmetic. The result should be written in decimal. Show your work.

3.12 [20] <§3.3> Using a table similar to that shown in Figure 3.6, calculate the product of the octal unsigned 6-bit integers 62 and 12 using the hardware described in Figure 3.3. You should show the contents of each register on each step.

Never give in, never give in, never, never, never—in nothing, great or small, large or petty—never give in.

Winston Churchill, address at Harrow School, 1941

3.13 [20] <§3.3> Using a table similar to that shown in Figure 3.6, calculate the product of the hexadecimal unsigned 8-bit integers 62 and 12 using the hardware described in Figure 3.5. You should show the contents of each register on each step.

3.14 [10] <§3.3> Calculate the time necessary to perform a multiply using the approach given in Figures 3.3 and 3.4 if an integer is 8 bits wide and each step of the operation takes four time units. Assume that in step 1a an addition is always performed—either the multiplicand will be added, or a zero will be. Also assume that the registers have already been initialized (you are just counting how long it takes to do the multiplication loop itself). If this is being done in hardware, the shifts of the multiplicand and multiplier can be done simultaneously. If this is being done in software, they will have to be done one after the other. Solve for each case.

3.15 [10] <§3.3> Calculate the time necessary to perform a multiply using the approach described in the text (31 adders stacked vertically) if an integer is 8 bits wide and an adder takes four time units.

3.16 [20] <§3.3> Calculate the time necessary to perform a multiply using the approach given in Figure 3.7 if an integer is 8 bits wide and an adder takes four time units.

3.17 [20] <§3.3> As discussed in the text, one possible performance enhancement is to do a shift and add instead of an actual multiplication. Since 9×6, for example, can be written $(2 \times 2 \times 2 + 1) \times 6$, we can calculate 9×6 by shifting 6 to the left three times and then adding 6 to that result. Show the best way to calculate $0 \times 33 \times 0 \times 55$ using shifts and adds/subtracts. Assume both inputs are 8-bit unsigned integers.

3.18 [20] <§3.4> Using a table similar to that shown in Figure 3.10, calculate 74 divided by 21 using the hardware described in Figure 3.8. You should show the contents of each register on each step. Assume both inputs are unsigned 6-bit integers.

3.19 [30] <§3.4> Using a table similar to that shown in Figure 3.10, calculate 74 divided by 21 using the hardware described in Figure 3.11. You should show the contents of each register on each step. Assume A and B are unsigned 6-bit integers. This algorithm requires a slightly different approach than that shown in Figure 3.9. You will want to think hard about this, do an experiment or two, or else go to the web to figure out how to make this work correctly. (Hint: one possible solution involves using the fact that Figure 3.11 implies the remainder register can be shifted either direction.)

3.20 [5] <§3.5> What decimal number does the bit pattern $0 \times 0C000000$ represent if it is a two's complement integer? An unsigned integer?

3.21 [10] <§3.5> If the bit pattern $0 \times 0C000000$ is placed into the Instruction Register, what MIPS instruction will be executed?

3.22 [10] <§3.5> What decimal number does the bit pattern $0 \times 0C000000$ represent if it is a floating point number? Use the IEEE 754 standard.

3.23 [10] <§3.5> Write down the binary representation of the decimal number 63.25 assuming the IEEE 754 single precision format.

3.24 [10] <§3.5> Write down the binary representation of the decimal number 63.25 assuming the IEEE 754 double precision format.

3.25 [10] <§3.5> Write down the binary representation of the decimal number 63.25 assuming it was stored using the single precision IBM format (base 16, instead of base 2, with 7 bits of exponent).

3.26 [20] <§3.5> Write down the binary bit pattern to represent -1.5625×10^{-1} assuming a format similar to that employed by the DEC PDP-8 (the leftmost 12 bits are the exponent stored as a two's complement number, and the rightmost 24 bits are the fraction stored as a two's complement number). No hidden 1 is used. Comment on how the range and accuracy of this 36-bit pattern compares to the single and double precision IEEE 754 standards.

3.27 [20] <§3.5> IEEE 754-2008 contains a half precision that is only 16 bits wide. The leftmost bit is still the sign bit, the exponent is 5 bits wide and has a bias of 15, and the mantissa is 10 bits long. A hidden 1 is assumed. Write down the bit pattern to represent -1.5625×10^{-1} assuming a version of this format, which uses an excess-16 format to store the exponent. Comment on how the range and accuracy of this 16-bit floating point format compares to the single precision IEEE 754 standard.

3.28 [20] <§3.5> The Hewlett-Packard 2114, 2115, and 2116 used a format with the leftmost 16 bits being the fraction stored in two's complement format, followed by another 16-bit field which had the leftmost 8 bits as an extension of the fraction (making the fraction 24 bits long), and the rightmost 8 bits representing the exponent. However, in an interesting twist, the exponent was stored in sign-magnitude format with the sign bit on the far right! Write down the bit pattern to represent -1.5625×10^{-1} assuming this format. No hidden 1 is used. Comment on how the range and accuracy of this 32-bit pattern compares to the single precision IEEE 754 standard.

3.29 [20] <§3.5> Calculate the sum of 2.6125×10^{1} and $4.150390625 \times 10^{-1}$ by hand, assuming A and B are stored in the 16-bit half precision described in Exercise 3.27. Assume 1 guard, 1 round bit, and 1 sticky bit, and round to the nearest even. Show all the steps.

3.30 [30] <§3.5> Calculate the product of -8.0546875×10^{0} and $-1.79931640625 \times 10^{-1}$ by hand, assuming A and B are stored in the 16-bit half precision format described in Exercise 3.27. Assume 1 guard, 1 round bit, and 1 sticky bit, and round to the nearest even. Show all the steps; however, as is done in the example in the text, you can do the multiplication in human-readable format instead of using the techniques described in Exercises 3.12 through 3.14. Indicate if there is overflow or underflow. Write your answer in both the 16-bit floating point format described in Exercise 3.27 and also as a decimal number. How accurate is your result? How does it compare to the number you get if you do the multiplication on a calculator?

3.31 [30] <§3.5> Calculate by hand 8.625×10^1 divided by -4.875×10^0. Show all the steps necessary to achieve your answer. Assume there is a guard, a round bit, and a sticky bit, and use them if necessary. Write the final answer in both the 16-bit floating point format described in Exercise 3.27 and in decimal and compare the decimal result to that which you get if you use a calculator.

3.32 [20] <§3.10> Calculate $(3.984375 \times 10^{-1} + 3.4375 \times 10^{-1}) + 1.771 \times 10^3$ by hand, assuming each of the values is stored in the 16-bit half precision format described in Exercise 3.27 (and also described in the text). Assume 1 guard, 1 round bit, and 1 sticky bit, and round to the nearest even. Show all the steps, and write your answer in both the 16-bit floating point format and in decimal.

3.33 [20] <§3.10> Calculate $3.984375 \times 10^{-1} + (3.4375 \times 10^{-1} + 1.771 \times 10^3)$ by hand, assuming each of the values is stored in the 16-bit half precision format described in Exercise 3.27 (and also described in the text). Assume 1 guard, 1 round bit, and 1 sticky bit, and round to the nearest even. Show all the steps, and write your answer in both the 16-bit floating point format and in decimal.

3.34 [10] <§3.10> Based on your answers to Exercises 3.32 and 3.33, does $(3.984375 \times 10^{-1} + 3.4375 \times 10^{-1}) + 1.771 \times 10^3 = 3.984375 \times 10^{-1} + (3.4375 \times 10^{-1} + 1.771 \times 10^3)$?

3.35 [30] <§3.10> Calculate $(3.41796875 \times 10^{-3} \times 6.34765625 \times 10^{-3}) \times 1.05625 \times 10^2$ by hand, assuming each of the values is stored in the 16-bit half precision format described in Exercise 3.27 (and also described in the text). Assume 1 guard, 1 round bit, and 1 sticky bit, and round to the nearest even. Show all the steps, and write your answer in both the 16-bit floating point format and in decimal.

3.36 [30] <§3.10> Calculate $3.41796875 \times 10^{-3} \times (6.34765625 \times 10^{-3} \times 1.05625 \times 10^2)$ by hand, assuming each of the values is stored in the 16-bit half precision format described in Exercise 3.27 (and also described in the text). Assume 1 guard, 1 round bit, and 1 sticky bit, and round to the nearest even. Show all the steps, and write your answer in both the 16-bit floating point format and in decimal.

3.37 [10] <§3.10> Based on your answers to Exercises 3.35 and 3.36, does $(3.41796875 \times 10^{-3} \times 6.34765625 \times 10^{-3}) \times 1.05625 \times 10^2 = 3.41796875 \times 10^{-3} \times (6.34765625 \times 10^{-3} \times 1.05625 \times 10^2)$?

3.38 [30] <§3.10> Calculate $1.666015625 \times 10^0 \times (1.9760 \times 10^4 + -1.9744 \times 10^4)$ by hand, assuming each of the values is stored in the 16-bit half precision format described in Exercise 3.27 (and also described in the text). Assume 1 guard, 1 round bit, and 1 sticky bit, and round to the nearest even. Show all the steps, and write your answer in both the 16-bit floating point format and in decimal.

3.39 [30] <§3.10> Calculate $(1.666015625 \times 10^0 \times 1.9760 \times 10^4) + (1.666015625 \times 10^0 \times -1.9744 \times 10^4)$ by hand, assuming each of the values is stored in the 16-bit half precision format described in Exercise 3.27 (and also described in the text). Assume 1 guard, 1 round bit, and 1 sticky bit, and round to the nearest even. Show all the steps, and write your answer in both the 16-bit floating point format and in decimal.

3.40 [10] <§3.10> Based on your answers to Exercises 3.38 and 3.39, does $(1.666015625 \times 10^0 \times 1.9760 \times 10^4) + (1.666015625 \times 10^0 \times -1.9744 \times 10^4) = 1.666015625 \times 10^0 \times (1.9760 \times 10^4 + -1.9744 \times 10^4)$?

3.41 [10] <§3.5> Using the IEEE 754 floating point format, write down the bit pattern that would represent $-1/4$. Can you represent $-1/4$ exactly?

3.42 [10] <§3.5> What do you get if you add $-1/4$ to itself four times? What is $-1/4 \times 4$? Are they the same? What should they be?

3.43 [10] <§3.5> Write down the bit pattern in the fraction of value 1/3 assuming a floating point format that uses binary numbers in the fraction. Assume there are 24 bits, and you do not need to normalize. Is this representation exact?

3.44 [10] <§3.5> Write down the bit pattern in the fraction of value 1/3 assuming a floating point format that uses Binary Coded Decimal (base 10) numbers in the fraction instead of base 2. Assume there are 24 bits, and you do not need to normalize. Is this representation exact?

3.45 [10] <§3.5> Write down the bit pattern assuming that we are using base 15 numbers in the fraction of value 1/3 instead of base 2. (Base 16 numbers use the symbols 0–9 and A–F. Base 15 numbers would use 0–9 and A–E.) Assume there are 24 bits, and you do not need to normalize. Is this representation exact?

3.46 [20] <§3.5> Write down the bit pattern assuming that we are using base 30 numbers in the fraction of value 1/3 instead of base 2. (Base 16 numbers use the symbols 0–9 and A–F. Base 30 numbers would use 0–9 and A–T.) Assume there are 20 bits, and you do not need to normalize. Is this representation exact?

3.47 [45] <§§3.6, 3.7> The following C code implements a four-tap FIR filter on input array sig_in. Assume that all arrays are 16-bit fixed-point values.

```
for (i = 3;i< 128;i+ +)
sig_out[i] = sig_in[i - 3] * f[0] + sig_in[i - 2] * f[1]
  + sig_in[i - 1] * f[2] + sig_in[i] * f[3];
```

Assume you are to write an optimized implementation of this code in assembly language on a processor that has SIMD instructions and 128-bit registers. Without knowing the details of the instruction set, briefly describe how you would implement this code, maximizing the use of sub-word operations and minimizing the amount of data that is transferred between registers and memory. State all your assumptions about the instructions you use.

§3.2, page 191: 2.
§3.5, page 229: 3.

**Answers to
Check Yourself**

4

The Processor

In a major matter, no details are small.

French Proverb

The Five Classic Components of a Computer

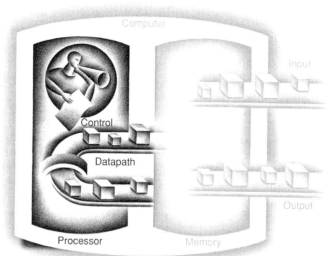

4.1 Introduction

Chapter 1 explains that the performance of a computer is determined by three key factors: instruction count, clock cycle time, and *clock cycles per instruction* (CPI). Chapter 2 explains that the compiler and the instruction set architecture determine the instruction count required for a given program. However, the implementation of the processor determines both the clock cycle time and the number of clock cycles per instruction. In this chapter, we construct the datapath and control unit for two different implementations of the LEGv8 instruction set.

This chapter contains an explanation of the principles and techniques used in implementing a processor, starting with a highly abstract and simplified overview in this section. It is followed by a section that builds up a datapath and constructs a simple version of a processor sufficient to implement an instruction set like LEGv8. The bulk of the chapter covers a more realistic **pipelined** LEGv8 implementation, followed by a section that develops the concepts necessary to implement more complex instruction sets, like the x86.

PIPELINING

For the reader interested in understanding the high-level interpretation of instructions and its impact on program performance, this initial section and Section 4.5 present the basic concepts of pipelining. Current trends are covered in Section 4.10, and Section 4.11 describes the recent Intel Core i7 and ARM Cortex-A53 architectures. Section 4.12 shows how to use instruction-level parallelism to more than double the performance of the matrix multiply from Section 3.9. These sections provide enough background to understand the pipeline concepts at a high level.

For the reader interested in understanding the processor and its performance in more depth, Sections 4.3, 4.4, and 4.6 will be useful. Those interested in learning how to build a processor should also cover Sections 4.2, 4.7, 4.8, and 4.9. For readers with an interest in modern hardware design, 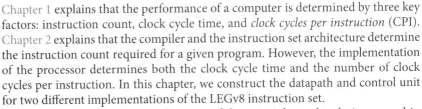 Section 4.13 describes how hardware design languages and CAD tools are used to implement hardware, and then how to use a hardware design language to describe a pipelined implementation. It also gives several more illustrations of how pipelining hardware executes. (The hardware models in this chapter have been sourced by the authors and do not imply ARM-endorsed architectures.)

A Basic LEGv8 Implementation

We will be examining an implementation that includes a subset of the core LEGv8 instruction set:

- The memory-reference instructions *load register unscaled* (LDUR) and *store register unscaled* (STUR)

- The arithmetic-logical instructions ADD, SUB, AND, and ORR

- The instructions *compare and branch on zero* (CBZ) and *branch* (B), which we add last

This subset does not include all the integer instructions (for example, shift, multiply, and divide are missing), nor does it include any floating-point instructions. However, it illustrates the key principles used in creating a datapath and designing the control. The implementation of the remaining instructions is similar.

In examining the implementation, we will have the opportunity to see how the instruction set architecture determines many aspects of the implementation, and how the choice of various implementation strategies affects the clock rate and CPI for the computer. Many of the key design principles introduced in Chapter 2 can be illustrated by looking at the implementation, such as *Simplicity favors regularity*. In addition, most concepts used to implement the LEGv8 subset in this chapter are the same basic ideas that are used to construct a broad spectrum of computers, from high-performance servers to general-purpose microprocessors to embedded processors.

An Overview of the Implementation

In Chapter 2, we looked at the core LEGv8 instructions, including the integer arithmetic-logical instructions, the memory-reference instructions, and the branch instructions. Much of what needs to be done to implement these instructions is the same, independent of the exact class of instruction. For every instruction, the first two steps are identical:

1. Send the *program counter* (PC) to the memory that contains the code and fetch the instruction from that memory.

2. Read one or two registers, using fields of the instruction to select the registers to read. For the LDUR and CBZ instructions, we need to read only one register, but most other instructions require reading two registers.

After these two steps, the actions required to complete the instruction depend on the instruction class. Fortunately, for each of the three instruction classes (memory-reference, arithmetic-logical, and branches), the actions are largely the same, independent of the exact instruction. The simplicity and regularity of the LEGv8 instruction set simplify the implementation by making the execution of many of the instruction classes similar.

For example, all instruction classes, except unconditional branch, use the arithmetic-logical unit (ALU) after reading the registers. The memory-reference instructions use the ALU for an address calculation, the arithmetic-logical instructions for the operation execution, and conditional branches for comparison to zero. After using the ALU, the actions required to complete various instruction classes differ. A memory-reference instruction will need to access the memory either to read data for a load or write data for a store. An arithmetic-logical or load instruction must write the data from the ALU or memory back into a register. Lastly, for a conditional branch instruction, we may need to change the next instruction address based on the comparison; otherwise, the PC should be incremented by four to get the address of the subsequent instruction.

Figure 4.1 shows the high-level view of an LEGv8 implementation, focusing on the various functional units and their interconnection. Although this figure shows most of the flow of data through the processor, it omits two important aspects of instruction execution.

First, in several places, Figure 4.1 shows data going to a particular unit as coming from two different sources. For example, the value written into the PC can come from one of two adders, the data written into the register file can come from either the ALU or the data memory, and the second input to the ALU can come from a register or the immediate field of the instruction. In practice, these data lines cannot simply be wired together; we must add a logic element that chooses from among the multiple sources and steers one of those sources to its destination. This selection is commonly done with a device called a *multiplexor*, although this device might better be called a *data selector*. Appendix A describes the multiplexor, which selects from among several inputs based on the setting of its control lines. The control lines are set based primarily on information taken from the instruction being executed.

The second omission in Figure 4.1 is that several of the units must be controlled depending on the type of instruction. For example, the data memory must read

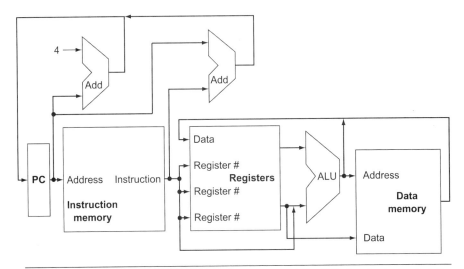

FIGURE 4.1   An abstract view of the implementation of the LEGv8 subset showing the major functional units and the major connections between them. All instructions start by using the program counter to supply the instruction address to the instruction memory. After the instruction is fetched, the register operands used by an instruction are specified by fields of that instruction. Once the register operands have been fetched, they can be operated on to compute a memory address (for a load or store), to compute an arithmetic result (for an integer arithmetic-logical instruction), or a compare to zero (for a branch). If the instruction is an arithmetic-logical instruction, the result from the ALU must be written to a register. If the operation is a load or store, the ALU result is used as an address to either store a value from the registers or load a value from memory into the registers. The result from the ALU or memory is written back into the register file. Branches require the use of the ALU output to determine the next instruction address, which comes either from the ALU (where the PC and branch offset are summed) or from an adder that increments the current PC by four. The thick lines interconnecting the functional units represent buses, which consist of multiple signals. The arrows are used to guide the reader in knowing how information flows. Since signal lines may cross, we explicitly show when crossing lines are connected by the presence of a dot where the lines cross.

on a load and write on a store. The register file must be written only on a load or an arithmetic-logical instruction. And, of course, the ALU must perform one of several operations. (Appendix A describes the detailed design of the ALU.) Like the multiplexors, control lines that are set based on various fields in the instruction direct these operations.

Figure 4.2 shows the datapath of Figure 4.1 with the three required multiplexors added, as well as control lines for the major functional units. A *control unit*, which has the instruction as an input, is used to determine how to set the control lines for the functional units and two of the multiplexors. The top multiplexor,

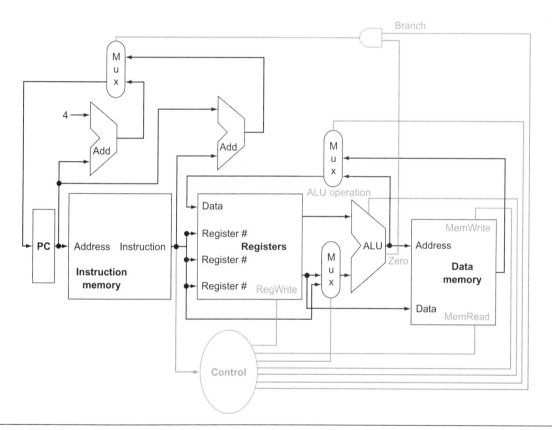

FIGURE 4.2 The basic implementation of the LEGv8 subset, including the necessary multiplexors and control lines. The top multiplexor ("Mux") controls what value replaces the PC (PC + 4 or the branch destination address); the multiplexor is controlled by the gate that "ANDs" together the Zero output of the ALU and a control signal that indicates that the instruction is a branch. The middle multiplexor, whose output returns to the register file, is used to steer the output of the ALU (in the case of an arithmetic-logical instruction) or the output of the data memory (in the case of a load) for writing into the register file. Finally, the bottom-most multiplexor is used to determine whether the second ALU input is from the registers (for an arithmetic-logical instruction or a branch) or from the offset field of the instruction (for a load or store). The added control lines are straightforward and determine the operation performed at the ALU, whether the data memory should read or write, and whether the registers should perform a write operation. The control lines are shown in color to make them easier to see.

which determines whether PC + 4 or the branch destination address is written into the PC, is set based on the Zero output of the ALU, which is used to perform the comparison of a `CBZ` instruction. The regularity and simplicity of the LEGv8 instruction set mean that a simple decoding process can be used to determine how to set the control lines.

In the remainder of the chapter, we refine this view to fill in the details, which requires that we add further functional units, increase the number of connections between units, and, of course, enhance a control unit to control what actions are taken for different instruction classes. Sections 4.3 and 4.4 describe a simple implementation that uses a single long clock cycle for every instruction and follows the general form of Figures 4.1 and 4.2. In this first design, every instruction begins execution on one clock edge and completes execution on the next clock edge.

While easier to understand, this approach is not practical, since the clock cycle must be severely stretched to accommodate the longest instruction. After designing the control for this simple computer, we will look at pipelined implementation with all its complexities, including exceptions.

Check Yourself How many of the five classic components of a computer—shown on page 255—do Figures 4.1 and 4.2 include?

4.2 Logic Design Conventions

To discuss the design of a computer, we must decide how the hardware logic implementing the computer will operate and how the computer is clocked. This section reviews a few key ideas in digital logic that we will use extensively in this chapter. If you have little or no background in digital logic, you will find it helpful to read Appendix A before continuing.

combinational element An operational element, such as an AND gate or an ALU

The datapath elements in the LEGv8 implementation consist of two different types of logic elements: elements that operate on data values and elements that contain state. The elements that operate on data values are all combinational, which means that their outputs depend only on the current inputs. Given the same input, a combinational element always produces the same output. The ALU shown in Figure 4.1 and discussed in Appendix A is an example of a combinational element. Given a set of inputs, it always produces the same output because it has no internal storage.

state element A memory element, such as a register or a memory.

Other elements in the design are not combinational, but instead contain *state*. An element contains state if it has some internal storage. We call these elements state elements because, if we pulled the power plug on the computer, we could restart it accurately by loading the state elements with the values they contained before we pulled the plug. Furthermore, if we saved and restored the state elements, it would be as if the computer had never lost power. Thus, these state elements completely characterize the computer. In Figure 4.1, the instruction and data memories, as well as the registers, are all examples of state elements.

A state element has at least two inputs and one output. The required inputs are the data value to be written into the element and the clock, which determines when the data value is written. The output from a state element provides the value that was written in an earlier clock cycle. For example, one of the logically simplest state elements is a D-type flip-flop (see Appendix A), which has exactly these two inputs (a value and a clock) and one output. In addition to flip-flops, our LEGv8 implementation uses two other types of state elements: memories and registers, both of which appear in Figure 4.1. The clock is used to determine when the state element should be written; a state element can be read at any time.

Logic components that contain state are also called *sequential*, because their outputs depend on both their inputs and the contents of the internal state. For example, the output from the functional unit representing the registers depends both on the register numbers supplied and on what was written into the registers previously. Appendix A discusses the operation of both the combinational and sequential elements and their construction in more detail.

Clocking Methodology

A clocking methodology defines when signals can be read and when they can be written. It is important to specify the timing of reads and writes, because if a signal is written at the same time that it is read, the value of the read could correspond to the old value, the newly written value, or even some mix of the two! Computer designs cannot tolerate such unpredictability. A clocking methodology is designed to make hardware predictable.

For simplicity, we will assume an edge-triggered clocking methodology. An edge-triggered clocking methodology means that any values stored in a sequential logic element are updated only on a clock edge, which is a quick transition from low to high or vice versa (see Figure 4.3). Because only state elements can store a data value, any collection of combinational logic must have its inputs come from a set of state elements and its outputs written into a set of state elements. The inputs are values that were written in a previous clock cycle, while the outputs are values that can be used in a following clock cycle.

clocking methodology The approach used to determine when data are valid and stable relative to the clock.

edge-triggered clocking A clocking scheme in which all state changes occur on a clock edge.

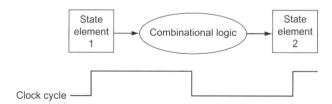

FIGURE 4.3   Combinational logic, state elements, and the clock are closely related. In a synchronous digital system, the clock determines when elements with state will write values into internal storage. Any inputs to a state element must reach a stable value (that is, have reached a value from which they will not change until after the clock edge) before the active clock edge causes the state to be updated. All state elements in this chapter, including memory, are assumed positive edge-triggered; that is, they change on the rising clock edge.

Figure 4.3 shows the two state elements surrounding a block of combinational logic, which operates in a single clock cycle: all signals must propagate from state element 1, through the combinational logic, and to state element 2 in the time of one clock cycle. The time necessary for the signals to reach state element 2 defines the length of the clock cycle.

For simplicity, we do not show a write control signal when a state element is written on every active clock edge. In contrast, if a state element is not updated on every clock, then an explicit write control signal is required. Both the clock signal and the write control signal are inputs, and the state element is changed only when the write control signal is asserted and a clock edge occurs.

We will use the word asserted to indicate a signal that is logically high and *assert* to specify that a signal should be driven logically high, and *deassert* or deasserted to represent logically low. We use the terms assert and deassert because when we implement hardware, at times 1 represents logically high and at times it can represent logically low.

An edge-triggered methodology allows us to read the contents of a register, send the value through some combinational logic, and write that register in the same clock cycle. Figure 4.4 gives a generic example. It doesn't matter whether we assume that all writes take place on the rising clock edge (from low to high) or on the falling clock edge (from high to low), since the inputs to the combinational logic block cannot change except on the chosen clock edge. In this book, we use the rising clock edge. With an edge-triggered timing methodology, there is *no* feedback within a single clock cycle, and the logic in Figure 4.4 works correctly. In Appendix A, we briefly discuss additional timing constraints (such as setup and hold times) as well as other timing methodologies.

For the 64-bit LEGv8 architecture, nearly all of these state and logic elements will have inputs and outputs that are 64 bits wide, since that is the width of most of the data handled by the processor. We will make it clear whenever a unit has an input or output that is other than 64 bits in width. The figures will indicate *buses*, which are signals wider than 1 bit, with thicker lines. At times, we will want to combine several buses to form a wider bus; for example, we may want to obtain a 64-bit bus by combining two 32-bit buses. In such cases, labels on the bus lines

control signal A signal used for multiplexor selection or for directing the operation of a functional unit; contrasts with a *data signal*, which contains information that is operated on by a functional unit.

asserted The signal is logically high or true.

deasserted The signal is logically low or false.

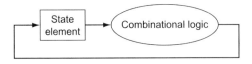

FIGURE 4.4   An edge-triggered methodology allows a state element to be read and written in the same clock cycle without creating a race that could lead to indeterminate data values. Of course, the clock cycle still must be long enough so that the input values are stable when the active clock edge occurs. Feedback cannot occur within one clock cycle because of the edge-triggered update of the state element. If feedback were possible, this design could not work properly. Our designs in this chapter and the next rely on the edge-triggered timing methodology and on structures like the one shown in this figure.

will make it clear that we are concatenating buses to form a wider bus. Arrows are also added to help clarify the direction of the flow of data between elements. Finally, color indicates a control signal contrary to a signal that carries data; this distinction will become clearer as we proceed through this chapter.

True or false: Because the register file is both read and written on the same clock cycle, any LEGv8 datapath using edge-triggered writes must have more than one copy of the register file.

Check Yourself

Elaboration: There is also a 32-bit version of the ARMv8 architecture, and, naturally enough, most paths in its implementation would be 32 bits wide.

4.3   Building a Datapath

A reasonable way to start a datapath design is to examine the major components required to execute each class of LEGv8 instructions. Let's start at the top by looking at which datapath elements each instruction needs, and then work our way down through the levels of **abstraction**. When we show the datapath elements, we will also show their control signals. We use abstraction in this explanation, starting from the bottom up.

Figure 4.5a shows the first element we need: a memory unit to store the instructions of a program and supply instructions given an address. Figure 4.5b also shows the program counter (PC), which as we saw in Chapter 2 is a register that holds the address of the current instruction. Lastly, we will need an adder to increment the PC to the address of the next instruction. This adder, which is combinational, can be built from the ALU described in detail in Appendix A simply by wiring the control lines so that the control always specifies an add operation. We will draw such an ALU with the label *Add*, as in Figure 4.5, to indicate that it has been permanently made an adder and cannot perform the other ALU functions.

To execute any instruction, we must start by fetching the instruction from memory. To prepare for executing the next instruction, we must also increment the program counter so that it points at the next instruction, 4 bytes later. Figure 4.6 shows how to combine the three elements from Figure 4.5 to form a datapath that fetches instructions and increments the PC to obtain the address of the next sequential instruction.

Now let's consider the R-format instructions (see Figure 2.21 on page 124). They all read two registers, perform an ALU operation on the contents of the registers, and write the result to a register. We call these instructions either *R-type instructions* or *arithmetic-logical instructions* (since they perform arithmetic or logical operations). This instruction class includes ADD, SUB, AND, and ORR, which

ABSTRACTION

datapath element A unit used to operate on or hold data within a processor. In the LEGv8 implementation, the datapath elements include the instruction and data memories, the register file, the ALU, and adders.

program counter (PC) The register containing the address of the instruction in the program being executed.

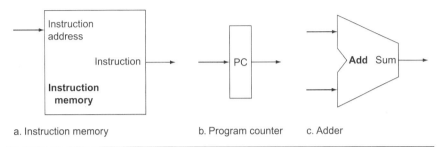

a. Instruction memory   b. Program counter   c. Adder

FIGURE 4.5 Two state elements are needed to store and access instructions, and an adder is needed to compute the next instruction address. The state elements are the instruction memory and the program counter. The instruction memory need only provide read access because the datapath does not write instructions. Since the instruction memory only reads, we treat it as combinational logic: the output at any time reflects the contents of the location specified by the address input, and no read control signal is needed. (We will need to write the instruction memory when we load the program; this is not hard to add, and we ignore it for simplicity.) The program counter is a 64-bit register that is written at the end of every clock cycle and thus does not need a write control signal. The adder is an ALU wired to always add its two 64-bit inputs and place the sum on its output.

were introduced in Chapter 2. Recall that a typical instance of such an instruction is ADD X1, X2, X3, which reads X2 and X3 and writes the sum into X1.

The processor's 32 general-purpose registers are stored in a structure called a register file. A register file is a collection of registers in which any register can be read or written by specifying the number of the register in the file. The register file contains the register state of the computer. In addition, we will need an ALU to operate on the values read from the registers.

R-format instructions have three register operands, so we will need to read two data words from the register file and write one data word into the register file for each instruction. For each data word to be read from the registers, we need an input to the register file that specifies the *register number* to be read and an output from the register file that will carry the value that has been read from the registers. To write a data word, we will need two inputs: one to specify the register number to be written and one to supply the *data* to be written into the register. The register file always outputs the contents of whatever register numbers are on the Read register inputs. Writes, however, are controlled by the write control signal, which must be asserted for a write to occur at the clock edge. Figure 4.7a shows the result; we need a total of four inputs (three for register numbers and one for data) and two outputs (both for data). The register number inputs are 5 bits wide to specify one of 32 registers ($32 = 2^5$), whereas the data input and two data output buses are each 64 bits wide.

Figure 4.7b shows the ALU, which takes two 64-bit inputs and produces a 64-bit result, as well as a 1-bit signal if the result is 0. The 4-bit control signal of the ALU is described in detail in Appendix A; we will review the ALU control shortly when we need to know how to set it.

register file A state element that consists of a set of registers that can be read and written by supplying a register number to be accessed.

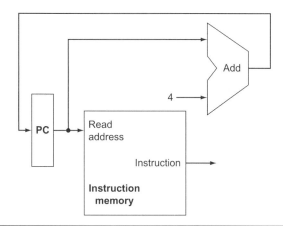

FIGURE 4.6 A portion of the datapath used for fetching instructions and incrementing the program counter. The fetched instruction is used by other parts of the datapath.

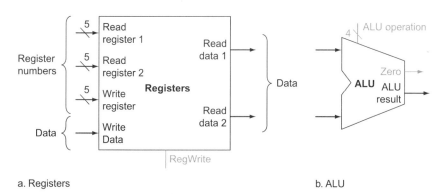

a. Registers

b. ALU

FIGURE 4.7 The two elements needed to implement R-format ALU operations are the register file and the ALU. The register file contains all the registers and has two read ports and one write port. The design of multiported register files is discussed in Section A.8 of Appendix A. The register file always outputs the contents of the registers corresponding to the Read register inputs on the outputs; no other control inputs are needed. In contrast, a register write must be explicitly indicated by asserting the write control signal. Remember that writes are edge-triggered, so that all the write inputs (i.e., the value to be written, the register number, and the write control signal) must be valid at the clock edge. Since writes to the register file are edge-triggered, our design can legally read and write the same register within a clock cycle: the read will get the value written in an earlier clock cycle, while the value written will be available to a read in a subsequent clock cycle. The inputs carrying the register number to the register file are all 5 bits wide, whereas the lines carrying data values are 64 bits wide. The operation to be performed by the ALU is controlled with the ALU operation signal, which will be 4 bits wide, using the ALU designed in Appendix A. We will use the Zero detection output of the ALU shortly to implement conditional branches.

Next, consider the LEGv8 load register and store register instructions, which have the general form `LDUR X1,[X2,offset_value]` or `STUR X1, [X2,offset_value]`. These instructions compute a memory address by adding the base register, which is X2, to the 9-bit signed offset field contained in the instruction. If the instruction is a store, the value to be stored must also be read from the register file where it resides in X1. If the instruction is a load, the value read from memory must be written into the register file in the specified register, which is X1. Thus, we will need both the register file and the ALU from Figure 4.7.

sign-extend To increase the size of a data item by replicating the high-order sign bit of the original data item in the high-order bits of the larger, destination data item.

In addition, we will need a unit to **sign-extend** the 9-bit offset field in the instruction to a 64-bit signed value, and a data memory unit to read from or write to. The data memory must be written on store instructions; hence, data memory has read and write control signals, an address input, and an input for the data to be written into memory. Figure 4.8 shows these two elements.

The CBZ instruction has two operands, a register that is tested for zero, and a 19-bit offset used to compute the **branch target address** relative to the branch instruction address. Its form is `CBZ X1,offset`. To implement this instruction, we must compute the branch target address by adding the sign-extended offset field of the instruction to the PC. There are two details in the definition of branch instructions (see Chapter 2) to which we must pay attention:

branch target address The address specified in a branch, which becomes the new program counter (PC) if the branch is taken. In the LEGv8 architecture, the branch target is given by the sum of the offset field of the instruction and the address of the branch.

- The instruction set architecture specifies that the base for the branch address calculation is the address of the branch instruction.

- The architecture also states that the offset field is shifted left 2 bits so that it is a word offset; this shift increases the effective range of the offset field by a factor of 4.

branch taken A branch where the branch condition is satisfied and the program counter (PC) becomes the branch target. All unconditional branches are taken branches.

To deal with the latter complication, we will need to shift the offset field by 2.

As well as computing the branch target address, we must also determine whether the next instruction is the instruction that follows sequentially or the instruction at the branch target address. When the condition is true (i.e., the operand is zero), the branch target address becomes the new PC, and we say that the branch is taken. If the operand is not zero, the incremented PC should replace the current PC (just as for any other normal instruction); in this case, we say that the branch is not taken.

branch not taken or (untaken branch) A branch where the branch condition is false and the program counter (PC) becomes the address of the instruction that sequentially follows the branch.

Thus, the branch datapath must do two operations: compute the branch target address and test the register contents. (Branches also affect the instruction fetch portion of the datapath, as we will deal with shortly.) Figure 4.9 shows the structure of the datapath segment that handles branches. To compute the branch target address, the branch datapath includes a sign extension unit, from Figure 4.8 and an adder. To perform the compare, we need to use the register file shown in Figure 4.7a to supply the register operand (although we will not need to write into the register file). In addition, the comparison can be done using the ALU we designed in Appendix A. Since that ALU provides an output signal that indicates whether

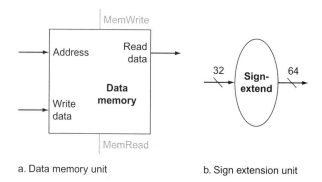

a. Data memory unit   b. Sign extension unit

FIGURE 4.8 The two units needed to implement loads and stores, in addition to the register file and ALU of Figure 4.7, are the data memory unit and the sign extension unit. The memory unit is a state element with inputs for the address and the write data, and a single output for the read result. There are separate read and write controls, although only one of these may be asserted on any given clock. The memory unit needs a read signal, since, unlike the register file, reading the value of an invalid address can cause problems, as we will see in Chapter 5. The sign extension unit has a 32-bit instruction as input that selects a 9-bit for load and store or a 19-bit field for compare and branch on zero that is sign-extended into a 64-bit result appearing on the output (see Chapter 2). We assume the data memory is edge-triggered for writes. Standard memory chips actually have a write enable signal that is used for writes. Although the write enable is not edge-triggered, our edge-triggered design could easily be adapted to work with real memory chips. See Section A.8 of Appendix A for further discussion of how real memory chips work.

the result was 0, we can send the register operand to the ALU with the control set to pass the register value. If the Zero signal out of the ALU unit is asserted, we know that the register value is zero. Although the Zero output always signals if the result is 0, we will be using it only to implement the compare test of conditional branches. Later, we will show exactly how to connect the control signals of the ALU for use in the datapath.

The branch instruction operates by adding the PC with the lower 26 bits of the instruction shifted left by 2 bits. Simply concatenating 00 to the branch offset accomplishes this shift, as described in Chapter 2.

Creating a Single Datapath

Now that we have examined the datapath components needed for the individual instruction classes, we can combine them into a single datapath and add the control to complete the implementation. This simplest datapath will attempt to execute all instructions in one clock cycle. This design means that no datapath resource can be used more than once per instruction, so any element needed more than once must be duplicated. We therefore need a memory for instructions separate from one for data. Although some of the functional units will need to be duplicated, many of the elements can be shared by different instruction flows.

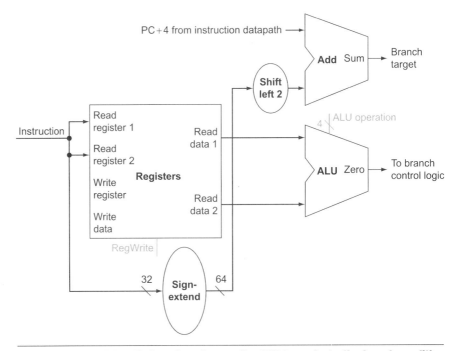

FIGURE 4.9 The datapath for a branch uses the ALU to evaluate the branch condition and a separate adder to compute the branch target as the sum of the PC and the sign-extended 19 bits of the instruction (the branch displacement), shifted left 2 bits. The unit labeled *Shift left 2* is simply a routing of the signals between input and output that adds 00_{two} to the low-order end of the sign-extended offset field; no actual shift hardware is needed, since the amount of the "shift" is constant. Since we know that the offset was sign-extended from 19 bits, the shift will throw away only "sign bits." Control logic is used to decide whether the incremented PC or branch target should replace the PC, based on the Zero output of the ALU.

To share a datapath element between two different instruction classes, we may need to allow multiple connections to the input of an element, using a multiplexor and control signal to select among the multiple inputs.

EXAMPLE

Building a Datapath

The operations of arithmetic-logical (or R-type) instructions and the memory instructions datapath are quite similar. The key differences are the following:

- The arithmetic-logical instructions use the ALU, with the inputs coming from the two registers. The memory instructions can also use the ALU to do the address calculation, although the second input is the sign-extended 9-bit offset field from the instruction.

■ The value stored into a destination register comes from the ALU (for an R-type instruction) or the memory (for a load).

Show how to build a datapath for the operational portion of the memory-reference and arithmetic-logical instructions that uses a single register file and a single ALU to handle both types of instructions, adding any necessary multiplexors.

To create a datapath with only a single register file and a single ALU, we must support two different sources for the second ALU input, as well as two different sources for the data stored into the register file. Thus, one multiplexor is placed at the ALU input and another at the data input to the register file. Figure 4.10 shows the operational portion of the combined datapath.

Now we can combine all the pieces to make a simple datapath for the core LEGv8 architecture by adding the datapath for instruction fetch (Figure 4.6), the datapath from R-type and memory instructions (Figure 4.10), and the datapath for branches (Figure 4.9). Figure 4.11 shows the datapath we obtain by composing the separate pieces. The branch instruction uses the main ALU to test the register operand, so we must keep the adder from Figure 4.9 for computing the branch target address. An additional multiplexor is required to select either the sequentially following instruction address (PC + 4) or the branch target address to be written into the PC.

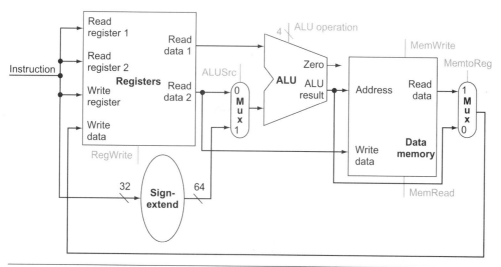

FIGURE 4.10 The datapath for the memory instructions and the R-type instructions. This example shows how a single datapath can be assembled from the pieces in Figures 4.7 and 4.8 by adding multiplexors. Two multiplexors are needed, as described in the example.

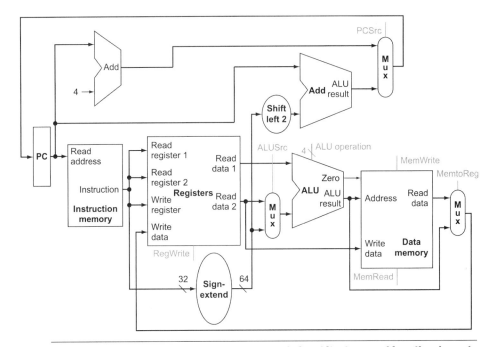

FIGURE 4.11   The simple datapath for the core LEGv8 architecture combines the elements required by different instruction classes. The components come from Figures 4.6, 4.9, and 4.10. This datapath can execute the basic instructions (load-store register, ALU operations, and branches) in a single clock cycle. Just one additional multiplexor is needed to integrate branches. The support for unconditional branches will be added later.

Check Yourself

I. Which of the following is correct for a load instruction? Refer to Figure 4.10.

  a. MemtoReg should be set to cause the data from memory to be sent to the register file.

  b. MemtoReg should be set to cause the correct register destination to be sent to the register file.

  c. We do not care about the setting of MemtoReg for loads.

II. The single-cycle datapath conceptually described in this section *must* have separate instruction and data memories, because

  a. the formats of data and instructions are different in LEGv8, and hence different memories are needed;

  b. having separate memories is less expensive;

  c. the processor operates in one cycle and cannot use a (single-ported) memory for two different accesses within that cycle.

Now that we have completed this simple datapath, we can add the control unit. The control unit must be able to take inputs and generate a write signal for each state element, the selector control for each multiplexor, and the ALU control. The ALU control is different in a number of ways, and it will be useful to design it first before we design the rest of the control unit.

Elaboration: The sign extension logic must choose between sign-extending a 9-bit field in instruction bits 20:12 for data transfer instructions or a 19-bit field (bits 23:5) for the conditional branch. Since the input is all 32 bits of the instruction, it can use the opcode bits of the instruction to select the proper field. LEGv8 opcode bit 26 happens to be 0 for data transfer instructions and 1 for conditional branch. Thus, bit 26 can control a 2:1 multiplexor inside the sign extension logic that selects the 9-bit field if it is 0 or the 19-bit field if it is 1.

4.4 A Simple Implementation Scheme

In this section, we look at what might be thought of as a simple implementation of our LEGv8 subset. We build this simple implementation using the datapath of the last section and adding a simple control function. This simple implementation covers *load register* (LDUR), *store register* (STUR), *compare and branch zero* (CBZ), and the arithmetic-logical instructions ADD, SUB, AND, and ORR. We will later enhance the design to include an unconditional branch instruction (B).

The ALU Control

The LEGv8 ALU in Appendix A defines the six following combinations of four control inputs:

ALU control lines	Function
0000	AND
0001	OR
0010	add
0110	subtract
0111	pass input b
1100	NOR

Depending on the instruction class, the ALU will need to perform one of these first five functions. (NOR can be used for other parts of the LEGv8 instruction set not found in the subset we are implementing.) For load register and store register instructions, we use the ALU to compute the memory address by addition. For the R-type instructions, the ALU needs to perform one of the four actions (AND, OR, subtract, or add), depending on the value of the 11-bit opcode field in the instruction (see Chapter 2). For compare and branch zero, the ALU just passes the register input value.

We can generate the 4-bit ALU control input using a small control unit that has as inputs the opcode field of the instruction and a 2-bit control field, which we call ALUOp. ALUOp indicates whether the operation to be performed should be add (00) for loads and stores, pass input b (01) for CBZ, or be determined by the operation encoded in the opcode field (10). The output of the ALU control unit is a 4-bit signal that directly controls the ALU by generating one of the 4-bit combinations shown previously.

In Figure 4.12, we show how to set the ALU control inputs based on the 2-bit ALUOp control and the 11-bit opcode. Later in this chapter, we will see how the ALUOp bits are generated from the main control unit.

This style of using multiple levels of decoding—that is, the main control unit generates the ALUOp bits, which then are used as input to the ALU control that generates the actual signals to control the ALU unit—is a common implementation technique. Using multiple levels of control can reduce the size of the main control unit. Using several smaller control units may also potentially reduce the latency of the control unit. Such optimizations are important, since the latency of the control unit is often a critical factor in determining the clock cycle time.

There are several different ways to implement the mapping from the 2-bit ALUOp field and the 11-bit opcode field to the four ALU operation control bits. Because only a small number of the 2048 possible values of the opcode field are of interest and the opcode field is used only when the ALUOp bits equal 10, we can use a small piece of logic that recognizes the subset of possible values and generates the appropriate ALU control signals.

As a step in designing this logic, it is useful to create a *truth table* for the interesting combinations of the opcode field and the ALUOp signals, as we've done

Instruction	ALUOp	Instruction operation	Opcode field	Desired ALU action	ALU control input
LDUR	00	load register	XXXXXXXXXX	add	0010
STUR	00	store register	XXXXXXXXXX	add	0010
CBZ	01	compare and branch on zero	XXXXXXXXXX	pass input b	0111
R-type	10	ADD	10001011000	add	0010
R-type	10	SUB	11001011000	subtract	0110
R-type	10	AND	10001010000	AND	0000
R-type	10	ORR	10101010000	OR	0001

FIGURE 4.12 How the ALU control bits are set depends on the ALUOp control bits and the different opcodes for the R-type instruction. The instruction, listed in the first column, determines the setting of the ALUOp bits. All the encodings are shown in binary. Notice that when the ALUOp code is 00 or 01, the desired ALU action does not depend on the opcode field; in this case, we say that we "don't care" about the value of the opcode, and the bits are shown as Xs. When the ALUOp value is 10, then the opcode is used to set the ALU control input. See Appendix A.

in Figure 4.13; this truth table shows how the 4-bit ALU control is set depending on these two input fields. Since the full truth table is very large (2^{13} = 8192 entries), and we don't care about the value of the ALU control for many of these input combinations, we show only the truth table entries for which the ALU control must have a specific value. Throughout this chapter, we will use this practice of showing only the truth table entries for outputs that must be asserted and not showing those that are all deasserted or don't care. (This practice has a disadvantage, which we discuss in Section C.2 of 🌐 Appendix C.)

Because in many instances we do not care about the values of some of the inputs, and because we wish to keep the tables compact, we also include don't-care terms. A don't-care term in this truth table (represented by an X in an input column) indicates that the output does not depend on the value of the input corresponding to that column. For example, when the ALUOp bits are 00, as in the first row of Figure 4.13, we always set the ALU control to 0010, independent of the opcode. In this case, then, the opcode inputs will be don't cares in this line of the truth table. Later, we will see examples of another type of don't-care term. If you are unfamiliar with the concept of don't-care terms, see Appendix A for more information.

Once the truth table has been constructed, it can be optimized and then turned into gates. This process is completely mechanical. Thus, rather than show the final steps here, we describe the process and the result in Section C.2 of 🌐 Appendix C.

truth table From logic, a representation of a logical operation by listing all the values of the inputs and then in each case showing what the resulting outputs should be.

don't-care term An element of a logical function in which the output does not depend on the values of all the inputs. Don't-care terms may be specified in different ways.

Designing the Main Control Unit

Now that we have described how to design an ALU that uses the opcode and a 2-bit signal as its control inputs, we can return to looking at the rest of the control. To start this process, let's identify the fields of an instruction and the control lines that are needed for the datapath we constructed in Figure 4.11. To understand how to connect the fields of an instruction to the datapath, it is useful to review the formats

ALUOp		Opcode field											Operation
ALUOp1	ALUOp0	I[31]	I[30]	I[29]	I[28]	I[27]	I[26]	I[25]	I[24]	I[23]	I[22]	I[21]	
0	0	X	X	X	X	X	X	X	X	X	X	X	0010
X	1	X	X	X	X	X	X	X	X	X	X	X	0111
1	X	1	0	0	0	1	0	1	1	0	0	0	0010
1	X	1	1	0	0	1	0	1	1	0	0	0	0110
1	X	1	0	0	0	1	0	1	0	0	0	0	0000
1	X	1	0	1	0	1	0	1	0	0	0	0	0001

FIGURE 4.13 The truth table for the 4 ALU control bits (called Operation). The inputs are the ALUOp and opcode field. Only the entries for which the ALU control is asserted are shown. Some don't-care entries have been added. For example, the ALUOp does not use the encoding 11, so the truth table can contain entries 1X and X1, rather than 10 and 01. While we show all 11 bits of the opcode, note that the only bits with different values for the four R-format instructions are bits 30, 29, and 24. Thus, we only need these three opcode bits as input for ALU control instead of all 11.

Field	opcode	Rm	shamt	Rn	Rd
Bit positions	31:21	20:16	15:10	9:5	4:0

a. R-type instruction

Field	1986 or 1984	address	0	Rn	Rt
Bit positions	31:21	20:12	11:10	9:5	4:0

b. Load or store instruction

Field	180	address	Rt
Bit positions	31:26	23:5	4:0

c. Conditional branch instruction

FIGURE 4.14   The three instruction classes (R-type, load and store, and conditional branch) use three different instruction formats. The unconditional branch instruction uses another format, which we will discuss shortly. (a) Instruction format for R-format instructions, have three register operands: Rn, Rm, and Rd. Fields Rn and Rm are sources, and Rd is the destination. The ALU function is in the opcode field and is decoded by the ALU control design in the previous section. The R-type instructions that we implement are ADD, SUB, AND, and ORR. The shamt field is used only for shifts; we will ignore it in this chapter. (b) Instruction format for load (opcode = 1986_{ten}) and store (opcode = 1984_{ten}) instructions. The register Rn is the base register that is added to the 9-bit address field to form the memory address. For loads, Rt is the destination register for the loaded value. For stores, Rt is the source register whose value should be stored into memory. (c) Instruction format for compare and branch on zero (opcode = 180). The register Rt is the source register that is tested for zero. The 19-bit address field is sign-extended, shifted, and added to the PC to compute the branch target address.

of the three instruction classes: the R-type, branch, and load-store instructions. Figure 4.14 shows these formats.

There are several major observations about this instruction format that we will rely on:

opcode The field that denotes the operation and format of an instruction.

- The opcode field, which as we saw in Chapter 2, is between 6 and 11 bits wide and found in bits 31:26 to 31:21.

- The first register operand is always in bit positions 9:5 (Rn) for both R-type instructions and for the base register for load and store instructions.

- The other register operand is in one of two places. It is in bit positions 20:16 (Rm) for R-type instructions and it is in bit positions 4:0 (Rt) for the register to be written by a load. That is also the field that specifies the register to be tested for zero for compare and branch on zero. Thus, we will need to add a multiplexor to select which field of the instruction is used to indicate the register number to be read.

- Another operand can also be a 19-bit offset for compare and branch on zero or a 9-bit offset for load and store.

- The destination register for R-type instructions (Rd) and for loads (Rt) is in bit positions 4:0.

The first design principle from Chapter 2—*simplicity favors regularity*—pays off here in specifying control.

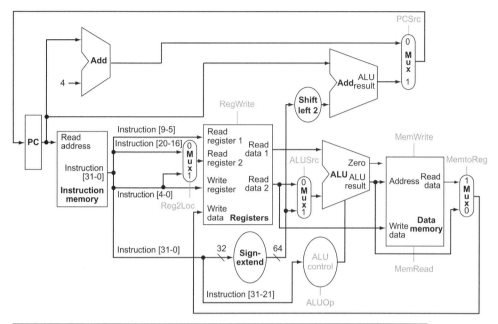

FIGURE 4.15   The datapath of Figure 4.11 with all necessary multiplexors and all control lines identified. The control lines are shown in color. The ALU control block has also been added. The PC does not require a write control, since it is written once at the end of every clock cycle; the branch control logic determines whether it is written with the incremented PC or the branch target address.

Using this information, we can add the instruction labels and extra multiplexor (for the Read register 2 number input of the register file) to the simple datapath. Figure 4.15 shows these additions plus the ALU control block, the write signals for state elements, the read signal for the data memory, and the control signals for the multiplexors. Since all the multiplexors have two inputs, they each require a single control line.

Figure 4.15 shows seven single-bit control lines plus the 2-bit ALUOp control signal. We have already defined how the ALUOp control signal works, and it is useful to define what the seven other control signals do informally before we determine how to set these control signals during instruction execution. Figure 4.16 describes the function of these seven control lines.

Now that we have looked at the function of each of the control signals, we can look at how to set them. The control unit can set all but one of the control signals based solely on the opcode field of the instruction. The PCSrc control line is the exception. That control line should be asserted if the instruction is compare and branch on zero (a decision that the control unit can make) *and* the Zero output of the ALU, which is used for the zero test, is asserted. To generate the PCSrc signal, we will need to AND together a signal from the control unit, which we call *Branch*, with the Zero signal out of the ALU.

Signal name	Effect when deasserted	Effect when asserted
Reg2Loc	The register number for Read register 2 comes from the Rm field (bits 20:16).	The register number for Read register 2 comes from the Rt field (bits 4:0).
RegWrite	None.	The register on the Write register input is written with the value on the Write data input.
ALUSrc	The second ALU operand comes from the second register file output (Read data 2).	The second ALU operand is the sign-extended, lower 16 bits of the instruction.
PCSrc	The PC is replaced by the output of the adder that computes the value of PC + 4.	The PC is replaced by the output of the adder that computes the branch target.
MemRead	None.	Data memory contents designated by the address input are put on the Read data output.
MemWrite	None.	Data memory contents designated by the address input are replaced by the value on the Write data input.
MemtoReg	The value fed to the register Write data input comes from the ALU.	The value fed to the register Write data input comes from the data memory.

FIGURE 4.16   The effect of each of the seven control signals. When the 1-bit control to a two-way multiplexor is asserted, the multiplexor selects the input corresponding to 1. Otherwise, if the control is deasserted, the multiplexor selects the 0 input. Remember that the state elements all have the clock as an implicit input and that the clock is used in controlling writes. Gating the clock externally to a state element can create timing problems. (See Appendix A for further discussion of this problem.)

These nine control signals (seven from Figure 4.16 and two for ALUOp) can now be set based on the input signals to the control unit, which are the opcode bits 31 to 21. Figure 4.17 shows the datapath with the control unit and the control signals.

Before we try to write a set of equations or a truth table for the control unit, it will be useful to try to define the control function informally. Because the setting of the control lines depends only on the opcode, we define whether each control signal should be 0, 1, or don't care (X) for each of the opcode values. Figure 4.18 defines how the control signals should be set for each opcode; this information follows directly from Figures 4.12, 4.16, and 4.17.

Operation of the Datapath

With the information contained in Figures 4.16 and 4.18, we can design the control unit logic, but before we do that, let's look at how each instruction uses the datapath. In the next few figures, we show the flow of three different instruction classes through the datapath. The asserted control signals and active datapath elements are highlighted in each of these. Note that a multiplexor whose control is 0 has a definite action, even if its control line is not highlighted. Multiple-bit control signals are highlighted if any constituent signal is asserted.

Figure 4.19 shows the operation of the datapath for an R-type instruction, such as ADD X1, X2, X3. Although everything occurs in one clock cycle, we can think

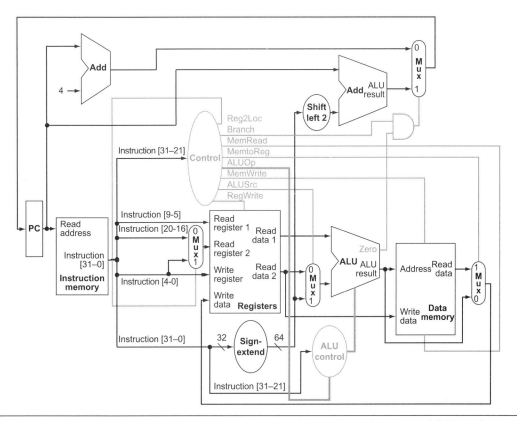

FIGURE 4.17 The simple datapath with the control unit. The input to the control unit is the 11-bit opcode field from the instruction. The outputs of the control unit consist of three 1-bit signals that are used to control multiplexors (Reg2Loc, ALUSrc, and MemtoReg), three signals for controlling reads and writes in the register file and data memory (RegWrite, MemRead, and MemWrite), a 1-bit signal used in determining whether to possibly branch (Branch), and a 2-bit control signal for the ALU (ALUOp). An AND gate is used to combine the branch control signal and the Zero output from the ALU; the AND gate output controls the selection of the next PC. Notice that PCSrc is now a derived signal, rather than one coming directly from the control unit. Thus, we drop the signal name in subsequent figures.

of four steps to execute the instruction; these steps are ordered by the flow of information:

1. The instruction is fetched, and the PC is incremented.

2. Two registers, X2 and X3, are read from the register file; also, the main control unit computes the setting of the control lines during this step.

3. The ALU operates on the data read from the register file, using portions of the opcode to generate the ALU function.

4. The result from the ALU is written into the destination register (X1) in the register file.

Instruction	Reg2Loc	ALUSrc	MemtoReg	RegWrite	MemRead	MemWrite	Branch	ALUOp1	ALUOp0
R-format	0	0	0	1	0	0	0	1	0
LDUR	X	1	1	1	1	0	0	0	0
STUR	1	1	X	0	0	1	0	0	0
CBZ	1	0	X	0	0	0	1	0	1

FIGURE 4.18   The setting of the control lines is completely determined by the opcode fields of the instruction. The first row of the table corresponds to the R-format instructions (ADD, SUB, AND, and ORR). For all these instructions, the source register fields are Rn and Rm, and the destination register field is Rd; this defines how the signals ALUSrc and Reg2Loc are set. Furthermore, an R-type instruction writes a register (RegWrite = 1), but neither reads nor writes data memory. When the Branch control signal is 0, the PC is unconditionally replaced with PC + 4; otherwise, the PC is replaced by the branch target if the Zero output of the ALU is also high. The ALUOp field for R-type instructions is set to 10 to indicate that the ALU control should be generated from the opcode field. The second and third rows of this table give the control signal settings for LDUR and STUR. These ALUSrc and ALUOp fields are set to perform the address calculation. The MemRead and MemWrite are set to perform the memory access. Finally, Reg2Loc and RegWrite are set for a load to cause the result to be stored in the Rt register. The ALUOp field for branch is set for a pass input b (ALUOp = 01), which is used to test for zero. Notice that the MemtoReg field is irrelevant when the RegWrite signal is 0: since the register is not being written, the value of the data on the register data write port is not used. Thus, the entry MemtoReg in the last two rows of the table is replaced with X for don't care. A don't care can also be added to Reg2Loc for LDUR, which doesn't use a second register. This type of don't care must be added by the designer, since it depends on knowledge of how the datapath works.

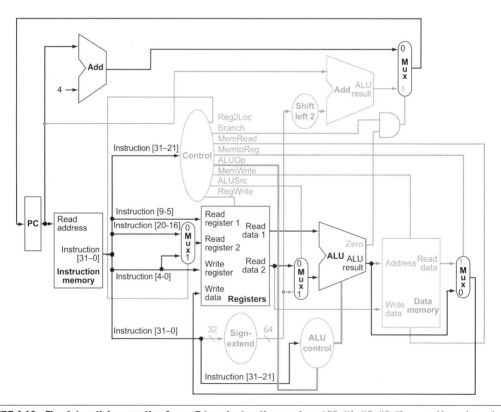

FIGURE 4.19   The datapath in operation for an R-type instruction, such as ADD X1, X2, X3. The control lines, datapath units, and connections that are active are highlighted.

Similarly, we can illustrate the execution of a load register, such as

```
LDUR  X1,  [X2,offset]
```

in a style similar to Figure 4.19. Figure 4.20 shows the active functional units and asserted control lines for a load. We can think of a load instruction as operating in five steps (similar to how the R-type executed in four):

1. An instruction is fetched from the instruction memory, and the PC is incremented.

2. A register (X2) value is read from the register file.

3. The ALU computes the sum of the value read from the register file and the sign-extended 9 bits of the instruction (offset).

4. The sum from the ALU is used as the address for the data memory.

5. The data from the memory unit is written into the register file (X1).

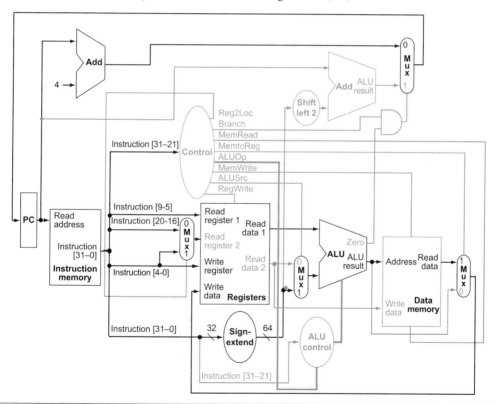

FIGURE 4.20 The datapath in operation for a load instruction. The control lines, datapath units, and connections that are active are highlighted. A store instruction would operate very similarly. The main difference would be that the memory control would indicate a write rather than a read, the second register value read would be used for the data to store, and the operation of writing the data memory value to the register file would not occur.

Finally, we can show the operation of the compare-and-branch-on-zero instruction, such as CBZ X1, offset, in the same fashion. It operates much like an R-format instruction, but the ALU output is used to determine whether the PC is written with PC + 4 or the branch target address. Figure 4.21 shows the four steps in execution:

1. An instruction is fetched from the instruction memory, and the PC is incremented.

2. The register, X1 is read from the register file using bits 4:0 of the instruction (Rt).

3. The ALU passes the data value read from the register file. The value of PC is added to the sign-extended, 19 bits of the instruction (offset) are shifted left by two; the result is the branch target address.

4. The Zero status information from the ALU is used to decide which adder result to store in the PC.

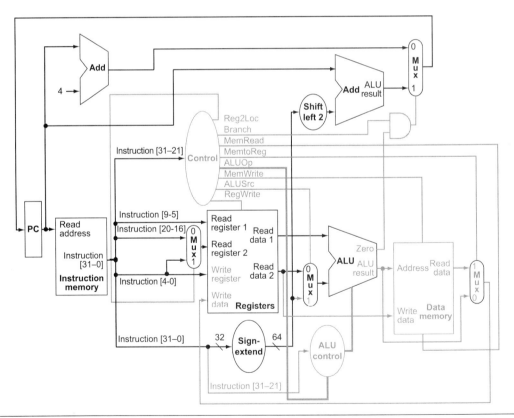

FIGURE 4.21 The datapath in operation for a compare-and-branch-on-zero instruction. The control lines, datapath units, and connections that are active are highlighted. After using the register file and ALU to perform the compare, the Zero output is used to select the next program counter from between the two candidates.

Finalizing Control

Now that we have seen how the instructions operate in steps, let's continue with the control implementation. The control function can be precisely defined using the contents of Figure 4.18. The outputs are the control lines, and the input is the opcode field. Thus, we can create a truth table for each of the outputs based on the binary encoding of the opcodes.

Figure 4.22 defines the logic in the control unit as one large truth table that combines all the outputs and that uses the opcode bits as inputs. It completely specifies the control function, and we can implement it directly in gates in an automated fashion. We show this final step in Section C.2 in 🌐 Appendix C.

Now that we have a single-cycle implementation of most of the LEGv8 core instruction set, let's add the unconditional branch instruction to show how the basic datapath and control can be extended to handle other instructions in the instruction set.

single-cycle implementation Also called **single clock cycle implementation**. An implementation in which an instruction is executed in one clock cycle. While easy to understand, it is too slow to be practical.

Input or output	Signal name	R-format	LDUR	STUR	CBZ
Inputs	I[31]	1	1	1	1
	I[30]	X	1	1	0
	I[29]	X	1	1	1
	I[28]	0	1	1	1
	I[27]	1	1	1	0
	I[26]	0	0	0	1
	I[25]	1	0	0	0
	I[24]	X	0	0	0
	I[23]	0	0	0	X
	I[22]	0	1	0	X
	I[21]	0	0	0	X
Outputs	Reg2Loc	0	X	1	1
	ALUSrc	0	1	1	0
	MemtoReg	0	1	X	X
	RegWrite	1	1	0	0
	MemRead	0	1	0	0
	MemWrite	0	0	1	0
	Branch	0	0	0	1
	ALUOp1	1	0	0	0
	ALUOp0	0	0	0	1

FIGURE 4.22 The control function for the simple single-cycle implementation is completely specified by this truth table. The top half of the table gives the combinations of input signals that correspond to the four instruction classes, one per column, that determine the control output settings. The bottom portion of the table gives the outputs for each of the four opcodes. Thus, the output RegWrite is asserted for two different combinations of the inputs. We simplified the truth table by using don't cares in the input portion to combine the four R-format instructions together in one column; we could have instead replaced that single column with four columns for the instructions ADD, SUB, AND, and ORR. The outputs would have been the same for all four of these R-format instructions.

EXAMPLE

Implementing Unconditional Branches

Figure 4.17 shows the implementation of many of the instructions we looked at in Chapter 2. One instruction that is missing is the unconditional branch instruction. Extend the datapath and control of Figure 4.17 to include the unconditional branch instruction. Describe how to set any new control lines.

ANSWER

The unconditional branch instruction looks like a branch instruction with a longer offset but it is not conditional. Like a branch, the low-order 2 bits of a branch address are always 00_{two}. The next 26 bits of this 64-bit address come from the 26-bit immediate field in the instruction and then are sign extended. Thus, we can implement a branch by storing into the PC sum of the PC and the sign extended and shifted 26-bit offset. Figure 4.23 shows the addition of the control for branch added to Figure 4.17. An additional OR-gate is used with a control signal to select the branch target PC always. This control signal, called *UncondBranch*, is asserted only when the instruction is an unconditional branch.

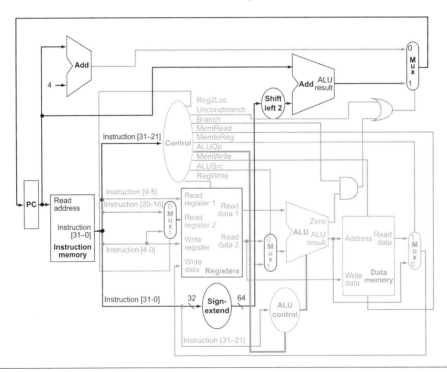

FIGURE 4.23 The simple control and datapath are extended to handle the unconditional branch instruction. An additional OR-gate (at the upper right) is used to control the multiplexor that chooses between the branch target and the sequential instruction following this one. One input to the OR-gate is the Uncondbranch control signal. Although not shown, the Sign-extend logic would recognize the unconditional branch opcode and sign-extend the lower 26 bits of the branch instruction to form a 64-bit address to be added to the PC.

Elaboration: We also need to modify the sign-extend unit to include the 26-bit address of this instruction. We solve the problem by expanding the 2:1 multiplexor that was controlled by opcode bit 26 mentioned in an earlier elaboration to include this address and then control it with the two opcode bits 31 and 26. The value 01 means select the address for B, 10 means address for LDUR or STUR, and 11 for CBZ.

Why a Single-Cycle Implementation is not Used Today

Although the single-cycle design will work correctly, it is too inefficient to be used in modern designs. To see why this is so, notice that the clock cycle must have the same length for every instruction in this single-cycle design. Of course, the longest possible path in the processor determines the clock cycle. This path is most likely a load instruction, which uses five functional units in series: the instruction memory, the register file, the ALU, the data memory, and the register file. Although the CPI is 1 (see Chapter 1), the overall performance of a single-cycle implementation is likely to be poor, since the clock cycle is too long.

The penalty for using the single-cycle design with a fixed clock cycle is significant, but might be considered acceptable for this small instruction set. Historically, early computers with very simple instruction sets did use this implementation technique. However, if we tried to implement the floating-point unit or an instruction set with more complex instructions, this single-cycle design wouldn't work well at all.

Because we must assume that the clock cycle is equal to the worst-case delay for all instructions, it's useless to try implementation techniques that reduce the delay of the common case but do not improve the worst-case cycle time. A single-cycle implementation thus violates the great idea from Chapter 1 of making the **common case fast**.

COMMON CASE FAST

In next section, we'll look at another implementation technique, called pipelining, that uses a datapath very similar to the single-cycle datapath but is much more efficient by having a much higher throughput. Pipelining improves efficiency by executing multiple instructions simultaneously.

Look at the control signals in Figure 4.22. Can you combine any together? Can any control signal output in the figure be replaced by the inverse of another? (Hint: take into account the don't cares.) If so, can you use one signal for the other without adding an inverter?

Check Yourself

4.5   An Overview of Pipelining

Pipelining is an implementation technique in which multiple instructions are overlapped in execution. Today, **pipelining** is nearly universal.

This section relies heavily on one analogy to give an overview of the pipelining terms and issues. If you are interested in just the big picture, you should concentrate on this section and then skip to Sections 4.10 and 4.11 to see an introduction to the advanced pipelining techniques used in recent processors such as the Intel Core i7 and ARM Cortex-A53. If you are curious about exploring the anatomy of a pipelined computer, this section is a good introduction to Sections 4.6 through 4.9.

Never waste time.
American proverb

pipelining An implementation technique in which multiple instructions are overlapped in execution, much like an assembly line.

PIPELINING

Anyone who has done a lot of laundry has intuitively used pipelining. The *non-pipelined* approach to laundry would be as follows:

1. Place one dirty load of clothes in the washer.

2. When the washer is finished, place the wet load in the dryer.

3. When the dryer is finished, place the dry load on a table and fold.

4. When folding is finished, ask your roommate to put the clothes away.

When your roommate is done, start over with the next dirty load.

The *pipelined* approach takes much less time, as Figure 4.24 shows. As soon as the washer is finished with the first load and placed in the dryer, you load the washer with the second dirty load. When the first load is dry, you place it on the table to start folding, move the wet load to the dryer, and put the next dirty load into the washer. Next, you have your roommate put the first load away, you start folding the second load, the dryer has the third load, and you put the fourth load into the washer. At this point all steps—called *stages* in pipelining—are operating concurrently. As long as we have separate resources for each stage, we can pipeline the tasks.

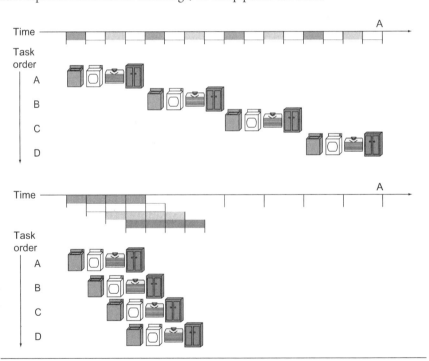

FIGURE 4.24 The laundry analogy for pipelining. Ann, Brian, Cathy, and Don each have dirty clothes to be washed, dried, folded, and put away. The washer, dryer, "folder," and "storer" each take 30 minutes for their task. Sequential laundry takes 8 hours for four loads of washing, while pipelined laundry takes just 3.5 hours. We show the pipeline stage of different loads over time by showing copies of the four resources on this two-dimensional time line, but we really have just one of each resource.

The pipelining paradox is that the time from placing a single dirty sock in the washer until it is dried, folded, and put away is not shorter for pipelining; the reason pipelining is faster for many loads is that everything is working in parallel, so more loads are finished per hour. Pipelining improves *throughput* of our laundry system. Hence, pipelining would not decrease the time to complete one load of laundry, but when we have many loads of laundry to do, the improvement in throughput decreases the total time to complete the work.

If all the stages take about the same amount of time and there is enough work to do, then the speed-up due to pipelining is equal to the number of stages in the pipeline, in this case four: washing, drying, folding, and putting away. Therefore, pipelined laundry is potentially four times faster than nonpipelined: 20 loads would take about five times as long as one load, while 20 loads of sequential laundry takes 20 times as long as one load. It's only 2.3 times faster in Figure 4.24, because we only show four loads. Notice that at the beginning and end of the workload in the pipelined version in Figure 4.24, the pipeline is not completely full; this start-up and wind-down affects performance when the number of tasks is not large compared to the number of stages in the pipeline. If the number of loads is much larger than four, then the stages will be full most of the time and the increase in throughput will be very close to four.

The same principles apply to processors where we pipeline instruction execution. LEGv8 instructions classically take five steps:

1. Fetch instruction from memory.

2. Read registers and decode the instruction.

3. Execute the operation or calculate an address.

4. Access an operand in data memory (if necessary).

5. Write the result into a register (if necessary).

Hence, the LEGv8 pipeline we explore in this chapter has five stages. The following example shows that pipelining speeds up instruction execution just as it speeds up the laundry.

Single-Cycle versus Pipelined Performance

EXAMPLE

To make this discussion concrete, let's create a pipeline. In this example, and in the rest of this chapter, we limit our attention to seven instructions: load register (LDUR), store register (STUR), add (ADD), subtract (SUB), AND (AND), OR (ORR), and compare and branch on zero (CBZ).

Contrast the average time between instructions of a single-cycle implementation, in which all instructions take one clock cycle, to a pipelined implementation. Assume that the operation times for the major functional units in this example are 200 ps for memory access for instructions or data, 200 ps for ALU operation, and 100 ps for register file read or write. In the single-cycle model, every instruction takes exactly one clock cycle, so the clock cycle must be stretched to accommodate the slowest instruction.

ANSWER

Figure 4.25 shows the time required for each of the seven instructions. The single-cycle design must allow for the slowest instruction—in Figure 4.25 it is LDUR—so the time required for every instruction is 800 ps. Similarly to Figure 4.24, Figure 4.26 compares nonpipelined and pipelined execution of three load register instructions. Thus, the time between the first and fourth instructions in the nonpipelined design is 3 × 800 ps or 2400 ps.

All the pipeline stages take a single clock cycle, so the clock cycle must be long enough to accommodate the slowest operation. Just as the single-cycle design must take the worst-case clock cycle of 800 ps, even though some instructions can be as fast as 500 ps, the pipelined execution clock cycle must have the worst-case clock cycle of 200 ps, even though some stages take only 100 ps. Pipelining still offers a fourfold performance improvement: the time between the first and fourth instructions is 3 × 200 ps or 600 ps.

We can turn the pipelining speed-up discussion above into a formula. If the stages are perfectly balanced, then the time between instructions on the pipelined processor—assuming ideal conditions—is equal to

$$\text{Time between instructions}_{\text{pipelined}} = \frac{\text{Time between instructions}_{\text{nonpipelined}}}{\text{Number of pipe stages}}$$

Under ideal conditions and with a large number of instructions, the speed-up from pipelining is approximately equal to the number of pipe stages; a five-stage pipeline is nearly five times faster.

The formula suggests that a five-stage pipeline should offer nearly a fivefold improvement over the 800 ps nonpipelined time, or a 160 ps clock cycle. The

Instruction class	Instruction fetch	Register read	ALU operation	Data access	Register write	Total time
Load register (LDUR)	200 ps	100 ps	200 ps	200 ps	100 ps	800 ps
Store register (STUR)	200 ps	100 ps	200 ps	200 ps		700 ps
R-format (ADD, SUB, AND, ORR)	200 ps	100 ps	200 ps		100 ps	600 ps
Branch (CBZ)	200 ps	100 ps	200 ps			500 ps

FIGURE 4.25   Total time for each instruction calculated from the time for each component. This calculation assumes that the multiplexors, control unit, PC accesses, and sign extension unit have no delay.

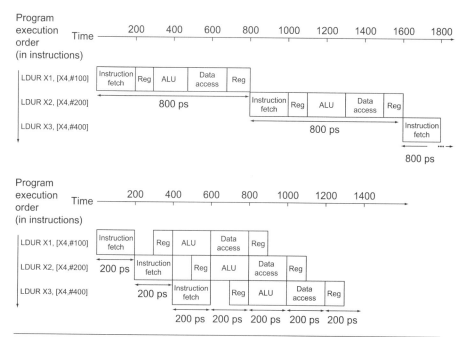

FIGURE 4.26 Single-cycle, nonpipelined execution (top) versus pipelined execution (bottom). Both use the same hardware components, whose time is listed in Figure 4.25. In this case, we see a fourfold speed-up on average time between instructions, from 800 ps down to 200 ps. Compare this figure to Figure 4.24. For the laundry, we assumed all stages were equal. If the dryer were slowest, then the dryer stage would set the stage time. The pipeline stage times of a computer are also limited by the slowest resource, either the ALU operation or the memory access. We assume the write to the register file occurs in the first half of the clock cycle and the read from the register file occurs in the second half. We use this assumption throughout this chapter.

example shows, however, that the stages may be imperfectly balanced. Moreover, pipelining involves some overhead, the source of which will be clearer shortly. Thus, the time per instruction in the pipelined processor will exceed the minimum possible, and speed-up will be less than the number of pipeline stages.

However, even our claim of fourfold improvement for our example is not reflected in the total execution time for the three instructions: it's 1400 ps versus 2400 ps. Of course, this is because the number of instructions is not large. What would happen if we increased the number of instructions? We could extend the previous figures to 1,000,003 instructions. We would add 1,000,000 instructions in the pipelined example; each instruction adds 200 ps to the total execution time. The total execution time would be 1,000,000 × 200 ps + 1400 ps, or 200,001,400 ps. In the nonpipelined example, we would add 1,000,000 instructions, each taking 800 ps, so total execution time would be 1,000,000 × 800 ps + 2400 ps, or 800,002,400 ps. Under these conditions, the ratio of total execution times for real programs on nonpipelined to pipelined processors is close to the ratio of times between instructions:

$$\frac{800,002,400 \text{ ps}}{200,001,400 \text{ ps}} \simeq \frac{800 \text{ ps}}{200 \text{ ps}} \simeq 4.00$$

Pipelining improves performance by *increasing instruction throughput, in contrast to decreasing the execution time of an individual instruction*, but instruction throughput is the important metric because real programs execute billions of instructions.

Designing Instruction Sets for Pipelining

Even with this simple explanation of pipelining, we can get insight into the design of the LEGv8 instruction set, which was designed for pipelined execution.

First, all LEGv8 instructions are the same length. This restriction makes it much easier to fetch instructions in the first pipeline stage and to decode them in the second stage. In an instruction set like the x86, where instructions vary from 1 byte to 15 bytes, pipelining is considerably more challenging. Modern implementations of the x86 architecture actually translate x86 instructions into simple operations that look like LEGv8 instructions and then pipeline the simple operations rather than the native x86 instructions! (See Section 4.10.)

Second, LEGv8 has just a few instruction formats, with the first source register and destination register fields being located in the same place in each instruction.

Third, memory operands only appear in loads or stores in LEGv8. This restriction means we can use the execute stage to calculate the memory address and then access memory in the following stage. If we could operate on the operands in memory, as in the x86, stages 3 and 4 would expand to an address stage, memory stage, and then execute stage. We will shortly see the downside of longer pipelines.

Pipeline Hazards

There are situations in pipelining when the next instruction cannot execute in the following clock cycle. These events are called *hazards*, and there are three different types.

Structural Hazard

structural hazard When a planned instruction cannot execute in the proper clock cycle because the hardware does not support the combination of instructions that are set to execute.

The first hazard is called a structural hazard. It means that the hardware cannot support the combination of instructions that we want to execute in the same clock cycle. A structural hazard in the laundry room would occur if we used a washer-dryer combination instead of a separate washer and dryer, or if our roommate was busy doing something else and wouldn't put clothes away. Our carefully scheduled pipeline plans would then be foiled.

As we said above, the LEGv8 instruction set was designed to be pipelined, making it fairly easy for designers to avoid structural hazards when designing a pipeline. Suppose, however, that we had a single memory instead of two memories. If the pipeline in Figure 4.26 had a fourth instruction, we would see that in the same clock cycle, the first instruction is accessing data from memory while the fourth instruction is fetching an instruction from that same memory. Without two memories, our pipeline could have a structural hazard.

Data Hazards

Data hazards occur when the pipeline must be stalled because one step must wait for another to complete. Suppose you found a sock at the folding station for which no match existed. One possible strategy is to run down to your room and search through your clothes bureau to see if you can find the match. Obviously, while you are doing the search, loads that have completed drying are ready to fold and those that have finished washing are ready to dry.

In a computer pipeline, data hazards arise from the dependence of one instruction on an earlier one that is still in the pipeline (a relationship that does not really exist when doing laundry). For example, suppose we have an add instruction followed immediately by a subtract instruction that uses that sum (X19):

```
ADD  X19, X0, X1
SUB  X2, X19, X3
```

Without intervention, a data hazard could severely stall the pipeline. The add instruction doesn't write its result until the fifth stage, meaning that we would have to waste three clock cycles in the pipeline.

Although we could try to rely on compilers to remove all such hazards, the results would not be satisfactory. These dependences happen just too often and the delay is far too long to expect the compiler to rescue us from this dilemma.

The primary solution is based on the observation that we don't need to wait for the instruction to complete before trying to resolve the data hazard. For the code sequence above, as soon as the ALU creates the sum for the add, we can supply it as an input for the subtract. Adding extra hardware to retrieve the missing item early from the internal resources is called forwarding or bypassing.

data hazard Also called a **pipeline data hazard**. When a planned instruction cannot execute in the proper clock cycle because data that are needed to execute the instruction are not yet available.

forwarding Also called **bypassing**. A method of resolving a data hazard by retrieving the missing data element from internal buffers rather than waiting for it to arrive from programmer-visible registers or memory.

Forwarding with Two Instructions

EXAMPLE

For the two instructions above, show what pipeline stages would be connected by forwarding. Use the drawing in Figure 4.27 to represent the datapath during the five stages of the pipeline. Align a copy of the datapath for each instruction, similar to the laundry pipeline in Figure 4.24.

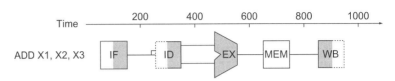

FIGURE 4.27 Graphical representation of the instruction pipeline, similar in spirit to the laundry pipeline in Figure 4.24. Here we use symbols representing the physical resources with the abbreviations for pipeline stages used throughout the chapter. The symbols for the five stages: *IF* for the instruction fetch stage, with the box representing instruction memory; *ID* for the instruction decode/register file read stage, with the drawing showing the register file being read; *EX* for the execution stage, with the drawing representing the ALU; *MEM* for the memory access stage, with the box representing data memory; and *WB* for the write-back stage, with the drawing showing the register file being written. The shading indicates the element is used by the instruction. Hence, MEM has a white background because ADD does not access the data memory. Shading on the right half of the register file or memory means the element is read in that stage, and shading of the left half means it is written in that stage. Hence the right half of ID is shaded in the second stage because the register file is read, and the left half of WB is shaded in the fifth stage because the register file is written.

ANSWER

Figure 4.28 shows the connection to forward the value in X1 after the execution stage of the ADD instruction as input to the execution stage of the SUB instruction.

In this graphical representation of events, forwarding paths are valid only if the destination stage is later in time than the source stage. For example, there cannot be a valid forwarding path from the output of the memory access stage in the first instruction to the input of the execution stage of the following, since that would mean going backward in time.

Forwarding works very well and is described in detail in Section 4.7. It cannot prevent all pipeline stalls, however. For example, suppose the first instruction was a load of X1 instead of an add. As we can imagine from looking at Figure 4.28, the

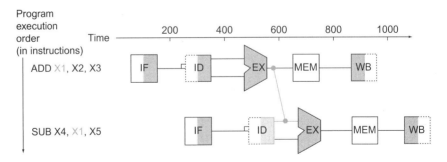

FIGURE 4.28 Graphical representation of forwarding. The connection shows the forwarding path from the output of the EX stage of ADD to the input of the EX stage for SUB, replacing the value from register X1 read in the second stage of SUB.

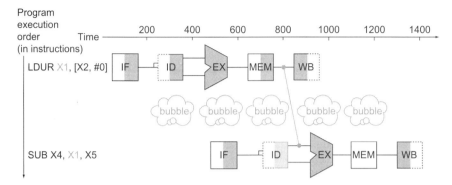

FIGURE 4.29   We need a stall even with forwarding when an R-format instruction following a load tries to use the data. Without the stall, the path from memory access stage output to execution stage input would be going backward in time, which is impossible. This figure is actually a simplification, since we cannot know until after the subtract instruction is fetched and decoded whether or not a stall will be necessary. Section 4.7 shows the details of what really happens in the case of a hazard.

desired data would be available only *after* the fourth stage of the first instruction in the dependence, which is too late for the *input* of the third stage of SUB. Hence, even with forwarding, we would have to stall one stage for a load-use data hazard, as Figure 4.29 shows. This figure shows an important pipeline concept, officially called a pipeline stall, but often given the nickname bubble. We shall see stalls elsewhere in the pipeline. Section 4.7 shows how we can handle hard cases like these, using either hardware detection and stalls or software that reorders code to try to avoid load-use pipeline stalls, as this example illustrates.

load-use data hazard A specific form of data hazard in which the data being loaded by a load instruction have not yet become available when they are needed by another instruction.

pipeline stall Also called bubble. A stall initiated in order to resolve a hazard.

Reordering Code to Avoid Pipeline Stalls

Consider the following code segment in C:

```
a = b + e;
c = b + f;
```

Here is the generated LEGv8 code for this segment, assuming all variables are in memory and are addressable as offsets from X0:

```
LDUR    X1, [X0,#0]    // Load b
LDUR    X2, [X0,#8]    // Load e
ADD     X3, X1,X2      // b + e
STUR    X3, [X0,#24]   // Store a
LDUR    X4, [X0,#16]   // Load f
ADD     X5, X1,X4      // b + f
STUR    X5, [X0,#32]   // Store c
```

EXAMPLE

Find the hazards in the preceding code segment and reorder the instructions to avoid any pipeline stalls.

ANSWER

Both ADD instructions have a hazard because of their respective dependence on the previous LDUR instruction. Notice that forwarding eliminates several other potential hazards, including the dependence of the first ADD on the first LDUR and any hazards for store instructions. Moving up the third LDUR instruction to become the third instruction eliminates both hazards:

```
LDUR    X1, [X0,#0]
LDUR    X2, [X0,#8]
LDUR    X4, [X0,#16]
ADD     X3, X1,X2
STUR    X3, [X0,#24]
ADD     X5, X1,X4
STUR    X5, [X0,#32]
```

On a pipelined processor with forwarding, the reordered sequence will complete in two fewer cycles than the original version.

Forwarding yields another insight into the LEGv8 architecture, in addition to the three mentioned on page 288. Each LEGv8 instruction writes at most one result and does this in the last stage of the pipeline. Forwarding is harder if there are multiple results to forward per instruction or if there is a need to write a result early on in instruction execution.

Elaboration: The name "forwarding" comes from the idea that the result is passed forward from an earlier instruction to a later instruction. "Bypassing" comes from passing the result around the register file to the desired unit.

Control Hazards

control hazard Also called branch hazard. When the proper instruction cannot execute in the proper pipeline clock cycle because the instruction that was fetched is not the one that is needed; that is, the flow of instruction addresses is not what the pipeline expected.

The third type of hazard is called a control hazard, arising from the need to make a decision based on the results of one instruction while others are executing.

Suppose our laundry crew was given the happy task of cleaning the uniforms of a football team. Given how filthy the laundry is, we need to determine whether the detergent and water temperature setting we select are strong enough to get the uniforms clean but not so strong that the uniforms wear out sooner. In our laundry pipeline, we have to wait until the second stage to examine the dry uniform to see if we need to change the washer setup or not. What to do?

Here is the first of two solutions to control hazards in the laundry room and its computer equivalent.

Stall: Just operate sequentially until the first batch is dry and then repeat until you have the right formula.

This conservative option certainly works, but it is slow.

The equivalent decision task in a computer is the conditional branch instruction. Notice that we must begin fetching the instruction following the branch on the following clock cycle. Nevertheless, the pipeline cannot possibly know what the next instruction should be, since it *only just received* the branch instruction from memory! Just as with laundry, one possible solution is to stall immediately after we fetch a branch, waiting until the pipeline determines the outcome of the branch and knows what instruction address to fetch from.

Let's assume that we put in enough extra hardware so that we can test a register, calculate the branch address, and update the PC during the second stage of the pipeline (see Section 4.8 for details). Even with this added hardware, the pipeline involving conditional branches would look like Figure 4.30. The instruction to be executed if the branch fails is stalled one extra 200 ps clock cycle before starting.

Performance of "Stall on Branch"

EXAMPLE

Estimate the impact on the *clock cycles per instruction* (CPI) of stalling on branches. Assume all other instructions have a CPI of 1.

Figure 3.28 in Chapter 3 shows that conditional branches are 17% of the instructions executed in SPECint2006. Since the other instructions run have a CPI of 1, and conditional branches took one extra clock cycle for the stall, then we would see a CPI of 1.17 and hence a slowdown of 1.17 versus the ideal case.

ANSWER

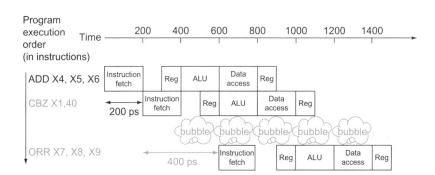

FIGURE 4.30 Pipeline showing stalling on every conditional branch as solution to control hazards. This example assumes the conditional branch is taken, and the instruction at the destination of the branch is the ORR instruction. There is a one-stage pipeline stall, or bubble, after the branch. In reality, the process of creating a stall is slightly more complicated, as we will see in Section 4.8. The effect on performance, however, is the same as would occur if a bubble were inserted.

If we cannot resolve the branch in the second stage, as is often the case for longer pipelines, then we'd see an even larger slowdown if we stall on conditional branches. The cost of this option is too high for most computers to use and motivates a second solution to the control hazard using one of our great ideas from Chapter 1:

Predict: If you're sure you have the right formula to wash uniforms, then just *predict* that it will work and wash the second load while waiting for the first load to dry.

This option does not slow down the pipeline when you are correct. When you are wrong, however, you need to redo the load that was washed while guessing the decision.

Computers do indeed use **prediction** to handle conditional branches. One simple approach is to predict always that conditional branches will be untaken. When you're right, the pipeline proceeds at full speed. Only when conditional branches are taken does the pipeline stall. Figure 4.31 shows such an example.

PREDICTION

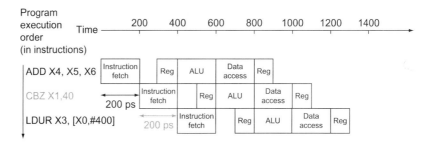

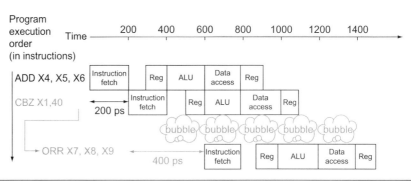

FIGURE 4.31   Predicting that branches are not taken as a solution to control hazard. The top drawing shows the pipeline when the branch is not taken. The bottom drawing shows the pipeline when the branch is taken. As we noted in Figure 4.30, the insertion of a bubble in this fashion simplifies what actually happens, at least during the first clock cycle immediately following the branch. Section 4.8 will reveal the details.

A more sophisticated version of branch prediction would have some conditional branches predicted as taken and some as untaken. In our analogy, the dark or home uniforms might take one formula while the light or road uniforms might take another. In the case of programming, at the bottom of loops are conditional branches that branch back to the top of the loop. Since they are likely to be taken and they branch backward, we could always predict taken for conditional branches that branch to an earlier address.

Such rigid approaches to branch prediction rely on stereotypical behavior and don't account for the individuality of a specific branch instruction. *Dynamic* hardware predictors, in stark contrast, make their guesses depending on the behavior of each conditional branch and may change predictions for a conditional branch over the life of a program. Following our analogy, in dynamic prediction a person would look at how dirty the uniform was and guess at the formula, adjusting the next **prediction** depending on the success of recent guesses.

One popular approach to dynamic prediction of conditional branches is keeping a history for each conditional branch as taken or untaken, and then using the recent past behavior to predict the future. As we will see later, the amount and type of history kept have become extensive, with the result being that dynamic branch predictors can correctly predict conditional branches with more than 90% accuracy (see Section 4.8). When the guess is wrong, the pipeline control must ensure that the instructions following the wrongly guessed conditional branch have no effect and must restart the pipeline from the proper branch address. In our laundry analogy, we must stop taking new loads so that we can restart the load that we incorrectly predicted.

As in the case of all other solutions to control hazards, longer pipelines exacerbate the problem, in this case by raising the cost of misprediction. Solutions to control hazards are described in more detail in Section 4.8.

branch prediction
A method of resolving a branch hazard that assumes a given outcome for the conditional branch and proceeds from that assumption rather than waiting to ascertain the actual outcome.

PREDICTION

Elaboration: There is a third approach to the control hazard, called a *delayed decision*. In our analogy, whenever you are going to make such a decision about laundry, just place a load of non-football clothes in the washer while waiting for football uniforms to dry. As long as you have enough dirty clothes that are not affected by the test, this solution works fine.

Called the *delayed branch* in computers, this is the solution actually used by the MIPS architecture. The delayed branch always executes the next sequential instruction, with the branch taking place *after* that one instruction delay. It is hidden from the MIPS assembly language programmer because the assembler can automatically arrange the instructions to get the branch behavior desired by the programmer. MIPS software will place an instruction immediately after the delayed branch instruction that is not affected by the branch, and a taken branch changes the address of the instruction that *follows* this safe instruction. In our example, the ADD instruction before the branch in Figure 4.30 does not affect the branch and can be moved after the branch to hide the branch delay fully. Since delayed branches are useful when the branches are short, it is rare to see a processor with a delayed branch of more than one cycle. For longer branch delays, hardware-based branch prediction is usually used.

PARALLELISM

PIPELINING

Pipeline Overview Summary

Pipelining is a technique that exploits **parallelism** between the instructions in a sequential instruction stream. It has the substantial advantage that, unlike programming a multiprocessor (see Chapter 6), it is fundamentally invisible to the programmer.

In the next few sections of this chapter, we cover the concept of pipelining using the LEGv8 instruction subset from the single-cycle implementation in Section 4.4 and show a simplified version of its pipeline. We then look at the problems that **pipelining** introduces and the performance attainable under typical situations.

If you wish to focus more on the software and the performance implications of pipelining, you now have sufficient background to skip to Section 4.10. Section 4.10 introduces advanced pipelining concepts, such as superscalar and dynamic scheduling, and Section 4.11 examines the pipelines of recent microprocessors.

Alternatively, if you are interested in understanding how pipelining is implemented and the challenges of dealing with hazards, you can proceed to examine the design of a pipelined datapath and the basic control, explained in Section 4.6. You can then use this understanding to explore the implementation of forwarding and stalls in Section 4.7. You can next read Section 4.8 to learn more about solutions to branch hazards, and finally see how exceptions are handled in Section 4.9.

Check Yourself

For each code sequence below, state whether it must stall, can avoid stalls using only forwarding, or can execute without stalling or forwarding.

Sequence 1	Sequence 2	Sequence 3
LDUR X0, [X0,#0]	ADD X1, X0, X0	ADDI X1, X0, #1
ADD X1, X0, X0	ADDI X2, X0, #5	ADDI X2, X0, #2
	ADDI X4, X1, #5	ADDI X3, X0, #3
		ADDI X4, X0, #4
		ADDI X5, X0, #5

Understanding Program Performance

Outside the memory system, the effective operation of the pipeline is usually the most important factor in determining the CPI of the processor and hence its performance. As we will see in Section 4.10, understanding the performance of a modern multiple-issue pipelined processor is complex and requires understanding more than just the issues that arise in a simple pipelined processor. Nonetheless, structural, data, and control hazards remain important in both simple pipelines and more sophisticated ones.

For modern pipelines, structural hazards usually revolve around the floating-point unit, which may not be fully pipelined, while control hazards are usually more of a problem in integer programs, which tend to have higher conditional branch frequencies as well as less predictable branches. Data hazards can be performance bottlenecks in both integer and floating-point programs. Often it is easier to deal with data hazards in floating-point programs because the lower conditional branch frequency and more regular memory access patterns allow the compiler to try to schedule instructions to avoid hazards. It is more difficult to perform such optimizations in integer programs that have less regular memory accesses, involving more use of pointers. As we will see in Section 4.10, there are more ambitious compiler and hardware techniques for reducing data dependences through scheduling.

PIPELINING

> **Pipelining** increases the number of simultaneously executing instructions and the rate at which instructions are started and completed. Pipelining does not reduce the time it takes to complete an individual instruction, also called the **latency**. For example, the five-stage pipeline still takes five clock cycles for the instruction to complete. In the terms used in Chapter 1, pipelining improves instruction *throughput* rather than individual instruction *execution time* or *latency*.
>
> Instruction sets can either make life harder or simpler for pipeline designers, who must already cope with structural, control, and data hazards. Branch **prediction** and forwarding help make a computer fast while still getting the right answers.

The **BIG** Picture

latency (pipeline) The number of stages in a pipeline or the number of stages between two instructions during execution.

PREDICTION

4.6   Pipelined Datapath and Control

Figure 4.32 shows the single-cycle datapath from Section 4.4 with the pipeline stages identified. The division of an instruction into five stages means a five-stage pipeline, which in turn means that up to five instructions will be in execution during any single clock cycle. Thus, we must separate the datapath into five pieces, with each piece named corresponding to a stage of instruction execution:

1. IF: Instruction fetch
2. ID: Instruction decode and register file read
3. EX: Execution or address calculation
4. MEM: Data memory access
5. WB: Write back

In Figure 4.32, these five components correspond roughly to the way the datapath is drawn; instructions and data move generally from left to right through the

There is less in this than meets the eye.

Tallulah Bankhead, remark to Alexander Woollcott, 1922

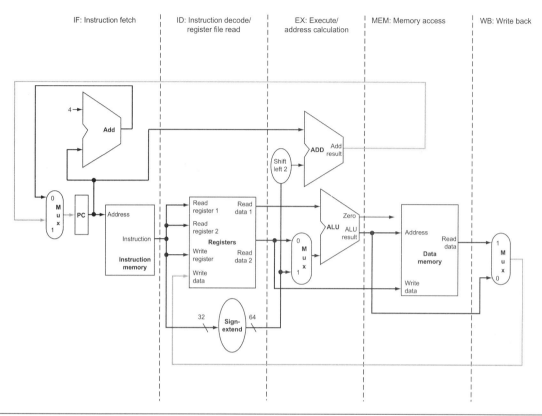

FIGURE 4.32   The single-cycle datapath from Section 4.4 (similar to Figure 4.17). Each step of the instruction can be mapped onto the datapath from left to right. The only exceptions are the update of the PC and the write-back step, shown in color, which sends either the ALU result or the data from memory to the left to be written into the register file. (Normally we use color lines for control, but these are data lines.)

five stages as they complete execution. Returning to our laundry analogy, clothes get cleaner, drier, and more organized as they move through the line, and they never move backward.

There are, however, two exceptions to this left-to-right flow of instructions:

■ The write-back stage, which places the result back into the register file in the middle of the datapath

■ The selection of the next value of the PC, choosing between the incremented PC and the branch address from the MEM stage

Data flowing from right to left do not affect the current instruction; these reverse data movements influence only later instructions in the pipeline. Note that the first right-to-left flow of data can lead to data hazards and the second leads to control hazards.

One way to show what happens in pipelined execution is to pretend that each instruction has its own datapath, and then to place these datapaths on a timeline to show their relationship. Figure 4.33 shows the execution of the instructions in Figure 4.26 by displaying their private datapaths on a common timeline. We use a stylized version of the datapath in Figure 4.32 to show the relationships in Figure 4.33.

Figure 4.33 seems to suggest that three instructions need three datapaths. Instead, we add registers to hold data so that portions of a single datapath can be shared during instruction execution.

For example, as Figure 4.33 shows, the instruction memory is used during only one of the five stages of an instruction, allowing it to be shared by following instructions during the other four stages. To retain the value of an individual instruction for its other four stages, the value read from instruction memory must be saved in a register. Similar arguments apply to every pipeline stage, so we must place registers wherever there are dividing lines between stages in Figure 4.32. Returning to our laundry analogy, we might have a basket between each pair of stages to hold the clothes for the next step.

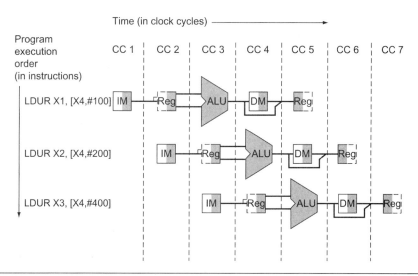

FIGURE 4.33 Instructions being executed using the single-cycle datapath in Figure 4.32, assuming pipelined execution. Similar to Figures 4.27 through 4.29, this figure pretends that each instruction has its own datapath, and shades each portion according to use. Unlike those figures, each stage is labeled by the physical resource used in that stage, corresponding to the portions of the datapath in Figure 4.32. *IM* represents the instruction memory and the PC in the instruction fetch stage, *Reg* stands for the register file and sign extender in the instruction decode/register file read stage (ID), and so on. To maintain proper time order, this stylized datapath breaks the register file into two logical parts: registers read during register fetch (ID) and registers written during write back (WB). This dual use is represented by drawing the unshaded left half of the register file using dashed lines in the ID stage, when it is not being written, and the unshaded right half in dashed lines in the WB stage, when it is not being read. As before, we assume the register file is written in the first half of the clock cycle and the register file is read during the second half.

Figure 4.34 shows the pipelined datapath with the pipeline registers high-lighted. All instructions advance during each clock cycle from one pipeline register to the next. The registers are named for the two stages separated by that register. For example, the pipeline register between the IF and ID stages is called IF/ID.

Notice that there is no pipeline register at the end of the write-back stage. All instructions must update some state in the processor—the register file, memory, or the PC—so a separate pipeline register is redundant to the state that is updated. For example, a load instruction will place its result in one of the 32 registers, and any later instruction that needs that data will simply read the appropriate register.

Of course, every instruction updates the PC, whether by incrementing it or by setting it to a branch destination address. The PC can be thought of as a pipeline register: one that feeds the IF stage of the pipeline. Unlike the shaded pipeline registers in Figure 4.34, however, the PC is part of the visible architectural state; its contents must be saved when an exception occurs, while the contents of the pipeline registers can be discarded. In the laundry analogy, you could think of the PC as corresponding to the basket that holds the load of dirty clothes before the wash step.

To show how the pipelining works, throughout this chapter we show sequences of figures to demonstrate operation over time. These extra pages would seem to require much more time for you to understand. Fear not; the sequences take much

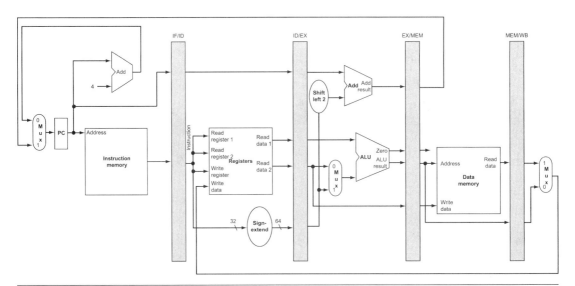

FIGURE 4.34 The pipelined version of the datapath in Figure 4.32. The pipeline registers, in color, separate each pipeline stage. They are labeled by the stages that they separate; for example, the first is labeled *IF/ID* because it separates the instruction fetch and instruction decode stages. The registers must be wide enough to store all the data corresponding to the lines that go through them. For example, the IF/ID register must be 96 bits wide, because it must hold both the 32-bit instruction fetched from memory and the incremented 64-bit PC address. We will expand these registers over the course of this chapter, but for now the other three pipeline registers contain 256, 193, and 128 bits, respectively.

less time than it might appear, because you can compare them to see what changes occur in each clock cycle. Section 4.7 describes what happens when there are data hazards between pipelined instructions; ignore them for now.

Figures 4.35 through 4.37, our first sequence, show the active portions of the datapath highlighted as a load instruction goes through the five stages of pipelined execution. We show a load first because it is active in all five stages. As in Figures 4.27 through 4.29, we highlight the *right half* of registers or memory when they are being *read* and highlight the *left half* when they are being *written*.

We show the instruction LDUR with the name of the pipe stage that is active in each figure. The five stages are the following:

1. *Instruction fetch:* The top portion of Figure 4.35 shows the instruction being read from memory using the address in the PC and then being placed in the IF/ID pipeline register. The PC address is incremented by 4 and then written back into the PC to be ready for the next clock cycle. This incremented address is also saved in the IF/ID pipeline register in case it is needed later for an instruction, such as CBZ. The computer cannot know which type of instruction is being fetched, so it must prepare for any instruction, passing potentially needed information down the pipeline.

2. *Instruction decode and register file read:* The bottom portion of Figure 4.35 shows the instruction portion of the IF/ID pipeline register supplying the immediate field, which is sign-extended to 64 bits, and the register numbers to read the two registers. All three values are stored in the ID/EX pipeline register, along with the incremented PC address. We again transfer everything that might be needed by any instruction during a later clock cycle.

3. *Execute or address calculation:* Figure 4.36 shows that the load instruction reads the contents of a register and the sign-extended immediate from the ID/EX pipeline register and adds them using the ALU. That sum is placed in the EX/MEM pipeline register.

4. *Memory access:* The top portion of Figure 4.37 shows the load instruction reading the data memory using the address from the EX/MEM pipeline register and loading the data into the MEM/WB pipeline register.

5. *Write-back:* The bottom portion of Figure 4.37 shows the final step: reading the data from the MEM/WB pipeline register and writing it into the register file in the middle of the figure.

This walk-through of the load instruction shows that any information needed in a later pipe stage must be passed to that stage via a pipeline register. Walking through a store instruction shows the similarity of instruction execution, as well as passing the information for later stages. Here are the five pipe stages of the store instruction:

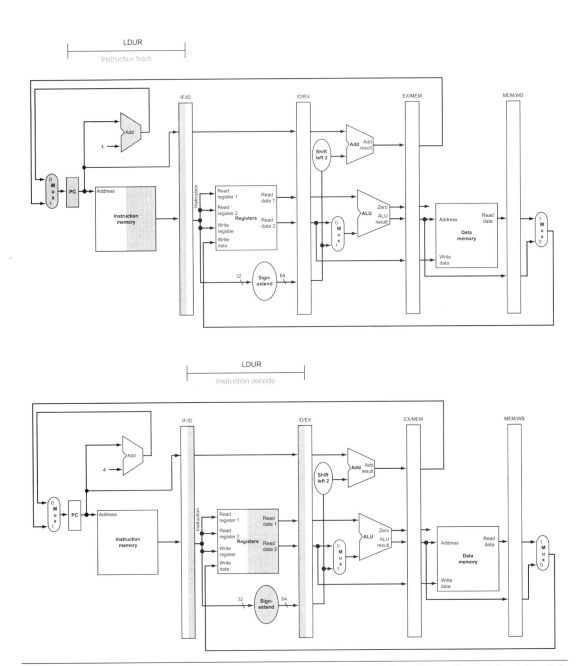

FIGURE 4.35 IF and ID: First and second pipe stages of an instruction, with the active portions of the datapath in Figure 4.34 highlighted. The highlighting convention is the same as that used in Figure 4.27. As in Section 4.2, there is no confusion when reading and writing registers, because the contents change only on the clock edge. Although the load needs only the top register in stage 2, it doesn't hurt to do potentially extra work, so it sign-extends the constant and reads both registers into the ID/EX pipeline register. We don't need all three operands, but it simplifies control to keep all three.

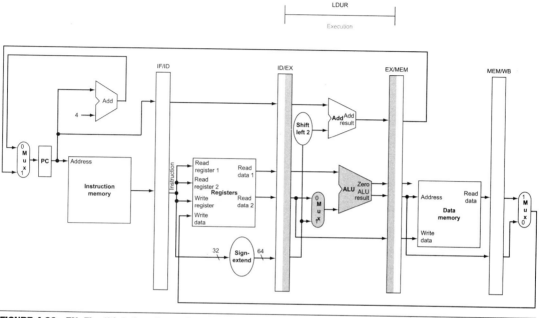

FIGURE 4.36 EX: The third pipe stage of a load instruction, highlighting the portions of the datapath in Figure 4.34 used in this pipe stage. The register is added to the sign-extended immediate, and the sum is placed in the EX/MEM pipeline register.

1. *Instruction fetch:* The instruction is read from memory using the address in the PC and then is placed in the IF/ID pipeline register. This stage occurs before the instruction is identified, so the top portion of Figure 4.35 works for store as well as load.

2. *Instruction decode and register file read:* The instruction in the IF/ID pipeline register supplies the register numbers for reading two registers and extends the sign of the immediate operand. These three 64-bit values are all stored in the ID/EX pipeline register. The bottom portion of Figure 4.35 for load instructions also shows the operations of the second stage for stores. These first two stages are executed by all instructions, since it is too early to know the type of the instruction. (While the store instruction uses the Rt field to read the second register in this pipe stage, that detail is not shown in this pipeline diagram, so we can use the same figure for both.)

3. *Execute and address calculation:* Figure 4.38 shows the third step; the effective address is placed in the EX/MEM pipeline register.

4. *Memory access:* The top portion of Figure 4.39 shows the data being written to memory. Note that the register containing the data to be stored was read in an earlier stage and stored in ID/EX. The only way to make the data available during the MEM stage is to place the data into the EX/MEM pipeline register in the EX stage, just as we stored the effective address into EX/MEM.

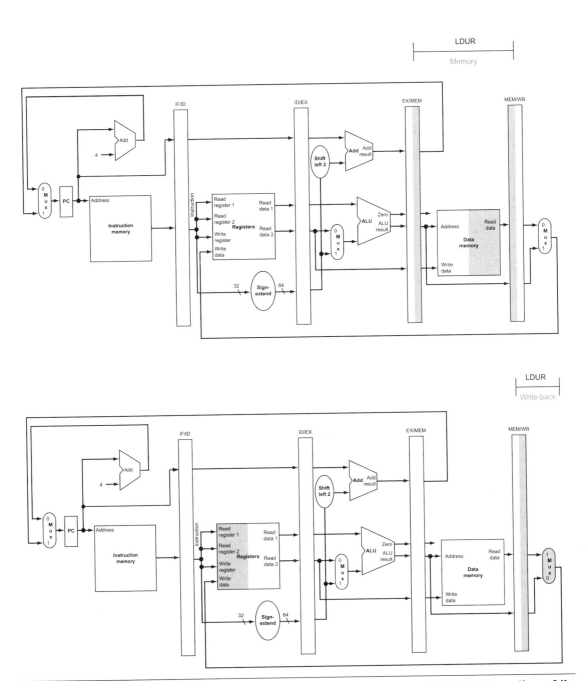

FIGURE 4.37 MEM and WB: The fourth and fifth pipe stages of a load instruction, highlighting the portions of the datapath in Figure 4.34 used in this pipe stage. Data memory is read using the address in the EX/MEM pipeline registers, and the data are placed in the MEM/WB pipeline register. Next, data are read from the MEM/WB pipeline register and written into the register file in the middle of the datapath. Note: there is a bug in this design that is repaired in Figure 4.40.

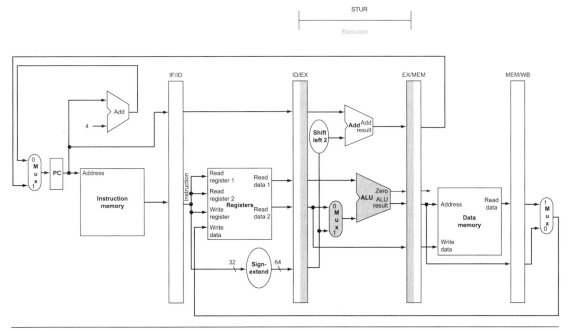

FIGURE 4.38 **EX: The third pipe stage of a store instruction.** Unlike the third stage of the load instruction in Figure 4.36, the second register value is loaded into the EX/MEM pipeline register to be used in the next stage. Although it wouldn't hurt to always write this second register into the EX/MEM pipeline register, we write the second register only on a store instruction to make the pipeline easier to understand.

5. *Write-back:* The bottom portion of Figure 4.39 shows the final step of the store. For this instruction, nothing happens in the write-back stage. Since every instruction behind the store is already in progress, we have no way to accelerate those instructions. Hence, an instruction passes through a stage even if there is nothing to do, because later instructions are already progressing at the maximum rate.

The store instruction again illustrates that to pass something from an early pipe stage to a later pipe stage, the information must be placed in a pipeline register; otherwise, the information is lost when the next instruction enters that pipeline stage. For the store instruction, we needed to pass one of the registers read in the ID stage to the MEM stage, where it is stored in memory. The data were first placed in the ID/EX pipeline register and then passed to the EX/MEM pipeline register.

Load and store illustrate a second key point: each logical component of the datapath—such as instruction memory, register read ports, ALU, data memory, and register write port—can be used only within a *single* pipeline stage. Otherwise, we would have a *structural hazard* (see page 288). Hence, these components, and their control, can be associated with a single pipeline stage.

Now we can uncover a bug in the design of the load instruction. Did you see it? Which register is changed in the final stage of the load? More specifically,

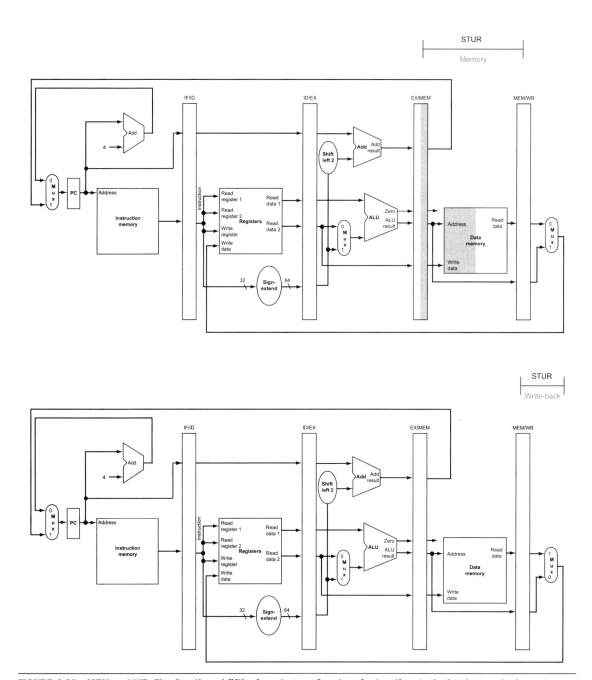

FIGURE 4.39   MEM and WB: The fourth and fifth pipe stages of a store instruction. In the fourth stage, the data are written into data memory for the store. Note that the data come from the EX/MEM pipeline register and that nothing is changed in the MEM/WB pipeline register. Once the data are written in memory, there is nothing left for the store instruction to do, so nothing happens in stage 5.

which instruction supplies the write register number? The instruction in the IF/ID pipeline register supplies the write register number, yet this instruction occurs considerably *after* the load instruction!

Hence, we need to preserve the destination register number in the load instruction. Just as store passed the register *value* from the ID/EX to the EX/MEM pipeline registers for use in the MEM stage, load must pass the register *number* from the ID/EX through EX/MEM to the MEM/WB pipeline register for use in the WB stage. Another way to think about the passing of the register number is that to share the pipelined datapath, we need to preserve the instruction read during the IF stage, so each pipeline register contains a portion of the instruction needed for that stage and later stages.

Figure 4.40 shows the correct version of the datapath, passing the write register number first to the ID/EX register, then to the EX/MEM register, and finally to the MEM/WB register. The register number is used during the WB stage to specify the register to be written. Figure 4.41 is a single drawing of the corrected datapath, highlighting the hardware used in all five stages of the load register instruction in Figures 4.35 through 4.37. See Section 4.8 for an explanation of how to make the branch instruction work as expected.

Graphically Representing Pipelines

Pipelining can be difficult to master, since many instructions are simultaneously executing in a single datapath in every clock cycle. To aid understanding, there are

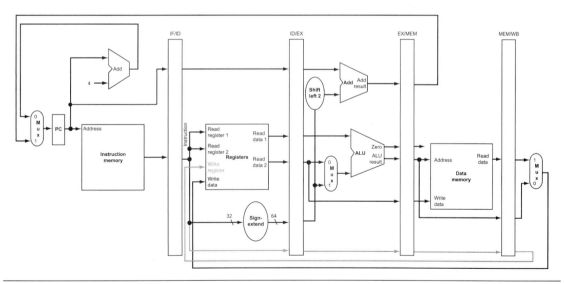

FIGURE 4.40 The corrected pipelined datapath to handle the load instruction properly. The write register number now comes from the MEM/WB pipeline register along with the data. The register number is passed from the ID pipe stage until it reaches the MEM/WB pipeline register, adding five more bits to the last three pipeline registers. This new path is shown in color.

two basic styles of pipeline figures: *multiple-clock-cycle pipeline diagrams*, such as Figure 4.33 on page 299, and *single-clock-cycle pipeline diagrams*, such as Figures 4.35 through 4.39. The multiple-clock-cycle diagrams are simpler but do not contain all the details. For example, consider the following five-instruction sequence:

```
LDUR     X10, [X1,#40]
SUB      X11, X2, X3
ADD      X12, X3, X4
LDUR     X13, [X1,#48]
ADD      X14, X5, X6
```

Figure 4.42 shows the multiple-clock-cycle pipeline diagram for these instructions. Time advances from left to right across the page in these diagrams, and instructions advance from the top to the bottom of the page, similar to the laundry pipeline in Figure 4.24. A representation of the pipeline stages is placed in each portion along the instruction axis, occupying the proper clock cycles. These stylized datapaths represent the five stages of our pipeline graphically, but a rectangle naming each pipe stage works just as well. Figure 4.43 shows the more traditional version of the multiple-clock-cycle pipeline diagram. Note that Figure 4.42 shows the physical resources used at each stage, while Figure 4.43 uses the *name* of each stage.

Single-clock-cycle pipeline diagrams show the state of the entire datapath during a single clock cycle, and usually all five instructions in the pipeline are identified by labels above their respective pipeline stages. We use this type of figure to show the details of what is happening within the pipeline during each clock

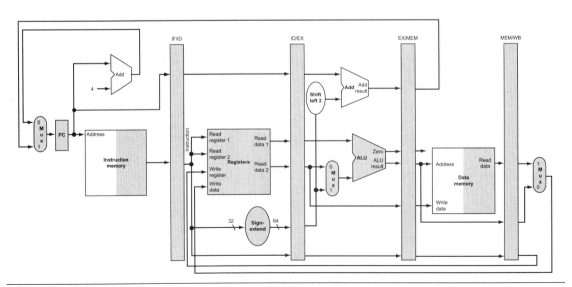

FIGURE 4.41 The portion of the datapath in Figure 4.40 that is used in all five stages of a load instruction.

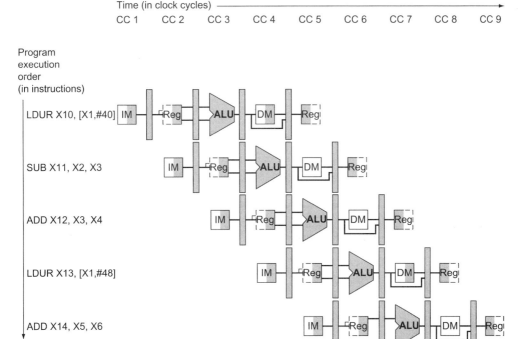

Time (in clock cycles)

CC 1 CC 2 CC 3 CC 4 CC 5 CC 6 CC 7 CC 8 CC 9

Program
execution
order
(in instructions)

LDUR X10, [X1,#40]

SUB X11, X2, X3

ADD X12, X3, X4

LDUR X13, [X1,#48]

ADD X14, X5, X6

FIGURE 4.42 Multiple-clock-cycle pipeline diagram of five instructions. This style of pipeline representation shows the complete execution of instructions in a single figure. Instructions are listed in instruction execution order from top to bottom, and clock cycles move from left to right. Unlike Figure 4.27, here we show the pipeline registers between each stage. Figure 4.43 shows the traditional way to draw this diagram.

cycle; typically, the drawings appear in groups to show pipeline operation over a sequence of clock cycles. We use multiple-clock-cycle diagrams to give overviews of pipelining situations. (⊞ Section 4.13 gives more illustrations of single-clock diagrams if you would like to see more details about Figure 4.42.) A single-clock-cycle diagram represents a vertical slice of one clock cycle through a set of multiple-clock-cycle diagrams, showing the usage of the datapath by each of the instructions in the pipeline at the designated clock cycle. For example, Figure 4.44 shows the single-clock-cycle diagram corresponding to clock cycle 5 of Figures 4.42 and 4.43. Obviously, the single-clock-cycle diagrams have more detail and take significantly more space to show the same number of clock cycles. The exercises ask you to create such diagrams for other code sequences.

Check Yourself

A group of students were debating the efficiency of the five-stage pipeline when one student pointed out that not all instructions are active in every stage of the pipeline. After deciding to ignore the effects of hazards, they made the following four statements. Which ones are correct?

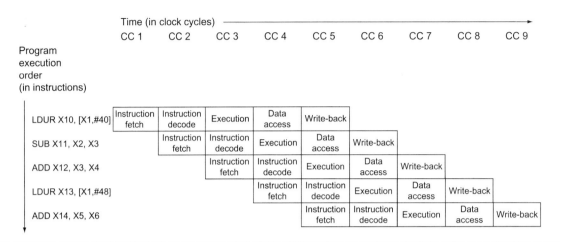

FIGURE 4.43   **Traditional multiple-clock-cycle pipeline diagram of five instructions in** Figure 4.42.

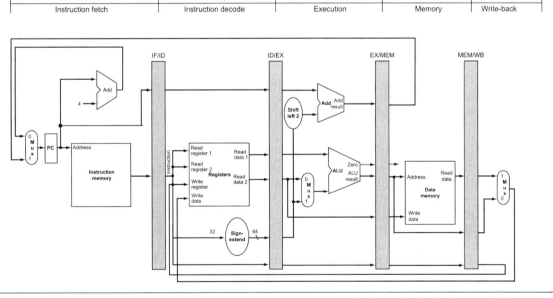

FIGURE 4.44   **The single-clock-cycle diagram corresponding to clock cycle 5 of the pipeline in** Figures 4.42 and 4.43.
As you can see, a single-clock-cycle figure is a vertical slice through a multiple-clock-cycle diagram.

1. Allowing branches and ALU instructions to take fewer stages than the five required by the load instruction will increase pipeline performance under all circumstances.

2. Trying to allow some instructions to take fewer cycles does not help, since the throughput is determined by the clock cycle; the number of pipe stages per instruction affects latency, not throughput.

3. You cannot make ALU instructions take fewer cycles because of the write-back of the result, but branches can take fewer cycles, so there is some opportunity for improvement.

4. Instead of trying to make instructions take fewer cycles, we should explore making the pipeline longer, so that instructions take more cycles, but the cycles are shorter. This could improve performance.

Pipelined Control

Just as we added control to the single-cycle datapath in Section 4.3, we now add control to the pipelined datapath. We start with a simple design that views the problem through rose-colored glasses.

The first step is to label the control lines on the existing datapath. Figure 4.45 shows those lines. We borrow as much as we can from the control for the simple

In the 6600 Computer, perhaps even more than in any previous computer, the control system is the difference.

James Thornton, *Design of a Computer: The Control Data 6600*, 1970

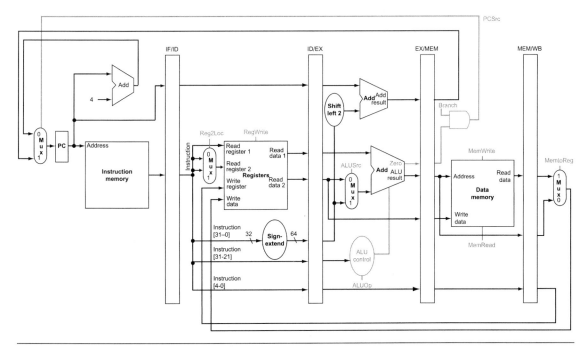

FIGURE 4.45 The pipelined datapath of Figure 4.40 with the control signals identified. This datapath borrows the control logic for PC source, register destination number, and ALU control from Section 4.4. Note that we now need the opcode field of the instruction in the EX stage as input to ALU control, so these bits must also be included in the ID/EX pipeline register.

datapath in Figure 4.17. In particular, we use the same ALU control logic, branch logic, read-register2-number multiplexor, and control lines. These functions are defined in Figures 4.12, 4.16, and 4.18. We reproduce the key information in Figures 4.46 through 4.48 on a single page to make the following discussion easier to absorb.

As was the case for the single-cycle implementation, we assume that the PC is written on each clock cycle, so there is no separate write signal for the PC. By the same argument, there are no separate write signals for the pipeline registers (IF/ID, ID/EX, EX/MEM, and MEM/WB), since the pipeline registers are also written during each clock cycle.

To specify control for the pipeline, we need only set the control values during each pipeline stage. Because each control line is associated with a component active in only a single pipeline stage, we can divide the control lines into five groups according to the pipeline stage.

1. *Instruction fetch:* The control signals to read instruction memory and to write the PC are always asserted, so there is nothing special to control in this pipeline stage.

2. *Instruction decode/register file read:* We need to select the correct register number for read register 2, so the signal Reg2Loc is set. The signal selects instruction bits 20:16 (Rm) or 4:0 (Rt).

3. *Execution/address calculation:* The signals to be set are ALUOp and ALUSrc (see Figures 4.46 and 4.47). The signals select the ALU operation and either Read data 2 or a sign-extended immediate as inputs to the ALU.

4. *Memory access:* The control lines set in this stage are Branch, MemRead, and MemWrite. The compare and branch on zero, load, and store instructions set these signals, respectively. Recall that PCSrc in Figure 4.47 selects the next sequential address unless control asserts Branch and the ALU result was 0.

5. *Write-back:* The two control lines are MemtoReg, which decides between sending the ALU result or the memory value to the register file, and RegWrite, which writes the chosen value.

Since pipelining the datapath leaves the meaning of the control lines unchanged, we can use the same control values. Figure 4.48 has the same values as in Section 4.4, but now the nine control lines are grouped by pipeline stage.

Implementing control means setting the nine control lines to these values in each stage for each instruction. We need to set Reg2Loc to read the right register while we are still decoding the instruction during the ID pipeline stage. We need to

Instruction	ALUOp	Instruction operation	Opcode	Desired ALU action	ALU control input
LDUR	00	load register	XXXXXXXXXX	add	0010
STUR	00	store register	XXXXXXXXXX	add	0010
CBZ	01	compare and branch on zero	XXXXXXXXXX	pass input b	0111
R-type	10	ADD	10001011000	add	0010
R-type	10	SUB	11001011000	subtract	0110
R-type	10	AND	10001010000	AND	0000
R-type	10	ORR	10101010000	OR	0001

FIGURE 4.46 A copy of Figure 4.12. This figure shows how the ALU control bits are set depending on the ALUOp control bits and the different opcodes for the R-type instruction.

Signal name	Effect when deasserted (0)	Effect when asserted (1)
Reg2Loc	The register number for Read register 2 comes from the Rm field (bits 20:16).	The register number for Read register 2 comes from the Rt field (bits 4:0).
RegWrite	None.	The register on the Write register input is written with the value on the Write data input.
ALUSrc	The second ALU operand comes from the second register file output (Read data 2).	The second ALU operand is the immediate field of the instruction.
PCSrc	The PC is replaced by the output of the adder that computes the value of PC + 4.	The PC is replaced by the output of the adder that computes the branch target.
MemRead	None.	Data memory contents designated by the address input are put on the Read data output.
MemWrite	None.	Data memory contents designated by the address input are replaced by the value on the Write data input.
MemtoReg	The value fed to the register Write data input comes from the ALU.	The value fed to the register Write data input comes from the data memory

FIGURE 4.47 A copy of Figure 4.16. The function of each of seven control signals is defined. The ALU control lines (ALUOp) are defined in the second column of Figure 4.46. When a 1-bit control to a two-way multiplexor is asserted, the multiplexor selects the input corresponding to 1. Otherwise, if the control is deasserted, the multiplexor selects the 0 input. Note that PCSrc is controlled by an AND gate in Figure 4.45. If the Branch signal and the ALU Zero signal are both set, then PCSrc is 1; otherwise, it is 0. Control sets the Branch signal only during a CBZ instruction; otherwise, PCSrc is set to 0.

Instruction	Instruction decode stage control lines	Execution/address calculation stage control lines			Memory access stage control lines			Write-back stage control lines	
Instruction	Reg2Loc	ALUOp1	ALUOp0	ALUSrc	Branch	MemRead	MemWrite	RegWrite	MemtoReg
R-format	0	1	0	0	0	0	0	1	0
LDUR	X	0	0	1	0	1	0	1	1
STUR	1	0	0	1	0	0	1	0	X
CBZ	1	0	1	0	1	0	0	0	X

FIGURE 4.48 The values of the control lines are the same as in Figure 4.18, but they have been shuffled into four groups corresponding to the last four pipeline stages.

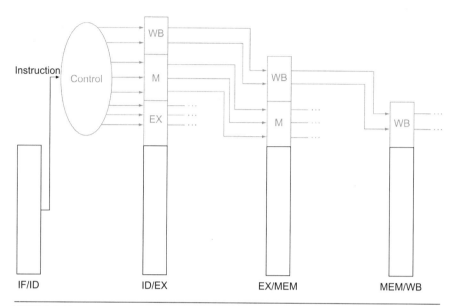

FIGURE 4.49   The eight control lines for the final three stages. Note that three of the eight control lines are used in the EX phase, with the remaining five control lines passed on to the EX/MEM pipeline register extended to hold the control lines; three are used during the MEM stage, and the last two are passed to MEM/WB for use in the WB stage.

use bits of the opcode to control this multiplexor, just as we did for sign extension as explained in the elaboration earlier. If you look at the opcodes carefully—such as in Figure 4.22—you can see that instruction bit 28 is a 0 for R-format instructions and a 1 for the rest. This is exactly what we need to control the register address multiplexor, so we simply connect instruction bit 28 to Reg2Loc.

Since the rest of the control lines starts with the EX stage, we can create the control information during instruction decode for the later stages. The simplest way to pass these control signals is to extend the pipeline registers to include control information. Figure 4.49 above shows that these control signals are then used in the appropriate pipeline stage as the instruction moves down the pipeline, just as the destination register number for loads moves down the pipeline in Figure 4.40. Figure 4.50 shows the full datapath with the extended pipeline registers and with the control lines connected to the proper stage along with the instruction bit 28 controlling the Reg2Loc register address multiplexor. (🌐 Section 4.13 gives more examples of LEGv8 code executing on pipelined hardware using single-clock diagrams, if you would like to see more details.)

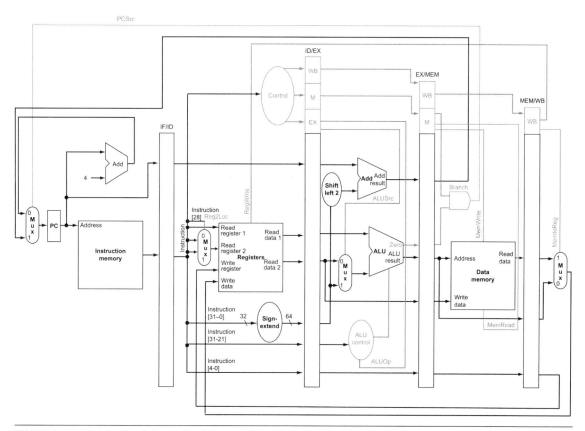

FIGURE 4.50 The pipelined datapath of Figure 4.45, with the control signals connected to the control portions of the pipeline registers. The control values for the last three stages are created during the instruction decode stage and then placed in the ID/EX pipeline register. The control lines for each pipe stage are used, and remaining control lines are then passed to the next pipeline stage.

Elaboration: Because one of the source registers for LEGv8 instructions is found in different places in different instructions, the LEGv8 computer designer has a more difficult challenge than the MIPS computer designer, where the source registers are always in the same location in the MIPS instruction formats. Thus, the MIPS decode stage can read two registers without doing any decoding. In the ID stage for LEGv8, either you need a three-ported register file to read the three possible registers, or you need to do some partial decoding, as we have done in this chapter.

4.7   Data Hazards: Forwarding versus Stalling

The examples in the previous section show the power of pipelined execution and how the hardware performs the task. It's now time to take off the rose-colored glasses and look at what happens with real programs. The LEGv8 instructions in Figures 4.42 through 4.44 were independent; none of them used the results calculated by any of the others. Yet, in Section 4.5, we saw that data hazards are obstacles to pipelined execution.

Let's look at a sequence with many dependences, shown in color:

```
SUB   X2, X1,X3      // Register X2 written by SUB
AND   X12,X2,X5      // 1st operand(X2) depends on SUB
OR    X13,X6,X2      // 2nd operand(X2) depends on SUB
ADD   X14,X2,X2      // 1st(X2) & 2nd(X2) depend on SUB
STUR  X15,[X2,#100]  // Base (X2) depends on SUB
```

The last four instructions are all dependent on the result in register X2 of the first instruction. If register X2 had the value 10 before the subtract instruction and −20 afterwards, the programmer intends that −20 will be used in the following instructions that refer to register X2.

How would this sequence perform with our pipeline? Figure 4.51 illustrates the execution of these instructions using a multiple-clock-cycle pipeline representation. To demonstrate the execution of this instruction sequence in our current pipeline, the top of Figure 4.51 shows the value of register X2, which changes during the middle of clock cycle 5, when the SUB instruction writes its result.

The last potential hazard can be resolved by the design of the register file hardware: What happens when a register is read and written in the same clock cycle? We assume that the write is in the first half of the clock cycle and the read is in the second half, so the read delivers what is written. As is the case for many implementations of register files, we have no data hazard in this case.

Figure 4.51 shows that the values read for register X2 would *not* be the result of the SUB instruction unless the read occurred during clock cycle 5 or later. Thus, the instructions that would get the correct value of −20 are ADD and STUR; the AND and ORR instructions would get the incorrect value 10! Using this style of drawing, such problems become apparent when a dependence line goes backward in time.

As mentioned in Section 4.5, the desired result is available at the end of the EX stage of the SUB instruction or clock cycle 3. When are the data actually needed by the AND and ORR instructions? The answer is at the beginning of the EX stage of the AND and ORR instructions, or clock cycles 4 and 5, respectively. Thus, we can execute this segment without stalls if we simply *forward* the data as soon as it is available to any units that need it before it is ready to read from the register file.

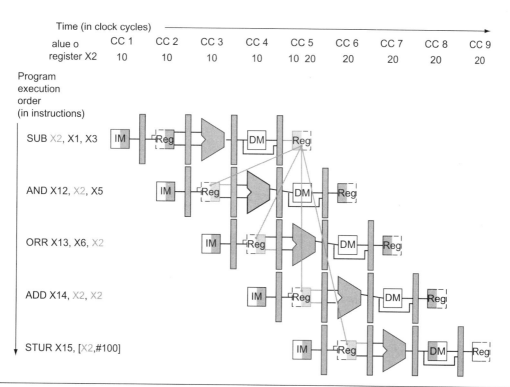

FIGURE 4.51 Pipelined dependences in a five-instruction sequence using simplified datapaths to show the dependences. All the dependent actions are shown in color, and "CC 1" at the top of the figure means clock cycle 1. The first instruction writes into X2, and all the following instructions read X2. This register is written in clock cycle 5, so the proper value is unavailable before clock cycle 5. (A read of a register during a clock cycle returns the value written at the end of the first half of the cycle, when such a write occurs.) The colored lines from the top datapath to the lower ones show the dependences. Those that must go backward in time are *pipeline data hazards*.

How does forwarding work? For simplicity in the rest of this section, we consider only the challenge of forwarding to an operation in the EX stage, which may be either an ALU operation or an effective address calculation. This means that when an instruction tries to use a register in its EX stage that an earlier instruction intends to write in its WB stage, we actually need the values as inputs to the ALU.

A notation that names the fields of the pipeline registers allows for a more precise notation of dependences. For example, "ID/EX.RegisterRn1" refers to the number of one register whose value is found in the pipeline register ID/EX; that is, the one from the first read port of the register file. The first part of the name, to the left of the period, is the name of the pipeline register; the second part is the name of the field in that register. Using this notation, the two pairs of hazard conditions are

1a. EX/MEM.RegisterRd = ID/EX.RegisterRn1

1b. EX/MEM.RegisterRd = ID/EX.RegisterRm2

2a. MEM/WB.RegisterRd = ID/EX.RegisterRn1

2b. MEM/WB.RegisterRd = ID/EX.RegisterRm2

The first hazard in the sequence on page 316 is on register X2, between the result of SUB X2,X1,X3 and the first read operand of AND X12,X2,X5. This hazard can be detected when the AND instruction is in the EX stage and the prior instruction is in the MEM stage, so this is hazard 1a:

EX/MEM.RegisterRd = ID/EX.RegisterRn1 = X2

EXAMPLE

Dependence Detection

Classify the dependences in this sequence from page 316:

```
SUB  X2,  X1, X3     // Register X2 set by SUB
AND  X12, X2, X5     // 1st operand(X2) set by SUB
OR   X13, X6, X2     // 2nd operand(X2) set by SUB
ADD  X14, X2, X2     // 1st(X2) & 2nd(X2) set by SUB
STUR X15, [X2,#100]  // Index(X2) set by SUB
```

ANSWER

As mentioned above, the SUB-AND is a type 1a hazard. The remaining hazards are as follows:

- The SUB-ORR is a type 2b hazard:

  MEM/WB.RegisterRd = ID/EX.RegisterRm2 = X2

- The two dependences on SUB-ADD are not hazards because the register file supplies the proper data during the ID stage of ADD.

- There is no data hazard between SUB and STUR because STUR reads X2 the clock cycle *after* SUB writes X2.

Because some instructions do not write registers, this policy is inaccurate; sometimes it would forward when it shouldn't. One solution is simply to check to see if the RegWrite signal will be active: examining the WB control field of the pipeline register during the EX and MEM stages determines whether RegWrite is asserted. Recall that LEGv8 requires that every use of XZR(X31) as an operand must yield an operand value of 0. If an instruction in the pipeline has XZR as its destination (for example, SUBS XZR,X1,2), we want to avoid forwarding its possibly nonzero result value. Not forwarding results destined for XZR frees the assembly programmer and the compiler of any requirement to avoid using XZR as a destination. The conditions above thus work properly as long as we add EX/ MEM.RegisterRd ≠ 31 to the first hazard condition and MEM/WB.RegisterRd ≠ 31 to the second.

The last detail is to recall that two instructions—STUR and CBZ—use Rt (bits 4:0) to specify the second register operand instead of the Rm field (bits 20:16). The pipeline resolves this decision in the ID stage using a 2:1 multiplexor and the Reg2Loc control signal. We can simplify pipeline control by using the output of this multiplexor instead of the actual Rm field so that control always sees the proper second register number. We'll still call the number RegisterRm, but remember it handles STUR and CBZ correctly too.

Now that we can detect hazards, half of the problem is resolved—but we must still forward the proper data.

Figure 4.52 shows the dependences between the pipeline registers and the inputs to the ALU for the same code sequence as in Figure 4.51. The change is that the

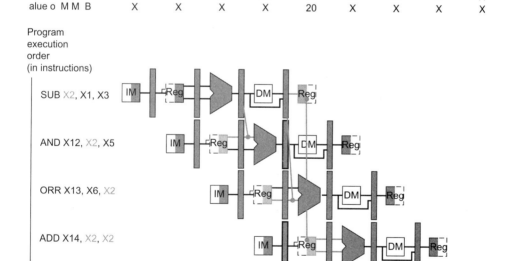

Time (in clock cycles)									
	CC 1	CC 2	CC 3	CC 4	CC 5	CC 6	CC 7	CC 8	CC 9
alue o register X2	10	10	10	10	10 20	20	20	20	20
alue o X M M	X	X	X	20	X	X	X	X	X
alue o M M B	X	X	X	X	20	X	X	X	X

Program
execution
order
(in instructions)

SUB X2, X1, X3

AND X12, X2, X5

ORR X13, X6, X2

ADD X14, X2, X2

STUR X15, [X2,#100]

FIGURE 4.52 The dependences between the pipeline registers move forward in time, so it is possible to supply the inputs to the ALU needed by the AND **instruction and** ORR **instruction by forwarding the results found in the pipeline registers.** The values in the pipeline registers show that the desired value is available before it is written into the register file. We assume that the register file forwards values that are read and written during the same clock cycle, so the ADD does not stall, but the values come from the register file instead of a pipeline register. Register file "forwarding"—that is, the read gets the value of the write in that clock cycle—is why clock cycle 5 shows register X2 having the value 10 at the beginning and −20 at the end of the clock cycle.

dependence begins from a *pipeline* register, rather than waiting for the WB stage to write the register file. Thus, the required data exist in time for later instructions, with the pipeline registers holding the data to be forwarded.

If we can take the inputs to the ALU from *any* pipeline register rather than just ID/EX, then we can forward the correct data. By adding multiplexors to the input of the ALU, and with the proper controls, we can run the pipeline at full speed in the presence of these data hazards.

For now, we will assume the only instructions we need to forward are the four R-format instructions: ADD, SUB, AND, and ORR. Figure 4.53 shows a close-up of the ALU and pipeline register before and after adding forwarding. Figure 4.54 shows the values of the control lines for the ALU multiplexors that select either the register file values or one of the forwarded values.

This forwarding control will be in the EX stage, because the ALU forwarding multiplexors are found in that stage. Thus, we must pass the operand register numbers from the ID stage via the ID/EX pipeline register to determine whether to forward values. Before forwarding, the ID/EX register had no need to include space to hold the Rn and Rm fields. Hence, they were added to ID/EX.

Let's now write both the conditions for detecting hazards, and the control signals to resolve them:

1. *EX hazard:*

```
if (EX/MEM.RegWrite
and (EX/MEM.RegisterRd ≠ 31)
and (EX/MEM.RegisterRd = ID/EX.RegisterRn1)) ForwardA = 10

if (EX/MEM.RegWrite
and (EX/MEM.RegisterRd ≠ 31)
and (EX/MEM.RegisterRd = ID/EX.RegisterRm2)) ForwardB = 10
```

This case forwards the result from the previous instruction to either input of the ALU. If the previous instruction is going to write to the register file, and the write register number matches the read register number of ALU inputs A or B, provided it is not register 31, then steer the multiplexor to pick the value instead from the pipeline register EX/MEM.

2. *MEM hazard:*

```
if (MEM/WB.RegWrite
and (MEM/WB.RegisterRd ≠ 31)
and (MEM/WB.RegisterRd = ID/EX.RegisterRn1)) ForwardA = 01
if (MEM/WB.RegWrite
and (MEM/WB.RegisterRd ≠ 31)
and (MEM/WB.RegisterRd = ID/EX.RegisterRm2)) ForwardB = 01
```

As mentioned above, there is no hazard in the WB stage, because we assume that the register file supplies the correct result if the instruction in the ID stage reads the same register written by the instruction in the WB stage. Such a register file performs another form of forwarding, but it occurs within the register file.

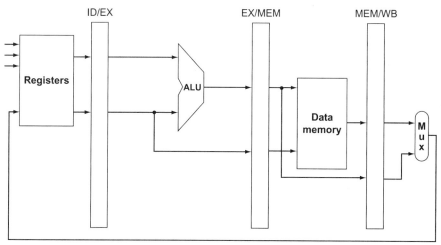

a. No forwarding

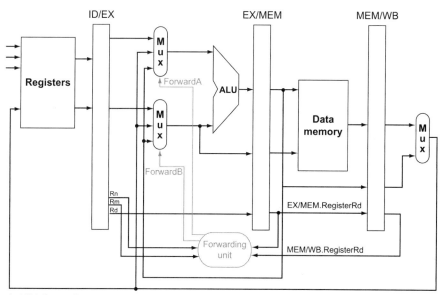

b. With forwarding

FIGURE 4.53 On the top are the ALU and pipeline registers before adding forwarding. On the bottom, the multiplexors have been expanded to add the forwarding paths, and we show the forwarding unit. The new hardware is shown in color. This figure is a stylized drawing, however, leaving out details from the full datapath such as the sign extension hardware.

Mux control	Source	Explanation
ForwardA = 00	ID/EX	The first ALU operand comes from the register file.
ForwardA = 10	EX/MEM	The first ALU operand is forwarded from the prior ALU result.
ForwardA = 01	MEM/WB	The first ALU operand is forwarded from data memory or an earlier ALU result.
ForwardB = 00	ID/EX	The second ALU operand comes from the register file.
ForwardB = 10	EX/MEM	The second ALU operand is forwarded from the prior ALU result.
ForwardB = 01	MEM/WB	The second ALU operand is forwarded from data memory or an earlier ALU result.

FIGURE 4.54   The control values for the forwarding multiplexors in Figure 4.53. The signed immediate that is another input to the ALU is described in the *Elaboration* at the end of this section.

One complication is potential data hazards between the result of the instruction in the WB stage, the result of the instruction in the MEM stage, and the source operand of the instruction in the ALU stage. For example, when summing a vector of numbers in a single register, a sequence of instructions will all read and write to the same register:

```
ADD X1,X1,X2
ADD X1,X1,X3
ADD X1,X1,X4
 . . .
```

In this case, the result should be forwarded from the MEM stage because the result in the MEM stage is the more recent result. Thus, the control for the MEM hazard would be (with the additions highlighted):

```
if  (MEM/WB.RegWrite
and (MEM/WB.RegisterRd ≠ 31)
and not(EX/MEM.RegWrite and (EX/MEM.RegisterRd ≠ 31)
        and (EX/MEM.RegisterRd ≠ ID/EX.RegisterRn1))
and (MEM/WB.RegisterRd = ID/EX.RegisterRn1)) ForwardA = 01

if  (MEM/WB.RegWrite
and (MEM/WB.RegisterRd ≠ 31)
and not(EX/MEM.RegWrite and (EX/MEM.RegisterRd ≠ 31)
        and (EX/MEM.RegisterRd ≠ ID/EX.RegisterRm2))
and (MEM/WB.RegisterRd = ID/EX.RegisterRm2)) ForwardB = 01
```

Figure 4.55 shows the hardware necessary to support forwarding for operations that use results during the EX stage. Note that the EX/MEM.RegisterRd field is the register destination for either an ALU instruction (which comes from the Rd field of the instruction) or a load (which comes from the Rt field, but we'll use the notation Rd in this section).

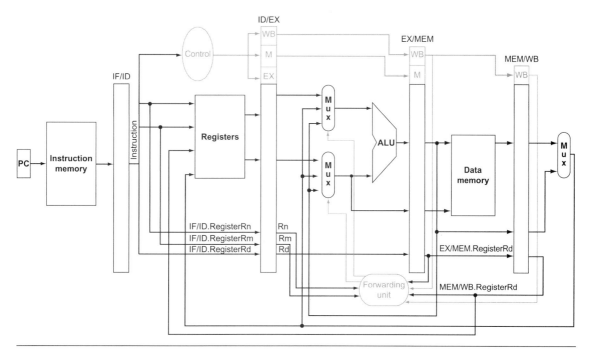

FIGURE 4.55   The datapath modified to resolve hazards via forwarding. Compared with the datapath in Figure 4.50, the additions are the multiplexors to the inputs to the ALU. This figure is a more stylized drawing, however, leaving out details from the full datapath, such as the branch hardware and the sign extension hardware.

If you would like to see more illustrated examples using single-cycle pipeline drawings, ⊞ Section 4.13 has figures that show two pieces of LEGv8 code with hazards that cause forwarding.

Elaboration: Forwarding can also help with hazards when store instructions are dependent on other instructions. Since they use just one data value during the MEM stage, forwarding is easy. However, consider loads immediately followed by stores, useful when performing memory-to-memory copies in the LEGv8 architecture. Since copies are frequent, we need to add more forwarding hardware to make them run faster. If we were to redraw Figure 4.52, replacing the SUB and AND instructions with LDUR and STUR, we would see that it is possible to avoid a stall, since the data exist in the MEM/WB register of a load instruction in time for its use in the MEM stage of a store instruction. We would need to add forwarding into the memory access stage for this option. We leave this modification as an exercise to the reader.

In addition, the signed-immediate input to the ALU, needed by loads and stores, is missing from the datapath in Figure 4.55. Since central control decides between register and immediate, and since the forwarding unit chooses the pipeline register for a register input to the ALU, the easiest solution is to add a 2:1 multiplexor that chooses between the ForwardB multiplexor output and the signed immediate. Figure 4.56 shows this addition.

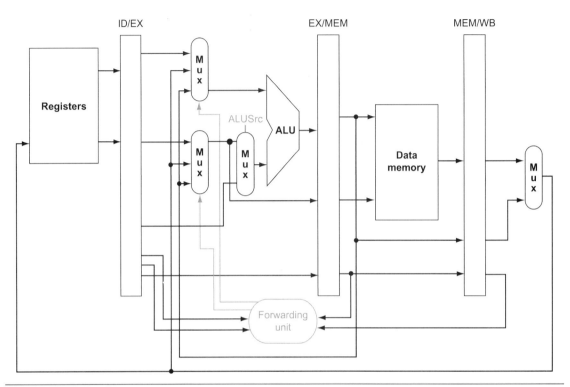

FIGURE 4.56   A close-up of the datapath in Figure 4.53 shows a 2:1 multiplexor, which has been added to select the signed immediate as an ALU input.

Data Hazards and Stalls

If at first you don't succeed, redefine success.

Anonymous

As we said in Section 4.5, one case where forwarding cannot save the day is when an instruction tries to read a register following a load instruction that writes the same register. Figure 4.57 illustrates the problem. The data are still being read from memory in clock cycle 4 while the ALU is performing the operation for the following instruction. Something must stall the pipeline for the combination of load followed by an instruction that reads its result.

Hence, in addition to a forwarding unit, we need a *hazard detection unit*. It operates during the ID stage so that it can insert the stall between the load and the instruction dependent on it. Checking for load instructions, the control for the hazard detection unit is this single condition:

```
if (ID/EX.MemRead and
    ((ID/EX.RegisterRd = IF/ID.RegisterRn1) or
     (ID/EX.RegisterRd = IF/ID.RegisterRm2)))
     stall the pipeline
```

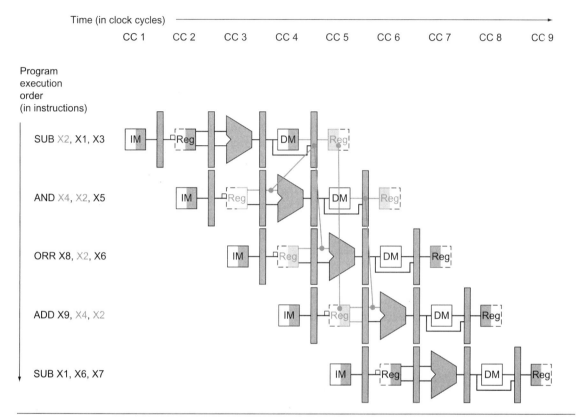

Time (in clock cycles)

CC 1   CC 2   CC 3   CC 4   CC 5   CC 6   CC 7   CC 8   CC 9

Program
execution
order
(in instructions)

SUB X2, X1, X3

AND X4, X2, X5

ORR X8, X2, X6

ADD X9, X4, X2

SUB X1, X6, X7

FIGURE 4.57 A pipelined sequence of instructions. Since the dependence between the load and the following instruction (and) goes backward in time, this hazard cannot be solved by forwarding. Hence, this combination must result in a stall by the hazard detection unit.

Recall that we are using the RegisterRd to refer the register specified in instruction bits 4:0 for both load and R-type instructions. The first line tests to see if the instruction is a load: the only instruction that reads data memory is a load. The next two lines check to see if the destination register field of the load in the EX stage matches either source register of the instruction in the ID stage. If the condition holds, the instruction stalls one clock cycle. After this one-cycle stall, the forwarding logic can handle the dependence and execution proceeds. (If there were no forwarding, then the instructions in Figure 4.57 would need another stall cycle.)

If the instruction in the ID stage is stalled, then the instruction in the IF stage must also be stalled; otherwise, we would lose the fetched instruction. Preventing these two instructions from making progress is accomplished simply by preventing the PC register and the IF/ID pipeline register from changing. Provided these registers are preserved, the instruction in the IF stage will continue to be read using the same PC, and the registers in the ID stage will continue to be read using

nop An instruction that does no operation to change state.

the same instruction fields in the IF/ID pipeline register. Returning to our favorite analogy, it's as if you restart the washer with the same clothes and let the dryer continue tumbling empty. Of course, like the dryer, the back half of the pipeline starting with the EX stage must be doing something; what it is doing is executing instructions that have no effect: **nops**.

How can we insert these nops, which act like bubbles, into the pipeline? In Figure 4.48, we see that deasserting all eight control signals (setting them to 0) in the EX, MEM, and WB stages will create a "do nothing" or nop instruction. By identifying the hazard in the ID stage, we can insert a bubble into the pipeline by changing the EX, MEM, and WB control fields of the ID/EX pipeline register to 0. These benign control values are percolated forward at each clock cycle with the proper effect: no registers or memories are written if the control values are all 0.

Figure 4.58 shows what really happens in the hardware: the pipeline execution slot associated with the AND instruction is turned into a nop and all instructions beginning with the AND instruction are delayed one cycle. Like an air bubble in

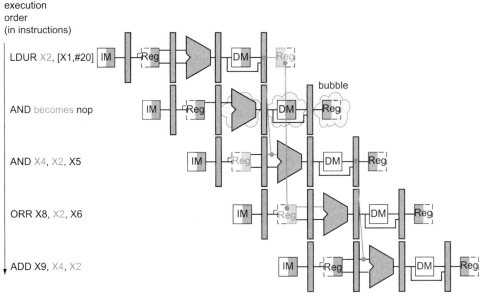

FIGURE 4.58 The way stalls are really inserted into the pipeline. A bubble is inserted beginning in clock cycle 4, by changing the AND instruction to a nop. Note that the AND instruction is really fetched and decoded in clock cycles 2 and 3, but its EX stage is delayed until clock cycle 5 (versus the unstalled position in clock cycle 4). Likewise, the ORR instruction is fetched in clock cycle 3, but its ID stage is delayed until clock cycle 5 (versus the unstalled clock cycle 4 position). After insertion of the bubble, all the dependences go forward in time and no further hazards occur.

a water pipe, a stall bubble delays everything behind it and proceeds down the instruction pipe one stage each clock cycle until it exits at the end. In this example, the hazard forces the AND and ORR instructions to repeat in clock cycle 4 what they did in clock cycle 3: AND reads registers and decodes, and ORR is refetched from instruction memory. Such repeated work is what a stall looks like, but its effect is to stretch the time of the AND and ORR instructions and delay the fetch of the ADD instruction.

Figure 4.59 highlights the pipeline connections for both the hazard detection unit and the forwarding unit. As before, the forwarding unit controls the ALU multiplexors to replace the value from a general-purpose register with the value from the proper pipeline register. The hazard detection unit controls the writing of the PC and IF/ID registers plus the multiplexor that chooses between the real control values and all 0s. The hazard detection unit stalls and deasserts the control fields if the load-use hazard test above is true. If you would like to see more details, Section 4.13 gives an example illustrated using single-clock pipeline diagrams of LEGv8 code with hazards that cause stalling.

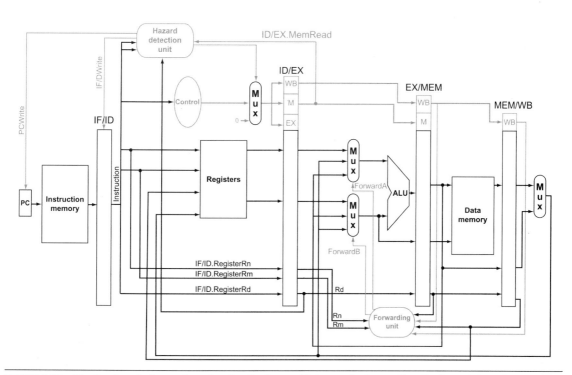

FIGURE 4.59 Pipelined control overview, showing the two multiplexors for forwarding, the hazard detection unit, and the forwarding unit. Although the ID and EX stages have been simplified—the sign-extended immediate and branch logic are missing—this drawing gives the essence of the forwarding hardware requirements.

The **BIG** Picture

Although the compiler generally relies upon the hardware to resolve hazards and thereby ensure correct execution, the compiler must understand the pipeline to achieve the best performance. Otherwise, unexpected stalls will reduce the performance of the compiled code.

Elaboration: Regarding the remark earlier about setting control lines to 0 to avoid writing registers or memory: only the signals RegWrite and MemWrite need be 0, while the other control signals can be don't cares.

There are a thousand hacking at the branches of evil to one who is striking at the root.

Henry David Thoreau, *Walden*, 1854

4.8 Control Hazards

Thus far, we have limited our concern to hazards involving arithmetic operations and data transfers. However, as we saw in Section 4.5, there are also pipeline hazards involving conditional branches. Figure 4.60 shows a sequence of instructions and indicates when the branch would occur in this pipeline. An instruction must be fetched at every clock cycle to sustain the pipeline, yet in our design the decision about whether to branch doesn't occur until the MEM pipeline stage. As mentioned in Section 4.5, this delay in determining the proper instruction to fetch is called a *control hazard* or *branch hazard*, in contrast to the *data hazards* we have just examined.

This section on control hazards is shorter than the previous sections on data hazards. The reasons are that control hazards are relatively simple to understand, they occur less frequently than data hazards, and there is nothing as effective against control hazards as forwarding is against data hazards. Hence, we use simpler schemes. We look at two schemes for resolving control hazards and one optimization to improve these schemes.

Assume Branch Not Taken

As we saw in Section 4.5, stalling until the branch is complete is too slow. One improvement over branch stalling is to **predict** that the conditional branch will not be taken and thus continue execution down the sequential instruction stream. If the conditional branch is taken, the instructions that are being fetched and decoded must be discarded. Execution continues at the branch target. If conditional branches are untaken half the time, and if it costs little to discard the instructions, this optimization halves the cost of control hazards.

PREDICTION

Time (in clock cycles)

CC 1   CC 2   CC 3   CC 4   CC 5   CC 6   CC 7   CC 8   CC 9

Program
execution
order
(in instructions)

40 CBZ X1, 8

44 AND X12, X2, X5

48 ORR X13, X6, X2

52 ADD X14, X2, X2

72 LDUR X4, [X7,#100]

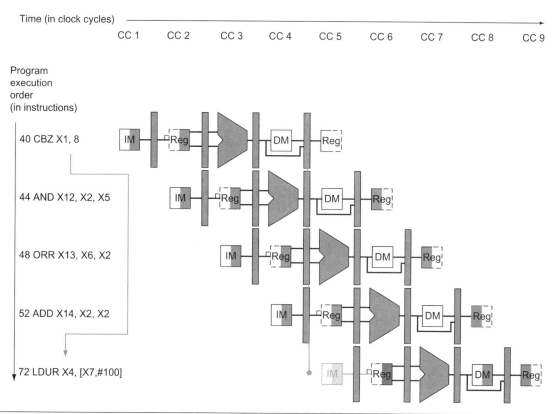

FIGURE 4.60 The impact of the pipeline on the branch instruction. The numbers to the left of the instruction (40, 44, …) are the addresses of the instructions. Since the branch instruction decides whether to branch in the MEM stage—clock cycle 4 for the CBZ instruction above—the three sequential instructions that follow the branch will be fetched and begin execution. Without intervention, those three following instructions will begin execution before CBZ branches to LDUR at location 72. (Figure 4.30 assumed extra hardware to reduce the control hazard to one clock cycle; this figure uses the nonoptimized datapath.)

To discard instructions, we merely change the original control values to 0s, much as we did to stall for a load-use data hazard. The difference is that we must also change the three instructions in the IF, ID, and EX stages when the branch reaches the MEM stage; for load-use stalls, we just change control to 0 in the ID stage and let them percolate through the pipeline. Discarding instructions, then, means we must be able to **flush** instructions in the IF, ID, and EX stages of the pipeline.

flush To discard instructions in a pipeline, usually due to an unexpected event.

Reducing the Delay of Branches

One way to improve conditional branch performance is to reduce the cost of the taken branch. Thus far, we have assumed the next PC for a branch is selected in the MEM stage, but if we move the conditional branch execution earlier in the pipeline, then fewer instructions need be flushed. Moving the branch decision up requires two actions to occur earlier: computing the branch target address and evaluating the branch decision. The easy part of this change is to move up the branch address calculation. We already have the PC value and the immediate field in the IF/ID pipeline register, so we just move the branch adder from the EX stage to the ID stage; of course, the address calculation for branch targets will be performed for all instructions, but only used when needed.

The harder part is the branch decision itself. For compare and branch zero, we would compare a register read during the ID stage to see if it is zero. Zero can be tested by ORing all 64 bits. Moving the branch test to the ID stage implies additional forwarding and hazard detection hardware, since a branch dependent on a result still in the pipeline must still work properly with this optimization. For example, to implement compare and branch on zero (and its inverse), we will need to forward results to the zero test logic that operates during ID. There are two complicating factors:

1. During ID, we must decode the instruction, decide whether a bypass to the zero test unit is needed, and complete the zero test so that if the instruction is a branch, we can set the PC to the branch target address. Forwarding for the operand of branches was formerly handled by the ALU forwarding logic, but the introduction of the zero test unit in ID will require new forwarding logic. Note that the bypassed source operands of a branch can come from either the ALU/MEM or MEM/WB pipeline latches.

2. Because the value in a branch comparison is needed during ID but may be produced later in time, it is possible that a data hazard can occur and a stall will be needed. For example, if an ALU instruction immediately preceding a branch produces the operand for the test in the conditional branch, a stall will be required, since the EX stage for the ALU instruction will occur after the ID cycle of the branch. By extension, if a load is immediately followed by a conditional branch that depends on the load result, two stall cycles will be needed, as the result from the load appears at the end of the MEM cycle but is needed at the beginning of ID for the branch.

Despite these difficulties, moving the conditional branch execution to the ID stage is an improvement, because it reduces the penalty of a branch to only one instruction if the branch is taken, namely, the one currently being fetched. The exercises explore the details of implementing the forwarding path and detecting the hazard.

To flush instructions in the IF stage, we add a control line, called IF.Flush, that zeros the instruction field of the IF/ID pipeline register. Clearing the register transforms the fetched instruction into a nop, an instruction that has no action and changes no state.

Pipelined Branch

Show what happens when the branch is taken in this instruction sequence, assuming the pipeline is optimized for branches that are not taken, and that we moved the branch execution to the ID stage:

```
36  SUB   X10, X4, X8
40  CBZ   X1,  X3, 8 //  PC-relative branch to 40+8*4=72
44  AND   X12, X2, X5
48  ORR   X13, X2, X6
52  ADD   X14, X4, X2
56  SUB   X15, X6, X7
    . . .
72  LDUR  X4,  [X7,#50]
```

Figure 4.61 shows what happens when a conditional branch is taken. Unlike Figure 4.60, there is only one pipeline bubble on a taken branch.

Dynamic Branch Prediction

Assuming a conditional branch is not taken is one simple form of *branch prediction*. In that case, we predict that conditional branches are untaken, flushing the pipeline when we are wrong. For the simple five-stage pipeline, such an approach, possibly coupled with compiler-based prediction, is probably adequate. With deeper pipelines, the branch penalty increases when measured in clock cycles. Similarly, with multiple issue (see Section 4.10), the branch penalty increases in terms of instructions lost. This combination means that in an aggressive pipeline, a simple static prediction scheme will probably waste too much performance. As we mentioned in Section 4.5, with more hardware it is possible to try to **predict** branch behavior during program execution.

One approach is to look up the address of the instruction to see if the conditional branch was taken the last time this instruction was executed, and, if so, to begin fetching new instructions from the same place as the last time. This technique is called dynamic branch prediction.

One implementation of that approach is a branch prediction buffer or branch history table. A branch prediction buffer is a small memory indexed by the lower portion of the address of the branch instruction. The memory contains a bit that says whether the branch was recently taken or not.

This prediction uses the simplest sort of buffer; we don't know, in fact, if the prediction is the right one—it may have been put there by another conditional branch that has the same low-order address bits. However, this doesn't affect correctness. Prediction is just a hint that we hope is correct, so fetching begins in the predicted direction. If the hint turns out to be wrong, the incorrectly predicted

PREDICTION

dynamic branch prediction Prediction of branches at runtime using runtime information.

branch prediction buffer Also called **branch history** table. A small memory that is indexed by the lower portion of the address of the branch instruction and that contains one or more bits indicating whether the branch was recently taken or not.

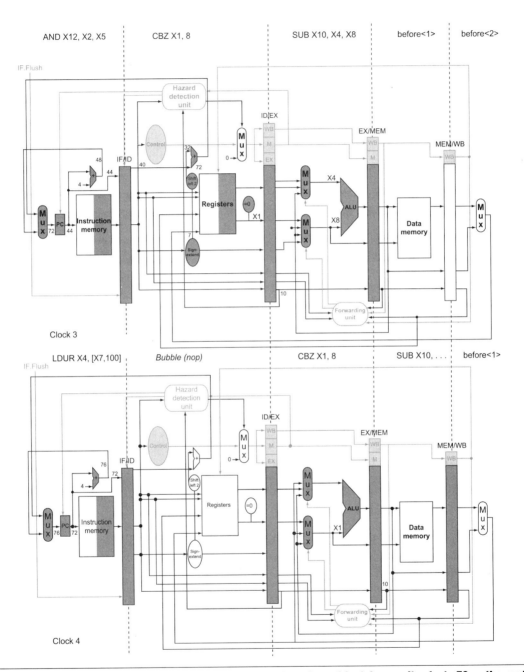

FIGURE 4.61 The ID stage of clock cycle 3 determines that a branch must be taken, so it selects 72 as the next PC address and zeros the instruction fetched for the next clock cycle. Clock cycle 4 shows the instruction at location 72 being fetched and the single bubble or nop instruction in the pipeline because of the taken branch.

instructions are deleted, the prediction bit is inverted and stored back, and the proper sequence is fetched and executed.

This simple 1-bit prediction scheme has a performance shortcoming: even if a conditional branch is almost always taken, we can predict incorrectly twice, rather than once, when it is not taken. The following example shows this dilemma.

Loops and Prediction

Consider a loop branch that branches nine times in a row, and then is not taken once. What is the prediction accuracy for this branch, assuming the prediction bit for this branch remains in the prediction buffer?

The steady-state prediction behavior will mispredict on the first and last loop iterations. Mispredicting the last iteration is inevitable since the prediction bit will indicate taken, as the branch has been taken nine times in a row at that point. The misprediction on the first iteration happens because the bit is flipped on prior execution of the last iteration of the loop, since the branch was not taken on that exiting iteration. Thus, the prediction accuracy for this branch that is taken 90% of the time is only 80% (two incorrect predictions and eight correct ones).

Ideally, the accuracy of the predictor would match the taken branch frequency for these highly regular branches. To remedy this weakness, 2-bit prediction schemes are often used. In a 2-bit scheme, a prediction must be wrong twice before it is changed. Figure 4.62 shows the finite-state machine for a 2-bit prediction scheme.

A branch prediction buffer can be implemented as a small, special buffer accessed with the instruction address during the IF pipe stage. If the instruction is predicted as taken, fetching begins from the target as soon as the PC is known; as mentioned on page 329, it can be as early as the ID stage. Otherwise, sequential fetching and executing continue. If the prediction turns out to be wrong, the prediction bits are changed as shown in Figure 4.62.

Elaboration: A branch predictor tells us whether a conditional branch is taken, but still requires the calculation of the branch target. In the five-stage pipeline, this calculation takes one cycle, meaning that taken branches will have a one-cycle penalty. One approach is to use a cache to hold the destination program counter or destination instruction using a **branch target buffer**.

branch target buffer
A structure that caches the destination PC or destination instruction for a branch. It is usually organized as a cache with tags, making it more costly than a simple prediction buffer.

correlating predictor
A branch predictor that combines local behavior of a particular branch and global information about the behavior of some recent number of executed branches.

The 2-bit dynamic prediction scheme uses only information about a particular branch. Researchers noticed that using information about both a local branch and the global behavior of recently executed branches together yields greater prediction accuracy for the same number of prediction bits. Such predictors are called **correlating predictors**. A typical correlating predictor might have two 2-bit

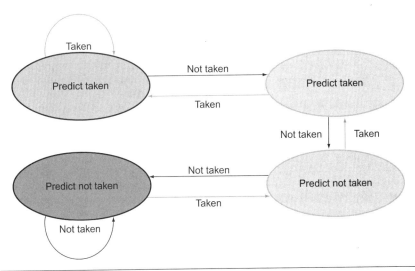

FIGURE 4.62 The states in a 2-bit prediction scheme. By using 2 bits rather than 1, a branch that strongly favors taken or not taken—as many branches do—will be mispredicted only once. The 2 bits are used to encode the four states in the system. The 2-bit scheme is a general instance of a counter-based predictor, which is incremented when the prediction is accurate and decremented otherwise, and uses the mid-point of its range as the division between taken and not taken.

predictors for each branch, with the choice between predictors made based on whether the last executed branch was taken or not taken. Thus, the global branch behavior can be thought of as adding additional index bits for the prediction lookup.

Another approach to branch prediction is the use of tournament predictors. A tournament branch predictor uses multiple predictors, tracking, for each branch, which predictor yields the best results. A typical tournament predictor might contain two predictions for each branch index: one based on local information and one based on global branch behavior. A selector would choose which predictor to use for any given prediction. The selector can operate similarly to a 1- or 2-bit predictor, favoring whichever of the two predictors has been more accurate. Some recent microprocessors use such ensemble predictors.

tournament branch predictor A branch predictor with multiple predictions for each branch and a selection mechanism that chooses which predictor to enable for a given branch.

Elaboration: One way to reduce the number of conditional branches is to add *conditional move* instructions. Instead of changing the PC with a conditional branch, the instruction conditionally changes the destination register of the move. For example, the full ARMv8 instruction set architecture has a conditional select instruction called CSEL. It specifies a destination register, two source registers, and a condition. The destination register gets a value of the first operand if the condition is true and the second operand otherwise. Thus, CSEL X8, X11, X4, NE copies the contents of register 11 into register 8 if the condition codes say the result of the operation was not equal zero or a copy of register 4 into register 11 if it was zero. Hence, programs using the full ARMv8 instruction set could have fewer conditional branches than in programs using just the LEGv8 core.

Pipeline Summary

We started in the laundry room, showing principles of pipelining in an everyday setting. Using that analogy as a guide, we explained instruction pipelining step-by-step, starting with the single-cycle datapath and then adding pipeline registers, forwarding paths, data hazard detection, branch prediction, and flushing instructions on mispredicted branches or load-use data hazards. Figure 4.63 shows the final evolved datapath and control. We now are ready for yet another control hazard: the sticky issue of exceptions.

Consider three branch prediction schemes: predict not taken, predict taken, and dynamic prediction. Assume that they all have zero penalty when they predict correctly and two cycles when they are wrong. Assume that the average predict accuracy of the dynamic predictor is 90%. Which predictor is the best choice for the following branches?

Check Yourself

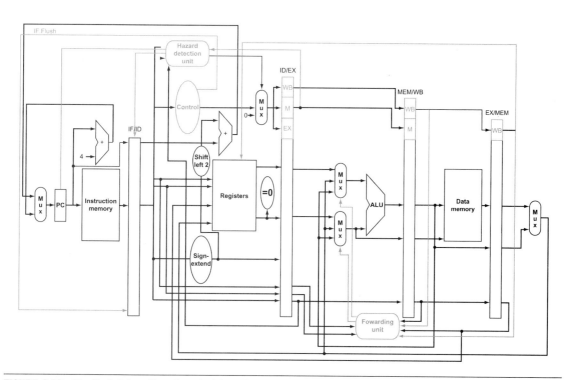

FIGURE 4.63 The final datapath and control for this chapter. Note that this is a stylized figure rather than a detailed datapath, so it's missing the ALUsrc Mux from Figure 4.56 and the multiplexor controls from Figure 4.50.

1. A conditional branch that is taken with 5% frequency

2. A conditional branch that is taken with 95% frequency

3. A conditional branch that is taken with 70% frequency

exception Also called **interrupt**. An unscheduled event that disrupts program execution; used to detect overflow.

interrupt An exception that comes from outside of the processor. (Some architectures use the term *interrupt* for all exceptions.)

4.9   Exceptions

Control is the most challenging aspect of processor design: it is both the hardest part to get right and the toughest part to make fast. One of the demanding tasks of control is implementing exceptions and interrupts—events other than branches that change the normal flow of instruction execution. They were initially created to handle unexpected events from within the processor, like floating-point overflow. The same basic mechanism was extended for I/O devices to communicate with the processor, as we will see in Chapter 5.

Many architectures and authors do not distinguish between interrupts and exceptions, often using either name to refer to both types of events. For example, the Intel x86 uses interrupt. We use the term *exception* to refer to *any* unexpected change in control flow without distinguishing whether the cause is internal or external; we use the term *interrupt* only when the event is externally caused. Here are examples showing whether the situation is internally generated by the processor or externally generated and the name that ARM uses:

Type of event	From where?	ARMv8 terminology
System reset	External	Exception
I/O device request	External	Interrupt
Invoke the operating system from user program	Internal	Exception
Floating-point arithmetic overflow or underflow	Internal	Exception
Using an undefined instruction	Internal	Exception
Hardware malfunctions	Either	Exception or interrupt

Many of the requirements to support exceptions come from the specific situation that causes an exception to occur. Accordingly, we will return to this topic in Chapter 5, when we will better understand the motivation for additional capabilities in the exception mechanism. In this section, we deal with the control implementation for detecting types of exceptions that arise from the portions of the instruction set and implementation that we have already discussed.

Detecting exceptional conditions and taking the appropriate action is often on the critical timing path of a processor, which determines the clock cycle time and thus performance. Without proper attention to exceptions during design of the control unit, attempts to add exceptions to an intricate implementation can significantly reduce performance, as well as complicate the task of getting the design correct.

How Exceptions are Handled in the LEGv8 Architecture

The types of exceptions that our current implementation can generate are execution of an undefined instruction, floating-point overflow and underflow, or a hardware malfunction. We'll assume a hardware malfunction occurs during the instruction ADD X1,X2,X1 as the example exception in the next few pages. The basic action that the processor must perform when an exception occurs is to save the address of the unfortunate instruction in the *exception link register* (ELR) and then transfer control to the operating system at some specified address.

The operating system can then take the appropriate action, which may involve providing some service to the user program, taking some predefined action in response to a malfunction, or stopping the execution of the program and reporting an error. After performing whatever action is required because of the exception, the operating system can terminate the program or may continue its execution, using the ELR to determine where to restart the execution of the program. In Chapter 5, we will look more closely at the issue of restarting the execution.

For the operating system to handle the exception, it must know the reason for the exception, in addition to the instruction that caused it. There are two main methods used to communicate the reason for an exception. The method used in the LEGv8 architecture is to include a register (called the *Exception Syndrome Register* or *ESR*), which holds a field that indicates the reason for the exception.

A second method is to use vectored interrupts. In a vectored interrupt, the address to which control is transferred is determined by the cause of the exception, possibly added to a base register that points to memory range for vectored interrupts. For example, we might define the following exception vector addresses to accommodate these exception types:

vectored interrupt An interrupt for which the address to which control is transferred is determined by the cause of the exception.

Exception type	Exception vector address to be added to a Vector Table Base Register
Unknown Reason	0000000_{two}
Floating-point arithmetic exception	101100_{two}
System Error (hardware malfunction)	101111_{two}

The operating system knows the reason for the exception by the address at which it is initiated. When the exception is not vectored, as in LEGv8, a single entry point for all exceptions can be used, and the operating system decodes the status register to find the cause. For architectures with vectored exceptions, the addresses might be separated by, say, 32 bytes or eight instructions, and the operating system must record the reason for the exception and may perform some limited processing in this sequence.

We can perform the processing required for exceptions by adding a few extra registers and control signals to our basic implementation and by slightly extending control. Let's assume that we are implementing the exception system with the single interrupt entry point being the address $0000\ 0000\ 1C09\ 0000_{hex}$. (Implementing vectored exceptions is no more difficult.) We will need to add two additional registers to our current LEGv8 implementation:

- *ELR:* A 64-bit register used to hold the address of the affected instruction. (Such a register is needed even when exceptions are vectored.)

- *ESR:* A register used to record the cause of the exception. In the LEGv8 architecture, this register is 32 bits, although some bits are currently unused. Assume there is a field that encodes the three possible exception sources mentioned above, with 8 representing an undefined instruction, 10 representing arithmetic overflow or underflow, and 12 representing hardware malfunction.

Exceptions in a Pipelined Implementation

A pipelined implementation treats exceptions as another form of control hazard. For example, suppose there is a hardware malfunction in an add instruction. Just as we did for the taken branch in the previous section, we must flush the instructions that follow the ADD instruction from the pipeline and begin fetching instructions from the new address. We will use the same mechanism we used for taken branches, but this time the exception causes the deasserting of control lines.

When we dealt with branch misprediction, we saw how to flush the instruction in the IF stage by turning it into a nop. To flush instructions in the ID stage, we use the multiplexor already in the ID stage that zeros control signals for stalls. A new control signal, called ID.Flush, is ORed with the stall signal from the hazard detection unit to flush during ID. To flush the instruction in the EX phase, we use a new signal called EX.Flush to cause new multiplexors to zero the control lines. To start fetching instructions from location 0000 0000 1C09 0000$_{hex}$, which we are using as the LEGv8 exception address, we simply add an additional input to the PC multiplexor that sends 0000 0000 1C09 0000$_{hex}$ to the PC. Figure 4.64 shows these changes.

This example points out a problem with exceptions: if we do not stop execution in the middle of the instruction, the programmer will not be able to see the original value of register X1 because it will be clobbered as the Destination register of the ADD instruction. If we assume the exception is detected during the EX stage, we can use the EX.Flush signal to prevent the instruction in the EX stage from writing its result in the WB stage. Many exceptions require that we eventually complete the instruction that caused the exception as if it executed normally. The easiest way to do this is to flush the instruction and restart it from the beginning after the exception is handled.

The final step is to save the address of the offending instruction in the *exception link register* (ELR). Figure 4.64 shows a stylized version of the datapath, including the branch hardware and necessary accommodations to handle exceptions.

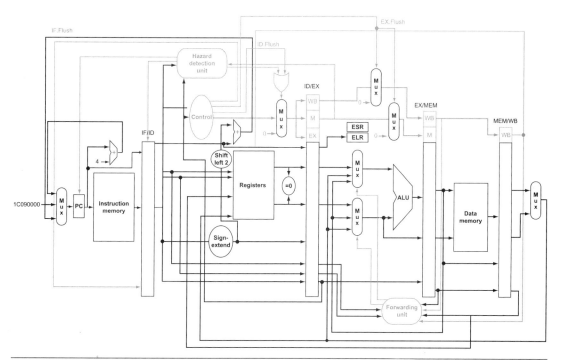

FIGURE 4.64   The datapath with controls to handle exceptions. The key additions include a new input with the value 0000 0000 1C09 0000$_{hex}$ in the multiplexor that supplies the new PC value; an ESR register to record the cause of the exception; and an ELR register to save the address of the instruction that caused the exception. The 0000 0000 1C09 0000$_{hex}$ input to the multiplexor is the initial address to begin fetching instructions in the event of an exception.

Exception in a Pipelined Computer

Given this instruction sequence,

```
40hex   SUB    X11, X2, X4
44hex   AND    X12, X2, X5
48hex   ORR    X13, X2, X6
4Chex   ADD    X1,  X2, X1
50hex   SUB    X15, X6, X7
54hex   LDUR   X16, [X7,#100]
.  .  .
```

assume the instructions to be invoked on an exception begin like this:

```
1C090000hex   STUR     X26, [X0,#1000]
1C090004hex   STUR     X27, [X0,#1008]
.  .  .
```

EXAMPLE

Show what happens in the pipeline if an exception occurs in the ADD instruction.

Figure 4.65 shows the events, starting with the add instruction in the EX stage. Assume the hardware malfunction is detected during that phase, and 0000 0000 1C09 0000$_{hex}$ is forced into the PC. Clock cycle 7 shows that the ADD and following instructions are flushed, and the first instruction of the exception code is fetched. Note that the address of the instruction *following* the ADD is saved: 4C$_{hex}$+4= 50$_{hex}$.

We mentioned several examples of exceptions on page 326, and we will see others in Chapter 5. With five instructions active in any clock cycle, the challenge is to associate an exception with the appropriate instruction. Moreover, multiple exceptions can occur simultaneously in a single clock cycle. The solution is to prioritize the exceptions so that it is easy to determine which is serviced first. In LEGv8 implementations, the hardware sorts exceptions so that the earliest instruction is interrupted.

I/O device requests and hardware malfunctions are not associated with a specific instruction, so the implementation has some flexibility as to when to interrupt the pipeline. Hence, the mechanism used for other exceptions works just fine.

The ELR captures the address of the interrupted instructions, and the ESR records the highest priority exception in a clock cycle if more than one exception occurs.

Hardware/ Software Interface

The hardware and the operating system must work in conjunction so that exceptions behave as you would expect. The hardware contract is normally to stop the offending instruction in midstream, let all prior instructions complete, flush all following instructions, set a register to show the cause of the exception, save the address of the offending instruction, and then branch to a prearranged address. The operating system contract is to look at the cause of the exception and act appropriately. For an undefined instruction or hardware failure, the operating system normally kills the program and returns an indicator of the reason. For an I/O device request or an operating system service call, the operating system saves the state of the program, performs the desired task, and, at some point in the future, restores the program to continue execution. In the case of I/O device requests, we may often choose to run another task before resuming the task that requested the I/O, since that task may often not be able to proceed until the I/O is complete. Exceptions are why the ability to save and restore the state of any task is critical. One of the most important and frequent uses of exceptions is handling page faults and TLB exceptions; Chapter 5 describes these exceptions and their handling in more detail.

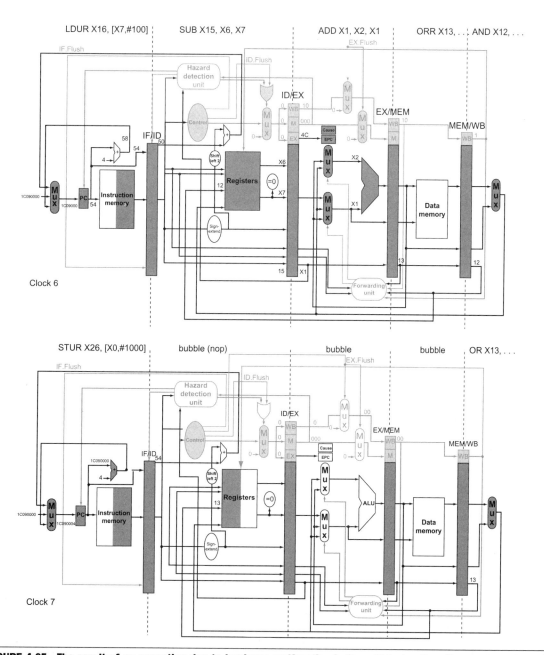

FIGURE 4.65 The result of an exception due to hardware malfunction in the add instruction. The exception is detected during the EX stage of clock 6, saving the address following the add in the ELR register ($4C + 4 = 50_{hex}$). It causes all the Flush signals to be set near the end of this clock cycle, deasserting control values (setting them to 0) for the ADD. Clock cycle 7 shows the instructions converted to bubbles in the pipeline plus the fetching of the first instruction of the exception routine—STUR X25,[X0,#1000]—from instruction location 0000 0000 1C09 0000$_{hex}$. Note that the AND and ORR instructions, which are prior to the ADD, still complete.

imprecise interrupt Also called **imprecise exception**. Interrupts or exceptions in pipelined computers that are not associated with the exact instruction that was the cause of the interrupt or exception.

precise interrupt Also called **precise exception**. An interrupt or exception that is always associated with the correct instruction in pipelined computers.

Elaboration: The difficulty of always associating the proper exception with the correct instruction in pipelined computers has led some computer designers to relax this requirement in noncritical cases. Such processors are said to have imprecise interrupts or imprecise exceptions. In the example above, PC would normally have 58_{hex} at the start of the clock cycle after the exception is detected, even though the offending instruction is at address $4C_{hex}$. A processor with imprecise exceptions might put 58_{hex} into ELR and leave it up to the operating system to determine which instruction caused the problem. LEGv8 and the vast majority of computers today support precise interrupts or precise exceptions. One reason is designers of a deeper pipeline processor might be tempted to record a different value in the ELR, which would create headaches for the OS. To prevent them, the deeper pipeline would likely be required to record the same PC that would have been recorded in the five-stage pipeline. It is simpler for everyone to just record the PC of the faulting instruction instead. (Another reason is to support virtual memory, which we shall see in Chapter 5.)

Elaboration: We show that LEGv8 uses the exception entry address 0000 0000 1C09 0000$_{hex}$, which is based on the ARMv8 Model Architecture. ARMv8 can have different exception entry addresses, depending on the platform.

Elaboration: The LEGv8 architecture has three levels of exception, each with their own ELR and ESR registers, as we'll see in Chapter 5.

Check Yourself

Which exception should be recognized first in this sequence?

1. `XXX X1, X2, X1    // undefined instruction`
2. `SUB X1, X2, X1    // hardware error`

PIPELINING

PARALLELISM

4.10 Parallelism via Instructions

Be forewarned: this section is a brief overview of fascinating but complex topics. If you want to learn more details, you should consult our more advanced book, *Computer Architecture: A Quantitative Approach*, fifth edition, where the material covered in these 13 pages is expanded to almost 200 pages (including appendices)!

Pipelining exploits the potential **parallelism** among instructions. This parallelism is called, naturally enough, instruction-level parallelism (ILP). There are two primary methods for increasing the potential amount of instruction-level parallelism. The first is increasing the depth of the pipeline to overlap more instructions. Using our laundry analogy and assuming that the washer cycle was longer than the others were, we could divide our washer into three machines that perform the wash, rinse, and spin steps of a traditional washer. We would then move from a four-stage to a six-stage pipeline. To get the full speed-up, we need to rebalance the remaining steps so they are the same length, in processors or in laundry. The amount of parallelism being exploited is higher, since there are more operations being overlapped. Performance is potentially greater since the clock cycle can be shorter.

Another approach is to replicate the internal components of the computer so that it can launch multiple instructions in every pipeline stage. The general name for this technique is multiple issue. A multiple-issue laundry would replace our household washer and dryer with, say, three washers and three dryers. You would also have to recruit more assistants to fold and put away three times as much laundry in the same amount of time. The downside is the extra work to keep all the machines busy and transferring the loads to the next pipeline stage.

Launching multiple instructions per stage allows the instruction execution rate to exceed the clock rate or, stated alternatively, the CPI to be less than 1. As mentioned in Chapter 1, it is sometimes useful to flip the metric and use *IPC*, or *instructions per clock cycle*. Hence, a 3-GHz four-way multiple-issue microprocessor can execute a peak rate of 12 billion instructions per second and have a best-case CPI of 0.33, or an IPC of 3. Assuming a five-stage pipeline, such a processor would have 15 instructions in execution at any given time. Today's high-end microprocessors attempt to issue from three to six instructions in every clock cycle. Even moderate designs will aim at a peak IPC of 2. There are typically, however, many constraints on what types of instructions may be executed simultaneously, and what happens when dependences arise.

There are two main ways to implement a multiple-issue processor, with the major difference being the division of work between the compiler and the hardware. Because the division of work dictates whether decisions are being made statically (that is, at compile time) or dynamically (that is, during execution), the approaches are sometimes called static multiple issue and dynamic multiple issue. As we will see, both approaches have other, more commonly used names, which may be less precise or more restrictive.

Two primary and distinct responsibilities must be dealt with in a multiple-issue pipeline:

1. Packaging instructions into issue slots: how does the processor determine how many instructions and which instructions can be issued in a given clock cycle? In most static issue processors, this process is at least partially handled by the compiler; in dynamic issue designs, it is normally dealt with at runtime by the processor, although the compiler will often have already tried to help improve the issue rate by placing the instructions in a beneficial order.

2. Dealing with data and control hazards: in static issue processors, the compiler handles some or all the consequences of data and control hazards statically. In contrast, most dynamic issue processors attempt to alleviate at least some classes of hazards using hardware techniques operating at execution time.

instruction-level parallelism The parallelism among instructions.

multiple issue A scheme whereby multiple instructions are launched in one clock cycle.

static multiple issue An approach to implementing a multiple-issue processor where many decisions are made by the compiler before execution.

dynamic multiple issue An approach to implementing a multiple-issue processor where many decisions are made during execution by the processor.

issue slots The positions from which instructions could issue in a given clock cycle; by analogy, these correspond to positions at the starting blocks for a sprint.

Although we describe these as distinct approaches, in reality, one approach often borrows techniques from the other, and neither approach can claim to be perfectly pure.

The Concept of Speculation

PREDICTION

speculation An approach whereby the compiler or processor guesses the outcome of an instruction to remove it as a dependence in executing other instructions.

One of the most important methods for finding and exploiting more ILP is speculation. Based on the great idea of **prediction**, speculation is an approach that allows the compiler or the processor to "guess" about the properties of an instruction, to enable execution to begin for other instructions that may depend on the speculated instruction. For example, we might speculate on the outcome of a branch, so that instructions after the branch could be executed earlier. Another example is that we might speculate that a store that precedes a load does not refer to the same address, which would allow the load to be executed before the store. The difficulty with speculation is that it may be wrong. So, any speculation mechanism must include both a method to check if the guess was right and a method to unroll or back out the effects of the instructions that were executed speculatively. The implementation of this back-out capability adds complexity.

Speculation may be done in the compiler or by the hardware. For example, the compiler can use speculation to reorder instructions, moving an instruction across a branch or a load across a store. The processor hardware can perform the same transformation at runtime using techniques we discuss later in this section.

The recovery mechanisms used for incorrect speculation are rather different. In the case of speculation in software, the compiler usually inserts additional instructions that check the accuracy of the speculation and provide a fix-up routine to use when the speculation is wrong. In hardware speculation, the processor usually buffers the speculative results until it knows they are no longer speculative. If the speculation is correct, the instructions are completed by allowing the contents of the buffers to be written to the registers or memory. If the speculation is incorrect, the hardware flushes the buffers and re-executes the correct instruction sequence. Misspeculation typically requires the pipeline to be flushed, or at least stalled, and thus further reduces performance.

Speculation introduces one other possible problem: speculating on certain instructions may introduce exceptions that were formerly not present. For example, suppose a load instruction is moved in a speculative manner, but the address it uses is not within bounds when the speculation is incorrect. The result would be that an exception that should not have occurred would occur. The problem is complicated by the fact that if the load instruction were not speculative, then the exception must occur! In compiler-based speculation, such problems are avoided by adding special speculation support that allows such exceptions to be ignored until it is clear that they really should occur. In hardware-based speculation, exceptions are simply buffered until it is clear that the instruction causing them is no longer speculative and is ready to complete; at that point, the exception is raised, and normal exception handling proceeds.

Since speculation can improve performance when done properly and decrease performance when done carelessly, significant effort goes into deciding when it is appropriate to speculate. Later in this section, we will examine both static and dynamic techniques for speculation.

Static Multiple Issue

Static multiple-issue processors all use the compiler to assist with packaging instructions and handling hazards. In a static issue processor, you can think of the set of instructions issued in a given clock cycle, which is called an issue packet, as one large instruction with multiple operations. This view is more than an analogy. Since a static multiple-issue processor usually restricts what mix of instructions can be initiated in a given clock cycle, it is useful to think of the issue packet as a single instruction allowing several operations in certain predefined fields. This view led to the original name for this approach: Very Long Instruction Word (VLIW).

Most static issue processors also rely on the compiler to take on some responsibility for handling data and control hazards. The compiler's responsibilities may include static branch prediction and code scheduling to reduce or prevent all hazards. Let's look at a simple static issue version of an LEGv8 processor, before we describe the use of these techniques in more aggressive processors.

An Example: Static Multiple Issue with the LEGv8 ISA

To give a flavor of static multiple issue, we consider a simple two-issue LEGv8 processor, where one of the instructions can be an integer ALU operation or branch and the other can be a load or store. Such a design is like that used in some embedded processors. Issuing two instructions per cycle will require fetching and decoding 64 bits of instructions. In many static multiple-issue processors, and essentially all VLIW processors, the layout of simultaneously issuing instructions is restricted to simplify the decoding and instruction issue. Hence, we will require that the instructions be paired and aligned on a 64-bit boundary, with the ALU or branch portion appearing first. Furthermore, if one instruction of the pair cannot be used, we require that it be replaced with a nop. Thus, the instructions always issue in pairs, possibly with a nop in one slot. Figure 4.66 shows how the instructions look as they go into the pipeline in pairs.

Static multiple-issue processors vary in how they deal with potential data and control hazards. In some designs, the compiler takes full responsibility for removing *all* hazards, scheduling the code, and inserting no-ops so that the code executes without any need for hazard detection or hardware-generated stalls. In others, the hardware detects data hazards and generates stalls between two issue packets, while requiring that the compiler avoid all dependences within an instruction packet. Even so, a hazard generally forces the entire issue packet containing the dependent instruction to stall. Whether the software must handle all hazards or only try to reduce the fraction of hazards between separate issue packets, the appearance of having a large single instruction with multiple operations is reinforced. We will assume the second approach for this example.

issue packet The set of instructions that issues together in one clock cycle; the packet may be determined statically by the compiler or dynamically by the processor.

Very Long Instruction Word (VLIW) A style of instruction set architecture that launches many operations that are defined to be independent in a single-wide instruction, typically with many separate opcode fields.

Instruction type	Pipe stages							
ALU or branch instruction	IF	ID	EX	MEM	WB			
Load or store instruction	IF	ID	EX	MEM	WB			
ALU or branch instruction		IF	ID	EX	MEM	WB		
Load or store instruction		IF	ID	EX	MEM	WB		
ALU or branch instruction			IF	ID	EX	MEM	WB	
Load or store instruction			IF	ID	EX	MEM	WB	
ALU or branch instruction				IF	ID	EX	MEM	WB
Load or store instruction				IF	ID	EX	MEM	WB

FIGURE 4.66 Static two-issue pipeline in operation. The ALU and data transfer instructions are issued at the same time. Here we have assumed the same five-stage structure as used for the single-issue pipeline. Although this is not strictly necessary, it does have some advantages. In particular, keeping the register writes at the end of the pipeline simplifies the handling of exceptions and the maintenance of a precise exception model, which become more difficult in multiple-issue processors.

To issue an ALU and a data transfer operation in parallel, the first need for additional hardware—beyond the usual hazard detection and stall logic—is extra ports in the register file (see Figure 4.67). In one clock cycle, we may need to read two registers for the ALU operation and two more for a store, and also one write port for an ALU operation and one write port for a load. Since the ALU is tied up for the ALU operation, we also need a separate adder to calculate the effective address for data transfers. Without these extra resources, our two-issue pipeline would be hindered by structural hazards.

use latency Number of clock cycles between a load instruction and an instruction that can use the result of the load without stalling the pipeline.

Clearly, this two-issue processor can improve performance by up to a factor of two! Doing so, however, requires that twice as many instructions be overlapped in execution, and this additional overlap increases the relative performance loss from data and control hazards. For example, in our simple five-stage pipeline, loads have a use latency of one clock cycle, which prevents one instruction from using the result without stalling. In the two-issue, five-stage pipeline the result of a load instruction cannot be used on the next *clock cycle*. This means that the next *two* instructions cannot use the load result without stalling. Furthermore, ALU instructions that had no use latency in the simple five-stage pipeline now have a one-instruction use latency, since the results cannot be used in the paired load or store. To effectively exploit the parallelism available in a multiple-issue processor, more ambitious compiler or hardware scheduling techniques are needed, and static multiple issue requires that the compiler take on this role.

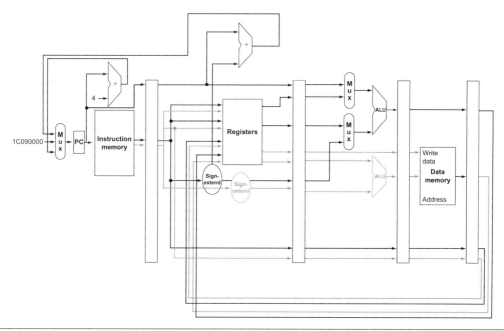

FIGURE 4.67   A static two-issue datapath. The additions needed for double issue are highlighted: another 32 bits from instruction memory, two more read ports and one more write port on the register file, and another ALU. Assume the bottom ALU handles address calculations for data transfers and the top ALU handles everything else.

Simple Multiple-Issue Code Scheduling

How would this loop be scheduled on a static two-issue pipeline for LEGv8?

```
Loop: LDUR X0, [X20,#0]   // X0=array element
      ADD  X0,X0,X21      // add scalar in X21
      STUR X0, [X20,#0]   // store result
      SUBI X20,X20,#8     // decrement pointer
      CMP  X20,X22        // compare to loop limit
      BGT  Loop           // branch if X20 > X22
```

Reorder the instructions to avoid as many pipeline stalls as possible. Assume branches are predicted, so that control hazards are handled by the hardware.

The first three instructions have data dependences, as do the next two. Figure 4.68 shows the best schedule for these instructions. Notice that just one pair of instructions has both issue slots used. It takes five clocks per loop iteration; at five clocks to execute six instructions, we get the disappointing CPI of 0.83 versus the best case of 0.5, or an IPC of 1.2 versus 2.0. Notice that in computing CPI or IPC, we do not count any nops executed as useful instructions. Doing so would improve CPI, but not performance!

	ALU or branch instruction	Data transfer instruction	Clock cycle
Loop:		LDUR X0,[X20,#0]	1
	SUBI X20, X20, #8		2
	ADD X0, X0, X21		3
	CMP X20, X22		4
	BGT Loop	STUR X0,[X20,#8]	5

FIGURE 4.68 The scheduled code as it would look on a two-issue LEGv8 pipeline. The empty slots are no-ops.

loop unrolling A technique to get more performance from loops that access arrays, in which multiple copies of the loop body are made and instructions from different iterations are scheduled together.

An important compiler technique to get more performance from loops is loop unrolling, where multiple copies of the loop body are made. After unrolling, there is more ILP available by overlapping instructions from different iterations.

EXAMPLE

ANSWER

register renaming The renaming of registers by the compiler or hardware to remove antidependences.

antidependence Also called name dependence An ordering forced by the reuse of a name, typically a register, rather than by a true dependence that carries a value between two instructions.

Loop Unrolling for Multiple-Issue Pipelines

See how well loop unrolling and scheduling work in the example above. For simplicity, assume that the loop index is a multiple of four.

To schedule the loop without any delays, it turns out that we need to make four copies of the loop body. After unrolling and eliminating the unnecessary loop overhead instructions, the loop will contain four copies each of LDUR, ADD, and STUR, plus one SUBI, one CMP, and one CBZ. Figure 4.69 shows the unrolled and scheduled code.

During the unrolling process, the compiler introduced additional registers (X1, X2, X3). The goal of this process, called register renaming, is to eliminate dependences that are not true data dependences, but could either lead to potential hazards or prevent the compiler from flexibly scheduling the code. Consider how the unrolled code would look using only X0. There would be repeated instances of LDUR X0,[X20,#0], ADD X0,X0,X21 followed by STUR X0,[X20,#8], but these sequences, despite using X0, are actually completely independent—no data values flow between one set of these instructions and the next set. This case is what is called an antidependence or name dependence, which is an ordering forced purely by the reuse of a name, rather than a real data dependence that is also called a true dependence.

Renaming the registers during the unrolling process allows the compiler to move these independent instructions subsequently to better schedule the code. The renaming process eliminates the name dependences, while preserving the true dependences.

	ALU or branch instruction		Data transfer instruction		Clock cycle
Loop:	SUBI	X20, X20,#32	LDUR	X0, [X20,#0]	1
			LDUR	X1, [X20,#24]	2
	ADD	X0, X0, X21	LDUR	X2, [X20,#16]	3
	ADD	X1, X1, X21	LDUR	X3, [X20,#8]	4
	ADD	X2, X2, X21	STUR	X0, [X20,#32]	5
	ADD	X3, X3, X21	STUR	X1, [X20,#24]	6
	CMP	X20, X22	STUR	X2, [X20,#16]	7
	BGT	Loop	STUR	X3, [X20,#8]	8

FIGURE 4.69 The unrolled and scheduled code of Figure 4.68 as it would look on a static two-issue LEGv8 pipeline. The empty slots are no-ops. Since the first instruction in the loop decrements X20 by 32, the addresses loaded are the original value of X20, then that address minus 8, minus 16, and minus 24.

Notice now that 14 of the 15 instructions in the loop execute as pairs. It takes eight clocks for four loop iterations, which yields an IPC of 15/8 = 1.88. Loop unrolling and scheduling more than doubled performance—8 versus 20 clock cycles for 4 iterations—partly from reducing the loop control instructions and partly from dual issue execution. The cost of this performance improvement is using four temporary registers rather than one, as well as more than doubling the code size.

Dynamic Multiple-Issue Processors

Dynamic multiple-issue processors are also known as superscalar processors, or simply superscalars. In the simplest superscalar processors, instructions issue in order, and the processor decides whether zero, one, or more instructions can issue in a given clock cycle. Obviously, achieving good performance on such a processor still requires the compiler to try to schedule instructions to move dependences apart and thereby improve the instruction issue rate. Even with such compiler scheduling, there is an important difference between this simple superscalar and a VLIW processor: the code, whether scheduled or not, is guaranteed by the hardware to execute correctly. Furthermore, compiled code will always run correctly independent of the issue rate or pipeline structure of the processor. In some VLIW designs, this has not been the case, and recompilation was required when moving across different processor models; in other static issue processors, code would run correctly across different implementations, but often so poorly as to make compilation effectively required.

Many superscalars extend the basic framework of dynamic issue decisions to include dynamic pipeline scheduling. Dynamic pipeline scheduling chooses which instructions to execute in a given clock cycle while trying to avoid hazards

superscalar An advanced pipelining technique that enables the processor to execute more than one instruction per clock cycle by selecting them during execution.

dynamic pipeline scheduling Hardware support for reordering the order of instruction execution to avoid stalls.

and stalls. Let's start with a simple example of avoiding a data hazard. Consider the following code sequence:

```
LDUR  X0,  [X21,#20]
ADD   X1,  X0, X2
SUB   X23, X23, X3
ANDI  X5,  X23, 20
```

PIPELINING

Even though the SUB instruction is ready to execute, it must wait for the LDUR and ADD to complete first, which might take many clock cycles if memory is slow. (Chapter 5 explains cache misses, the reason that memory accesses are sometimes very slow.) Dynamic **pipeline** scheduling allows such hazards to be avoided either fully or partially.

Dynamic Pipeline Scheduling

Dynamic pipeline scheduling chooses which instructions to execute next, possibly reordering them to avoid stalls. In such processors, the pipeline is divided into three major units: an instruction fetch and issue unit, multiple functional units (a dozen or more in high-end designs in 2015), and a commit unit. Figure 4.70 shows the model. The first unit fetches instructions, decodes them, and sends each instruction to a corresponding functional unit for execution. Each functional unit has buffers, called reservation stations, which hold the operands and the operation. (In the next section, we will discuss an alternative to reservation stations used by many recent processors.) As soon as the buffer contains all its operands and the functional unit is ready to execute, the result is calculated. When the result is completed, it is sent to any reservation stations waiting for this particular result as well as to the commit unit, which buffers the result until it is safe to put the result into the register file or, for a store, into memory. The buffer in the commit unit, often called the reorder buffer, is also used to supply operands, in much the same way as forwarding logic does in a statically scheduled pipeline. Once a result is committed to the register file, it can be fetched directly from there, just as in a normal pipeline.

The combination of buffering operands in the reservation stations and results in the reorder buffer provides a form of register renaming, just like that used by the compiler in our earlier loop-unrolling example on page 348. To see how this conceptually works, consider the following steps:

1. When an instruction issues, it is copied to a reservation station for the appropriate functional unit. Any operands that are available in the register file or reorder buffer are also immediately copied into the reservation station. The instruction is buffered in the reservation station until all the operands and the functional unit are available. For the issuing instruction, the register copy of the operand is no longer required, and if a write to that register occurred, the value could be overwritten.

commit unit The unit in a dynamic or out-of-order execution pipeline that decides when it is safe to release the result of an operation to programmer-visible registers and memory.

reservation station A buffer within a functional unit that holds the operands and the operation.

reorder buffer The buffer that holds results in a dynamically scheduled processor until it is safe to store the results to memory or a register.

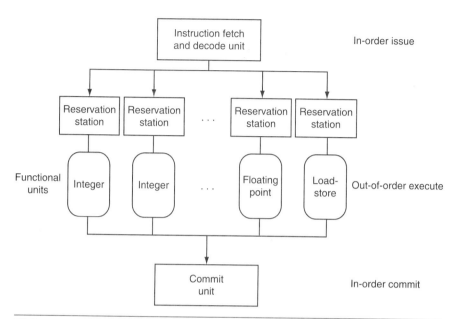

FIGURE 4.70 The three primary units of a dynamically scheduled pipeline. The final step of updating the state is also called retirement or graduation.

2. If an operand is not in the register file or reorder buffer, it must be waiting to be produced by a functional unit. The name of the functional unit that will produce the result is tracked. When that unit eventually produces the result, it is copied directly into the waiting reservation station from the functional unit bypassing the registers.

These steps effectively use the reorder buffer and the reservation stations to implement register renaming.

Conceptually, you can think of a dynamically scheduled pipeline as analyzing the data flow structure of a program. The processor then executes the instructions in some order that preserves the data flow order of the program. This style of execution is called an out-of-order execution, since the instructions can be executed in a different order than they were fetched.

To make programs behave as if they were running on a simple in-order pipeline, the instruction fetch and decode unit is required to issue instructions in order, which allows dependences to be tracked, and the commit unit is required to write results to registers and memory in program fetch order. This conservative mode is called in-order commit. Hence, if an exception occurs, the computer can point to the last instruction executed, and the only registers updated will be those written

out-of-order execution A situation in pipelined execution when an instruction blocked from executing does not cause the following instructions to wait.

in-order commit A commit in which the results of pipelined execution are written to the programmer visible state in the same order that instructions are fetched.

by instructions before the instruction causing the exception. Although the front end (fetch and issue) and the back end (commit) of the pipeline run in order, the functional units are free to initiate execution whenever the data they need are available. Today, all dynamically scheduled pipelines use in-order commit.

Dynamic scheduling is often extended by including hardware-based speculation, especially for branch outcomes. By predicting the direction of a branch, a dynamically scheduled processor can continue to fetch and execute instructions along the predicted path. Because the instructions are committed in order, we know whether the branch was correctly predicted before any instructions from the predicted path are committed. A speculative, dynamically scheduled pipeline can also support speculation on load addresses, allowing load-store reordering, and using the commit unit to avoid incorrect speculation. In the next section, we will look at the use of dynamic scheduling with speculation in the Intel Core i7 design.

Understanding Program Performance

HIERARCHY

PREDICTION

Given that compilers can also schedule code around data dependences, you might ask why a superscalar processor would use dynamic scheduling. There are three major reasons. First, not all stalls are predictable. In particular, cache misses (see Chapter 5) in the **memory hierarchy** cause unpredictable stalls. Dynamic scheduling allows the processor to hide some of those stalls by continuing to execute instructions while waiting for the stall to end.

Second, if the processor speculates on branch outcomes using dynamic branch **prediction**, it cannot know the exact order of instructions at compile time, since it depends on the predicted and actual behavior of branches. Incorporating dynamic speculation to exploit more *instruction-level parallelism* (ILP) without incorporating dynamic scheduling would significantly restrict the benefits of speculation.

Third, as the pipeline latency and issue width change from one implementation to another, the best way to compile a code sequence also changes. For example, how to schedule a sequence of dependent instructions is affected by both issue width and latency. The pipeline structure affects both the number of times a loop must be unrolled to avoid stalls as well as the process of compiler-based register renaming. Dynamic scheduling allows the hardware to hide most of these details. Thus, users and software distributors do not need to worry about having multiple versions of a program for different implementations of the same instruction set. Similarly, old legacy code will get much of the benefit of a new implementation without the need for recompilation.

Both **pipelining** and multiple-issue execution increase peak instruction throughput and attempt to exploit instruction-level **parallelism** (ILP). Data and control dependences in programs, however, offer an upper limit on sustained performance because the processor must sometimes wait for a dependence to be resolved. Software-centric approaches to exploiting ILP rely on the ability of the compiler to find and reduce the effects of such dependences, while hardware-centric approaches rely on extensions to the pipeline and issue mechanisms. Speculation, performed by the compiler or the hardware, can increase the amount of ILP that can be exploited via **prediction**, although care must be taken since speculating incorrectly is likely to reduce performance.

The **BIG** Picture

PIPELINING

PARALLELISM

PREDICTION

Modern, high-performance microprocessors are capable of issuing several instructions per clock; unfortunately, sustaining that issue rate is very difficult. For example, despite the existence of processors with four to six issues per clock, very few applications can sustain more than two instructions per clock. There are two primary reasons for this.

Hardware/ Software Interface

First, within the pipeline, the major performance bottlenecks arise from dependences that cannot be alleviated, thus reducing the parallelism among instructions and the sustained issue rate. Although little can be done about true data dependences, often the compiler or hardware does not know precisely whether a dependence exists or not, and so must conservatively assume the dependence exists. For example, code that makes use of pointers, particularly in ways that may lead to aliasing, will lead to more implied potential dependences. In contrast, the greater regularity of array accesses often allows a compiler to deduce that no

HIERARCHY

dependences exist. Similarly, branches that cannot be accurately predicted whether at runtime or compile time will limit the ability to exploit ILP. Often, additional ILP is available, but the ability of the compiler or the hardware to find ILP that may be widely separated (sometimes by the execution of thousands of instructions) is limited.

Second, losses in the **memory hierarchy** (the topic of Chapter 5) also limit the ability to keep the pipeline full. Some memory system stalls can be hidden, but limited amounts of ILP also limit the extent to which such stalls can be hidden.

Energy Efficiency and Advanced Pipelining

The downside to the increasing exploitation of instruction-level parallelism via dynamic multiple issue and speculation is potential energy inefficiency. Each innovation was able to turn more transistors into performance, but they often did so very inefficiently. Now that we have collided with the power wall, we are seeing designs with multiple processors per chip where the processors are not as deeply pipelined or as aggressively speculative as its predecessors.

The belief is that while the simpler processors are not as fast as their sophisticated brethren, they deliver better performance per Joule, so that they can deliver more performance per chip when designs are constrained more by energy than they are by the number of transistors.

Figure 4.71 shows the number of pipeline stages, the issue width, speculation level, clock rate, cores per chip, and power of several past and recent Intel microprocessors. Note the drop in pipeline stages and power as companies switch to multicore designs.

Microprocessor	Year	Clock Rate	Pipeline Stages	Issue Width	Out-of-Order/ Speculation	Cores/ Chip	Power	
Intel 486	1989	25 MHz	5	1	No	1	5	W
Intel Pentium	1993	66 MHz	5	2	No	1	10	W
Intel Pentium Pro	1997	200 MHz	10	3	Yes	1	29	W
Intel Pentium 4 Willamette	2001	2000 MHz	22	3	Yes	1	75	W
Intel Pentium 4 Prescott	2004	3600 MHz	31	3	Yes	1	103	W
Intel Core	2006	2930 MHz	14	4	Yes	2	75	W
Intel Core i5 Nehalem	2010	3300 MHz	14	4	Yes	2–4	87	W
Intel Core i5 Ivy Bridge	2012	3400 MHz	14	4	Yes	8	77	W

FIGURE 4.71 Record of Intel Microprocessors in terms of pipeline complexity, number of cores, and power. The Pentium 4 pipeline stages do not include the commit stages. If we included them, the Pentium 4 pipelines would be even deeper.

Elaboration: A commit unit controls updates to the register file *and* memory. Some dynamically scheduled processors update the register file immediately during execution, using extra registers to implement the renaming function and preserving the older copy of a register until the instruction updating the register is no longer speculative. Other processors buffer the result, which, as mentioned above, is typically in a structure called a reorder buffer, and the actual update to the register file occurs later as part of the commit. Stores to memory must be buffered until commit time either in a *store buffer* (see Chapter 5) or in the reorder buffer. The commit unit allows the store to write to memory from the buffer when the buffer has a valid address and valid data, and when the store is no longer dependent on predicted branches.

Elaboration: Memory accesses benefit from *nonblocking caches*, which continue servicing cache accesses during a cache miss (see Chapter 5). Out-of-order execution processors need the cache to allow instructions to execute during a miss.

State whether the following techniques or components are associated primarily with a software- or hardware-based approach to exploiting ILP. In some cases, the answer may be both.

Check Yourself

1. Branch prediction

2. Multiple issue

3. VLIW

4. Superscalar

5. Dynamic scheduling

6. Out-of-order execution

7. Speculation

8. Reorder buffer

9. Register renaming

4.11   Real Stuff: The ARM Cortex-A53 and Intel Core i7 Pipelines

Figure 4.72 describes the two microprocessors we examine in this section, whose targets are the two endpoints of the post-PC era.

The ARM Cortex-A53

The ARM Corxtex-A53 runs at 1.5 GHz with an eight-stage pipeline and executes the ARMv8 instruction set. It uses dynamic multiple issue, with two instructions per clock cycle. It is a static in-order pipeline, in that instructions issue, execute, and commit in order. The pipeline consists of three sections for instruction fetch, instruction decode, and execute. Figure 4.73 shows the overall pipeline.

Processor	ARM A53	Intel Core i7 920
Market	Personal Mobile Device	Server, Cloud
Thermal design power	100 milliWatts (1 core @ 1 GHz)	130 Watts
Clock rate	1.5 GHz	2.66 GHz
Cores/Chip	4 (configurable)	4
Floating point?	Yes	Yes
Multiple Issue?	Dynamic	Dynamic
Peak instructions/clock cycle	2	4
Pipeline Stages	8	14
Pipeline schedule	Static In-order	Dynamic Out-of-order with Speculation
Branch prediction	Hybrid	2-level
1st level caches/core	16-64 KiB I, 16-64 KiB D	32 KiB I, 32 KiB D
2nd level cache/core	128–2048 KiB (shared)	256 KiB (per core)
3rd level cache (shared)	(platform dependent)	2–8 MiB

FIGURE 4.72   Specification of the ARM Cortex-A53 and the Intel Core i7 920.

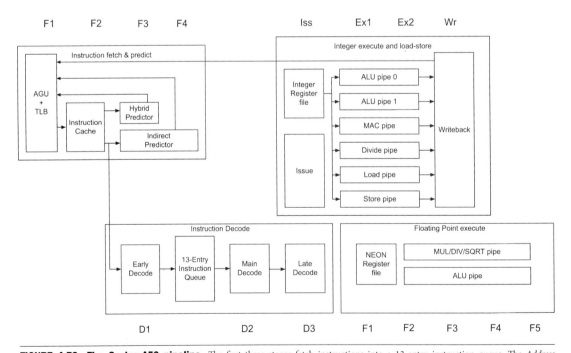

FIGURE 4.73   The Cortex-A53 pipeline. The first three stages fetch instructions into a 13-entry instruction queue. The *Address Generation Unit* (AGU) uses a *Hybrid Predictor, Indirect Predictor,* and a *Return Stack* to predict branches to try to keep the instruction queue full. Instruction decode is three stages and instruction execution is three stages. With two additional stages for floating point and SIMD operations.

The first three stages fetch two instructions at a time and try to keep a 13-entry instruction queue full. It uses a 6k-bit hybrid conditional branch predictor, a 256-entry indirect branch predictor, and an 8-entry return address stack to predict future function returns. The prediction of indirect branches takes an additional pipeline stage. This design choice will incur extra latency if the instruction queue cannot decouple the decode and execute stages from the fetch stage, primarily in the case of a branch misprediction or an instruction cache miss. When the branch prediction is wrong, it empties the pipeline, resulting in an eight-clock cycle misprediction penalty.

The decode stages of the pipeline determine if there are dependences between a pair of instructions, which would force sequential execution, and in which pipeline of the execution stages to send the instructions.

The instruction execution section primarily occupies three pipeline stages and provides one pipeline for load instructions, one pipeline for store instructions, two pipelines for integer arithmetic operations, and separate pipelines for integer multiply and divide operations. Either instruction from the pair can be issued to the load or store pipelines. The execution stages have full forwarding between the pipelines.

Floating-point and SIMD operations add a two more pipeline stages to the instruction execution section and feature one pipeline for multiply/divide/square root operations and one pipeline for other arithmetic operations.

Figure 4.74 shows the CPI of the Cortex-A53 using the SPEC2006 benchmarks. While the ideal CPI is 0.5, the best case achieved is 1.0, the median case is 1.3, and

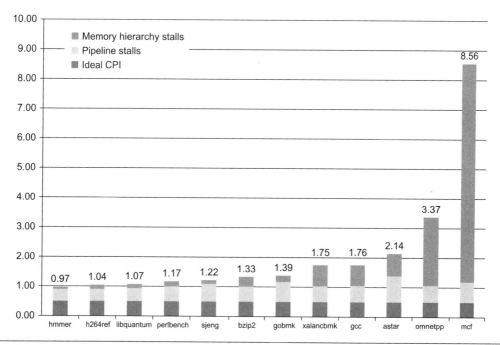

FIGURE 4.74 CPI on ARM Cortex-A53 for the SPEC2006 integer benchmarks.

the worst case is 8.6. For the median case, 60% of the stalls are due to the pipelining hazards and 40% are stalls due to the memory hierarchy. Pipeline stalls are caused by branch mispredictions, structural hazards, and data dependencies between pairs of instructions. Given the static pipeline of the Cortex-A53, it is up to the compiler to try to avoid structural hazards and data dependences.

Elaboration: The Cortex-A53 is a configurable core that supports the ARMv8 instruction set architecture. It is delivered as an IP (*Intellectual Property*) core. IP cores are the dominant form of technology delivery in the embedded, personal mobile device, and related markets; billions of ARM and MIPS processors have been created from these IP cores.

Note that IP cores are different than the cores in the Intel i7 multicore computers. An IP core (which may itself be a multicore) is designed to be incorporated with other logic (hence it is the "core" of a chip), including application-specific processors (such as an encoder or decoder for video), I/O interfaces, and memory interfaces, and then fabricated to yield a processor optimized for a particular application. Although the processor core is almost identical logically, the resultant chips have many differences. One parameter is the size of the L2 cache, which can vary by a factor of 16.

The Intel Core i7 920

x86 microprocessors employ sophisticated pipelining approaches, using both dynamic multiple issue and dynamic pipeline scheduling with out-of-order execution and speculation for their pipelines. These processors, however, are still faced with the challenge of implementing the complex x86 instruction set, described in Chapter 2. Intel fetches x86 instructions and translates them into internal ARMv8-like instructions, which Intel calls *micro-operations*. The micro-operations are then executed by a sophisticated, dynamically scheduled, speculative pipeline capable of sustaining an execution rate of up to six micro-operations per clock cycle. This section focuses on that micro-operation pipeline.

microarchitecture The organization of the processor, including the major functional units, their interconnection, and control.

When we consider the design of such processors, the design of the functional units, the cache and register file, instruction issue, and overall pipeline control become intermingled, making it difficult to separate the datapath from the pipeline. Because of this, many engineers and researchers have adopted the term microarchitecture to refer to the detailed internal architecture of a processor.

architectural registers The instruction set of visible registers of a processor; for example, in LEGv8, these are the 32 integer and 32 floating-point registers.

The Intel Core i7 uses a scheme for resolving antidependences and incorrect speculation that uses a reorder buffer together with register renaming. Register renaming explicitly renames the architectural registers in a processor (16 in the case of the 64-bit version of the x86 architecture) to a larger set of physical registers. The Core i7 uses register renaming to remove antidependences. Register renaming requires the processor to maintain a map between the architectural registers and the physical registers, indicating which physical register is the most current copy of an architectural register. By keeping track of the renamings that have occurred, register renaming offers another approach to recovery in the event of incorrect speculation: simply undo the mappings that have occurred since the first incorrectly speculated instruction. This undo will cause the state of the processor to return to the last correctly executed instruction, keeping the correct mapping between the architectural and physical registers.

Figure 4.75 shows the overall organization and pipeline of the Core i7. Below are the eight steps an x86 instruction goes through for execution.

1. Instruction fetch—The processor uses a multilevel branch target buffer to achieve a balance between speed and prediction accuracy. There is also a return address stack to speed up function return. Mispredictions cause a penalty of about 15 cycles. Using the predicted address, the instruction fetch unit fetches 16 bytes from the instruction cache.

2. The 16 bytes are placed in the predecode instruction buffer—The predecode stage transforms the 16 bytes into individual x86 instructions. This predecode is nontrivial since the length of an x86 instruction can be from 1 to 15 bytes

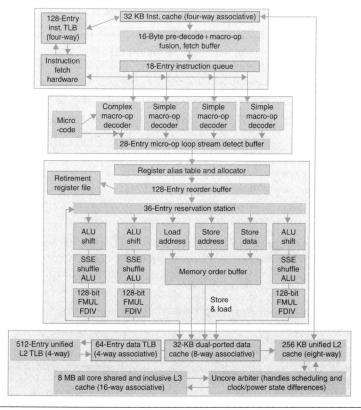

FIGURE 4.75 The Core i7 pipeline with memory components. The total pipeline depth is 14 stages, with branch mispredictions costing 17 clock cycles. This design can buffer 48 loads and 32 stores. The six independent units can begin execution of a ready micro-operation each clock cycle.

and the predecoder must look through a number of bytes before it knows the instruction length. Individual x86 instructions are placed into the 18-entry instruction queue.

3. Micro-op decode—Individual x86 instructions are translated into micro-operations (micro-ops). Three of the decoders handle x86 instructions that translate directly into one micro-op. For x86 instructions that have more complex semantics, there is a microcode engine that is used to produce the micro-op sequence; it can produce up to four micro-ops every cycle and continues until the necessary micro-op sequence has been generated. The micro-ops are placed according to the order of the x86 instructions in the 28-entry micro-op buffer.

4. The micro-op buffer performs *loop stream detection*—If there is a small sequence of instructions (less than 28 instructions or 256 bytes in length) that comprises a loop, the loop stream detector will find the loop and directly issue the micro-ops from the buffer, eliminating the need for the instruction fetch and instruction decode stages to be activated.

5. Perform the basic instruction issue—Looking up the register location in the register tables, renaming the registers, allocating a reorder buffer entry, and fetching any results from the registers or reorder buffer before sending the micro-ops to the reservation stations.

6. The i7 uses a 36-entry centralized reservation station shared by six functional units. Up to six micro-ops may be dispatched to the functional units every clock cycle.

7. The individual function units execute micro-ops and then results are sent back to any waiting reservation station as well as to the register retirement unit, where they will update the register state, once it is known that the instruction is no longer speculative. The entry corresponding to the instruction in the reorder buffer is marked as complete.

8. When one or more instructions at the head of the reorder buffer have been marked as complete, the pending writes in the register retirement unit are executed, and the instructions are removed from the reorder buffer.

Elaboration: Hardware in the second and fourth steps can combine or *fuse* operations together to reduce the number of operations that must be performed. *Macro-op fusion* in the second step takes x86 instruction combinations, such as compare followed by a branch, and fuses them into a single operation. *Microfusion* in the fourth step combines micro-operation pairs such as load/ALU operation and ALU operation/store and issues them to a single reservation station (where they can still issue independently), thus increasing the usage of the buffer. In a study of the Intel Core architecture, which also incorporated microfusion and macrofusion, Bird et al. [2007] discovered that microfusion had little impact on performance, while macrofusion appears to have a modest positive impact on integer performance and little impact on floating-point performance.

Performance of the Intel Core i7 920

Figure 4.76 shows the CPI of the Intel Core i7 for each of the SPEC2006 benchmarks. While the ideal CPI is 0.25, the best case achieved is 0.44, the median case is 0.79, and the worst case is 2.67.

Although it is difficult to differentiate between pipeline stalls and memory stalls in a dynamic out-of-order execution pipeline, we can show the effectiveness of branch prediction and speculation. Figure 4.77 shows the percentage of branches mispredicted and the percentage of the work (measured by the numbers of micro-ops dispatched into the pipeline) that does not retire (that is, their results are annulled) relative to all micro-op dispatches. The min, median, and max of branch mispredictions are 0%, 2%, and 10%. For wasted work, they are 1%, 18%, and 39%.

The wasted work in some cases closely matches the branch misprediction rates, such as for gobmk and astar. In several instances, such as mcf, the wasted work seems relatively larger than the misprediction rate. This divergence is likely due to the memory behavior. With very high data cache miss rates, mcf will dispatch many instructions during an incorrect speculation as long as sufficient reservation stations are available for the stalled memory references. When a branch among the many speculated instructions is finally mispredicted, the micro-ops corresponding to all these instructions will be flushed.

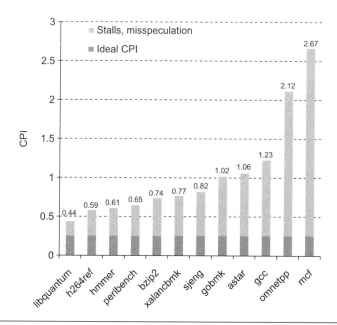

FIGURE 4.76   CPI of Intel Core i7 920 running SPEC2006 integer benchmarks.

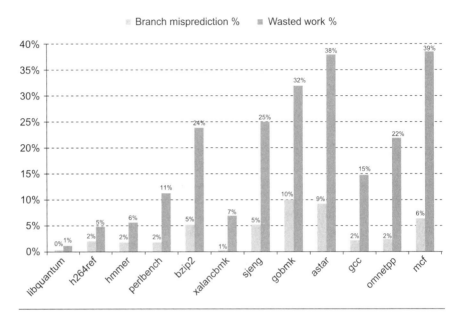

FIGURE 4.77 Percentage of branch mispredictions and wasted work due to unfruitful speculation of Intel Core i7 920 running SPEC2006 integer benchmarks.

Understanding Program Performance

HIERARCHY

The Intel Core i7 combines a 14-stage pipeline and aggressive multiple issue to achieve high performance. By keeping the latencies for back-to-back operations low, the impact of data dependences is reduced. What are the most serious potential performance bottlenecks for programs running on this processor? The following list includes some possible performance problems, the last three of which can apply in some form to any high-performance pipelined processor.

- The use of x86 instructions that do not map to a few simple micro-operations

- Branches that are difficult to predict, causing misprediction stalls and restarts when speculation fails

- Long dependences—typically caused by long-running instructions or the **memory hierarchy**—that lead to stalls

- Performance delays arising in accessing memory (see Chapter 5) that cause the processor to stall

4.12 Going Faster: Instruction-Level Parallelism and Matrix Multiply

Returning to the DGEMM example from Chapter 3, we can see the impact of instruction-level parallelism by unrolling the loop so that the multiple-issue, out-of-order execution processor has more instructions to work with. Figure 4.78 shows the unrolled version of Figure 3.23, which contains the C intrinsics to produce the AVX instructions.

Like the unrolling example in Figure 4.69 above, we are going to unroll the loop four times. Rather than manually unrolling the loop in C by making four copies of each of the intrinsics in Figure 3.23, we can rely on the gcc compiler to do the unrolling at −O3 optimization. (We use the constant UNROLL in the C code to control the amount of unrolling in case we want to try other values.) We surround each intrinsic with a simple *for* loop with four iterations (lines 9, 15, and 20) and replace the scalar C0 in Figure 3.23 with a four-element array c[] (lines 8, 10, 16, and 21).

```
1  //include <x86intrin.h>
2  //define UNROLL (4)
3
4  void dgemm (int n, double* A, double* B, double* C)
5  {
6      for ( int i = 0; i < n; i+=UNROLL*4 )
7          for ( int j = 0; j < n; j++ ) {
8              __m256d c[4];
9              for ( int x = 0; x < UNROLL; x++ )
10                 c[x] = _mm256_load_pd(C+i+x*4+j*n);
11
12             for( int k = 0; k < n; k++ )
13             {
14                 __m256d b = _mm256_broadcast_sd(B+k+j*n);
15                 for (int x = 0; x < UNROLL; x++)
16                 c[x] = _mm256_add_pd(c[x],
17                     _mm256_mul_pd(_mm256_load_pd(A+n*k+x*4+i), b));
18             }
19
20             for ( int x = 0; x < UNROLL; x++ )
21                 _mm256_store_pd(C+i+x*4+j*n, c[x]);
22         }
23  }
```

FIGURE 4.78 Optimized C version of DGEMM using C intrinsics to generate the AVX subword-parallel instructions for the x86 (Figure 3.23) and loop unrolling to create more opportunities for instruction-level parallelism. Figure 4.79 shows the assembly language produced by the compiler for the inner loop, which unrolls the three for-loop bodies to expose instruction-level parallelism.

Figure 4.79 shows the assembly language output of the unrolled code. As expected, in Figure 4.79 there are four versions of each of the AVX instructions in Figure 3.24, with one exception. We only need one copy of the vbroadcastsd instruction, since we can use the four copies of the B element in register %ymm0 repeatedly throughout the loop. Thus, the five AVX instructions in Figure 3.24 become 17 in Figure 4.79, and the seven integer instructions appear in both, although the constants and addressing changes to account for the unrolling. Hence, despite unrolling four times, the number of instructions in the body of the loop only doubles: from 12 to 24.

```
 1   vmovapd (%r11),%ymm4                // Load 4 elements of C into %ymm4
 2   mov     %rbx,%rax                   // register %rax = %rbx
 3   xor     %ecx,%ecx                   // register %ecx = 0
 4   vmovapd 0x20(%r11),%ymm3            // Load 4 elements of C into %ymm3
 5   vmovapd 0x40(%r11),%ymm2            // Load 4 elements of C into %ymm2
 6   vmovapd 0x60(%r11),%ymm1            // Load 4 elements of C into %ymm1
 7   vbroadcastsd (%rcx,%r9,1),%ymm0     // Make 4 copies of B element
 8   add     $0x8,%rcx                   // register %rcx = %rcx + 8
 9   vmulpd (%rax),%ymm0,%ymm5           // Parallel mul %ymm1,4 A
10   vaddpd %ymm5,%ymm4,%ymm4            // Parallel add %ymm5, %ymm4
11   vmulpd 0x20(%rax),%ymm0,%ymm5       // Parallel mul %ymm1,4 A
12   vaddpd %ymm5,%ymm3,%ymm3            // Parallel add %ymm5, %ymm3
13   vmulpd 0x40(%rax),%ymm0,%ymm5       // Parallel mul %ymm1,4 A
14   vmulpd 0x60(%rax),%ymm0,%ymm0       // Parallel mul %ymm1,4 A
15   add     %r8,%rax                    // register %rax = %rax + %r8
16   cmp     %r10,%rcx                   // compare %r8 to %rax
17   vaddpd %ymm5,%ymm2,%ymm2            // Parallel add %ymm5, %ymm2
18   vaddpd %ymm0,%ymm1,%ymm1            // Parallel add %ymm0, %ymm1
19   jne     68 <dgemm+0x68>             // branch if %r8 !=   %rax
20   add     $0x1,%esi                   // register % esi = % esi + 1
21   vmovapd %ymm4,(%r11)                // Store %ymm4 into 4 C elements
22   vmovapd %ymm3,0x20(%r11)            // Store %ymm3 into 4 C elements
23   vmovapd %ymm2,0x40(%r11)            // Store %ymm2 into 4 C elements
24   vmovapd %ymm1,0x60(%r11)            // Store %ymm1 into 4 C elements
```

FIGURE 4.79 **The x86 assembly language for the body of the nested loops generated by compiling the unrolled C code in Figure 4.78.**

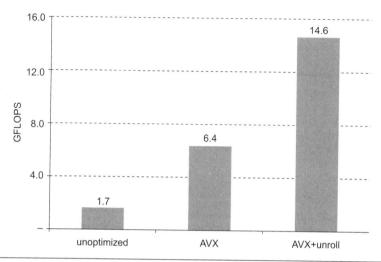

FIGURE 4.80 Performance of three versions of DGEMM for 32 × 32 matrices. Subword parallelism and instruction-level parallelism have led to speedup of almost a factor of 9 over the unoptimized code in Figure 3.21.

Figure 4.80 shows the performance increase DGEMM for 32 × 32 matrices in going from unoptimized to AVX and then to AVX with unrolling. Unrolling more than doubles performance, going from 6.4 GFLOPS to 14.6 GFLOPS. Optimizations for **subword parallelism** and **instruction-level parallelism** result in an overall speedup of 8.59 versus the unoptimized DGEMM in Figure 3.21.

Elaboration: As mentioned in the Elaboration in Section 3.8, these results are with Turbo mode turned off. If we turn it on, like in Chapter 3, we improve all the results by the temporary increase in the clock rate of 3.3/2.6 = 1.27 to 2.1 GFLOPS for unoptimized DGEMM, 8.1 GFLOPS with AVX, and 18.6 GFLOPS with unrolling and AVX. As mentioned in Section 3.8, Turbo mode works particularly well in this case because it is using only a single core of an eight-core chip.

Elaboration: There are no pipeline stalls despite the reuse of register %ymm5 in lines 9 to 17 of Figure 4.79 because the Intel Core i7 pipeline renames the registers.

PARALLELISM

Are the following statements true or false?

Check Yourself

1. The Intel Core i7 uses a multiple-issue pipeline to directly execute x86 instructions.

2. Both the A8 and the Core i7 use dynamic multiple issue.

3. The Core i7 microarchitecture has many more registers than x86 requires.

4. The Intel Core i7 uses less than half the pipeline stages of the earlier Intel Pentium 4 Prescott (see Figure 4.71).

Advanced Topic: An Introduction to Digital Design Using a Hardware Design Language to Describe and Model a Pipeline and More Pipelining Illustrations

Modern digital design is done using hardware description languages and modern computer-aided synthesis tools that can create detailed hardware designs from the descriptions using both libraries and logic synthesis. Entire books are written on such languages and their use in digital design. This section, which appears online, gives a brief introduction and shows how a hardware design language, Verilog in this case, can be used to describe the processor control both behaviorally and in a form suitable for hardware synthesis. It then provides a series of behavioral models in Verilog of the five-stage pipeline. The initial model ignores hazards, and additions to the model highlight the changes for forwarding, data hazards, and branch hazards.

We then provide about a dozen illustrations using the single-cycle graphical pipeline representation for readers who want to see more detail on how pipelines work for a few sequences of LEGv8 instructions.

4.14   Fallacies and Pitfalls

Fallacy: Pipelining is easy.

Our books testify to the subtlety of correct pipeline execution. Our advanced book had a pipeline bug in its first edition, despite its being reviewed by more than 100 people and being class-tested at 18 universities. The bug was uncovered only when someone tried to build the computer in that book. The fact that the Verilog to describe a pipeline like that in the Intel Core i7 will be hundreds of thousands of lines is an indication of the complexity. Beware!

Fallacy: Pipelining ideas can be implemented independent of technology.

When the number of transistors on-chip and the speed of transistors made a five-stage pipeline the best solution, then the delayed branch (see the *Elaboration* on page 295) was a simple solution to control hazards. With longer pipelines, superscalar execution, and dynamic branch prediction, it is now redundant. In the early 1990s, dynamic pipeline scheduling took too many resources and was not required for high performance, but as transistor budgets continued to double due to **Moore's Law** and logic became much faster than memory, then multiple functional units and dynamic pipelining made more sense. Today, concerns about power are leading to less aggressive and more efficient designs.

MOORE'S LAW

Pitfall: Failure to consider instruction set design can adversely impact pipelining.

Many of the difficulties of pipelining arise because of instruction set complications. Here are some examples:

- Widely variable instruction lengths and running times can lead to imbalance among pipeline stages and severely complicate hazard detection in a design pipelined at the instruction set level. This problem was overcome, initially in the DEC VAX 8500 in the late 1980s, using the micro-operations and micropipelined scheme that the Intel Core i7 employs today. Of course, the overhead of translation and maintaining correspondence between the micro-operations and the actual instructions remains.

- Sophisticated-addressing modes can lead to different sorts of problems. Addressing modes that update registers complicate hazard detection. Other addressing modes that require multiple memory accesses substantially complicate pipeline control and make it difficult to keep the pipeline flowing smoothly.

- Perhaps the best example is the DEC Alpha and the DEC NVAX. In comparable technology, the newer instruction set architecture of the Alpha allowed an implementation whose performance is more than twice as fast as NVAX. In another example, Bhandarkar and Clark [1991] compared the MIPS M/2000 and the DEC VAX 8700 by counting clock cycles of the SPEC benchmarks; they concluded that although the MIPS M/2000 executes more instructions, the VAX on average executes 2.7 times as many clock cycles, so the MIPS is faster.

4.15 Concluding Remarks

Nine-tenths of wisdom consists of being wise in time.

American proverb

As we have seen in this chapter, both the datapath and control for a processor can be designed starting with the instruction set architecture and an understanding of the basic characteristics of the technology. In Section 4.3, we saw how the datapath for an LEGv8 processor could be constructed based on the architecture and the decision to build a single-cycle implementation. Of course, the underlying technology also affects many design decisions by dictating what components can be used in the datapath, as well as whether a single-cycle implementation even makes sense.

Pipelining improves throughput but not the inherent execution time, or instruction latency, of instructions; for some instructions, the latency is similar in length to the single-cycle approach. Multiple instruction issue adds additional datapath hardware to allow multiple instructions to begin every clock cycle, but at an increase in effective latency. Pipelining was presented as reducing the clock cycle time of the simple single-cycle datapath. Multiple instruction issue, in comparison, clearly focuses on reducing *clock cycles per instruction* (CPI).

PIPELINING

instruction latency The inherent execution time for an instruction.

Pipelining and multiple issue both attempt to exploit instruction-level parallelism. The presence of data and control dependences, which can become hazards, are the primary limitations on how much parallelism can be exploited. Scheduling and speculation via **prediction**, both in hardware and in software, are the primary techniques used to reduce the performance impact of dependences.

We showed that unrolling the DGEMM loop four times exposed more instructions that could take advantage of the out-of-order execution engine of the Core i7 to more than double performance.

The switch to longer pipelines, multiple instruction issue, and dynamic scheduling in the mid-1990s helped sustain the 60% per year processor performance increase that started in the early 1980s. As mentioned in Chapter 1, these microprocessors preserved the sequential programming model, but they eventually ran into the power wall. Thus, the industry was forced to switch to multiprocessors, which exploit parallelism at much coarser levels (the subject of Chapter 6). This trend has also caused designers to reassess the energy-performance implications of some of the inventions since the mid-1990s, resulting in a simplification of pipelines in the more recent versions of microarchitectures.

To sustain the advances in processing performance via parallel processors, Amdahl's law suggests that another part of the system will become the bottleneck. That bottleneck is the topic of the next chapter: the **memory hierarchy**.

Historical Perspective and Further Reading

This section, which appears online, discusses the history of the first pipelined processors, the earliest superscalars, and the development of out-of-order and speculative techniques, as well as important developments in the accompanying compiler technology.

Exercises

4.1 Consider the following instruction:

Instruction: `AND Rd, Rn, Rm`

Interpretation: `Reg[Rd] = Reg[Rn] AND Reg[Rm]`

4.1.1 [5] <§4.3> What are the values of control signals generated by the control in Figure 4.10 for this instruction?

4.1.2 [5] <§4.3> Which resources (blocks) perform a useful function for this instruction?

4.1.3 [10] <§4.3> Which resources (blocks) produce no output for this instruction? Which resources produce output that is not used?

4.2 [10] <§4.4> Explain each of the "don't cares" in Figure 4.18.

4.3 Consider the following instruction mix:

R-type	I-Type	LDUR	STUR	CBZ	B
24%	28%	25%	10%	11%	2%

4.3.1 [5] <§4.4> What fraction of all instructions use data memory?

4.3.2 [5] <§4.4> What fraction of all instructions use instruction memory?

4.3.3 [5] <§4.4> What fraction of all instructions use the sign extend?

4.3.4 [5] <§4.4> What is the sign extend doing during cycles in which its output is not needed?

4.4 When silicon chips are fabricated, defects in materials (e.g., silicon) and manufacturing errors can result in defective circuits. A very common defect is for one signal wire to get "broken" and always register a logical 0. This is often called a "stuck-at-0" fault.

4.4.1 [5] <§4.4> Which instructions fail to operate correctly if the MemToReg wire is stuck at 0?

4.4.2 [5] <§4.4> Which instructions fail to operate correctly if the ALUSrc wire is stuck at 0?

4.4.3 [5] <§4.4> Which instructions fail to operate correctly if the Reg2Loc wire is stuck at 0?

4.5 In this exercise, we examine in detail how an instruction is executed in a single-cycle datapath. Problems in this exercise refer to a clock cycle in which the processor fetches the following instruction word: 0xf8014062.

4.5.1 [5] <§4.4> What are the outputs of the sign-extend and the "shift left 2" unit (near the top of Figure 4.23) for this instruction word?

4.5.2 [10] <§4.4> What are the values of the ALU control unit's inputs for this instruction?

4.5.3 [5] <§4.4> What is the new PC address after this instruction is executed? Highlight the path through which this value is determined.

4.5.4 [10] <§4.4> For each mux, show the values of its inputs and outputs during the execution of this instruction. List values that are register outputs at Reg [Xn].

4.5.5 [10] <§4.4> What are the input values for the ALU and the two add units?

4.5.6 [10] <§4.4> What are the values of all inputs for the registers unit?

4.6 Section 4.4 does not discuss I-type instructions like ADDI or ANDI.

4.6.1 [5] <§4.4> What additional logic blocks, if any, are needed to add I-type instructions to the CPU shown in Figure 4.23? Add any necessary logic blocks to Figure 4.23 and explain their purpose.

4.6.2 [10] <§4.4> List the values of the signals generated by the control unit for ADDI. Explain the reasoning for any "don't care" control signals.

4.7 Problems in this exercise assume that the logic blocks used to implement a processor's datapath have the following latencies:

I-Mem / D-Mem	Register File	Mux	ALU	Adder	Single gate	Register Read	Register Setup	Sign extend	Control
250 ps	150 ps	25 ps	200 ps	150 ps	5 ps	30 ps	20 ps	50 ps	50 ps

"Register read" is the time needed after the rising clock edge for the new register value to appear on the output. This value applies to the PC only. "Register setup" is the amount of time a register's data input must be stable before the rising edge of the clock. This value applies to both the PC and Register File.

4.7.1 [20] <§4.4> Although the control unit as a whole requires 50 ps, it so happens that we can extract the correct value of the Reg2Loc control wire directly from the instruction. Thus, the value of this control wire is available at the same time as the instruction. Explain how we can extract this value directly from the instruction. Hints: Carefully examine the opcodes shown in Figure 2.20. Also, remember that LSR and LSL do not use the Rm field. Finally, ignore STXR.

4.7.2 [5] <§4.4> What is the latency of an R-type instruction (i.e., how long must the clock period be to ensure that this instruction works correctly)?

4.7.3 [10] <§4.4> What is the latency of LDUR? (Check your answer carefully. Many students place extra muxes on the critical path.)

4.7.4 [10] <§4.4> What is the latency of STUR? (Check your answer carefully. Many students place extra muxes on the critical path.)

4.7.5 [5] <§4.4> What is the latency of CBZ?

4.7.6 [5] <§4.4> What is the latency of B?

4.7.7 [5] <§4.4> What is the latency of an I-type instruction?

4.7.8 [5] <§4.4> What is the minimum clock period for this CPU?

4.8 [10] <§4.4> Suppose you could build a CPU where the clock cycle time was different for each instruction. What would the speedup of this new CPU be over the CPU presented in Figure 4.23 given the instruction mix below?

R-type/I-Type	LDUR	STUR	CBZ	B
52%	25%	10%	11%	2%

4.9 Consider the addition of a multiplier to the CPU shown in Figure 4.23. This addition will add 300 ps to the latency of the ALU, but will reduce the number of instructions by 5% (because there will no longer be a need to emulate the multiply instruction).

4.9.1 [5] <§4.4> What is the clock cycle time with and without this improvement?

4.9.2 [10] <§4.4> What is the speedup achieved by adding this improvement?

4.9.3 [10] <§4.4> What is the slowest the new ALU can be and still result in improved performance?

4.10 When processor designers consider a possible improvement to the processor datapath, the decision usually depends on the cost/performance trade-off. In the following three problems, assume that we are beginning with the datapath from Figure 4.23, the latencies from Exercise 4.7, and the following costs:

I-Mem	Register File	Mux	ALU	Adder	D-Mem	Single Register	Sign extend	Single gate	Control
1000	200	10	100	30	2000	5	100	1	500

Suppose doubling the number of general purpose registers from 32 to 64 would reduce the number of LDUR and STUR instruction by 12%, but increase the latency of the register file from 150 ps to 160 ps and double the cost from 200 to 400. (Use the instruction mix from Exercise 4.8 and ignore the other effects on the ISA discussed in Exercise 2.18.)

4.10.1 [5] <§4.4> What is the speedup achieved by adding this improvement?

4.10.2 [10] <§4.4> Compare the change in performance to the change in cost.

4.10.3 [10] <§4.4> Given the cost/performance ratios you just calculated, describe a situation where it makes sense to add more registers and describe a situation where it doesn't make sense to add more registers.

4.11 Examine the difficulty of adding a proposed LWI Rd, Rm(Rn) ("*Load With Increment*") instruction to LEGv8.

Interpretation: Reg[Rd]=Mem[Reg[Rm]+Reg[Rn]]

4.11.1 [5] <§4.4> Which new functional blocks (if any) do we need for this instruction?

4.11.2 [5] <§4.4> Which existing functional blocks (if any) require modification?

4.11.3 [5] <§4.4> Which new data paths (if any) do we need for this instruction?

4.11.4 [5] <§4.4> What new signals do we need (if any) from the control unit to support this instruction?

4.12 Examine the difficulty of adding a proposed `swap Rd, Rn` instruction to LEGv8.

Interpretation: `Reg[Rd]=Reg[Rn]; Reg[Rn]=Reg[Rd]`

4.12.1 [5] <§4.4> Which new functional blocks (if any) do we need for this instruction?

4.12.2 [10] <§4.4> Which existing functional blocks (if any) require modification?

4.12.3 [5] <§4.4> What new data paths do we need (if any) to support this instruction?

4.12.4 [5] <§4.4> What new signals do we need (if any) from the control unit to support this instruction?

4.12.5 [5] <§4.4> Modify Figure 4.23 to demonstrate an implementation of this new instruction.

4.13 Examine the difficulty of adding a proposed `ss Rd,Rm,Rn` (Store Sum) instruction to LEGv8.

Interpretation: `Mem[Reg[Rd]]=Reg[Rn]+immediate`

4.13.1 [10] <§4.4> Which new functional blocks (if any) do we need for this instruction?

4.13.2 [10] <§4.4> Which existing functional blocks (if any) require modification?

4.13.3 [5] <§4.4> What new data paths do we need (if any) to support this instruction?

4.13.4 [5] <§4.4> What new signals do we need (if any) from the control unit to support this instruction?

4.13.5 [5] <§4.4> Modify Figure 4.23 to demonstrate an implementation of this new instruction.

4.14 [5] <§4.4> For which instructions (if any) is the sign-extend block on the critical path?

4.15 LDUR is instruction with the longest latency on the CPU from Section 4.4. If we modified LDUR and STUR so that there was no offset (i.e., the address to be loaded from/stored to must be calculated and placed in Rd before calling LDUR/ STUR), then no instruction would use both the ALU and Data memory. This would allow us to reduce the clock cycle time. However, it would also increase the number of instructions, because many LDUR and STUR instructions would need to be replaced with LDUR/ADD or STUR/ADD combinations.

4.15.1 [5] <§4.4> What would the new clock cycle time be?

4.15.2 [10] <§4.4> Would a program with the instruction mix presented in Exercise 4.7 run faster or slower on this new CPU? By how much? (For simplicity, assume every LDUR and STUR instruction is replaced with a sequence of two instructions.)

4.15.3 [5] <§4.4> What is the primary factor that influences whether a program will run faster or slower on the new CPU?

4.15.4 [5] <§4.4> Do you consider the original CPU (as shown in Figure 4.23) a better overall design; or do you consider the new CPU a better overall design? Why?

4.16 In this exercise, we examine how pipelining affects the clock cycle time of the processor. Problems in this exercise assume that individual stages of the datapath have the following latencies:

IF	ID	EX	MEM	WB
250 ps	350 ps	150 ps	300 ps	200 ps

Also, assume that instructions executed by the processor are broken down as follows:

ALU/Logic	Jump/Branch	LDUR	STUR
45%	20%	20%	15%

4.16.1 [5] <§4.5> What is the clock cycle time in a pipelined and non-pipelined processor?

4.16.2 [10] <§4.5> What is the total latency of an LDUR instruction in a pipelined and non-pipelined processor?

4.16.3 [10] <§4.5> If we can split one stage of the pipelined datapath into two new stages, each with half the latency of the original stage, which stage would you split and what is the new clock cycle time of the processor?

4.16.4 [10] <§4.5> Assuming there are no stalls or hazards, what is the utilization of the data memory?

4.16.5 [10] <§4.5> Assuming there are no stalls or hazards, what is the utilization of the write-register port of the "Registers" unit?

4.17 [10] <§4.5> What is the minimum number of cycles needed to completely execute n instructions on a CPU with a k stage pipeline? Justify your formula.

4.18 [5] <§4.5> Assume that X1 is initialized to 11 and X2 is initialized to 22. Suppose you executed the code below on a version of the pipeline from Section 4.5 that does not handle data hazards (i.e., the programmer is responsible for addressing data hazards by inserting NOP instructions where necessary). What would the final values of registers X3 and X4 be?

```
ADDI  X1, X2, #5
ADD   X3, X1, X2
ADDI  X4, X1, #15
```

4.19 [10] <§4.5> Assume that X1 is initialized to 11 and X2 is initialized to 22. Suppose you executed the code below on a version of the pipeline from Section 4.5 that does not handle data hazards (i.e., the programmer is responsible for addressing data hazards by inserting NOP instructions where necessary). What would the final values of register X5 be? Assume the register file is written at the beginning of the cycle and read at the end of a cycle. Therefore, an ID stage will return the results of a WB state occurring during the same cycle. See Section 4.7 and Figure 4.51 for details.

```
ADDI  X1, X2, #5
ADD   X3, X1, X2
ADDI  X4, X1, #15
ADD   X5, X1, X1
```

4.20 [5] <§4.5> Add NOP instructions to the code below so that it will run correctly on a pipeline that does not handle data hazards.

```
ADDI  X1, X2, #5
ADD   X3, X1, X2
ADDI  X4, X1, #15
ADD   X5, X3, X2
```

4.21 Consider a version of the pipeline from Section 4.5 that does not handle data hazards (i.e., the programmer is responsible for addressing data hazards by inserting NOP instructions where necessary). Suppose that (after optimization) a typical n-instruction program requires an additional .4*n NOP instructions to correctly handle data hazards.

4.21.1 [5] <§4.5> Suppose that the cycle time of this pipeline without forwarding is 250 ps. Suppose also that adding forwarding hardware will reduce the number of NOPs from .4*n to .05*n, but increase the cycle time to 300 ps. What is the speedup of this new pipeline compared to the one without forwarding?

4.21.2 [10] <§4.5> Different programs will require different amounts of NOPs. How many NOPs (as a percentage of code instructions) can remain in the typical program before that program runs slower on the pipeline with forwarding?

4.21.3 [10] <§4.5> Repeat 4.21.2; however, this time let x represent the number of NOP instructions relative to n. (In 4.21.2, x was equal to .4.) Your answer will be with respect to x.

4.21.4 [10] <§4.5> Can a program with only .075*n NOPs possibly run faster on the pipeline with forwarding? Explain why or why not.

4.21.5 [10] <§4.5> At minimum, how many NOPs (as a percentage of code instructions) must a program have before it can possibly run faster on the pipeline with forwarding?

4.22 [5] <§4.5> Consider the fragment of LEGv8 assembly below:

```
STUR X16, [X6, #12]
LDUR X16, [X6, #8]
SUB  X7, X5, X4
CBZ  X7, Label
ADD  X5, X1, X4
SUB  X5, X15, X4
```

Suppose we modify the pipeline so that it has only one memory (that handles both instructions and data). In this case, there will be a structural hazard every time a program needs to fetch an instruction during the same cycle in which another instruction accesses data.

4.22.1 [5] <§4.5> Draw a pipeline diagram to show were the code above will stall.

4.22.2 [5] <§4.5> In general, is it possible to reduce the number of stalls/NOPs resulting from this structural hazard by reordering code?

4.22.3 [5] <§4.5> Must this structural hazard be handled in hardware? We have seen that data hazards can be eliminated by adding NOPs to the code. Can you do the same with this structural hazard? If so, explain how. If not, explain why not.

4.22.4 [5] <§4.5> Approximately how many stalls would you expect this structural hazard to generate in a typical program? (Use the instruction mix from Exercise 4.8.)

4.23 If we change load/store instructions to use a register (without an offset) as the address, these instructions no longer need to use the ALU. (See Exercise 4.15.) As a result, the MEM and EX stages can be overlapped and the pipeline has only four stages.

4.23.1 [10] <§4.5> How will the reduction in pipeline depth affect the cycle time?

4.23.2 [5] <§4.5> How might this change improve the performance of the pipeline?

4.23.3 [5] <§4.5> How might this change degrade the performance of the pipeline?

4.24 [10] <§4.7> Which of the two pipeline diagrams below better describes the operation of the pipeline's hazard detection unit? Why?

Choice 1:

```
LDUR X1, [X2, #0]:    IF ID EX ME WB
ADD X3, X1, X4:          IF ID EX..ME WB
ORR X5, X6, X7:             IF ID..EX ME WB
```

Choice 2:

```
LDUR X1, [X2, #0]:    IF ID EX ME WB
ADD X3, X1, X4:          IF ID..EX ME WB
ORR X5, X6, X7:             IF..ID EX ME WB
```

4.25 Consider the following loop.

```
LOOP: LDUR X10, [X1, #0]
      LDUR X11, [X1, #8]
      ADD  X12, X10, X11
      SUBI X1, X1, #16
      CBNZ X12, LOOP
```

Assume that perfect branch prediction is used (no stalls due to control hazards), that there are no delay slots, that the pipeline has full forwarding support, and that branches are resolved in the EX (as opposed to the ID) stage.

4.25.1 [10] <§4.7> Show a pipeline execution diagram for the first two iterations of this loop.

4.25.2 [10] <§4.7> Mark pipeline stages that do not perform useful work. How often while the pipeline is full do we have a cycle in which all five pipeline stages are doing useful work? (Begin with the cycle during which the SUBI is in the IF stage. End with the cycle during which the CBNZ is in the IF stage.)

4.26 This exercise is intended to help you understand the cost/complexity/ performance trade-offs of forwarding in a pipelined processor. Problems in this exercise refer to pipelined datapaths from Figure 4.53. These problems assume that, of all the instructions executed in a processor, the following fraction of these instructions has a particular type of RAW data dependence. The type of RAW data dependence is identified by the stage that produces the result (EX or MEM) and the next instruction that consumes the result (1st instruction that follows the one that produces the result, 2nd instruction that follows, or both). We assume that the register write is done in the first half of the clock cycle and that register reads are done in the second half of the cycle, so "EX to 3rd" and "MEM to 3rd" dependences are not counted because they cannot result in data hazards. We also assume that branches are resolved in the EX stage (as opposed to the ID stage), and that the CPI of the processor is 1 if there are no data hazards.

EX to 1st Only	MEM to 1st Only	EX to 2nd Only	MEM to 2nd Only	EX to 1st and EX to 2nd
5%	20%	5%	10%	10%

Assume the following latencies for individual pipeline stages. For the EX stage, latencies are given separately for a processor without forwarding and for a processor with different kinds of forwarding.

IF	ID	EX (no FW)	EX (full FW)	EX (FW from EX/ MEM only)	EX (FW from MEM/ WB only)	MEM	WB
120 ps	100 ps	110 ps	130 ps	120 ps	120 ps	120 ps	100 ps

4.26.1 [5] <§4.7> For each RAW dependency listed above, give a sequence of at least three assembly statements that exhibits that dependency.

4.26.2 [5] <§4.7> For each RAW dependency above, how many NOPs would need to be inserted to allow your code from 4.26.1 to run correctly on a pipeline with no forwarding or hazard detection? Show where the NOPs could be inserted.

4.26.3 [10] <§4.7> Analyzing each instruction independently will over-count the number of NOPs needed to run a program on a pipeline with no forwarding or hazard detection. Write a sequence of three assembly instructions so that, when you consider each instruction in the sequence independently, the sum of the stalls is larger than the number of stalls the sequence actually needs to avoid data hazards.

4.26.4 [5] <§4.7> Assuming no other hazards, what is the CPI for the program described by the table above when run on a pipeline with no forwarding? What percent of cycles are stalls? (For simplicity, assume that all necessary cases are listed above and can be treated independently.)

4.26.5 [5] <§4.7> What is the CPI if we use full forwarding (forward all results that can be forwarded)? What percent of cycles are stalls?

4.26.6 [10] <§4.7> Let us assume that we cannot afford to have three-input multiplexors that are needed for full forwarding. We have to decide if it is better to forward only from the EX/MEM pipeline register (next-cycle forwarding) or only from the MEM/WB pipeline register (two-cycle forwarding). What is the CPI for each option?

4.26.7 [5] <§4.7> For the given hazard probabilities and pipeline stage latencies, what is the speedup achieved by each type of forwarding (EX/MEM, MEM/WB, for full) as compared to a pipeline that has no forwarding?

4.26.8 [5] <§4.7> What would be the additional speedup (relative to the fastest processor from 4.26.7) be if we added "time-travel" forwarding that eliminates all data hazards? Assume that the yet-to-be-invented time-travel circuitry adds 100 ps to the latency of the full-forwarding EX stage.

4.26.9 [5] <§4.7> The table of hazard types has separate entries for "EX to 1st" and "EX to 1st and EX to 2nd". Why is there no entry for "MEM to 1st and MEM to 2nd"?

4.27 Problems in this exercise refer to the following sequence of instructions, and assume that it is executed on a five-stage pipelined datapath:

```
ADD    X5, X2, X1
LDUR   X3, [X5, #4]
LDUR   X2, [X2, #0]
ORR    X3, X5, X3
STUR   X3, [X5, #0]
```

4.27.1 [5] <§4.7> If there is no forwarding or hazard detection, insert NOPs to ensure correct execution.

4.27.2 [10] <§4.7> Now, change and/or rearrange the code to minimize the number of NOPs needed. You can assume register X7 can be used to hold temporary values in your modified code.

4.27.3 [10] <§4.7> If the processor has forwarding, but we forgot to implement the hazard detection unit, what happens when the original code executes?

4.27.4 [20] <§4.7> If there is forwarding, for the first seven cycles during the execution of this code, specify which signals are asserted in each cycle by hazard detection and forwarding units in Figure 4.59.

4.27.5 [10] <§4.7> If there is no forwarding, what new input and output signals do we need for the hazard detection unit in Figure 4.59? Using this instruction sequence as an example, explain why each signal is needed.

4.27.6 [20] <§4.7> For the new hazard detection unit from 4.26.5, specify which output signals it asserts in each of the first five cycles during the execution of this code.

4.28 The importance of having a good branch predictor depends on how often conditional branches are executed. Together with branch predictor accuracy, this will determine how much time is spent stalling due to mispredicted branches. In this exercise, assume that the breakdown of dynamic instructions into various instruction categories is as follows:

R-Type	CBZ/CBNZ	B	LDUR	STUR
40%	25%	5%	25%	5%

Also, assume the following branch predictor accuracies:

Always-Taken	Always-Not-Taken	2-Bit
45%	55%	85%

4.28.1 [10] <§4.8> Stall cycles due to mispredicted branches increase the CPI. What is the extra CPI due to mispredicted branches with the always-taken predictor? Assume that branch outcomes are determined in the ID stage and applied in the EX stage that there are no data hazards, and that no delay slots are used.

4.28.2 [10] <§4.8> Repeat 4.28.1 for the "always-not-taken" predictor.

4.28.3 [10] <§4.8> Repeat 4.28.1 for the 2-bit predictor.

4.28.4 [10] <§4.8> With the 2-bit predictor, what speedup would be achieved if we could convert half of the branch instructions to some ALU instruction? Assume that correctly and incorrectly predicted instructions have the same chance of being replaced.

4.28.5 [10] <§4.8> With the 2-bit predictor, what speedup would be achieved if we could convert half of the branch instructions in a way that replaced each branch instruction with two ALU instructions? Assume that correctly and incorrectly predicted instructions have the same chance of being replaced.

4.28.6 [10] <§4.8> Some branch instructions are much more predictable than others. If we know that 80% of all executed branch instructions are easy-to-predict loop-back branches that are always predicted correctly, what is the accuracy of the 2-bit predictor on the remaining 20% of the branch instructions?

4.29 This exercise examines the accuracy of various branch predictors for the following repeating pattern (e.g., in a loop) of branch outcomes: T, NT, T, T, NT.

4.29.1 [5] <§4.8> What is the accuracy of always-taken and always-not-taken predictors for this sequence of branch outcomes?

4.29.2 [5] <§4.8> What is the accuracy of the 2-bit predictor for the first four branches in this pattern, assuming that the predictor starts off in the bottom left state from Figure 4.62 (predict not taken)?

4.29.3 [10] <§4.8> What is the accuracy of the 2-bit predictor if this pattern is repeated forever?

4.29.4 [30] <§4.8> Design a predictor that would achieve a perfect accuracy if this pattern is repeated forever. You predictor should be a sequential circuit with one output that provides a prediction (1 for taken, 0 for not taken) and no inputs other than the clock and the control signal that indicates that the instruction is a conditional branch.

4.29.5 [10] <§4.8> What is the accuracy of your predictor from 4.29.4 if it is given a repeating pattern that is the exact opposite of this one?

4.29.6 [20] <§4.8> Repeat 4.29.4, but now your predictor should be able to eventually (after a warm-up period during which it can make wrong predictions) start perfectly predicting both this pattern and its opposite. Your predictor should have an input that tells it what the real outcome was. Hint: this input lets your predictor determine which of the two repeating patterns it is given.

4.30 This exercise explores how exception handling affects pipeline design. The first three problems in this exercise refer to the following two instructions:

Instruction 1	Instruction 2
CBZ X1, LABEL	LDUR X1, [X2,#0]

4.30.1 [5] <§4.9> Which exceptions can each of these instructions trigger? For each of these exceptions, specify the pipeline stage in which it is detected.

4.30.2 [10] <§4.9> If there is a separate handler address for each exception, show how the pipeline organization must be changed to be able to handle this exception. You can assume that the addresses of these handlers are known when the processor is designed.

4.30.3 [10] <§4.9> If the second instruction is fetched immediately after the first instruction, describe what happens in the pipeline when the first instruction causes the first exception you listed in Exercise 4.30.1. Show the pipeline execution diagram from the time the first instruction is fetched until the time the first instruction of the exception handler is completed.

4.30.4 [20] <§4.9> In vectored exception handling, the table of exception handler addresses is in data memory at a known (fixed) address. Change the pipeline to implement this exception handling mechanism. Repeat Exercise 4.30.3 using this modified pipeline and vectored exception handling.

4.30.5 [15] <§4.9> We want to emulate vectored exception handling (described in Exercise 4.30.4) on a machine that has only one fixed handler address. Write the code that should be at that fixed address. Hint: this code should identify the exception, get the right address from the exception vector table, and transfer execution to that handler.

4.31 In this exercise we compare the performance of 1-issue and 2-issue processors, taking into account program transformations that can be made to optimize for 2-issue execution. Problems in this exercise refer to the following loop (written in C):

```
for(i=0;i!=j;i+=2)
    b[i]=a[i]-a[i+1];
```

A compiler doing little or no optimization might produce the following LEGv8 assembly code:

```
        MOV   X5, XZR
        B     ENT
TOP:    LSL   X10, X5, #3
        ADD   X11, X1, X10
        LDUR  X12, [X11, #0]
        LDUR  X13, [X11, #8]
        SUB   X14, X12, X13
        ADD   X15, X2, X10
        STUR  X14, [X15, #0]
        ADDI  X5,  X5, #2
ENT:    CMP   X5,  X6
        B.NE  TOP
```

The code above uses the following registers:

i	j	a	b	Temporary values
X5	X6	X1	X2	X10–X15

Assume the two-issue, statically scheduled processor for this exercise has the following properties:

1. One instruction must be a memory operation; the other must be an arithmetic/logic instruction or a branch.
2. The processor has all possible forwarding paths between stages (including paths to the ID stage for branch resolution).
3. The processor has perfect branch prediction.
4. Two instruction may not issue together in a packet if one depends on the other. (See page 345.)
5. If a stall is necessary, both instructions in the issue packet must stall. (See page 345.)

As you complete these exercises, notice how much effort goes into generating code that will produce a near-optimal speedup.

4.31.1 [30] <§4.10> Draw a pipeline diagram showing how LEGv8 code given above executes on the two-issue processor. Assume that the loop exits after two iterations.

4.31.2 [10] <§4.10> What is the speedup of going from a one-issue to a two-issue processor? (Assume the loop runs thousands of iterations.)

4.31.3 [10] <§4.10> Rearrange/rewrite the LEGv8 code given above to achieve better performance on the one-issue processor. Hint: Use the instruction "CBZ X6, XZR, DONE" to skip the loop entirely if j = 0.

4.31.4 [20] <§4.10> Rearrange/rewrite the LEGv8 code given above to achieve better performance on the two-issue processor. (Do not unroll the loop, however.)

4.31.5 [30] <§4.10> Repeat Exercise 4.31.1, but this time use your optimized code from Exercise 4.31.4.

4.31.6 [10] <§4.10> What is the speedup of going from a one-issue processor to a two-issue processor when running the optimized code from Exercises 4.31.3 and 4.31.4.

4.31.7 [10] <§4.10> Unroll the LEGv8 code from Exercise 4.31.3 so that each iteration of the unrolled loop handles two iterations of the original loop. Then, rearrange/rewrite your unrolled code to achieve better performance on the one-issue processor. You may assume that j is a multiple of 4.

4.31.8 [20] <§4.10> Unroll the LEGv8 code from Exercise 4.31.4 so that each iteration of the unrolled loop handles two iterations of the original loop. Then, rearrange/rewrite your unrolled code to achieve better performance on the two-issue processor. You may assume that j is a multiple of 4. (Hint: Re-organize the loop so that some calculations appear both outside the loop and at the end of the loop. You may assume that the values in temporary registers are not needed after the loop.)

4.31.9 [10] <§4.10> What is the speedup of going from a one-issue processor to a two-issue processor when running the unrolled, optimized code from Exercises 4.31.7 and 4.31.8?

4.31.10 [30] <§4.10> Repeat Exercises 4.31.8 and 4.31.9, but this time assume the two-issue processor can run two arithmetic/logic instructions together. (In other words, the first instruction in a packet can be any type of instruction, but the second must be an arithmetic or logic instruction. Two memory operations cannot be scheduled at the same time.)

4.32 This exercise explores energy efficiency and its relationship with performance. Problems in this exercise assume the following energy consumption for activity in Instruction memory, Registers, and Data memory. You can assume that the other components of the datapath consume a negligible amount of energy. ("Register Read" and "Register Write" refer to the register file only.)

I-Mem	1 Register Read	Register Write	D-Mem Read	D-Mem Write
140pJ	70pJ	60pJ	140pJ	120pJ

Assume that components in the datapath have the following latencies. You can assume that the other components of the datapath have negligible latencies.

I-Mem	Control	Register Read or Write	ALU	D-Mem Read or Write
200 ps	150 ps	90 ps	90 ps	250 ps

4.32.1 [5] <§§4.3, 4.6, 4.14> How much energy is spent to execute an ADD instruction in a single-cycle design and in the five-stage pipelined design?

4.32.2 [10] <§§4.6, 4.14> What is the worst-case ARMv8 instruction in terms of energy consumption? What is the energy spent to execute it?

4.32.3 [10] <§§4.6, 4.14> If energy reduction is paramount, how would you change the pipelined design? What is the percentage reduction in the energy spent by an LDUR instruction after this change?

4.32.4 [10] <§§4.6, 4.14> What other instructions can potentially benefit from the change discussed in Exercise 4.32.2?

4.32.5 [10] <§§4.6, 4.14> How do your changes from Exercise 4.32.3 affect the performance of a pipelined CPU?

4.32.6 [10] <§§4.6, 4.14> We can eliminate the MemRead control signal and have the data memory be read in every cycle, i.e., we can permanently have MemRead=1. Explain why the processor still functions correctly after this change. If 25% of instructions are loads, what is the effect of this change on clock frequency and energy consumption?

4.33 When silicon chips are fabricated, defects in materials (e.g., silicon) and manufacturing errors can result in defective circuits. A very common defect is for one wire to affect the signal in another. This is called a "cross-talk fault". A special class of cross-talk faults is when a signal is connected to a wire that has a constant logical value (e.g., a power supply wire). These faults, where the affected signal always has a logical value of either 0 or 1 are called "stuck-at-0" or "stuck-at-1" faults. The following problems refer to bit 0 of the Write Register input on the register file in Figure 4.23.

4.33.1 [10] <§§4.3, 4.4> Let us assume that processor testing is done by (1) filling the PC, registers, and data and instruction memories with some values (you can choose which values), (2) letting a single instruction execute, then (3) reading the PC, memories, and registers. These values are then examined to determine if a particular fault is present. Can you design a test (values for PC, memories, and registers) that would determine if there is a stuck-at-0 fault on this signal?

4.33.2 [10] <§§4.3, 4.4> Repeat Exercise 4.33.1 for a stuck-at-1 fault. Can you use a single test for both stuck-at-0 and stuck-at-1? If yes, explain how; if no, explain why not.

4.33.3 [10] <§§4.3, 4.4> If we know that the processor has a stuck-at-1 fault on this signal, is the processor still usable? To be usable, we must be able to convert any program that executes on a normal LEGv8 processor into a program that works on this processor. You can assume that there is enough free instruction memory and data memory to let you make the program longer and store additional data.

4.33.4 [10] <§§4.3, 4.4> Repeat Exercise 4.33.1; but now the fault to test for is whether the MemRead control signal becomes 0 if the branch control signal is 0, no fault otherwise.

4.33.5 [10] <§§4.3, 4.4> Repeat Exercise 4.33.1; but now the fault to test for is whether the MemRead control signal becomes 1 if RegRt control signal is 1, no fault otherwise. Hint: This problem requires knowledge of operating systems. Consider what causes segmentation faults.

4.33.6 [10] <§§4.3, 4.4> Repeat Exercise 4.33.1; but now the fault to test for is whether the Branch control signal becomes 0 if Reg2Loc control signal is 0, no fault otherwise.

Answers to Check Yourself

§4.1, page 260: 3 of 5: Control, Datapath, Memory. Input and Output are missing.
§4.2, page 263: false. Edge-triggered state elements make simultaneous reading and writing both possible and unambiguous.
§4.3, page 270: I. a. II. c.
§4.4, page 283: Yes, Branch and ALUOp0 are identical. In addition, you can use the flexibility of the don't care bits to combine other signals together. ALUSrc and MemtoReg can be made the same by setting the two don't care bits of MemtoReg to 1 and 0. Reg2Loc and RegWrite can be made to be inverses of one another by setting the don't care bit of Reg2Loc to 0. You don't need an inverter; simply use the other signal and flip the order of the inputs to the Reg2Loc multiplexor!
§4.5, page 296: 1. Stall due to a load-use data hazard of the LDUR result. II. Avoid stalling in the third instruction for the read-after-write data hazard on X1 by forwarding the ADD result. 3. It need not stall, even without forwarding.
§4.6, page 309: Statements 2 and 4 are correct; the rest are incorrect.
§4.8, page 335: 1. Predict not taken. 2. Predict taken. 3. Dynamic prediction.
§4.9, page 342: The first instruction, since it is logically executed before the others.
§4.10, page 355: 1. Both. 2. Both. 3. Software. 4. Hardware. 5. Hardware. 6. Hardware. 7. Both. 8. Hardware. 9. Both.
§4.12, page 365: First two are false and the last two are true.
§4.13, page 4.13-4: Only statement #3 is completely accurate.
§4.13, page 4.13-6: Statements #1 and #4 are true.

5

Large and Fast: Exploiting Memory Hierarchy

Ideally one would desire an indefinitely large memory capacity such that any particular ... word would be immediately available. ... We are ... forced to recognize the possibility of constructing a hierarchy of memories, each of which has greater capacity than the preceding but which is less quickly accessible.

A. W. Burks, H. H. Goldstine, and J. von Neumann,
Preliminary Discussion of the Logical Design of an Electronic Computing Instrument, 1946

The Five Classic Components of a Computer

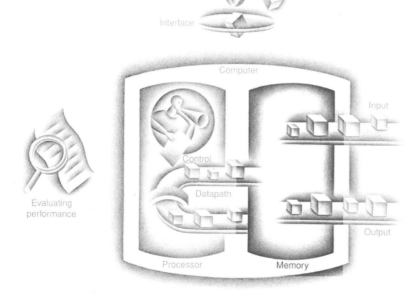

5.1   Introduction

From the earliest days of computing, programmers have wanted unlimited amounts of fast memory. The topics in this chapter aid programmers by creating that illusion. Before we look at creating the illusion, let's consider a simple analogy that illustrates the key principles and mechanisms that we use.

Suppose you were a student writing a term paper on important historical developments in computer hardware. You are sitting at a desk in a library with a collection of books that you have pulled from the shelves and are examining. You find that several of the important computers that you need to write about are described in the books you have, but there is nothing about the EDSAC. Therefore, you go back to the shelves and look for an additional book. You find a book on early British computers that covers the EDSAC. Once you have a good selection of books on the desk in front of you, there is a high probability that many of the topics you need can be found in them, and you may spend most of your time just using the books on the desk without returning to the shelves. Having several books on the desk in front of you saves time compared to having only one book there and constantly having to go back to the shelves to return it and take out another.

The same principle allows us to create the illusion of a large memory that we can access as fast as a very small memory. Just as you did not need to access all the books in the library at once with equal probability, a program does not access all of its code or data at once with equal probability. Otherwise, it would be impossible to make most memory accesses fast and still have large memory in computers, just as it would be impossible for you to fit all the library books on your desk and still find what you wanted quickly.

This *principle of locality* underlies both the way in which you did your work in the library and the way that programs operate. The principle of locality states that programs access a relatively small portion of their address space at any instant of time, just as you accessed a very small portion of the library's collection. There are two different types of locality:

temporal locality The locality principle stating that if a data location is referenced then it will tend to be referenced again soon.

- **Temporal locality** (locality in time): if an item is referenced, it will tend to be referenced again soon. If you recently brought a book to your desk to look at, you will probably need to look at it again soon.

spatial locality The locality principle stating that if a data location is referenced, data locations with nearby addresses will tend to be referenced soon.

- **Spatial locality** (locality in space): if an item is referenced, items whose addresses are close by will tend to be referenced soon. For example, when you brought out the book on early English computers to learn about the EDSAC, you also noticed that there was another book shelved next to it about early mechanical computers, so you likewise brought back that book and, later on, found something useful in that book. Libraries put books on the same topic together on the same shelves to increase spatial locality. We'll see how memory hierarchies use spatial locality a little later in this chapter.

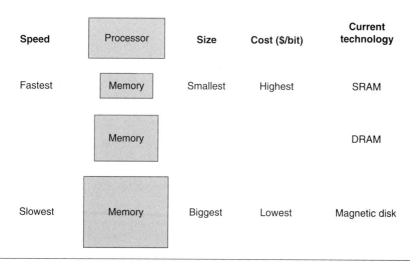

Speed	Processor	Size	Cost ($/bit)	Current technology
Fastest	Memory	Smallest	Highest	SRAM
	Memory			DRAM
Slowest	Memory	Biggest	Lowest	Magnetic disk

FIGURE 5.1   The basic structure of a memory hierarchy. By implementing the memory system as a hierarchy, the user has the illusion of a memory that is as large as the largest level of the hierarchy, but can be accessed as if it were all built from the fastest memory. Flash memory has replaced disks in many personal mobile devices, and may lead to a new level in the storage hierarchy for desktop and server computers; see Section 5.2.

Just as accesses to books on the desk naturally exhibit locality, locality in programs arises from simple and natural program structures. For example, most programs contain loops, so instructions and data are likely to be accessed repeatedly, showing large temporal locality. Since instructions are normally accessed sequentially, programs also show high spatial locality. Accesses to data also exhibit a natural spatial locality. For example, sequential accesses to elements of an array or a record will naturally have high degrees of spatial locality.

We take advantage of the principle of locality by implementing the memory of a computer as a memory hierarchy. A memory hierarchy consists of multiple levels of memory with different speeds and sizes. The faster memories are more expensive per bit than the slower memories and thus are smaller.

Figure 5.1 shows the faster memory is close to the processor and the slower, less expensive memory is below it. The goal is to present the user with as much memory as is available in the cheapest technology, while providing access at the speed offered by the fastest memory.

The data are similarly hierarchical: a level closer to the processor is generally a subset of any level further away, and all the data are stored at the lowest level. By analogy, the books on your desk form a subset of the library you are working in, which is in turn a subset of all the libraries on campus. Furthermore, as we move away from the processor, the levels take progressively longer to access, just as we might encounter in a hierarchy of campus libraries.

A memory hierarchy can consist of multiple levels, but data are copied between only two adjacent levels at a time, so we can focus our attention on just two levels.

memory hierarchy
A structure that uses multiple levels of memories; as the distance from the processor increases, the size of the memories and the access time both increase.

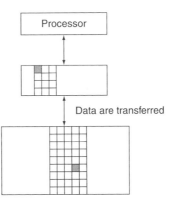

FIGURE 5.2   Every pair of levels in the memory hierarchy can be thought of as having an upper and lower level. Within each level, the unit of information that is present or is not is called a *block* or a *line*. Usually we transfer an entire block when we copy something between levels.

block (or line) The minimum unit of information that can be either present or not present in a cache.

hit rate The fraction of memory accesses found in a level of the memory hierarchy.

miss rate The fraction of memory accesses not found in a level of the memory hierarchy.

hit time The time required to access a level of the memory hierarchy, including the time needed to determine whether the access is a hit or a miss.

miss penalty The time required to fetch a block into a level of the memory hierarchy from the lower level, including the time to access the block, transmit it from one level to the other, insert it in the level that experienced the miss, and then pass the block to the requestor.

The upper level—the one closer to the processor—is smaller and faster than the lower level, since the upper level uses technology that is more expensive. Figure 5.2 shows that the minimum unit of information that can be either present or not present in the two-level hierarchy is called a block or a line; in our library analogy, a block of information is one book.

If the data requested by the processor appear in some block in the upper level, this is called a *hit* (analogous to your finding the information in one of the books on your desk). If the data are not found in the upper level, the request is called a *miss*. The lower level in the hierarchy is then accessed to retrieve the block containing the requested data. (Continuing our analogy, you go from your desk to the shelves to find the desired book.) The hit rate, or *hit ratio*, is the fraction of memory accesses found in the upper level; it is often used as a measure of the performance of the memory hierarchy. The miss rate (1−hit rate) is the fraction of memory accesses not found in the upper level.

Since performance is the major reason for having a memory hierarchy, the time to service hits and misses is important. Hit time is the time to access the upper level of the memory hierarchy, which includes the time needed to determine whether the access is a hit or a miss (that is, the time needed to look through the books on the desk). The miss penalty is the time to replace a block in the upper level with the corresponding block from the lower level, plus the time to deliver this block to the processor (or the time to get another book from the shelves and place it on the desk). Because the upper level is smaller and built using faster memory parts, the hit time will be much smaller than the time to access the next level in the hierarchy, which is the major component of the miss penalty. (The time to examine the books on the desk is much smaller than the time to get up and get a new book from the shelves.)

As we will see in this chapter, the concepts used to build memory systems affect many other aspects of a computer, including how the operating system manages memory and I/O, how compilers generate code, and even how applications use the computer. Of course, because all programs spend much of their time accessing memory, the memory system is necessarily a major factor in determining performance. The reliance on memory hierarchies to achieve performance has meant that programmers, who used to be able to think of memory as a flat, random access storage device, now need to understand that memory is a hierarchy to get good performance. We show how important this understanding is in later examples, such as Figure 5.18 on page 422, and Section 5.14, which shows how to double matrix multiply performance.

Since memory systems are critical to performance, computer designers devote a great deal of attention to these systems and develop sophisticated mechanisms for improving the performance of the memory system. In this chapter, we discuss the major conceptual ideas, although we use many simplifications and abstractions to keep the material manageable in length and complexity. (The hardware models in this chapter have been sourced by the authors and do not imply ARM-endorsed architectures.)

Programs exhibit both temporal locality, the tendency to reuse recently accessed data items, and spatial locality, the tendency to reference data items that are close to other recently accessed items. Memory hierarchies take advantage of temporal locality by keeping more recently accessed data items closer to the processor. Memory hierarchies take advantage of spatial locality by moving blocks consisting of multiple contiguous words in memory to upper levels of the hierarchy.

Figure 5.3 shows that a memory hierarchy uses smaller and faster memory technologies close to the processor. Thus, accesses that hit in the highest level of the hierarchy can be processed quickly. Accesses that miss go to lower levels of the hierarchy, which are larger but slower. If the hit rate is high enough, the memory hierarchy has an effective access time close to that of the highest (and fastest) level and a size equal to that of the lowest (and largest) level.

In most systems, the memory is a true hierarchy, meaning that data cannot be present in level i unless they are also present in level $i + 1$.

The BIG Picture

Which of the following statements are generally true?

1. Memory hierarchies take advantage of temporal locality.
2. On a read, the value returned depends on which blocks are in the cache.
3. Most of the cost of the memory hierarchy is at the highest level.
4. Most of the capacity of the memory hierarchy is at the lowest level.

Check Yourself

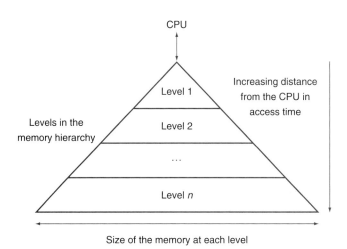

FIGURE 5.3   This diagram shows the structure of a memory hierarchy: as the distance from the processor increases, so does the size. This structure, with the appropriate operating mechanisms, allows the processor to have an access time that is determined primarily by level 1 of the hierarchy and yet have a memory as large as level *n*. Maintaining this illusion is the subject of this chapter. Although the local disk is normally the bottom of the hierarchy, some systems use tape or a file server over a local area network as the next levels of the hierarchy.

5.2   Memory Technologies

There are four primary technologies used today in memory hierarchies. Main memory is implemented from DRAM (*dynamic random access memory*), while levels closer to the processor (caches) use SRAM (*static random access memory*). DRAM is less costly per bit than SRAM, although it is substantially slower. The price difference arises because DRAM uses significantly less area per bit of memory, and DRAMs thus have larger capacity for the same amount of silicon; the speed difference arises from several factors described in Section A.9 of Appendix A. The third technology is flash memory. This nonvolatile memory is the secondary memory in Personal Mobile Devices. The fourth technology, used to implement the largest and slowest level in the hierarchy in servers, is magnetic disk. The access time and price per bit vary widely among these technologies, as the table below shows, using typical values for 2012.

Memory technology	Typical access time	$ per GiB in 2012
SRAM semiconductor memory	0.5–2.5 ns	$500–$1000
DRAM semiconductor memory	50–70 ns	$10–$20
Flash semiconductor memory	5,000–50,000 ns	$0.75–$1.00
Magnetic disk	5,000,000–20,000,000 ns	$0.05–$0.10

We describe each memory technology in the remainder of this section.

SRAM Technology

SRAMs are simply integrated circuits that are memory arrays with (usually) a single access port that can provide either a read or a write. SRAMs have a fixed access time to any datum, though the read and write access times may differ.

SRAMs don't need to refresh and so the access time is very close to the cycle time. SRAMs typically use six to eight transistors per bit to prevent the information from being disturbed when read. SRAM needs only minimal power to retain the charge in standby mode.

In the past, most PCs and server systems used separate SRAM chips for either their primary, secondary, or even tertiary caches. Today, thanks to **Moore's Law**, all levels of caches are integrated onto the processor chip, so the market for independent SRAM chips has nearly evaporated.

MOORE'S LAW

DRAM Technology

In a SRAM, as long as power is applied, the value can be kept indefinitely. In a *dynamic RAM* (DRAM), the value kept in a cell is stored as a charge in a capacitor. A single transistor is then used to access this stored charge, either to read the value or to overwrite the charge stored there. Because DRAMs use only one transistor per bit of storage, they are much denser and cheaper per bit than SRAM. As DRAMs store the charge on a capacitor, it cannot be kept indefinitely and must periodically be refreshed. That is why this memory structure is called dynamic, in contrast to the static storage in an SRAM cell.

To refresh the cell, we merely read its contents and write it back. The charge can be kept for several milliseconds. If every bit had to be read out of the DRAM and then written back individually, we would constantly be refreshing the DRAM, leaving no time for accessing it. Fortunately, DRAMs use a two-level decoding structure, and this allows us to refresh an entire *row* (which shares a word line) with a read cycle followed immediately by a write cycle.

Figure 5.4 shows the internal organization of a DRAM, and Figure 5.5 shows how the density, cost, and access time of DRAMs have changed over the years.

The row organization that helps with refresh also helps with performance. To improve performance, DRAMs buffer rows for repeated access. The buffer acts like an SRAM; by changing the address, random bits can be accessed in the buffer until the next row access. This capability improves the access time significantly, since the access time to bits in the row is much lower. Making the chip wider also improves the memory bandwidth of the chip. When the row is in the buffer, it can be transferred by successive addresses at whatever the width of the DRAM is (typically 4, 8, or 16 bits), or by specifying a block transfer and the starting address within the buffer.

To improve the interface to processors further, DRAMs added clocks and are properly called synchronous DRAMs or SDRAMs. The advantage of SDRAMs is that the use of a clock eliminates the time for the memory and processor to synchronize. The speed advantage of synchronous DRAMs comes from the ability to transfer the bits in the burst without having to specify additional address bits.

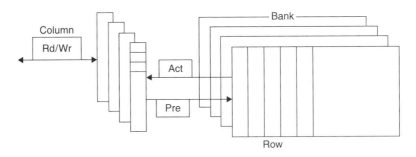

FIGURE 5.4   Internal organization of a DRAM. Modern DRAMs are organized in banks, typically four for DDR3. Each bank consists of a series of rows. Sending a PRE (precharge) command opens or closes a bank. A row address is sent with an Act (activate), which causes the row to transfer to a buffer. When the row is in the buffer, it can be transferred by successive column addresses at whatever the width of the DRAM is (typically 4, 8, or 16 bits in DDR3) or by specifying a block transfer and the starting address. Each command, as well as block transfers, is synchronized with a clock.

Year introduced	Chip size	$ per GiB	Total access time to a new row/column	Average column access time to existing row
1980	64 Kibibit	$1,500,000	250 ns	150 ns
1983	256 Kibibit	$500,000	185 ns	100 ns
1985	1 Mebibit	$200,000	135 ns	40 ns
1989	4 Mebibit	$50,000	110 ns	40 ns
1992	16 Mebibit	$15,000	90 ns	30 ns
1996	64 Mebibit	$10,000	60 ns	12 ns
1998	128 Mebibit	$4,000	60 ns	10 ns
2000	256 Mebibit	$1,000	55 ns	7 ns
2004	512 Mebibit	$250	50 ns	5 ns
2007	1 Gibibit	$50	45 ns	1.25 ns
2010	2 Gibibit	$30	40 ns	1 ns
2012	4 Gibibit	$1	35 ns	0.8 ns

FIGURE 5.5   DRAM size increased by multiples of four approximately once every 3 years until 1996, and thereafter considerably slower. The improvements in access time have been slower but continuous, and cost roughly tracks density improvements, although cost is often affected by other issues, such as availability and demand. The cost per gibibyte is not adjusted for inflation.

Instead, the clock transfers the successive bits in a burst. The fastest version is called *Double Data Rate (DDR)* SDRAM. The name means data transfers on both the rising *and* falling edge of the clock, thereby getting twice as much bandwidth as you might expect based on the clock rate and the data width. The latest version of this technology is called DDR4. A DDR4-3200 DRAM can do 3200 million transfers per second, which means it has a 1600-MHz clock.

Sustaining that much bandwidth requires clever organization *inside* the DRAM. Instead of just a faster row buffer, the DRAM can be internally organized to read or write from multiple *banks*, with each having its own row buffer. Sending an address to several banks permits them all to read or write simultaneously. For example, with four banks, there is just one access time and then accesses rotate between the four banks to supply four times the bandwidth. This rotating access scheme is called *address interleaving*.

Although personal mobile devices like the iPad (see Chapter 1) use individual DRAMs, memory for servers is commonly sold on small boards called *dual inline memory modules* (DIMMs). DIMMs typically contain 4–16 DRAMs, and they are normally organized to be 8 bytes wide for server systems. A DIMM using DDR4-3200 SDRAMs could transfer at $8 \times 3200 = 25,600$ megabytes per second. Such DIMMs are named after their bandwidth: PC25600. Since a DIMM can have so many DRAM chips that only a portion of them are used for a particular transfer, we need a term to refer to the subset of chips in a DIMM that share common address lines. To avoid confusion with the internal DRAM names of row and banks, we use the term *memory rank* for such a subset of chips in a DIMM.

Elaboration: One way to measure the performance of the memory system behind the caches is the Stream benchmark [McCalpin, 1995]. It measures the performance of long vector operations. They have no temporal locality and they access arrays that are larger than the cache of the computer being tested.

Flash Memory

Flash memory is a type of *electrically erasable programmable read-only memory* (EEPROM).

Unlike disks and DRAM, but like other EEPROM technologies, writes can wear out flash memory bits. To cope with such limits, most flash products include a controller to spread the writes by remapping blocks that have been written many times to less trodden blocks. This technique is called *wear leveling*. With wear leveling, personal mobile devices are very unlikely to exceed the write limits in the flash. Such wear leveling lowers the potential performance of flash, but it is needed unless higher-level software monitors block wear. Flash controllers that perform wear leveling can also improve yield by mapping out memory cells that were manufactured incorrectly.

Disk Memory

As Figure 5.6 shows, a magnetic hard disk consists of a collection of platters, which rotate on a spindle at 5400 to 15,000 revolutions per minute. The metal platters are covered with magnetic recording material on both sides, similar to the material found on a cassette or videotape. To read and write information on a hard disk, a movable *arm* containing a small electromagnetic coil called a *read-write head* is located just above each surface. The entire drive is permanently sealed to control the environment inside the drive, which, in turn, allows the disk heads to be much closer to the drive surface.

Each disk surface is divided into concentric circles, called tracks. There are typically tens of thousands of tracks per surface. Each track is in turn divided into

track One of thousands of concentric circles that make up the surface of a magnetic disk.

sector One of the segments that make up a track on a magnetic disk; a sector is the smallest amount of information that is read or written on a disk.

sectors that contain the information; each track may have thousands of sectors. Sectors are typically 512 to 4096 bytes in size. The sequence recorded on the magnetic media is a sector number, a gap, the information for that sector including error correction code (see Section 5.5), a gap, the sector number of the next sector, and so on.

The disk heads for each surface are connected together and move in conjunction, so that every head is over the same track of every surface. The term *cylinder* is used to refer to all the tracks under the heads at a given point on all surfaces.

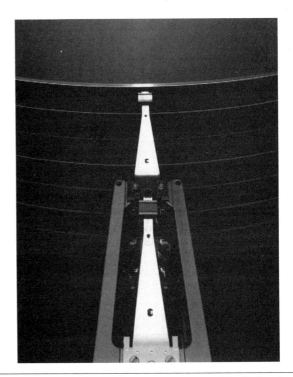

FIGURE 5.6   A disk showing 10 disk platters and the read/write heads. The diameter of today's disks is 2.5 or 3.5 inches, and there are typically one or two platters per drive today.

To access data, the operating system must direct the disk through a three-stage process. The first step is to position the head over the proper track. This operation is called a **seek**, and the time to move the head to the desired track is called the *seek time*.

seek The process of positioning a read/write head over the proper track on a disk.

Disk manufacturers report minimum seek time, maximum seek time, and average seek time in their manuals. The first two are easy to measure, but the average is open to wide interpretation because it depends on the seek distance. The industry calculates average seek time as the sum of the time for all possible seeks divided by the number of possible seeks. Average seek times are usually advertised as 3 ms to 13 ms, but, depending on the application and scheduling of disk requests, the actual average seek time may be only 25% to 33% of the advertised number because of the locality of disk references. This locality arises both because of successive accesses to the same file and because the operating system tries to schedule such accesses together.

Once the head has reached the correct track, we must wait for the desired sector to rotate under the read/write head. This time is called the rotational latency or rotational delay. The average latency to the desired information is halfway around the disk. Disks rotate at 5400 RPM to 15,000 RPM. The average rotational latency at 5400 RPM is

rotational latency Also called **rotational delay**. The time required for the desired sector of a disk to rotate under the read/write head; usually assumed to be half the rotation time.

$$\text{Average rotational latency} = \frac{0.5 \text{ rotation}}{5400 \text{ RPM}} = \frac{0.5 \text{ rotation}}{5400 \text{ RPM}/\left(60\dfrac{\text{seconds}}{\text{minute}}\right)}$$

$$= 0.0056 \text{ seconds} = 5.6 \text{ ms}$$

The last component of a disk access, *transfer time*, is the time to transfer a block of bits. The transfer time is a function of the sector size, the rotation speed, and the recording density of a track. Transfer rates in 2012 were between 100 and 200 MB/sec.

One complication is that most disk controllers have a built-in cache that stores sectors as they are passed over; transfer rates from the cache are typically higher, and were up to 750 MB/sec (6 Gbit/sec) in 2012.

Alas, where block numbers are located is no longer intuitive. The assumptions of the sector-track-cylinder model above are that nearby blocks are on the same track, blocks in the same cylinder take less time to access since there is no seek time, and some tracks are closer than others. The reason for the change was the raising of the level of the disk interfaces. To speed-up sequential transfers, these higher-level interfaces organize disks more like tapes than like random access devices. The logical blocks are ordered in serpentine fashion across a single surface, trying to capture all the sectors that are recorded at the same bit density to try to get best performance. Hence, sequential blocks may be on different tracks.

In summary, the two primary differences between magnetic disks and semiconductor memory technologies are that disks have a slower access time because they are mechanical devices—flash is 1000 times as fast and DRAM is 100,000 times as fast—yet they are cheaper per bit because they have very high storage capacity at a modest cost—disks are 10 to 100 times cheaper. Magnetic disks are nonvolatile like flash, but unlike flash there is no write wear-out problem. However, flash is much more rugged and hence a better match to the jostling inherent in personal mobile devices.

5.3 The Basics of Caches

Cache: a safe place for hiding or storing things.

Webster's New World Dictionary of the American Language, Third College Edition, 1988

In our library example, the desk acted as a cache—a safe place to store things (books) that we needed to examine. *Cache* was the name chosen to represent the level of the memory hierarchy between the processor and main memory in the first commercial computer to have this extra level. The memories in the datapath in Chapter 4 are simply replaced by caches. Today, although this remains the dominant use of the word *cache*, the term is also used to refer to any storage managed to take advantage of locality of access. Caches first appeared in research computers in the early 1960s and in production computers later in that same decade; every general-purpose computer built now from servers to low-power embedded processors, includes caches.

In this section, we begin by looking at a very simple cache in which the processor requests are each one word, and the blocks also consist of a single word. (Readers already familiar with cache basics may want to skip to Section 5.4.) Figure 5.7 shows such a simple cache, before and after requesting a data item that is not initially in the cache. Before the request, the cache contains a collection of recent references $X_1, X_2, \ldots, X_{n-1}$, and the processor requests a word X_n that is not in the cache. This request results in a miss, and the word X_n is brought from memory into the cache.

In looking at the scenario in Figure 5.7, there are two questions to answer: How do we know if a data item is in the cache? Moreover, if it is, how do we find it? The answers are related. If each word can go in exactly one place in the cache, then it is straightforward to find the word if it is in the cache. The simplest way to assign a location in the cache for each word in memory is to assign the cache location based on the *address* of the word in memory. This cache structure is called direct mapped, since each memory location is mapped directly to exactly one location in the cache. The typical mapping between addresses and cache locations for a direct-mapped cache is usually simple. For example, almost all direct-mapped caches use this mapping to find a block:

direct-mapped cache A cache structure in which each memory location is mapped to exactly one location in the cache.

(Block address) modulo (Number of blocks in the cache)

If the number of entries in the cache is a power of 2, then modulo can be computed simply by using the low-order $\log_2$ (cache size in blocks) bits of the address. Thus, an 8-block cache uses the three lowest bits ($8 = 2^3$) of the block address. For example, Figure 5.8 shows how the memory addresses between 1_{ten} (00001_{two}) and 29_{ten} (11101_{two}) map to locations 1_{ten} (001_{two}) and 5_{ten} (101_{two}) in a direct-mapped cache of eight words.

Because each cache location can contain the contents of a number of different memory locations, how do we know whether the data in the cache corresponds to a requested word? That is, how do we know whether a requested word is in the cache or not? We answer this question by adding a set of tags to the cache. The tags contain the address information required to identify whether a word in the cache corresponds to the requested word. The tag needs just to contain the upper portion of the address, corresponding to the bits that are not used as an index into the cache. For example, in Figure 5.8 we need only have the upper two of the five address bits in the tag, since the lower 3-bit index field of the address selects the block. Architects omit the index bits because they are redundant, since by definition, the index field of any address of a cache block must be that block number.

tag A field in a table used for a memory hierarchy that contains the address information required to identify whether the associated block in the hierarchy corresponds to a requested word.

We also need a way to recognize that a cache block does not have valid information. For instance, when a processor starts up, the cache does not have good data, and the tag fields will be meaningless. Even after executing many instructions,

| X_4 |
| X_1 |
| X_{n-2} |
| |
| X_{n-1} |
| X_2 |
| |
| X_3 |

| X_4 |
| X_1 |
| X_{n-2} |
| |
| X_{n-1} |
| X_2 |
| X_n |
| X_3 |

a. Before the reference to X_n b. After the reference to X_n

FIGURE 5.7 The cache just before and just after a reference to a word X_n that is not initially in the cache. This reference causes a miss that forces the cache to fetch X_n from memory and insert it into the cache.

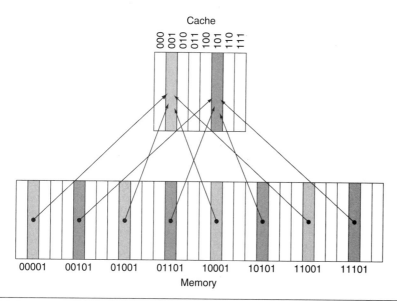

FIGURE 5.8 A direct-mapped cache with eight entries showing the addresses of memory words between 0 and 31 that map to the same cache locations. Because there are eight words in the cache, an address X maps to the direct-mapped cache word X modulo 8. That is, the low-order $\log_2(8)$ = 3 bits are used as the cache index. Thus, addresses 00001_{two}, 01001_{two}, 10001_{two}, and 11001_{two} all map to entry 001_{two} of the cache, while addresses 00101_{two}, 01101_{two}, 10101_{two}, and 11101_{two} all map to entry 101_{two} of the cache.

valid bit A field in the tables of a memory hierarchy that indicates that the associated block in the hierarchy contains valid data.

some of the cache entries may still be empty, as in Figure 5.7. Thus, we need to know that the tag should be ignored for such entries. The most common method is to add a **valid bit** to indicate whether an entry contains a valid address. If the bit is not set, there cannot be a match for this block.

For the rest of this section, we will focus on explaining how a cache deals with reads. In general, handling reads is a little simpler than handling writes, since reads do not have to change the contents of the cache. After seeing the basics of how reads work and how cache misses can be handled, we'll examine the cache designs for real computers and detail how these caches handle writes.

The **BIG** Picture

PREDICTION

Caching is perhaps the most important example of the big idea of **prediction**. It relies on the principle of locality to try to find the desired data in the higher levels of the memory hierarchy, and provides mechanisms to ensure that when the prediction is wrong it finds and uses the proper data from the lower levels of the memory hierarchy. The hit rates of the cache prediction on modern computers are often above 95% (see Figure 5.46).

Accessing a Cache

Below is a sequence of nine memory references to an empty eight-block cache, including the action for each reference. Figure 5.9 shows how the contents of the cache change on each miss. Since there are eight blocks in the cache, the low-order 3 bits of an address give the block number:

Decimal address of reference	Binary address of reference	Hit or miss in cache	Assigned cache block (where found or placed)
22	10110_{two}	miss (5.9b)	$(10110_{two} \bmod 8) = 110_{two}$
26	11010_{two}	miss (5.9c)	$(11010_{two} \bmod 8) = 010_{two}$
22	10110_{two}	hit	$(10110_{two} \bmod 8) = 110_{two}$
26	11010_{two}	hit	$(11010_{two} \bmod 8) = 010_{two}$
16	10000_{two}	miss (5.9d)	$(10000_{two} \bmod 8) = 000_{two}$
3	00011_{two}	miss (5.9e)	$(00011_{two} \bmod 8) = 011_{two}$
16	10000_{two}	hit	$(10000_{two} \bmod 8) = 000_{two}$
18	10010_{two}	miss (5.9f)	$(10010_{two} \bmod 8) = 010_{two}$
16	10000_{two}	hit	$(10000_{two} \bmod 8) = 000_{two}$

Since the cache is empty, several of the first references are misses; the caption of Figure 5.9 describes the actions for each memory reference. On the eighth reference

Index	V	Tag	Data
000	N		
001	N		
010	N		
011	N		
100	N		
101	N		
110	N		
111	N		

a. The initial state of the cache after power-on

Index	V	Tag	Data
000	N		
001	N		
010	N		
011	N		
100	N		
101	N		
110	Y	10_{two}	Memory (10110_{two})
111	N		

b. After handling a miss of address (10110_{two})

Index	V	Tag	Data
000	N		
001	N		
010	Y	11_{two}	Memory (11010_{two})
011	N		
100	N		
101	N		
110	Y	10_{two}	Memory (10110_{two})
111	N		

c. After handling a miss of address (11010_{two})

Index	V	Tag	Data
000	Y	10_{two}	Memory (10000_{two})
001	N		
010	Y	11_{two}	Memory (11010_{two})
011	N		
100	N		
101	N		
110	Y	10_{two}	Memory (10110_{two})
111	N		

d. After handling a miss of address (10000_{two})

Index	V	Tag	Data
000	Y	10_{two}	Memory (10000_{two})
001	N		
010	Y	11_{two}	Memory (11010_{two})
011	Y	00_{two}	Memory (00011_{two})
100	N		
101	N		
110	Y	10_{two}	Memory (10110_{two})
111	N		

e. After handling a miss of address (00011_{two})

Index	V	Tag	Data
000	Y	10_{two}	Memory (10000_{two})
001	N		
010	Y	10_{two}	Memory (10010_{two})
011	Y	00_{two}	Memory (00011_{two})
100	N		
101	N		
110	Y	10_{two}	Memory (10110_{two})
111	N		

f. After handling a miss of address (10010_{two})

FIGURE 5.9 The cache contents are shown after each reference request that misses, with the index and tag fields shown in binary for the sequence of addresses on page 400. The cache is initially empty, with all valid bits (V entry in cache) turned off (N). The processor requests the following addresses: 10110_{two} (miss), 11010_{two} (miss), 10110_{two} (hit), 11010_{two} (hit), 10000_{two} (miss), 00011_{two} (miss), 10000_{two} (hit), 10010_{two} (miss), and 10000_{two} (hit). The figures show the cache contents after each miss in the sequence has been handled. When address 10010_{two} (18) is referenced, the entry for address 11010_{two} (26) must be replaced, and a reference to 11010_{two} will cause a subsequent miss. The tag field will contain only the upper portion of the address. The full address of a word contained in cache block i with tag field j for this cache is $j \times 8 + i$, or equivalently the concatenation of the tag field j and the index i. For example, in cache f above, index 010_{two} has tag 10_{two} and corresponds to address 10010_{two}.

we have conflicting demands for a block. The word at address 18 (10010_{two}) should be brought into cache block 2 (010_{two}). Hence, it must replace the word at address 26 (11010_{two}), which is already in cache block 2 (010_{two}). This behavior allows a cache to take advantage of temporal locality: recently referenced words replace less recently referenced words.

This situation is directly analogous to needing a book from the shelves and having no more space on your desk—some book already on your desk must be returned to the shelves. In a direct-mapped cache, there is only one place to put the newly requested item and hence just one choice of what to replace.

We know where to look in the cache for each possible address: the low-order bits of an address can be used to find the unique cache entry to which the address could map. Figure 5.10 shows how a referenced address is divided into

- A *tag field*, which is used to compare with the value of the tag field of the cache

- A *cache index*, which is used to select the block

The index of a cache block, together with the tag contents of that block, uniquely specifies the memory address of the word contained in the cache block. Because the index field is used as an address to reference the cache, and because an n-bit field has 2^n values, the total number of entries in a direct-mapped cache must be a power of 2. Since words are aligned to multiples of four bytes, the least significant two bits of every address specify a byte within a word. Hence, if the words are aligned in memory, the least significant two bits can be ignored when selecting a word in the block. For this chapter, we'll assume that data are aligned in memory, and discuss how to handle unaligned cache accesses in an Elaboration.

The total number of bits needed for a cache is a function of the cache size and the address size, because the cache includes both the storage for the data and the tags. The size of the block above was one word (4 bytes), but normally it is several. For the following situation:

- 64-bit addresses

- A direct-mapped cache

- The cache size is 2^n blocks, so n bits are used for the index

- The block size is 2^m words (2^{m+2} bytes), so m bits are used for the word within the block, and two bits are used for the byte part of the address

The size of the tag field is

$$64 - (n + m + 2).$$

The total number of bits in a direct-mapped cache is

$$2^n \times (\text{block size} + \text{tag size} + \text{valid field size}).$$

Address (showing bit positions)

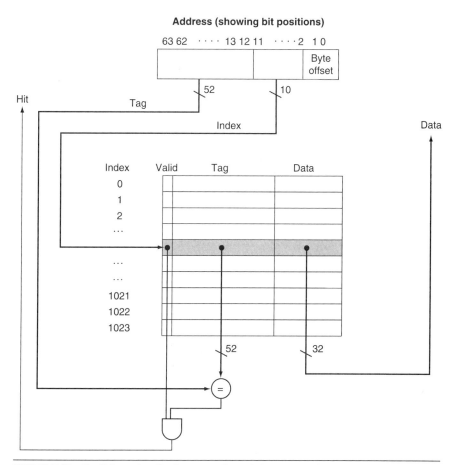

FIGURE 5.10 For this cache, the lower portion of the address is used to select a cache entry consisting of a data word and a tag. This cache holds 1024 words or 4 KiB. Unless noted otherwise, we assume 64-bit addresses in this chapter. The tag from the cache is compared against the upper portion of the address to determine whether the entry in the cache corresponds to the requested address. Because the cache has 2^{10} (or 1024) words and a block size of one word, 10 bits are used to index the cache, leaving $64 - 10 - 2 = 52$ bits to be compared against the tag. If the tag and upper 52 bits of the address are equal and the valid bit is on, then the request hits in the cache, and the word is supplied to the processor. Otherwise, a miss occurs.

Since the block size is 2^m words (2^{m+5} bits), and we need 1 bit for the valid field, the number of bits in such a cache is

$$2^n \times (2^m \times 32 + (64 - n - m - 2) + 1) = 2^n \times (2^m \times 32 + 63 - n - m).$$

Although this is the actual size in bits, the naming convention is to exclude the size of the tag and valid field and to count only the size of the data. Thus, the cache in Figure 5.10 is called a 4 KiB cache.

Bits in a Cache

EXAMPLE

How many total bits are required for a direct-mapped cache with 16 KiB of data and four-word blocks, assuming a 64-bit address?

ANSWER

We know that 16 KiB is 4096 (2^{12}) words. With a block size of four words (2^2), there are 1024 (2^{10}) blocks. Each block has 4×32 or 128 bits of data plus a tag, which is $64 - 10 - 2 - 2$ bits, plus a valid bit. Thus, the complete cache size is

$$2^{10} \times (4 \times 32 + (64 - 10 - 2 - 2) + 1) = 2^{10} \times 179 = 179 \, \text{Kibibits}$$

or 22.4 KiB for a 16 KiB cache. For this cache, the total number of bits in the cache is about 1.4 times as many as needed just for the storage of the data.

Mapping an Address to a Multiword Cache Block

EXAMPLE

Consider a cache with 64 blocks and a block size of 16 bytes. To what block number does byte address 1200 map?

ANSWER

We saw the formula on page 398. The block is given by

(Block address) modulo (Number of blocks in the cache)

where the address of the block is

$$\frac{\text{Byte address}}{\text{Bytes per block}}$$

Notice that this block address is the block containing all addresses between

$$\left\lfloor \frac{\text{Byte address}}{\text{Bytes per block}} \right\rfloor \times \text{Bytes per block}$$

and

$$\left\lceil \frac{\text{Byte address}}{\text{Bytes per block}} \right\rceil \times \text{Bytes per block} + (\text{Bytes per block} - 1)$$

Thus, with 16 bytes per block, byte address 1200 is block address

$$\left\lceil \frac{1200}{6} \right\rceil = 75$$

which maps to cache block number (75 modulo 64) = 11. In fact, this block maps all addresses between 1200 and 1215.

Larger blocks exploit spatial locality to lower miss rates. As Figure 5.11 shows, increasing the block size usually decreases the miss rate. The miss rate may go up eventually if the block size becomes a significant fraction of the cache size, because the number of blocks that can be held in the cache will become small, and there will be a great deal of competition for those blocks. As a result, a block will be bumped out of the cache before many of its words are accessed. Stated alternatively, spatial locality among the words in a block decreases with a very large block; consequently, the benefits to the miss rate become smaller.

A more serious problem associated with just increasing the block size is that the cost of a miss rises. The miss penalty is determined by the time required to fetch

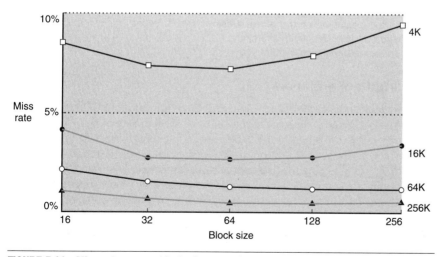

FIGURE 5.11 Miss rate versus block size. Note that the miss rate actually goes up if the block size is too large relative to the cache size. Each line represents a cache of different size. (This figure is independent of associativity, discussed soon.) Unfortunately, SPEC CPU2000 traces would take too long if block size were included, so these data are based on SPEC92.

the block from the next lower level of the hierarchy and load it into the cache. The time to fetch the block has two parts: the latency to the first word and the transfer time for the rest of the block. Clearly, unless we change the memory system, the transfer time—and hence the miss penalty—will likely increase as the block size expands. Furthermore, the improvement in the miss rate starts to decrease as the blocks become larger. The result is that the increase in the miss penalty overwhelms the decrease in the miss rate for blocks that are too large, and cache performance thus decreases. Of course, if we design the memory to transfer larger blocks more efficiently, we can increase the block size and obtain further improvements in cache performance. We discuss this topic in the next section.

Elaboration: Although it is hard to do anything about the longer latency component of the miss penalty for large blocks, we may be able to hide some of the transfer time so that the miss penalty is effectively smaller. The easiest method for doing this, called *early restart*, is simply to resume execution as soon as the requested word of the block is returned, rather than wait for the entire block. Many processors use this technique for instruction access, where it works best. Instruction accesses are largely sequential, so if the memory system can deliver a word every clock cycle, the processor may be able to restart operation when the requested word is returned, with the memory system delivering new instruction words just in time. This technique is usually less effective for data caches because it is likely that the words will be requested from the block in a less predictable way, and the probability that the processor will need another word from a different cache block before the transfer completes is high. If the processor cannot access the data cache because a transfer is ongoing, then it must stall.

An even more sophisticated scheme is to organize the memory so that the requested word is transferred from the memory to the cache first. The remainder of the block is then transferred, starting with the address after the requested word and wrapping around to the beginning of the block. This technique, called *requested word first* or *critical word first*, can be slightly faster than early restart, but it is limited by the same properties that restrain early restart.

Handling Cache Misses

cache miss A request for data from the cache that cannot be filled because the data are not present in the cache.

Before we look at the cache of a real system, let's see how the control unit deals with cache misses. (We describe a cache controller in detail in Section 5.9.) The control unit must detect a miss and process the miss by fetching the requested data from memory (or, as we shall see, a lower-level cache). If the cache reports a hit, the computer continues using the data as if nothing happened.

Modifying the control of a processor to handle a hit is trivial; misses, however, require some extra work. The cache miss handling is done in collaboration with the processor control unit and with a separate controller that initiates the memory access and refills the cache. The processing of a cache miss creates a pipeline stall (Chapter 4) in constrast to an exception or interrupt, which would require saving the state of all registers. For a cache miss, we can stall the entire processor, essentially freezing the contents of the temporary and programmer-visible registers, while we wait for memory. More sophisticated out-of-order processors can allow execution of instructions while waiting for a cache miss, but we'll assume in-order processors that stall on cache misses in this section.

Let's look a little more closely at how instruction misses are handled; the same approach can be easily extended to handle data misses. If an instruction access results in a miss, then the content of the Instruction register is invalid. To get the proper instruction into the cache, we must be able to tell the lower level in the memory hierarchy to perform a read. Since the program counter is incremented in the first clock cycle of execution, the address of the instruction that generates an instruction cache miss is equal to the value of the program counter minus 4. Once we have the address, we need to instruct the main memory to perform a read. We wait for the memory to respond (since the access will take multiple clock cycles), and then write the words containing the desired instruction into the cache.

We can now define the steps to be taken on an instruction cache miss:

1. Send the original PC value to the memory.

2. Instruct main memory to perform a read and wait for the memory to complete its access.

3. Write the cache entry, putting the data from memory in the data portion of the entry, writing the upper bits of the address (from the ALU) into the tag field, and turning the valid bit on.

4. Restart the instruction execution at the first step, which will refetch the instruction, this time finding it in the cache.

The control of the cache on a data access is essentially identical: on a miss, we simply stall the processor until the memory responds with the data.

Handling Writes

Writes work somewhat differently. Suppose on a store instruction, we wrote the data into only the data cache (without changing main memory); then, after the write into the cache, memory would have a different value from that in the cache. In such a case, the cache and memory are said to be *inconsistent*. The simplest way to keep the main memory and the cache consistent is always to write the data into both the memory and the cache. This scheme is called write-through.

The other key aspect of writes is what occurs on a write miss. We first fetch the words of the block from memory. After the block is fetched and placed into the cache, we can overwrite the word that caused the miss into the cache block. We also write the word to main memory using the full address.

Although this design handles writes very simply, it would not provide good performance. With a write-through scheme, every write causes the data to be written to main memory. These writes will take a long time, likely at least 100 processor clock cycles, and could slow down the processor considerably. For example, suppose 10% of the instructions are stores. If the CPI without cache

write-through
A scheme in which writes always update both the cache and the next lower level of the memory hierarchy, ensuring that data are always consistent between the two.

write buffer A queue that holds data while the data are waiting to be written to memory.

write-back A scheme that handles writes by updating values only to the block in the cache, then writing the modified block to the lower level of the hierarchy when the block is replaced.

misses was 1.0, spending 100 extra cycles on every write would lead to a CPI of $1.0 + 100 \times 10\% = 11$, reducing performance by more than a factor of 10.

One solution to this problem is to use a write buffer. A write buffer stores the data while they are waiting to be written to memory. After writing the data into the cache and into the write buffer, the processor can continue execution. When a write to main memory completes, the entry in the write buffer is freed. If the write buffer is full when the processor reaches a write, the processor must stall until there is an empty position in the write buffer. Of course, if the rate at which the memory can complete writes is less than the rate at which the processor is generating writes, no amount of buffering can help, because writes are being generated faster than the memory system can accept them.

The rate at which writes are generated may also be *less* than the rate at which the memory can accept them, and yet stalls may still occur. This can happen when the writes occur in bursts. To reduce the occurrence of such stalls, processors usually increase the depth of the write buffer beyond a single entry.

The alternative to a write-through scheme is a scheme called write-back. In a write-back scheme, when a write occurs, the new value is written only to the block in the cache. The modified block is written to the lower level of the hierarchy when it is replaced. Write-back schemes can improve performance, especially when processors can generate writes as fast or faster than the writes can be handled by main memory; a write-back scheme is, however, more complex to implement than write-through.

In the rest of this section, we describe caches from real processors, and we examine how they handle both reads and writes. In Section 5.8, we will describe the handling of writes in more detail.

Elaboration: Writes introduce several complications into caches that are not present for reads. Here we discuss two of them: the policy on write misses and efficient implementation of writes in write-back caches.

Consider a miss in a write-through cache. The most common strategy is to allocate a block in the cache, called *write allocate*. The block is fetched from memory and then the appropriate portion of the block is overwritten. An alternative strategy is to update the portion of the block in memory but not put it in the cache, called *no write allocate*. The motivation is that sometimes programs write entire blocks of data, such as when the operating system zeros a page of memory. In such cases, the fetch associated with the initial write miss may be unnecessary. Some computers allow the write allocation policy to be changed on a per-page basis.

Actually implementing stores efficiently in a cache that uses a write-back strategy is more complex than in a write-through cache. A write-through cache can write the data into the cache and read the tag; if the tag mismatches, then a miss occurs. Because the cache is write-through, the overwriting of the block in the cache is not catastrophic, since memory has the correct value. In a write-back cache, we must first write the block back to memory if the data in the cache are modified and we have a cache miss. If we simply overwrote the block on a store instruction before we knew whether the store had hit in the cache (as we could for a write-through cache), we would destroy the contents of the block, which is not backed up in the next lower level of the memory hierarchy.

In a write-back cache, because we cannot overwrite the block, stores either require two cycles (a cycle to check for a hit followed by a cycle to actually perform the write) or require a write buffer to hold that data—effectively allowing the store to take only one cycle by pipelining it. When a store buffer is used, the processor does the cache lookup and places the data in the store buffer during the normal cache access cycle. Assuming a cache hit, the new data are written from the store buffer into the cache on the next unused cache access cycle.

By comparison, in a write-through cache, writes can always be done in one cycle. We read the tag and write the data portion of the selected block. If the tag matches the address of the block being written, the processor can continue normally, since the correct block has been updated. If the tag does not match, the processor generates a write miss to fetch the rest of the block corresponding to that address.

Many write-back caches also include write buffers that are used to reduce the miss penalty when a miss replaces a modified block. In such a case, the modified block is moved to a write-back buffer associated with the cache while the requested block is read from memory. The write-back buffer is later written back to memory. Assuming another miss does not occur immediately, this technique halves the miss penalty when a dirty block must be replaced.

An Example Cache: The Intrinsity FastMATH Processor

The Intrinsity FastMATH is an embedded microprocessor that uses the MIPS architecture and a simple cache implementation. Near the end of the chapter, we will examine the more complex cache designs of ARM and Intel microprocessors, but we start with this simple, yet real, example for pedagogical reasons. Figure 5.12 shows the organization of the Intrinsity FastMATH data cache. Note that the address size for this computer is just 32 bits, not 64 as in the rest of the book.

This processor has a 12-stage pipeline. When operating at peak speed, the processor can request both an instruction word and a data word on every clock. To satisfy the demands of the pipeline without stalling, separate instruction and data caches are used. Each cache is 16 KiB, or 4096 words, with 16-word blocks.

Read requests for the cache are straightforward. Because there are separate data and instruction caches, we need separate control signals to read and write each cache. (Remember that we need to update the instruction cache when a miss occurs.) Thus, the steps for a read request to either cache are as follows:

1. Send the address to the appropriate cache. The address comes either from the PC (for an instruction) or from the ALU (for data).

2. If the cache signals hit, the requested word is available on the data lines. Since there are 16 words in the desired block, we need to select the right one. A block index field is used to control the multiplexor (shown at the bottom of the figure), which selects the requested word from the 16 words in the indexed block.

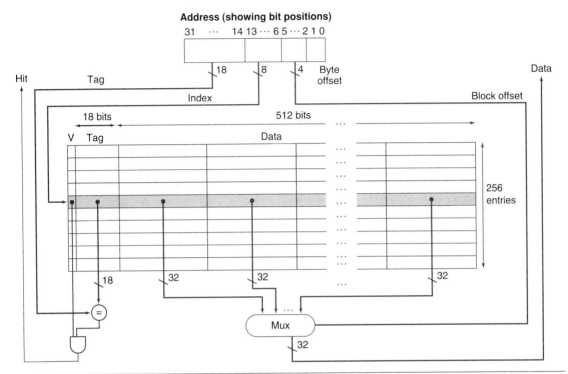

FIGURE 5.12 The 16 KiB caches in the Intrinsity FastMATH each contain 256 blocks with 16 words per block. Note that the address size for this computer is just 32 bits. The tag field is 18 bits wide and the index field is 8 bits wide, while a 4-bit field (bits 5–2) is used to index the block and select the word from the block using a 16-to-1 multiplexor. In practice, to eliminate the multiplexor, caches use a separate large RAM for the data and a smaller RAM for the tags, with the block offset supplying the extra address bits for the large data RAM. In this case, the large RAM is 32 bits wide and must have 16 times as many words as blocks in the cache.

3. If the cache signals miss, we send the address to the main memory. When the memory returns with the data, we write it into the cache and then read it to fulfill the request.

For writes, the Intrinsity FastMATH offers both write-through and write-back, leaving it up to the operating system to decide which strategy to use for an application. It has a one-entry write buffer.

What cache miss rates are attained with a cache structure like that used by the Intrinsity FastMATH? Figure 5.13 shows the miss rates for the instruction and data caches. The combined miss rate is the effective miss rate per reference for each program after accounting for the differing frequency of instruction and data accesses.

Instruction miss rate	Data miss rate	Effective combined miss rate
0.4%	11.4%	3.2%

FIGURE 5.13 Approximate instruction and data miss rates for the Intrinsity FastMATH processor for SPEC CPU2000 benchmarks. The combined miss rate is the effective miss rate seen for the combination of the 16 KiB instruction cache and 16 KiB data cache. It is obtained by weighting the instruction and data individual miss rates by the frequency of instruction and data references.

Although miss rate is an important characteristic of cache designs, the ultimate measure will be the effect of the memory system on program execution time; we'll see how miss rate and execution time are related shortly.

Elaboration: A combined cache with a total size equal to the sum of the two split caches will usually have a better hit rate. This higher rate occurs because the combined cache does not rigidly divide the number of entries that may be used by instructions from those that may be used by data. Nonetheless, almost all processors today use split instruction and data caches to increase cache *bandwidth* to match what modern pipelines expect. (There may also be fewer conflict misses; see Section 5.8.)

Here are miss rates for caches the size of those found in the Intrinsity FastMATH processor, and for a combined cache whose size is equal to the sum of the two caches:

split cache A scheme in which a level of the memory hierarchy is composed of two independent caches that operate in parallel with each other, with one handling instructions and one handling data.

- Total cache size: 32 KiB
- Split cache effective miss rate: 3.24%
- Combined cache miss rate: 3.18%

The miss rate of the split cache is only slightly worse.

The advantage of doubling the cache bandwidth, by supporting both an instruction and data access simultaneously, easily overcomes the disadvantage of a slightly increased miss rate. This observation cautions us that we cannot use miss rate as the sole measure of cache performance, as Section 5.4 shows.

Summary

We began the previous section by examining the simplest of caches: a direct-mapped cache with a one-word block. In such a cache, both hits and misses are simple, since a word can go in exactly one location and there is a separate tag for every word. To keep the cache and memory consistent, a write-through scheme can be used, so that every write into the cache also causes memory to be updated. The alternative to write-through is a write-back scheme that copies a block back to memory when it is replaced; we'll discuss this scheme further in upcoming sections.

To take advantage of spatial locality, a cache must have a block size larger than one word. The use of a bigger block decreases the miss rate and improves the efficiency of the cache by reducing the amount of tag storage relative to the amount of data storage in the cache. Although a larger block size decreases the miss rate, it can also increase the miss penalty. If the miss penalty increased linearly with the block size, larger blocks could easily lead to lower performance.

To avoid performance loss, the bandwidth of main memory is increased to transfer cache blocks more efficiently. Common methods for increasing bandwidth external to the DRAM are making the memory wider and interleaving. DRAM designers have steadily improved the interface between the processor and memory to increase the bandwidth of burst mode transfers to reduce the cost of larger cache block sizes.

Check Yourself

The speed of the memory system affects the designer's decision on the size of the cache block. Which of the following cache designer guidelines is generally valid?

1. The shorter the memory latency, the smaller the cache block

2. The shorter the memory latency, the larger the cache block

3. The higher the memory bandwidth, the smaller the cache block

4. The higher the memory bandwidth, the larger the cache block

5.4 Measuring and Improving Cache Performance

In this section, we begin by examining ways to measure and analyze cache performance. We then explore two different techniques for improving cache performance. One focuses on reducing the miss rate by reducing the probability that two distinct memory blocks will contend for the same cache location. The second technique reduces the miss penalty by adding an additional level to the hierarchy. This technique, called *multilevel caching*, first appeared in high-end computers selling for more than $100,000 in 1990; since then it has become common on personal mobile devices selling for a few hundred dollars!

CPU time can be divided into the clock cycles that the CPU spends executing the program and the clock cycles that the CPU spends waiting for the memory system. Normally, we assume that the costs of cache accesses that are hits are part of the normal CPU execution cycles. Thus,

$$\text{CPU time} = (\text{CPU execution clock cycles} + \text{Memory-stall clock cycles}) \times \text{Clock cycle time}$$

The memory-stall clock cycles come primarily from cache misses, and we make that assumption here. We also restrict the discussion to a simplified model of the memory system. In real processors, the stalls generated by reads and writes can be quite complex, and accurate performance prediction usually requires very detailed simulations of the processor and memory system.

Memory-stall clock cycles can be defined as the sum of the stall cycles coming from reads plus those coming from writes:

$$\text{Memory-stall clock cycles} = (\text{Read-stall cycles} + \text{Write-stall cycles})$$

The read-stall cycles can be defined in terms of the number of read accesses per program, the miss penalty in clock cycles for a read, and the read miss rate:

$$\text{Read-stall cycles} = \frac{\text{Reads}}{\text{Program}} \times \text{Read miss rate} \times \text{Read miss penalty}$$

Writes are more complicated. For a write-through scheme, we have two sources of stalls: write misses, which usually require that we fetch the block before continuing the write (see the *Elaboration* on page 408 for more details on dealing with writes), and write buffer stalls, which occur when the write buffer is full when a write happens. Thus, the cycles stalled for writes equal the sum of these two:

$$\text{Write-stall cycles} = \left(\frac{\text{Writes}}{\text{Program}} \times \text{Write miss rate} \times \text{Write miss penalty} \right) + \text{Write buffer stalls}$$

Because the write buffer stalls depend on the proximity of writes, and not just the frequency, it is impossible to give a simple equation to compute such stalls. Fortunately, in systems with a reasonable write buffer depth (e.g., four or more words) and a memory capable of accepting writes at a rate that significantly exceeds the average write frequency in programs (e.g., by a factor of 2), the write buffer stalls will be small, and we can safely ignore them. If a system did not meet these criteria, it would not be well designed; instead, the designer should have used either a deeper write buffer or a write-back organization.

Write-back schemes also have potential additional stalls arising from the need to write a cache block back to memory when the block is replaced. We will discuss this more in Section 5.8.

In most write-through cache organizations, the read and write miss penalties are the same (the time to fetch the block from memory). If we assume that the write buffer stalls are negligible, we can combine the reads and writes by using a single miss rate and the miss penalty:

$$\text{Memory-stall clock cycles} = \frac{\text{Memory accesses}}{\text{Program}} \times \text{Miss rate} \times \text{Miss penalty}$$

We can also factor this as

$$\text{Memory-stall clock cycles} = \frac{\text{Instructions}}{\text{Program}} \times \frac{\text{Misses}}{\text{Instruction}} \times \text{Miss penalty}$$

Let's consider a simple example to help us understand the impact of cache performance on processor performance.

EXAMPLE

Calculating Cache Performance

Assume the miss rate of an instruction cache is 2% and the miss rate of the data cache is 4%. If a processor has a CPI of 2 without any memory stalls, and the miss penalty is 100 cycles for all misses, determine how much faster a processor would run with a perfect cache that never missed. Assume the frequency of all loads and stores is 36%.

ANSWER

The number of memory miss cycles for instructions in terms of the Instruction count (I) is

$$\text{Instruction miss cycles} = I \times 2\% \times 100 = 2.00 \times I$$

As the frequency of all loads and stores is 36%, we can find the number of memory miss cycles for data references:

$$\text{Data miss cycles} = I \times 36\% \times 4\% \times 100 = 1.44 \times I$$

The total number of memory-stall cycles is 2.00 I + 1.44 I = 3.44 I. This is more than three cycles of memory stall per instruction. Accordingly, the total CPI including memory stalls is 2 + 3.44 = 5.44. Since there is no change in instruction count or clock rate, the ratio of the CPU execution times is

$$\frac{\text{CPU time with stalls}}{\text{CPU time with perfect cache}} = \frac{I \times \text{CPI}_{\text{stall}} \times \text{Clock cycle}}{I \times \text{CPI}_{\text{perfect}} \times \text{Clock cycle}}$$

$$= \frac{\text{CPI}_{\text{stall}}}{\text{CPI}_{\text{perfect}}} = \frac{5.44}{2}$$

The performance with the perfect cache is better by $\dfrac{5.44}{2} = 2.72$.

What happens if the processor is made faster, but the memory system is not? The amount of time spent on memory stalls will take up an increasing fraction of the execution time; Amdahl's Law, which we examined in Chapter 1, reminds us of this fact. A few simple examples show how serious this problem can be. Suppose we speed-up the computer in the previous example by reducing its CPI from 2 to 1 without changing the clock rate, which might be done with an improved pipeline. The system with cache misses would then have a CPI of 1 + 3.44 = 4.44, and the system with the perfect cache would be

$$\frac{4.44}{1} = 4.44 \text{ times as fast.}$$

The amount of execution time spent on memory stalls would have risen from

$$\frac{3.44}{5.44} = 63\%$$

to

$$\frac{3.44}{4.44} = 77\%$$

Similarly, increasing the clock rate without changing the memory system also increases the performance lost due to cache misses.

The previous examples and equations assume that the hit time is not a factor in determining cache performance. Clearly, if the hit time increases, the total time to access a word from the memory system will increase, possibly causing an increase in the processor cycle time. Although we will see additional examples of what can raise hit time shortly, one example is increasing the cache size. A larger cache could clearly have a bigger access time, just as, if your desk in the library was very large (say, 3 square meters), it would take longer to locate a book on the desk. An increase in hit time likely adds another stage to the pipeline, since it may take multiple cycles for a cache hit. Although it is more complex to calculate the performance impact of a deeper pipeline, at some point the increase in hit time for a larger cache could dominate the improvement in hit rate, leading to a decrease in processor performance.

To capture the fact that the time to access data for both hits and misses affects performance, designers sometime use *average memory access time* (AMAT) as a way to examine alternative cache designs. Average memory access time is the average time to access memory considering both hits and misses and the frequency of different accesses; it is equal to the following:

$$\text{AMAT} = \text{Time for a hit} + \text{Miss rate} \times \text{Miss penalty}$$

Calculating Average Memory Access Time

EXAMPLE

Find the AMAT for a processor with a 1 ns clock cycle time, a miss penalty of 20 clock cycles, a miss rate of 0.05 misses per instruction, and a cache access time (including hit detection) of 1 clock cycle. Assume that the read and write miss penalties are the same and ignore other write stalls.

ANSWER

The average memory access time per instruction is

$$
\begin{aligned}
\text{AMAT} &= \text{Time for a hit} + \text{Miss rate} \times \text{Miss penalty} \\
&= 1 + 0.05 \times 20 \\
&= 2 \text{ clock cycles}
\end{aligned}
$$

or 2 ns.

The next subsection discusses alternative cache organizations that decrease miss rate but may sometimes increase hit time; additional examples appear in Section 5.16.

Reducing Cache Misses by More Flexible Placement of Blocks

So far, when we put a block in the cache, we have used a simple placement scheme: A block can go in exactly one place in the cache. As mentioned earlier, it is called *direct mapped* because there is a direct mapping from any block address in memory to a single location in the upper level of the hierarchy. However, there is actually a whole range of schemes for placing blocks. Direct mapped, where a block can be placed in exactly one location, is at one extreme.

At the other extreme is a scheme where a block can be placed in *any* location in the cache. Such a scheme is called fully associative, because a block in memory may be associated with any entry in the cache. To find a given block in a fully associative cache, all the entries in the cache must be searched because a block can be placed in any one. To make the search practical, it is done in parallel with a comparator associated with each cache entry. These comparators significantly increase the hardware cost, effectively making fully associative placement practical only for caches with small numbers of blocks.

The middle range of designs between direct mapped and fully associative is called set associative. In a set-associative cache, there are a fixed number of locations where each block can be placed. A set-associative cache with n locations for a block is called an n-way set-associative cache. An n-way set-associative cache consists of a number of sets, each of which consists of n blocks. Each block in the memory maps to a unique *set* in the cache given by the index field, and a block can be placed in *any* element of that set. Thus, a set-associative placement combines direct-mapped placement and fully associative placement: a block is directly mapped into a set, and then all the blocks in the set are searched for a match. For example, Figure 5.14 shows where block 12 may be put in a cache with eight blocks total, according to the three block placement policies.

Remember that in a direct-mapped cache, the position of a memory block is given by

(Block number) modulo (Number of *blocks* in the cache)

fully associative cache A cache structure in which a block can be placed in any location in the cache.

set-associative cache A cache that has a fixed number of locations (at least two) where each block can be placed.

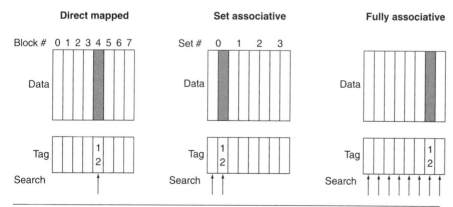

FIGURE 5.14 The location of a memory block whose address is 12 in a cache with eight blocks varies for direct-mapped, set-associative, and fully associative placement. In direct-mapped placement, there is only one cache block where memory block 12 can be found, and that block is given by (12 modulo 8) = 4. In a two-way set-associative cache, there would be four sets, and memory block 12 must be in set (12 mod 4) = 0; the memory block could be in either element of the set. In a fully associative placement, the memory block for block address 12 can appear in any of the eight cache blocks.

In a set-associative cache, the set containing a memory block is given by

(Block number) modulo (Number of *sets* in the cache)

Since the block may be placed in any element of the set, *all the tags of all the elements of the set* must be searched. In a fully associative cache, the block can go anywhere, and *all tags of all the blocks in the cache* must be searched.

We can also think of all block placement strategies as a variation on set associativity. Figure 5.15 shows the possible associativity structures for an eight-block cache. A direct-mapped cache is just a one-way set-associative cache: each cache entry holds one block and each set has one element. A fully associative cache with *m* entries is simply an *m*-way set-associative cache; it has one set with *m* blocks, and an entry can reside in any block within that set.

The advantage of increasing the degree of associativity is that it usually decreases the miss rate, as the next example shows. The main disadvantage, which we discuss in more detail shortly, is a potential increase in the hit time.

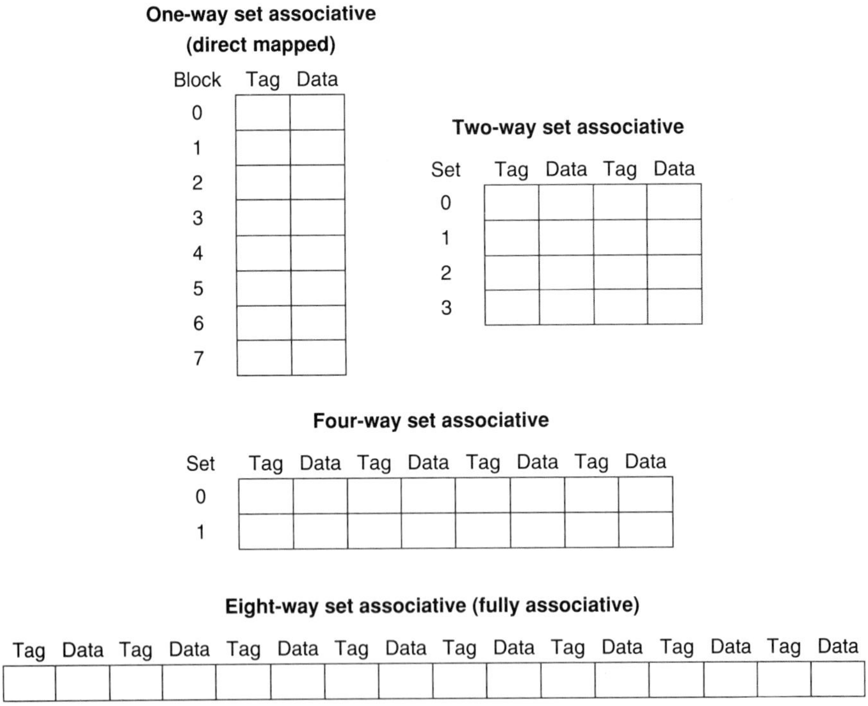

FIGURE 5.15   An eight-block cache configured as direct-mapped, two-way set associative, four-way set associative, and fully associative. The total size of the cache in blocks is equal to the number of sets times the associativity. Thus, for a fixed cache size, increasing the associativity decreases the number of sets while increasing the number of elements per set. With eight blocks, an eight-way set-associative cache is the same as a fully associative cache.

Misses and Associativity in Caches

Assume there are three small caches, each consisting of four one-word blocks. One cache is fully associative, a second is two-way set-associative, and the third is direct-mapped. Find the number of misses for each cache organization given the following sequence of block addresses: 0, 8, 0, 6, and 8.

EXAMPLE

The direct-mapped case is easiest. First, let's determine to which cache block each block address maps:

ANSWER

Block address	Cache block
0	(0 modulo 4) = 0
6	(6 modulo 4) = 2
8	(8 modulo 4) = 0

Now we can fill in the cache contents after each reference, using a blank entry to mean that the block is invalid, colored text to show a new entry added to the cache for the associated reference, and plain text to show an old entry in the cache:

Address of memory block accessed	Hit or miss	Contents of cache blocks after reference			
		0	1	2	3
0	miss	Memory[0]			
8	miss	Memory[8]			
0	miss	Memory[0]			
6	miss	Memory[0]		Memory[6]	
8	miss	Memory[8]		Memory[6]	

The direct-mapped cache generates five misses for the five accesses.

The set-associative cache has two sets (with indices 0 and 1) with two elements per set. Let's first determine to which set each block address maps:

Block address	Cache set
0	(0 modulo 2) = 0
6	(6 modulo 2) = 0
8	(8 modulo 2) = 0

Because we have a choice of which entry in a set to replace on a miss, we need a replacement rule. Set-associative caches usually replace the least recently used block within a set; that is, the block that was used furthest in the past is replaced. (We will discuss other replacement rules in more detail shortly.)

Using this replacement rule, the contents of the set-associative cache after each reference look like this:

Address of memory block accessed	Hit or miss	Contents of cache blocks after reference			
		Set 0	Set 0	Set 1	Set 1
0	miss	Memory[0]			
8	miss	Memory[0]	Memory[8]		
0	hit	Memory[0]	Memory[8]		
6	miss	Memory[0]	Memory[6]		
8	miss	Memory[8]	Memory[6]		

Notice that when block 6 is referenced, it replaces block 8, since block 8 has been less recently referenced than block 0. The two-way set-associative cache has four misses, one less than the direct-mapped cache.

The fully associative cache has four cache blocks (in a single set); any memory block can be stored in any cache block. The fully associative cache has the best performance, with only three misses:

Address of memory block accessed	Hit or miss	Contents of cache blocks after reference			
		Block 0	Block 1	Block 2	Block 3
0	miss	Memory[0]			
8	miss	Memory[0]	Memory[8]		
0	hit	Memory[0]	Memory[8]		
6	miss	Memory[0]	Memory[8]	Memory[6]	
8	hit	Memory[0]	Memory[8]	Memory[6]	

For this series of references, three misses is the best we can do, because three unique block addresses are accessed. Notice that if we had eight blocks in the cache, there would be no replacements in the two-way set-associative cache (check this for yourself), and it would have the same number of misses as the fully associative cache. Similarly, if we had 16 blocks, all three caches would have the identical number of misses. Even this trivial example shows that cache size and associativity are not independent in determining cache performance.

How much of a reduction in the miss rate is achieved by associativity? Figure 5.16 shows the improvement for a 64 KiB data cache with a 16-word block, and associativity ranging from direct-mapped to eight-way. Going from one-way to two-way associativity decreases the miss rate by about 15%, but there is little further improvement in going to higher associativity.

Associativity	Data miss rate
1	10.3%
2	8.6%
4	8.3%
8	8.1%

FIGURE 5.16 The data cache miss rates for an organization like the Intrinsity FastMATH processor for SPEC CPU2000 benchmarks with associativity varying from one-way to eight-way. These results for 10 SPEC CPU2000 programs are from Hennessy and Patterson (2003).

Tag	Index	Block offset

FIGURE 5.17 The three portions of an address in a set-associative or direct-mapped cache. The index is used to select the set, then the tag is used to choose the block by comparison with the blocks in the selected set. The block offset is the address of the desired data within the block.

Locating a Block in the Cache

Now, let's consider the task of finding a block in a cache that is set-associative. Just as in a direct-mapped cache, each block in a set-associative cache includes an address tag that gives the block address. The tag of every cache block within the appropriate set is checked to see if it matches the block address from the processor. Figure 5.17 decomposes the address. The index value is used to select the set containing the address of interest, and the tags of all the blocks in the set must be searched. Because speed is of the essence, all the tags in the selected set are searched in parallel. As in a fully associative cache, a sequential search would make the hit time of a set-associative cache too slow.

If the total cache size is kept the same, increasing the associativity raises the number of blocks per set, which is the number of simultaneous compares needed to perform the search in parallel: each increase by a factor of 2 in associativity doubles the number of blocks per set and halves the number of sets. Accordingly, each factor-of-2 increase in associativity decreases the size of the index by 1 bit and expands the size of the tag by 1 bit. In a fully associative cache, there is effectively only one set, and all the blocks must be checked in parallel. Thus, there is no index, and the entire address, excluding the block offset, is compared against the tag of every block. In other words, we search the full cache without any indexing.

In a direct-mapped cache, only a single comparator is needed, because the entry can be in only one block, and we access the cache simply by indexing. Figure 5.18 shows that in a four-way set-associative cache, four comparators are needed, together with a 4-to-1 multiplexor to choose among the four potential members of the selected set. The cache access consists of indexing the appropriate set and then searching the tags of the set. The costs of an associative cache are the extra comparators and any delay imposed by having to do the compare and select from among the elements of the set.

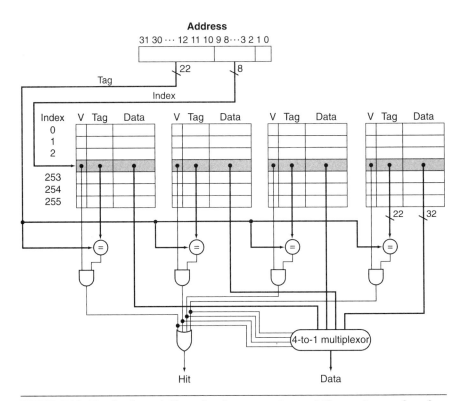

FIGURE 5.18   The implementation of a four-way set-associative cache requires four comparators and a 4-to-1 multiplexor. The comparators determine which element of the selected set (if any) matches the tag. The output of the comparators is used to select the data from one of the four blocks of the indexed set, using a multiplexor with a decoded select signal. In some implementations, the Output enable signals on the data portions of the cache RAMs can be used to select the entry in the set that drives the output. The Output enable signal comes from the comparators, causing the element that matches to drive the data outputs. This organization eliminates the need for the multiplexor.

The choice among direct-mapped, set-associative, or fully associative mapping in any memory hierarchy will depend on the cost of a miss versus the cost of implementing associativity, both in time and in extra hardware.

Elaboration: A *Content Addressable Memory* (CAM) is a circuit that combines comparison and storage in a single device. Instead of supplying an address and reading a word like a RAM, you send the data and the CAM looks to see if it has a copy and returns the index of the matching row. CAMs mean that cache designers can afford to implement much higher set associativity than if they needed to build the hardware out of SRAMs and comparators. In 2013, the greater size and power of CAM generally leads to two-way and four-way set associativity being built from standard SRAMs and comparators, with eight-way and above built using CAMs.

Choosing Which Block to Replace

When a miss occurs in a direct-mapped cache, the requested block can go in exactly one position, and the block occupying that position must be replaced. In an associative cache, we have a choice of where to place the requested block, and hence a choice of which block to replace. In a fully associative cache, all blocks are candidates for replacement. In a set-associative cache, we must choose among the blocks in the selected set.

The most commonly used scheme is least recently used (LRU), which we used in the previous example. In an LRU scheme, the block replaced is the one that has been unused for the longest time. The set-associative example on page 419 uses LRU, which is why we replaced Memory(0) instead of Memory(6).

LRU replacement is implemented by keeping track of when each element in a set was used relative to the other elements in the set. For a two-way set-associative cache, tracking when the two elements were used can be implemented by keeping a single bit in each set and setting the bit to indicate an element whenever that element is referenced. As associativity increases, implementing LRU gets harder; in Section 5.8, we will see an alternative scheme for replacement.

least recently used (LRU) A replacement scheme in which the block replaced is the one that has been unused for the longest time.

Size of Tags versus Set Associativity

Increasing associativity requires more comparators and more tag bits per cache block. Assuming a cache of 4096 blocks, a four-word block size, and a 64-bit address, find the total number of sets and the total number of tag bits for caches that are direct-mapped, two-way and four-way set-associative, and fully associative.

EXAMPLE

Since there are 16 $(= 2^4)$ bytes per block, a 64-bit address yields $64 - 4 = 60$ bits to be used for index and tag. The direct-mapped cache has the same number of sets as blocks, and hence 12 bits of index, since $\log_2(4096) = 12$; hence, the total number is $(60 - 12) \times 4096 = 48 \times 4096 = 197\,\text{K}$ tag bits.

Each degree of associativity decreases the number of sets by a factor of 2 and thus decreases the number of bits used to index the cache by 1 and increases the number of bits in the tag by 1. Thus, for a two-way set-associative cache, there are 2048 sets, and the total number of tag bits is $(60 - 11) \times 2 \times 2048 = 98 \times 2048 = 401$ Kbits. For a four-way set-associative cache, the total number of sets is 1024, and the total number is $(60 - 10) \times 4 \times 1024 = 100 \times 1024 = 205\,\text{K}$ tag bits.

For a fully associative cache, there is only one set with 4096 blocks, and the tag is 60 bits, leading to $60 \times 4096 \times 1 = 246\,\text{K}$ tag bits.

ANSWER

Reducing the Miss Penalty Using Multilevel Caches

All modern computers make use of caches. To close the gap further between the fast clock rates of modern processors and the increasingly long time required to access DRAMs, most microprocessors support an additional level of caching. This second-level cache is normally on the same chip and is accessed whenever a miss occurs in the primary cache. If the second-level cache contains the desired data, the miss penalty for the first-level cache will be essentially the access time of the second-level cache, which will be much less than the access time of main memory. If neither the primary nor the secondary cache contains the data, a main memory access is required, and a larger miss penalty is incurred.

How significant is the performance improvement from the use of a secondary cache? The next example shows us.

EXAMPLE

Performance of Multilevel Caches

Suppose we have a processor with a base CPI of 1.0, assuming all references hit in the primary cache, and a clock rate of 4 GHz. Assume a main memory access time of 100 ns, including all the miss handling. Suppose the miss rate per instruction at the primary cache is 2%. How much faster will the processor be if we add a secondary cache that has a 5-ns access time for either a hit or a miss and is large enough to reduce the miss rate to main memory to 0.5%?

ANSWER

The miss penalty to main memory is

$$\frac{100 \text{ ns}}{0.25 \dfrac{\text{ns}}{\text{clock cycle}}} = 400 \text{ clock cycles}$$

The effective CPI with one level of caching is given by

$$\text{Total CPI} = \text{Base CPI} + \text{Memory-stall cycles per instruction}$$

For the processor with one level of caching,

$$\text{Total CPI} = 1.0 + \text{Memory-stall cycles per instruction} = 1.0 + 2\% \times 400 = 9$$

With two levels of caching, a miss in the primary (or first-level) cache can be satisfied either by the secondary cache or by main memory. The miss penalty for an access to the second-level cache is

$$\frac{5 \text{ ns}}{0.25 \dfrac{\text{ns}}{\text{clock cycle}}} = 20 \text{ clock cycles}$$

If the miss is satisfied in the secondary cache, then this is the entire miss penalty. If the miss needs to go to main memory, then the total miss penalty is the sum of the secondary cache access time and the main memory access time.

Thus, for a two-level cache, total CPI is the sum of the stall cycles from both levels of cache and the base CPI:

Total CPI $= 1 +$ Primary stalls per instruction $+$ Secondary stalls per instruction
$$= 1 + 2\% \times 20 + 0.5\% \times 400 = 1 + 0.4 + 2.0 = 3.4$$

Thus, the processor with the secondary cache is faster by

$$\frac{9.0}{3.4} = 2.6$$

Alternatively, we could have computed the stall cycles by summing the stall cycles of those references that hit in the secondary cache $((2\% - 0.5\%) \times 20 = 0.3)$. Those references that go to main memory, which must include the cost to access the secondary cache as well as the main memory access time, are $(0.5\% \times (20 + 400) = 2.1)$. The sum, $1.0 + 0.3 + 2.1$, is again 3.4.

The design considerations for a primary and secondary cache are significantly different, because the presence of the other cache changes the best choice versus a single-level cache. In particular, a two-level cache structure allows the primary cache to focus on minimizing hit time to yield a shorter clock cycle or fewer pipeline stages, while allowing the secondary cache to focus on miss rate to reduce the penalty of long memory access times.

The effect of these changes on the two caches can be seen by comparing each cache to the optimal design for a single level of cache. In comparison to a single-level cache, the primary cache of a multilevel cache is often smaller. Furthermore, the primary cache may use a smaller block size, to go with the smaller cache size and also to reduce the miss penalty. In comparison, the secondary cache will be much larger than in a single-level cache, since the access time of the secondary cache is less critical. With a larger total size, the secondary cache may use a larger block size than appropriate with a single-level cache. It often uses higher associativity than the primary cache given the focus of reducing miss rates.

multilevel cache
A memory hierarchy with multiple levels of caches, rather than just a cache and main memory.

Sorting has been exhaustively analyzed to find better algorithms: Bubble Sort, Quicksort, Radix Sort, and so on. Figure 5.19(a) shows instructions executed by item searched for Radix Sort versus Quicksort. As expected, for large arrays, Radix Sort has an algorithmic advantage over Quicksort in terms of number of operations. Figure 5.19(b) shows time per key instead of instructions executed. We see that the lines start on the same trajectory as in Figure 5.19(a), but then the Radix Sort line

Understanding Program Performance

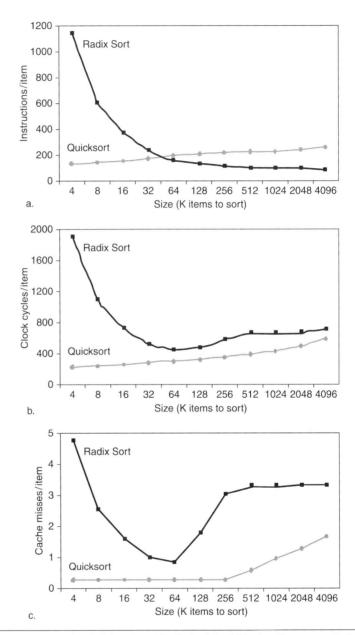

FIGURE 5.19 Comparing Quicksort and Radix Sort by (a) instructions executed per item sorted, (b) time per item sorted, and (c) cache misses per item sorted. These data are from a paper by LaMarca and Ladner [1996]. Due to such results, new versions of Radix Sort have been invented that take memory hierarchy into account, to regain its algorithmic advantages (see Section 5.15). The basic idea of cache optimizations is to use all the data in a block repeatedly before they are replaced on a miss.

diverges as the data to sort increase. What is going on? Figure 5.19(c) answers by looking at the cache misses per item sorted: Quicksort consistently has many fewer misses per item to be sorted.

Alas, standard algorithmic analysis often ignores the impact of the memory hierarchy. As faster clock rates and **Moore's Law** allow architects to squeeze all the performance out of a stream of instructions, using the memory hierarchy well is vital to high performance. As we said in the introduction, understanding the behavior of the memory hierarchy is critical to understanding the performance of programs on today's computers.

MOORE'S LAW

Software Optimization via Blocking

Given the importance of the memory hierarchy to program performance, not surprisingly many software optimizations were invented that can dramatically improve performance by reusing data within the cache and hence lower miss rates due to improved temporal locality.

When dealing with arrays, we can get good performance from the memory system if we store the array in memory so that accesses to the array are sequential in memory. Suppose that we are dealing with multiple arrays, however, with some arrays accessed by rows and some by columns. Storing the arrays row-by-row (called *row major order*) or column-by-column (*column major order*) does not solve the problem because both rows and columns are used in every loop iteration.

Instead of operating on entire rows or columns of an array, *blocked* algorithms operate on submatrices or *blocks*. The goal is to maximize accesses to the data loaded into the cache before the data are replaced; that is, improve temporal locality to reduce cache misses.

For example, the inner loops of DGEMM (lines 4 through 9 of Figure 3.22 in Chapter 3) are

```
for (int j = 0; j < n; ++j)
    {
      double cij = C[i+j*n]; /* cij = C[i][j] */
      for( int k = 0; k < n; k++ )
       cij += A[i+k*n] * B[k+j*n]; /* cij += A[i][k]*B[k][j] */
      C[i+j*n] = cij; /* C[i][j] = cij */
    }
```

It reads all N-by-N elements of B, reads the same N elements in what corresponds to one row of A repeatedly, and writes what corresponds to one row of N elements of C. (The comments make the rows and columns of the matrices easier to identify.) Figure 5.20 gives a snapshot of the accesses to the three arrays. A dark shade indicates a recent access, a light shade indicates an older access, and white means not yet accessed.

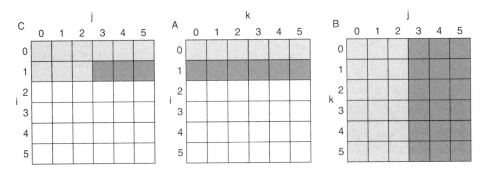

FIGURE 5.20   A snapshot of the three arrays C, A, **and** B **when** N=6 **and** i=**1.** The age of accesses to the array elements is indicated by shade: white means not yet touched, light means older accesses, and dark means newer accesses. Compared to Figure 5.22, elements of A and B are read repeatedly to calculate new elements of C. The variables i, j, and k are shown along the rows or columns used to access the arrays.

The number of capacity misses clearly depends on N and the size of the cache. If it can hold all three N-by-N matrices, then all is well, provided there are no cache conflicts. We purposely picked the matrix size to be 32 by 32 in DGEMM for Chapters 3 and 4 so that this would be the case. Each matrix is $32 \times 32 = 1024$ elements and each element is 8 bytes, so the three matrices occupy 24 KiB, which comfortably fit in the 32 KiB data cache of the Intel Core i7 (Sandy Bridge).

If the cache can hold one N-by-N matrix and one row of N, then at least the ith row of A and the array B may stay in the cache. Less than that and misses may occur for both B and C. In the worst case, there would be 2 N^3 + N^2 memory words accessed for N^3 operations.

To ensure that the elements being accessed can fit in the cache, the original code is changed to compute on a submatrix. Hence, we essentially invoke the version of DGEMM from Figure 4.78 in Chapter 4 repeatedly on matrices of size BLOCKSIZE by BLOCKSIZE. BLOCKSIZE is called the *blocking factor*.

Figure 5.21 shows the blocked version of DGEMM. The function do_block is DGEMM from Figure 3.22 with three new parameters si, sj, and sk to specify the starting position of each submatrix of of A, B, and C. The two inner loops of the do_block now compute in steps of size BLOCKSIZE rather than the full length of B and C. The gcc optimizer removes any function call overhead by "inlining" the function; that is, it inserts the code directly to avoid the conventional parameter passing and return address bookkeeping instructions.

Figure 5.22 illustrates the accesses to the three arrays using blocking. Looking only at capacity misses, the total number of memory words accessed is 2 N^3/BLOCKSIZE + N^2. This total is an improvement by about a factor of BLOCKSIZE. Hence, blocking exploits a combination of spatial and temporal locality, since A benefits from spatial locality and B benefits from temporal locality.

```
1   #define BLOCKSIZE 32
2   void do_block (int n, int si, int sj, int sk, double *A, double
3   *B, double *C)
4   {
5     for (int i = si; i < si+BLOCKSIZE; ++i)
6       for (int j = sj; j < sj+BLOCKSIZE; ++j)
7         {
8             double cij = C[i+j*n];/* cij = C[i][j] */
9             for( int k = sk; k < sk+BLOCKSIZE; k++ )
10               cij += A[i+k*n] * B[k+j*n];/* cij+=A[i][k]*B[k][j] */
11           C[i+j*n] = cij;/* C[i][j] = cij */
12        }
13  }
14  void dgemm (int n, double* A, double* B, double* C)
15  {
16    for ( int sj = 0; sj < n; sj += BLOCKSIZE )
17      for ( int si = 0; si < n; si += BLOCKSIZE )
18        for ( int sk = 0; sk < n; sk += BLOCKSIZE )
19          do_block(n, si, sj, sk, A, B, C);
20  }
```

FIGURE 5.21   Cache blocked version of DGEMM in Figure 3.22. Assume C is initialized to zero. The do_block function is basically DGEMM from Chapter 3 with new parameters to specify the starting positions of the submatrices of BLOCKSIZE. The gcc optimizer can remove the function overhead instructions by inlining the do_block function.

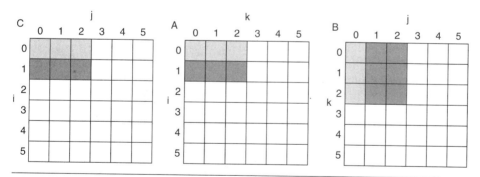

FIGURE 5.22   The age of accesses to the arrays C, A, and B when *BLOCKSIZE* = 3. Note that, in contrast to Figure 5.20, fewer elements are accessed.

Although we have aimed at reducing cache misses, blocking can also be used to help register allocation. By taking a small blocking size, such that the block can be held in registers, we can minimize the number of loads and stores in the program, which again improves performance.

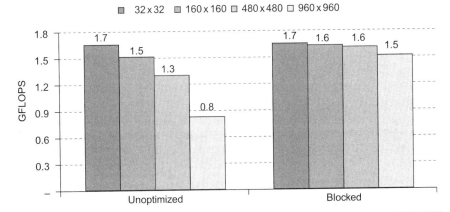

FIGURE 5.23 Performance of unoptimized DGEMM (Figure 3.22) versus cache blocked DGEMM (Figure 5.21) as the matrix dimension varies from 32 × 32 (where all three matrices fit in the cache) to 960 × 960.

Figure 5.23 shows the impact of cache blocking on the performance of the unoptimized DGEMM as we increase the matrix size beyond where all three matrices fit in the cache. The unoptimized performance is halved for the largest matrix. The cache-blocked version is less than 10% slower even at matrices that are 960 × 960, or 900 times larger than the 32 × 32 matrices in Chapters 3 and 4.

global miss rate The fraction of references that miss in all levels of a multilevel cache.

local miss rate The fraction of references to one level of a cache that miss; used in multilevel hierarchies.

Elaboration: Multilevel caches create many complications. First, there are now several different types of misses and corresponding miss rates. In the example on pages 424–425, we saw the primary cache miss rate and the global miss rate—the fraction of references that missed in all cache levels. There is also a miss rate for the secondary cache, which is the ratio of all misses in the secondary cache divided by the number of accesses to it. This miss rate is called the local miss rate of the secondary cache. Because the primary cache filters accesses, especially those with good spatial and temporal locality, the local miss rate of the secondary cache is much higher than the global miss rate. For the example on pages 424–425, we can compute the local miss rate of the secondary cache as 0.5%/2% = 25%! Luckily, the global miss rate dictates how often we must access the main memory.

Elaboration: With out-of-order processors (see Chapter 4), performance is more complex, since they execute instructions during the miss penalty. Instead of instruction miss rates and data miss rates, we use misses per instruction, and this formula:

$$\frac{\text{Memory} - \text{stall cycles}}{\text{Instruction}} = \frac{\text{Misses}}{\text{Instruction}} \times (\text{Total miss latency} - \text{Overlapped miss latency})$$

There is no general way to calculate overlapped miss latency, so evaluations of memory hierarchies for out-of-order processors inevitably require simulation of the processor and the memory hierarchy. Only by seeing the execution of the processor during each miss can we see if the processor stalls waiting for data or simply finds other work to do. A guideline is that the processor often hides the miss penalty for an L1 cache miss that hits in the L2 cache, but it rarely hides a miss to the L2 cache.

Elaboration: The performance challenge for algorithms is that the memory hierarchy varies between different implementations of the same architecture in cache size, associativity, block size, and number of caches. To cope with such variability, some recent numerical libraries parameterize their algorithms and then search the parameter space at runtime to find the best combination for a particular computer. This approach is called *autotuning*.

Which of the following is generally true about a design with multiple levels of caches?   **Check Yourself**

1. First-level caches are more concerned about hit time, and second-level caches are more concerned about miss rate.

2. First-level caches are more concerned about miss rate, and second-level caches are more concerned about hit time.

Summary

In this section, we focused on four topics: cache performance, using associativity to reduce miss rates, the use of multilevel cache hierarchies to reduce miss penalties, and software optimizations to improve effectiveness of caches.

The memory system has a significant effect on program execution time. The number of memory-stall cycles depends on both the miss rate and the miss penalty. The challenge, as we will see in Section 5.8, is to reduce one of these factors without significantly affecting other critical factors in the memory hierarchy.

To reduce the miss rate, we examined the use of associative placement schemes. Such schemes can reduce the miss rate of a cache by allowing more flexible placement of blocks within the cache. Fully associative schemes allow blocks to be placed anywhere, but also require that every block in the cache be searched to satisfy a request. The higher costs make large fully associative caches impractical. Set-associative caches are a practical alternative, since we need only search among the elements of a unique set that is chosen by indexing. Set-associative caches have higher miss rates but are faster to access. The amount of associativity that yields the best performance depends on both the technology and the details of the implementation.

We looked at multilevel caches as a technique to reduce the miss penalty by allowing a larger secondary cache to handle misses to the primary cache. Second-level caches have become commonplace as designers find that limited silicon and the goals of high clock rates prevent primary caches from becoming large. The secondary cache, which is often 10 or more times larger than the primary cache, handles many accesses that miss in the primary cache. In such cases, the miss penalty is that of the access time to the secondary cache (typically <10 processor cycles) versus the access time to memory (typically > 100 processor cycles). As with associativity, the design tradeoffs between the size of the secondary cache and its access time depend on a number of aspects of the implementation.

Finally, given the importance of the memory hierarchy in performance, we looked at how to change algorithms to improve cache behavior, with blocking being an important technique when dealing with large arrays.

5.5   Dependable Memory Hierarchy

DEPENDABILITY

Implicit in all the prior discussion is that the memory hierarchy doesn't forget. Fast but undependable is not very attractive. As we learned in Chapter 1, the one great idea for **dependability** is redundancy. In this section we'll first go over the terms to define terms and measures associated with failure, and then show how redundancy can make nearly unforgettable memories.

Defining Failure

We start with an assumption that you have a specification of proper service. Users can then see a system alternating between two states of delivered service with respect to the service specification:

1. *Service accomplishment*, where the service is delivered as specified

2. *Service interruption*, where the delivered service is different from the specified service

Transitions from state 1 to state 2 are caused by *failures*, and transitions from state 2 to state 1 are called *restorations*. Failures can be permanent or intermittent. The latter is the more difficult case; it is harder to diagnose the problem when a system oscillates between the two states. Permanent failures are far easier to diagnose.

This definition leads to two related terms: reliability and availability.

Reliability is a measure of the continuous service accomplishment—or, equivalently, of the time to failure—from a reference point. Hence, *mean time to failure* (MTTF) is a reliability measure. A related term is *annual failure rate* (AFR), which is just the percentage of devices that would be expected to fail in a year for a given MTTF. When MTTF gets large it can be misleading, while AFR leads to better intuition.

MTTF vs. AFR of Disks

EXAMPLE

Some disks today are quoted to have a 1,000,000-hour MTTF. As 1,000,000 hours is $1,000,000/(365 \times 24) = 114$ years, it would seem like they practically never fail. Warehouse-scale computers that run Internet services such as Search might have 50,000 servers. Assume each server has two disks. Use AFR to calculate how many disks we would expect to fail per year.

One year is $365 \times 24 = 8760$ hours. A 1,000,000-hour MTTF means an AFR of $8760/1,000,000 = 0.876\%$. With 100,000 disks, we would expect 876 disks to fail per year, or on average more than two disk failures per day!

Service interruption is measured as *mean time to repair* (MTTR). *Mean time between failures* (MTBF) is simply the sum of MTTF + MTTR. Although MTBF is widely used, MTTF is often the more appropriate term. *Availability* is then a measure of service accomplishment with respect to the alternation between the two states of accomplishment and interruption. Availability is statistically quantified as

$$\text{Availability} = \frac{\text{MTTF}}{(\text{MTTF} + \text{MTTR})}$$

Note that reliability and availability are actually quantifiable measures, rather than just synonyms for dependability. Shrinking MTTR can help availability as much as increasing MTTF. For example, tools for fault detection, diagnosis, and repair can help reduce the time to repair faults and thereby improve availability.

We want availability to be very high. One shorthand is to quote the number of "nines of availability" per year. For instance, a very good Internet service today offers 4 or 5 nines of availability. Given 365 days per year, which is $365 \times 24 \times 60 = 526{,}000$ minutes, then the shorthand is decoded as follows:

One nine: 90%   => 36.5 days of repair/year
Two nines: 99%   => 3.65 days of repair/year
Three nines: 99.9%   => 526 minutes of repair/year
Four nines: 99.99%   => 52.6 minutes of repair/year
Five nines: 99.999%   => 5.26 minutes of repair/year

and so on.

To increase MTTF, you can improve the quality of the components or design systems to continue operation in the presence of components that have failed. Hence, failure needs to be defined with respect to a context, as failure of a component may not lead to a failure of the system. To make this distinction clear, the term *fault* is used to mean failure of a component. Here are three ways to improve MTTF:

1. *Fault avoidance:* Preventing fault occurrence by construction.

2. *Fault tolerance:* Using redundancy to allow the service to comply with the service specification despite faults occurring.

3. *Fault forecasting:* **Predicting** the presence and creation of faults, allowing the component to be replaced *before* it fails.

PREDICTION

The Hamming Single Error Correcting, Double Error Detecting Code (SEC/DED)

Richard Hamming invented a popular redundancy scheme for memory, for which he received the Turing Award in 1968. To invent redundant codes, it is helpful to talk about how "close" correct bit patterns can be. What we call the *Hamming distance* is just the minimum number of bits that are different between any two correct bit patterns. For example, the distance between $0\underline{11}0\underline{1}1$ and $00\underline{1}\underline{1}11$ is two. What happens if the minimum distance between members of a code is two, and we get a one-bit error? It will turn a valid pattern in a code to an invalid one. Thus, if we can detect whether members of a code are accurate or not, we can detect single bit errors, and can say we have a single bit error detection code.

error detection code A code that enables the detection of an error in data, but not the precise location and, hence, correction of the error.

Hamming used a *parity code* for error detection. In a parity code, the number of 1s in a word is counted; the word has odd parity if the number of 1s is odd and even otherwise. When a word is written into memory, the parity bit is also written (1 for odd, 0 for even). That is, the parity of the N+1 bit word should always be even. Then, when the word is read out, the parity bit is read and checked. If the parity of the memory word and the stored parity bit do not match, an error has occurred.

EXAMPLE

Calculate the parity of a byte with the value 31_{ten} and show the pattern stored to memory. Assume the parity bit is on the right. Suppose the most significant bit was inverted in memory, and then you read it back. Did you detect the error? What happens if the two most significant bits are inverted?

ANSWER

31_{ten} is 00011111_{two}, which has five 1s. To make parity even, we need to write a 1 in the parity bit, or 000111111_{two}. If the most significant bit is inverted when we read it back, we would see $\underline{1}00111111_{two}$ which has seven 1s. Since we expect even parity and calculated odd parity, we would signal an error. If the *two* most significant bits are inverted, we would see $\underline{11}0111111_{two}$ which has eight 1s or even parity, and we would *not* signal an error.

If there are 2 bits of error, then a 1-bit parity scheme will not detect any errors, since the parity will match the data with two errors. (Actually, a 1-bit parity scheme can detect any odd number of errors; however, the probability of having three errors is much lower than the probability of having two, so, in practice, a 1-bit parity code is limited to detecting a single bit of error.)

Of course, a parity code cannot correct errors, which Hamming wanted to do as well as detect them. If we used a code that had a minimum distance of 3, then any single bit error would be closer to the correct pattern than to any other valid pattern. He came up with an easy to understand mapping of data into a distance 3 code that we call *Hamming Error Correction Code* (ECC) in his honor. We use extra parity bits to allow the position identification of a single error. Here are the steps to calculate Hamming ECC

1. Start numbering bits from 1 on the left, contrary to the traditional numbering of the rightmost bit being 0.

2. Mark all bit positions that are powers of 2 as parity bits (positions 1, 2, 4, 8, 16, …).

3. All other bit positions are used for data bits (positions 3, 5, 6, 7, 9, 10, 11, 12, 13, 14, 15, …).

4. The position of parity bit determines sequence of data bits that it checks (Figure 5.24 shows this coverage graphically) is:

   ■ Bit 1 (0001_{two}) checks bits (1,3,5,7,9,11,...), which are bits where rightmost bit of address is 1 (0001_{two}, 0011_{two}, 0101_{two}, 0111_{two}, 1001_{two}, 1011_{two},…).

   ■ Bit 2 (0010_{two}) checks bits (2,3,6,7,10,11,14,15,…), which are the bits where the second bit to the right in the address is 1.

   ■ Bit 4 (0100_{two}) checks bits (4–7, 12–15, 20–23,…), which are the bits where the third bit to the right in the address is 1.

   ■ Bit 8 (1000_{two}) checks bits (8–15, 24–31, 40–47,...), which are the bits where the fourth bit to the right in the address is 1.

   Note that each data bit is covered by two or more parity bits.

5. Set parity bits to create even parity for each group.

Bit position		1	2	3	4	5	6	7	8	9	10	11	12
Encoded data bits		p1	p2	d1	p4	d2	d3	d4	p8	d5	d6	d7	d8
	p1	X		X		X		X		X		X	
Parity bit coverage	p2		X	X			X	X			X	X	
	p4				X	X	X	X					X
	p8								X	X	X	X	X

FIGURE 5.24 Parity bits, data bits, and field coverage in a Hamming ECC code for eight data bits.

In what seems like a magic trick, you can determine whether bits are incorrect by looking at the parity bits. Using the 12 bit code in Figure 5.24, if the value of the four parity calculations (p8,p4,p2,p1) was 0000, then there was no error. However, if the pattern was, say, 1010, which is 10_{ten}, then Hamming ECC tells us that bit 10 (d6) is an error. Since the number is binary, we can correct the error just by inverting the value of bit 10.

EXAMPLE

Assume one byte data value is 10011010_{two}. First show the Hamming ECC code for that byte, and then invert bit 10 and show that the ECC code finds and corrects the single bit error.

Leaving spaces for the parity bits, the 12 bit pattern is__ 1_ 0 0 1_ 1 0 1 0.

ANSWER

Position 1 checks bits 1,3,5,7,9, and 11, which we highlight:__ **1**_ 0 0 **1**_ **1** 0 **1** 0. To make the group even parity, we should set bit 1 to 0.

Position 2 checks bits 2,3,6,7,10,11, which is 0_ **1**_ 0 0 **1**_ **1 0 1** 0 or odd parity, so we set position 2 to a 1.

Position 4 checks bits 4,5,6,7,12, which is 0 1 1_ **0 0 1**_ 1 0 1, so we set it to a 1.

Position 8 checks bits 8,9,10,11,12, which is 0 1 1 1 0 0 1_ **1 0 1 0**, so we set it to a 0.

The final code word is 011100101010. Inverting bit 10 changes it to 011100101110.

Parity bit 1 is 0 (**0**1**1**1**0**0**1**0**1**1**1**0 is four 1s, so even parity; this group is OK).

Parity bit 2 is 1 (0**11**1**00**1**0**1**1**1**0 is five 1s, so odd parity; there is an error somewhere).

Parity bit 4 is 1 (011**100**101110 is two 1s, so even parity; this group is OK).

Parity bit 8 is 1 (0111001**01110** is three 1s, so odd parity; there is an error somewhere).

Parity bits 2 and 8 are incorrect. As $2 + 8 = 10$, bit 10 must be wrong. Hence, we can correct the error by inverting bit 10: 011100101**0**10. Voila!

Hamming did not stop at single bit error correction code. At the cost of one more bit, we can make the minimum Hamming distance in a code be 4. This means we can correct single bit errors *and detect double bit errors*. The idea is to add a parity bit that is calculated over the whole word. Let's use a 4-bit data word as an example, which would only need 7 bits for single bit error detection. Hamming parity bits H (p1 p2 p3) are computed (even parity as usual) plus the even parity over the entire word, p4:

$$1 \quad 2 \quad 3 \quad 4 \quad 5 \quad 6 \quad 7 \quad \underline{8}$$
$$p_1 \quad p_2 \quad d_1 \quad p_3 \quad d_2 \quad d_3 \quad d_4 \quad \underline{\mathbf{p_4}}$$

Then the algorithm to correct one error and detect two is just to calculate parity over the ECC groups (H) as before plus one more over the whole group (p_4). There are four cases:

1. H is even and p_4 is even, so no error occurred.

2. H is odd and p_4 is odd, so a correctable single error occurred. (p_4 should calculate odd parity if one error occurred.)

3. H is even and p_4 is odd, a single error occurred in p_4 bit, not in the rest of the word, so correct the p_4 bit.

4. H is odd and p$_4$ is even, a double error occurred. (p$_4$ should calculate even parity if two errors occurred.)

Single Error Correcting/Double Error Detecting (SEC/DED) is common in memory for servers today. Conveniently, 8-byte data blocks can get SEC/DED with just one more byte, which is why many DIMMs are 72 bits wide.

Elaboration: To calculate how many bits are needed for SEC, let p be total number of parity bits and d number of data bits in $p + d$ bit word. If p error correction bits are to point to error bit ($p + d$ cases) plus one case to indicate that no error exists, we need:

$$2^p \geq p + d + 1 \text{ bits, and thus } p \geq \log(p + d + 1).$$

For example, for 8 bits data means $d = 8$ and $2^p \geq p + 8 + 1$, so $p = 4$. Similarly, $p = 5$ for 16 bits of data, 6 for 32 bits, 7 for 64 bits, and so on.

Elaboration: In very large systems, the possibility of multiple errors as well as complete failure of a single wide memory chip becomes significant. IBM introduced *chipkill* to solve this problem, and many big systems use this technology. (Intel calls their version SDDC.) Similar in nature to the RAID approach used for disks (see ⊞ Section 5.11), Chipkill distributes the data and ECC information, so that the complete failure of a single memory chip can be handled by supporting the reconstruction of the missing data from the remaining memory chips. Assuming a 10,000-processor cluster with 4 GiB per processor, IBM calculated the following rates of *unrecoverable* memory errors in 3 years of operation:

- Parity only—about 90,000, or one unrecoverable (or undetected) failure every 17 minutes.
- SEC/DED only—about 3500, or about one undetected or unrecoverable failure every 7.5 hours.
- Chipkill—6, or about one undetected or unrecoverable failure every 2 months.

Hence, Chipkill is a requirement for warehouse-scale computers.

Elaboration: While single or double bit errors are typical for memory systems, networks can have bursts of bit errors. One solution is called *Cyclic Redundancy Check*. For a block of k bits, a transmitter generates an n-k bit frame check sequence. It transmits n bits exactly divisible by some number. The receiver divides the frame by that number. If there is no remainder, it assumes there is no error. If there is, the receiver rejects the message, and asks the transmitter to send again. As you might guess from Chapter 3, it is easy to calculate division for some binary numbers with a shift register, which made CRC codes popular even when hardware was more precious. Going even further, Reed-Solomon codes use Galois fields to *correct* multibit transmission errors, but now data are considered coefficients of a polynomial and the code space is values of a polynomial. The Reed-Solomon calculation is considerably more complicated than binary division!

5.6   Virtual Machines

Virtual machines (VM) were first developed in the mid-1960s, and they have remained an important part of mainframe computing over the years. Although largely ignored in the single-user PC era in the 1980s and 1990s, they have recently gained popularity due to

- The increasing importance of isolation and security in modern systems

- The failures in security and reliability of standard operating systems

- The sharing of a single computer among many unrelated users, in particular for Cloud computing

- The dramatic increases in raw speed of processors over the decades, which makes the overhead of VMs more acceptable

The broadest definition of VMs includes basically all emulation methods that provide a standard software interface, such as the Java VM. In this section, we are interested in VMs that provide a complete system-level environment at the binary *instruction set architecture* (ISA) level. Although some VMs run different ISAs in the VM from the native hardware, we assume they always match the hardware. Such VMs are called (Operating) *System Virtual Machines*. IBM VM/370, VirtualBox, VMware ESX Server, and Xen are examples.

System virtual machines present the illusion that the users have an entire computer to themselves, including a copy of the operating system. A single computer runs multiple VMs and can support a number of different *operating systems* (OSes). On a conventional platform, a single OS "owns" all the hardware resources, but with a VM, multiple OSes all share the hardware resources.

The software that supports VMs is called a *virtual machine monitor* (VMM) or *hypervisor*; the VMM is the heart of virtual machine technology. The underlying hardware platform is called the *host*, and its resources are shared among the *guest* VMs. The VMM determines how to map virtual resources to physical resources: a physical resource may be time-shared, partitioned, or even emulated in software. The VMM is much smaller than a traditional OS; the isolation portion of a VMM is perhaps only 10,000 lines of code.

Although our interest here is in VMs for improving protection, VMs provide two other benefits that are commercially significant:

1. *Managing software.* VMs provide an abstraction that can run the complete software stack, even including old operating systems like DOS. A typical deployment might be some VMs running legacy OSes, many running the current stable OS release, and a few testing the next OS release.

2. *Managing hardware.* One reason for multiple servers is to have each application running with the compatible version of the operating system on separate computers, as this separation can improve dependability. VMs

allow these separate software stacks to run independently yet share hardware, thereby consolidating the number of servers. Another example is that some VMMs support migration of a running VM to a different computer, either to balance load or to evacuate from failing hardware.

Amazon Web Services (AWS) uses the virtual machines in its Cloud computing offering EC2 for five reasons:

Hardware/ Software Interface

1. It allows AWS to protect users from each other while sharing the same server.

2. It simplifies software distribution within a warehouse-scale computer. A customer installs a virtual machine image configured with the appropriate software, and AWS distributes it to all the instances a customer wants to use.

3. Customers (and AWS) can reliably "kill" a VM to control resource usage when customers complete their work.

4. Virtual machines hide the identity of the hardware on which the customer is running, which means AWS can keep using old servers *and* introduce new, more efficient servers. The customer expects performance for instances to match their ratings in "EC2 Compute Units," which AWS defines: to "provide the equivalent CPU capacity of a 1.0–1.2 GHz 2007 AMD Opteron or 2007 Intel Xeon processor." Thanks to **Moore's Law**, newer servers clearly offer more EC2 Compute Units than older ones, but AWS can keep renting old servers as long as they are economical.

MOORE'S LAW

5. Virtual machine monitors can control the rate that a VM uses the processor, the network, and disk space, which allows AWS to offer many price points of instances of different types running on the same underlying servers. For example, in 2012 AWS offered 14 instance types, from small standard instances at $0.08 per hour to high I/O quadruple extra large instances at $3.10 per hour.

In general, the cost of processor virtualization depends on the workload. User-level processor-bound programs have zero virtualization overhead, because the OS is rarely invoked, so everything runs at native speeds. I/O-intensive workloads are generally also OS-intensive, executing many system calls and privileged instructions that can result in high virtualization overhead. On the other hand, if the I/O-intensive workload is also *I/O-bound*, the cost of processor virtualization can be completely hidden, since the processor is often idle waiting for I/O.

The overhead is determined by both the number of instructions that must be emulated by the VMM and by how much time each takes to emulate. Hence, when the guest VMs run the same ISA as the host, as we assume here, the goal of the architecture and the VMM is to run almost all instructions directly on the native hardware.

Requirements of a Virtual Machine Monitor

What must a VM monitor do? It presents a software interface to guest software, it must isolate the state of guests from each other, and it must protect itself from guest software (including guest OSes). The qualitative requirements are:

- Guest software should behave on a VM exactly as if it were running on the native hardware, except for performance-related behavior or limitations of fixed resources shared by multiple VMs.

- Guest software should not be able to change the allocation of real system resources directly.

To "virtualize" the processor, the VMM must control just about everything—access to privileged state, I/O, exceptions, and interrupts—even though the guest VM and OS presently running are temporarily using them.

For example, in the case of a timer interrupt, the VMM would suspend the currently running guest VM, save its state, handle the interrupt, determine which guest VM to run next, and then load its state. Guest VMs that rely on a timer interrupt are provided with a virtual timer and an emulated timer interrupt by the VMM.

To be in charge, the VMM must be at a higher privilege level than the guest VM, which generally runs in user mode; this also ensures that the execution of any privileged instruction will be handled by the VMM. The basic system requirements to support VMMs are:

- At least two processor modes, system and user.

- A privileged subset of instructions that is available only in system mode, resulting in a trap if executed in user mode; all system resources must be controllable just via these instructions.

(Lack of) Instruction Set Architecture Support for Virtual Machines

If VMs are planned for during the design of the ISA, it's relatively easy to reduce both the number of instructions that must be executed by a VMM and improve their emulation speed. An architecture that allows the VM to execute directly on the hardware earns the title *virtualizable*, and the IBM 370 and the ARMv8 architectures proudly bear that label.

Alas, since VMs have been considered for PC and server applications only fairly recently, most instruction sets were created without virtualization in mind. These culprits include x86 and most RISC architectures, including ARMv7 and MIPS.

Because the VMM must ensure that the guest system only interacts with virtual resources, a conventional guest OS runs as a user mode program on top of the VMM. Then, if a guest OS attempts to access or modify information related to hardware resources via a privileged instruction—for example, reading or writing a status bit that enables interrupts—it will trap to the VMM. The VMM can then affect the appropriate changes to corresponding real resources.

Hence, if any instruction that tries to read or write such sensitive information traps when executed in user mode, the VMM can intercept it and support a virtual version of the sensitive information, as the guest OS expects.

In the absence of such support, other measures must be taken. A VMM must take special precautions to locate all problematic instructions and ensure that they behave correctly when executed by a guest OS, thereby increasing the complexity of the VMM and reducing the performance of running the VM.

Protection and Instruction Set Architecture

Protection is a joint effort of architecture and operating systems, but architects had to modify some awkward details of existing instruction set architectures when virtual memory became popular.

For example, the x86 instruction POPF loads the flag registers from the top of the stack in memory. One of the flags is the *Interrupt Enable* (IE) flag. If you run the POPF instruction in user mode, rather than trap it, it simply changes all the flags except IE. In system mode, it does change the IE. Since a guest OS runs in user mode inside a VM, this is a problem, as it expects to see a changed IE.

Historically, IBM mainframe hardware and VMM took three steps to improve the performance of virtual machines:

1. Reduce the cost of processor virtualization.

2. Reduce interrupt overhead cost due to the virtualization.

3. Reduce interrupt cost by steering interrupts to the proper VM without invoking VMM.

AMD and Intel tried to address the first point in 2006 by reducing the cost of processor virtualization. It will be interesting to see how many generations of architecture and VMM modifications it will take to address all three points, and how long before virtual machines of the 21st century for x86 will be as efficient as the IBM mainframes and VMMs of the 1970s.

Elaboration: ARMv8 provides a third state (EL2) specifically to allow the VMM to run at a higher privilege level than the guest operating system.

… a system has been devised to make the core drum combination appear to the programmer as a single level store, the requisite transfers taking place automatically.

Kilburn et al., One-level storage system, 1962

5.7 Virtual Memory

In earlier sections, we saw how caches provided fast access to recently-used portions of a program's code and data. Similarly, the main memory can act as a "cache" for the

virtual memory
A technique that uses main memory as a "cache" for secondary storage.

physical address An address in main memory.

protection A set of mechanisms for ensuring that multiple processes sharing the processor, memory, or I/O devices cannot interfere, intentionally or unintentionally, with one another by reading or writing each other's data. These mechanisms also isolate the operating system from a user process.

page fault An event that occurs when an accessed page is not present in main memory.

virtual address
An address that corresponds to a location in virtual space and is translated by address mapping to a physical address when memory is accessed.

secondary storage, traditionally implemented with magnetic disks. This technique is called *virtual memory*. Historically, there were two major motivations for virtual memory: to allow efficient and safe sharing of memory among several programs, such as for the memory needed by multiple virtual machines for Cloud computing, and to remove the programming burdens of a small, limited amount of main memory. Five decades after its invention, it's the former reason that reigns today.

Of course, to allow multiple virtual machines to share the same memory, we must be able to protect the virtual machines from each other, ensuring that a program can just read and write the portions of main memory that have been assigned to it. Main memory need contain only the active portions of the many virtual machines, just as a cache contains only the active portion of one program. Thus, the principle of locality enables virtual memory as well as caches, and virtual memory allows us to share the processor efficiently as well as the main memory.

We cannot know which virtual machines will share the memory with other virtual machines when we compile them. In fact, the virtual machines sharing the memory change dynamically while they are running. Because of this dynamic interaction, we would like to compile each program into its own *address space*—a separate range of memory locations accessible only to this program. Virtual memory implements the translation of a program's address space to *physical addresses*. This translation process enforces *protection* of a program's address space from other virtual machines.

The second motivation for virtual memory is to allow a single-user program to exceed the size of primary memory. Formerly, if a program became too large for memory, it was up to the programmer to make it fit. Programmers divided programs into pieces and then identified the pieces that were mutually exclusive. These *overlays* were loaded or unloaded under user program control during execution, with the programmer ensuring that the program at no time tried to access an overlay that was not loaded and that the overlays loaded never exceeded the total size of the memory. Overlays were traditionally organized as modules, each containing both code and data. Calls between procedures in different modules would lead to overlaying of one module with another.

As you can well imagine, this responsibility was a substantial burden on programmers. Virtual memory, which was invented to relieve programmers of this difficulty, automatically manages the two levels of the memory hierarchy represented by main memory (sometimes called *physical memory* to distinguish it from virtual memory) and secondary storage.

Although the concepts at work in virtual memory and in caches are the same, their differing historical roots have led to the use of different terminology. A virtual memory block is called a *page*, and a virtual memory miss is called a *page fault*. With virtual memory, the processor produces a *virtual address*, which is translated by a combination of hardware and software to a *physical address*, which in turn can be used to access main memory. Figure 5.25 shows the virtually addressed memory with pages mapped to main memory. This process is called *address mapping* or address translation. Today, the two memory hierarchy levels controlled by virtual memory are usually DRAMs and flash memory in personal mobile devices and DRAMs and magnetic disks in servers (see Section 5.2). If we return to our library analogy, we can think of a virtual address as the title of a book and a physical address as the location of that book in the library, such as might be given by the Library of Congress call number.

Virtual memory also simplifies loading the program for execution by providing *relocation*. Relocation maps the virtual addresses used by a program to different physical addresses before the addresses are used to access memory. This relocation allows us to load the program anywhere in main memory. Furthermore, all virtual memory systems in use today relocate the program as a set of fixed-size blocks (pages), thereby eliminating the need to find a contiguous block of memory to allocate to a program; instead, the operating system needs only to find enough pages in main memory.

In virtual memory, the address is broken into a *virtual page number* and a *page offset*. Figure 5.26 shows the translation of the virtual page number to a *physical page number*. While ARMv8 has a 64-bit address, the upper 16 bits are not used, so the address to be mapped is 48 bits. This figure assumes the physical memory is 1 TiB, or 2^{40} bytes, which needs a 40-bit address. The physical page number constitutes the upper portion of the physical address, while the page offset, which is not changed, constitutes the lower portion. The number of bits in the page offset field determines the page size. The number of pages addressable with the virtual address can be different than the number of pages addressable with the physical address. Having a larger number of virtual pages than physical pages is the basis for the illusion of an essentially unbounded amount of virtual memory.

address translation Also called **address mapping**. The process by which a virtual address is mapped to an address used to access memory.

FIGURE 5.25 In virtual memory, blocks of memory (called *pages*) are mapped from one set of addresses (called *virtual addresses*) to another set (called *physical addresses*). The processor generates virtual addresses while the memory is accessed using physical addresses. Both the virtual memory and the physical memory are broken into pages, so that a virtual page is mapped to a physical page. Of course, it is also possible for a virtual page to be absent from main memory and not be mapped to a physical address; in that case, the page resides on disk. Physical pages can be shared by having two virtual addresses point to the same physical address. This capability is used to allow two different programs to share data or code.

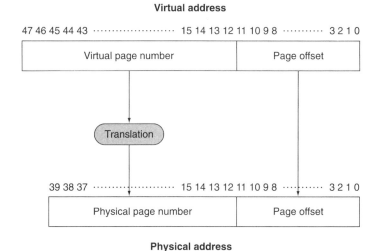

FIGURE 5.26 Mapping from a virtual to a physical address. The page size is $2^{12} = 4$ KiB. The number of physical pages allowed in memory is 2^{28}, since the physical page number has 28 bits in it. Thus, main memory can have at most 1 TiB, while the virtual address space is 256 TiB. ARMv8 allows physical memory to be up to 256 TiB; we chose 1 TiB since that matches the maximum of some ARMv8 computers in 2015.

Many design choices in virtual memory systems are motivated by the high cost of a page fault. A page fault to disk will take millions of clock cycles to process. (The table on page 392 shows that main memory latency is about 100,000 times quicker than disk.) This enormous miss penalty, dominated by the time to get the first word for typical page sizes, leads to several key decisions in designing virtual memory systems:

■ Pages should be large enough to try to amortize the high access time. Sizes from 4 KiB to 64 KiB are typical today. New desktop and server systems are being developed to support 32 KiB and 64 KiB pages, but new embedded systems are going in the other direction, to 1 KiB pages.

■ Organizations that reduce the page fault rate are attractive. The primary technique used here is to allow fully associative placement of pages in memory.

■ Page faults can be handled in software because the overhead will be small compared to the disk access time. In addition, software can afford to use clever algorithms for choosing how to place pages because even little reductions in the miss rate will pay for the cost of such algorithms.

■ Write-through will not work for virtual memory, since writes take too long. Instead, virtual memory systems use write-back.

The next few subsections address these factors in virtual memory design.

Elaboration: We present the motivation for virtual memory as many virtual machines sharing the same memory, but virtual memory was originally invented so that many programs could share a computer as part of a timesharing system. Since many readers today have no experience with time-sharing systems, we use virtual machines to motivate this section.

Elaboration: ARMv8 uses the term *granule* instead of page, and it calls a page fault a *Memory Management Unit (MMU) exception*. While the maximum virtual address is 48 bits, ARMv8 allows implementations with a smaller virtual address. It also allows physical addresses as large as 48 bits. It supports three options for minimum page or granule size: 4, 16, and 64 Kibibyte.

Elaboration: For servers and even PCs, 32-bit address processors are problematic. Although we normally think of virtual addresses as much larger than physical addresses, the opposite can occur when the processor address size is small relative to the state of the memory technology. No single program or virtual machine can benefit, but a collection of programs or virtual machines running at the same time can benefit from not having to be swapped out of main memory or by running on parallel processors.

Elaboration: The discussion of virtual memory in this book focuses on paging, which uses fixed-size blocks. There is also a variable-size block scheme called segmentation. In segmentation, an address consists of two parts: a segment number and a segment offset. The segment number is mapped to a physical address, and the offset is *added* to find the actual physical address. Because the segment can vary in size, a bounds check is also needed to make sure that the offset is within the segment. The major use of segmentation is to support more powerful methods of protection and sharing in an address space. Most operating system textbooks contain extensive discussions of segmentation compared to paging and of the use of segmentation to share the address space logically. The major disadvantage of segmentation is that it splits the address space into logically separate pieces that must be manipulated as a two-part address: the segment number and the offset. Paging, in contrast, makes the boundary between page number and offset invisible to programmers and compilers.

Segments have also been used as a method to extend the address space without changing the word size of the computer. Such attempts have been unsuccessful because of the awkwardness and performance penalties inherent in a two-part address, of which programmers and compilers must be aware.

Many architectures divide the address space into large fixed-size blocks that simplify protection between the operating system and user programs and increase the efficiency of implementing paging. Although these divisions are often called "segments," this mechanism is much simpler than variable block size segmentation and is not visible to user programs; we discuss it in more detail shortly.

segmentation A variable-size address mapping scheme in which an address consists of two parts: a segment number, which is mapped to a physical address, and a segment offset.

Placing a Page and Finding It Again

Because of the incredibly high penalty for a page fault, designers reduce page fault frequency by optimizing page placement. If we allow a virtual page to be mapped to any physical page, the operating system can then choose to replace any page it wants when a page fault occurs. For example, the operating system can use a sophisticated algorithm and complex data structures that track page usage to try to choose a page that will not be needed for a long time. The ability to use a clever and flexible replacement scheme reduces the page fault rate and simplifies the use of fully associative placement of pages.

As mentioned in Section 5.4, the difficulty in using fully associative placement is in locating an entry, since it can be anywhere in the upper level of the hierarchy. A full search is impractical. In virtual memory systems, we locate pages by using a table that indexes the main memory; this structure is called a page table, and it resides in main memory. A page table is indexed by the page number from the virtual address to discover the corresponding physical page number. Each program has its own page table, which maps the virtual address space of that program to main memory. In our library analogy, the page table corresponds to a mapping between book titles and library locations. Just as the card catalog may contain entries for books in another library on campus rather than the local branch library, we will see that the page table may contain entries for pages not present in memory. To indicate the location of the page table in memory, the hardware includes a register that points to the start of the page table; we call this the *page table register*. Assume for now that the page table is in a fixed and contiguous area of memory.

page table The table containing the virtual to physical address translations in a virtual memory system. The table, which is stored in memory, is typically indexed by the virtual page number; each entry in the table contains the physical page number for that virtual page if the page is currently in memory.

Hardware/ Software Interface

The page table, together with the program counter and the registers, specifies the *state* of a virtual machine. If we want to allow another virtual machine to use the processor, we must save this state. Later, after restoring this state, the virtual machine can continue execution. We often refer to this state as a *process*. The process is considered *active* when it is in possession of the processor; otherwise, it is considered *inactive*. The operating system can make a process active by loading the process's state, including the program counter, which will initiate execution at the value of the saved program counter.

The process's address space, and hence all the data it can access in memory, is defined by its page table, which resides in memory. Rather than save the entire page table, the operating system simply loads the page table register to point to the page table of the process it wants to make active. Each process has its own page table, since different processes use the same virtual addresses. The operating system is responsible for allocating the physical memory and updating the page tables, so that the virtual address spaces of distinct processes do not collide. As we will see shortly, the use of separate page tables also provides protection of one process from another.

Figure 5.27 uses the page table register, the virtual address, and the indicated page table to show how the hardware can form a physical address. A valid bit is used in each page table entry, just as we did in a cache. If the bit is off, the page is not present in main memory and a page fault occurs. If the bit is on, the page is in memory and the entry contains the physical page number.

Because the page table contains a mapping for every possible virtual page, no tags are required. In cache terminology, the index that is used to access the page table consists of the full block address, which in this case is the virtual page number.

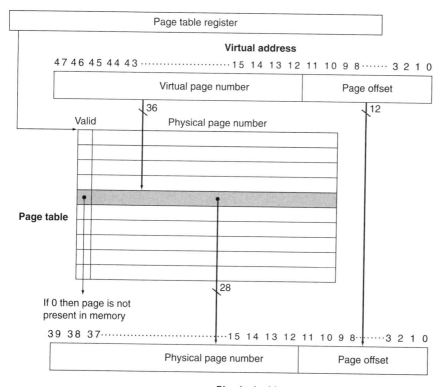

Physical address

FIGURE 5.27 The page table is indexed with the virtual page number to obtain the corresponding portion of the physical address. We assume a 48-bit address. The page table pointer gives the starting address of the page table. In this figure, the page size is 2^{12} bytes, or 4 KiB. The virtual address space is 2^{48} bytes, or 256 TiB, and the physical address space is 2^{40} bytes, which allows main memory of up to 1 TiB. If ARMv8 used a single page table as shown in this figure, the number of entries in the page table would be 2^{36}, or about 64 billion entries. (We'll see what ARMv8 does to reduce the number of entries shortly.) The valid bit for each entry indicates whether the mapping is legal. If it is off, then the page is not present in memory. Although the page table entry shown here need only be 29 bits wide, it would typically be rounded up to a power of 2 bits for ease of indexing. The page table entries in ARMv8 are 64 bits. The extra bits would be used to store additional information that needs to be kept on a per-page basis, such as protection.

Page Faults

If the valid bit for a virtual page is off, a page fault occurs. The operating system must be given control. This transfer is done with the exception mechanism, which we saw in Chapter 4 and will discuss again later in this section. Once the operating system gets control, it must find the page in the next level of the hierarchy (usually flash memory or magnetic disk) and decide where to place the requested page in the main memory.

The virtual address alone does not immediately tell us where the page is in secondary memory. Returning to our library analogy, we cannot find the location of a library book on the shelves just by knowing its title. Instead, we go to the catalog and look up the book, obtaining an address for the location on the shelves, such as the Library of Congress call number. Likewise, in a virtual memory system, we must keep track of the location in secondary memory of each page in virtual address space.

swap space The space on the disk reserved for the full virtual memory space of a process.

Because we do not know ahead of time when a page in memory will be replaced, the operating system usually creates the space on flash memory or disk for all the pages of a process when it creates the process. This space is called the swap space. At that time, it also creates a data structure to record where each virtual page is stored on disk. This data structure may be part of the page table or may be an auxiliary data structure indexed in the same way as the page table. Figure 5.28 shows the organization when a single table holds either the physical page number or the secondary memory address.

The operating system also creates a data structure that tracks which processes and which virtual addresses use each physical page. When a page fault occurs, if all the pages in main memory are in use, the operating system must choose a page to replace. Because we want to minimize the number of page faults, most operating systems try to choose a page that they hypothesize will not be needed soon. Using the past to predict the future, operating systems follow the *least recently used* (LRU) replacement scheme, which we mentioned in Section 5.4. The operating system searches for the least recently used page, assuming that a page that has not been used in a long time is less likely to be needed than a more recently accessed page. The replaced pages are written to swap space in secondary memory. In case you are wondering, the operating system is just another process, and these tables controlling memory are in memory; the details of this seeming contradiction will be explained shortly.

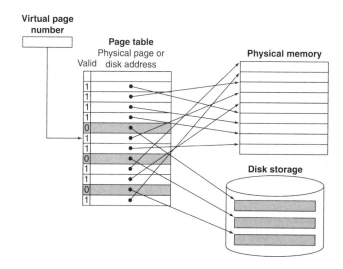

FIGURE 5.28   The page table maps each page in virtual memory to either a page in main memory or a page stored on disk, which is the next level in the hierarchy. The virtual page number is used to index the page table. If the valid bit is on, the page table supplies the physical page number (i.e., the starting address of the page in memory) corresponding to the virtual page. If the valid bit is off, the page currently resides only on disk, at a specified disk address. In many systems, the table of physical page addresses and disk page addresses, while logically one table, is stored in two separate data structures. Dual tables are justified in part because we must keep the disk addresses of all the pages, even if they are currently in main memory. Remember that the pages in main memory and the pages on disk are the same size.

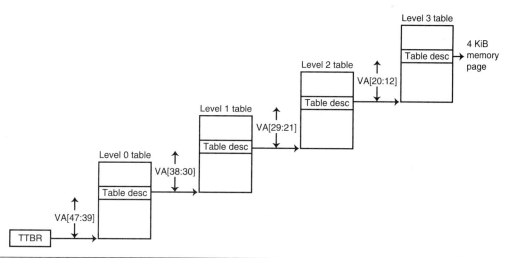

FIGURE 5.29   ARMv8 uses four levels of tables to translate a 48-bit virtual address into a 40-bit physical address. Rather than needing 64 billion page table entries for the single page table in Figure 5.27, this hierarchical approach needs just a tiny fraction. Each step of the translation uses 9 bits of the virtual address to find the next level table, until the upper 36 bits of the virtual address are mapped to the physical address of the desired 4 KiB page. Each ARMv8 page table entry is 8 bytes, so the 512 entries of a table fill a single 4 KiB page. The *Translation Table Base Register* (TTBR) gives the starting address of the first page table.

**Hardware/
Software
Interface**

reference bit Also called
use bit or access bit. A
field that is set whenever
a page is accessed and
that is used to implement
LRU or other replacement
schemes.

Implementing a completely accurate LRU scheme is too expensive, since it requires updating a data structure on *every* memory reference. Thus, most operating systems approximate LRU by keeping track of which pages have and which pages have not been recently used. To help the operating system estimate the LRU pages, some computers provide a reference bit or use bit, which is set whenever a page is accessed. (ARMv8 calls it an access bit.) The operating system periodically clears the reference bits and later records them so it can determine which pages were touched during a particular time period. With this usage information, the operating system can select a page that is among the least recently referenced (detected by having its reference bit off). If this bit is not provided by the hardware, the operating system must find another way to estimate which pages have been accessed.

Virtual Memory for Large Virtual Addresses

The caption in Figure 5.27 points out that with a single level page table for a 48-bit address with 4 KiB pages, we need 64 billion table entries. As each page table entry is 8 bytes for ARMv8, it would require 0.5 TiB just to map the virtual addresses to physical addresses! Moreover, there could be hundreds of processes running, each with its own page table. That much memory for translation would be unaffordable even for the largest systems.

A range of techniques is used to reduce the amount of storage required for the page table. The five techniques below aim at reducing the total maximum storage required as well as minimizing the main memory dedicated to page tables:

1. The simplest technique is to keep a limit register that restricts the size of the page table for a given process. If the virtual page number becomes larger than the contents of the limit register, entries must be added to the page table. This technique allows the page table to grow as a process consumes more space. Thus, the page table will only be large if the process is using many pages of virtual address space. This technique requires that the address space expand in just one direction.

2. Allowing growth in only one direction is not sufficient, since most languages require two areas whose size is expandable: one area holds the stack, and the other area holds the heap. Because of this duality, it is convenient to divide the page table and let it grow from the highest address down, as well as from the lowest address up. This means that there will be two separate page tables and two separate limits. The use of two page tables breaks the address space into two segments. The high-order bit of an address usually determines which segment and thus which page table to use for that address. Since the high-order address bit specifies the segment, each segment can be as large as one-half of the address space. A limit register for each segment specifies the current size of the segment, which grows in units of pages.

This type of segmentation is used by many architectures, including ARMv8 and MIPS. Unlike the type of segmentation discussed in the second elaboration on page 445, this form of segmentation is invisible to the application program, although not to the operating system. The major disadvantage of this scheme is that it does not work well when the address space is used in a sparse fashion rather than as a contiguous set of virtual addresses.

3. Another approach to reducing the page table size is to apply a hashing function to the virtual address so that the page table need be only the size of the number of *physical* pages in main memory. Such a structure is called an *inverted page table*. Of course, the lookup process is slightly more complex with an inverted page table, because we can no longer just index the page table.

4. To reduce the actual main memory tied up in page tables, most modern systems also allow the page tables to be paged. Although this sounds tricky, it works by using the same basic ideas of virtual memory and simply allowing the page tables to reside in the virtual address space. In addition, there are some small but critical problems, such as a never-ending series of page faults, which must be avoided. How these problems are overcome is both very detailed and typically highly processor-specific. In brief, these problems are avoided by placing all the page tables in the address space of the operating system and placing at least some of the page tables for the operating system in a portion of main memory that is physically addressed and is always present and thus never in secondary memory.

5. Multiple levels of page tables can also be used to reduce the total amount of page table storage, and this is the solution that ARMv8 uses to reduce the memory footprint of address translation. Figure 5.29 above shows the four levels of address translation to go from a 48-bit virtual address to a 40-bit physical address of a 4 KiB page. Address translation happens by first looking in the level 0 table, using the highest-order bits of the address. If the address in this table is valid, the next set of high-order bits is used to index the page table indicated by the segment table entry, and so on. Thus, the level 0 table maps the virtual address to a 512 GB (2^{39} bytes) region. The level 1 table in turn maps the virtual address to a 1 GB (2^{30}) region. The next level maps this down to a 2 MB (2^{21}) region. The final table maps the virtual address to the 4 KiB (2^{12}) memory page. This scheme allows the address space to be used in a sparse fashion (multiple noncontiguous segments can be active) without having to allocate the entire page table. Such schemes are particularly useful with very large address spaces and in software systems that require noncontiguous allocation. The primary disadvantage of this multi-level mapping is the more complex process for address translation.

Elaboration: An earlier elaboration mentioned that ARMv8 offers three choices for the minimum page size: 4, 16, and 64 Kibibyte. The caption of Figure 5.29 notes that the page table at each level is exactly one page, translating 9 bits of the virtual address per level. If a system uses the larger minimum page sizes, it can map more bits of the virtual address in a single page as well as the page itself being bigger, and thus systems with a 64 KiB page require just three levels of page tables.

What about Writes?

The difference between the access time to the cache and main memory is tens to hundreds of cycles, and write-through schemes can be used, although we need a write buffer to hide the latency of the write from the processor. In a virtual memory system, writes to the next level of the hierarchy (disk) can take millions of processor clock cycles; therefore, building a write buffer to allow the system to write-through to disk would be completely impractical. Instead, virtual memory systems must use write-back, performing the individual writes into the page in memory, and copying the page back to secondary memory when it is replaced in the main memory.

Hardware/ Software Interface

A write-back scheme has another major advantage in a virtual memory system. Because the disk transfer time is small compared with its access time, copying back an entire page is much more efficient than writing individual words back to the disk. A write-back operation, although faster than transferring separate words, is still costly. Thus, we would like to know whether a page *needs* to be copied back when we choose to replace it. To track whether a page has been written since it was read into the memory, a *dirty bit* is added to the page table. The dirty bit is set when any word in a page is written. If the operating system chooses to replace the page, the dirty bit indicates whether the page needs to be written out before its location in memory can be given to another page. Hence, a modified page is often called a *dirty* page.

Making Address Translation Fast: the TLB

Since the page tables are stored in main memory, every memory access by a program can take at least twice as long: one memory access to obtain the physical address and a second access to get the data. The key to improving access performance is to rely on locality of reference to the page table. When a translation for a virtual page number is used, it will probably be needed again soon, because the references to the words on that page have both temporal and spatial locality.

Accordingly, modern processors include a special cache that keeps track of recently used translations. This special address translation cache is traditionally referred to as a translation-lookaside buffer (TLB), although it would be more accurate to call it a translation cache. The TLB corresponds to that little piece of paper we typically use to record the location of a set of books we look up in the card catalog; rather than continually searching the entire catalog, we record the location of several books and use the scrap of paper as a cache of Library of Congress call numbers.

translation-lookaside buffer (TLB) A cache that keeps track of recently used address mappings to try to avoid an access to the page table.

Figure 5.30 shows that each tag entry in the TLB holds a portion of the virtual page number, and each data entry of the TLB holds a physical page number. Because we

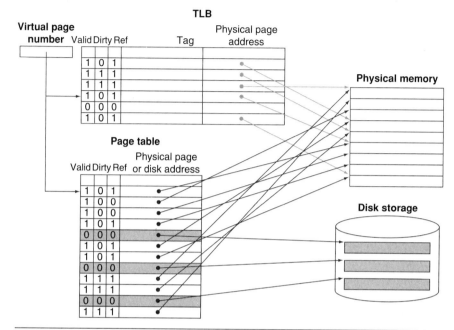

FIGURE 5.30 The TLB acts as a cache of the page table for the entries that map to physical pages only. The TLB contains a subset of the virtual-to-physical page mappings that are in the page table. The TLB mappings are shown in color. Because the TLB is a cache, it must have a tag field. If there is no matching entry in the TLB for a page, the page table must be examined. The page table either supplies a physical page number for the page (which can then be used to build a TLB entry) or indicates that the page resides on disk, in which case a page fault occurs. Since the page table has an entry for every virtual page, no tag field is needed; in other words, unlike a TLB, a page table is *not* a cache.

access the TLB instead of the page table on every reference, the TLB will need to include other status bits, such as the dirty and the reference bits. Although Figure 5.30 shows a single page table, TLBs work fine with multi-level page tables as well. The TLB simply loads the physical address and protection tags from the last level page table.

On every reference, we look up the virtual page number in the TLB. If we get a hit, the physical page number is used to form the address, and the corresponding reference bit is turned on. If the processor is performing a write, the dirty bit is also turned on. If a miss in the TLB occurs, we must determine whether it is a page fault or merely a TLB miss. If the page exists in memory, then the TLB miss indicates only that the translation is missing. In such cases, the processor can handle the TLB miss by loading the translation from the (last-level) page table into the TLB and then trying the reference again. If the page is not present in memory, then the TLB miss indicates a true page fault. In this case, the processor invokes the operating system using an exception. Because the TLB has many fewer entries than the number of pages in main memory, TLB misses will be much more frequent than true page faults.

TLB misses can be handled either in hardware or in software. In practice, with care there can be little performance difference between the two approaches, because the basic operations are the same in either case.

After a TLB miss occurs and the missing translation has been retrieved from the page table, we will need to select a TLB entry to replace. Because the reference and dirty bits are contained in the TLB entry, we need to copy these bits back to the page table entry when we replace an entry. These bits are the only portion of the TLB entry that can be changed. Using write-back—that is, copying these entries back at miss time rather than when they are written—is very efficient, since we expect the TLB miss rate to be small. Some systems use other techniques to approximate the reference and dirty bits, eliminating the need to write into the TLB except to load a new table entry on a miss.

Some typical values for a TLB might be

- TLB size: 16–512 entries
- Block size: 1–2 page table entries (typically 4–8 bytes each)
- Hit time: 0.5–1 clock cycle
- Miss penalty: 10–100 clock cycles
- Miss rate: 0.01%–1%

Designers have used a wide variety of associativities in TLBs. Some systems use small, fully associative TLBs because a fully associative mapping has a lower miss rate; furthermore, since the TLB is small, the cost of a fully associative mapping is not too high. Other systems use large TLBs, often with small associativity. With a fully associative mapping, choosing the entry to replace becomes tricky since implementing a hardware LRU scheme is too expensive. Furthermore, since TLB misses are much more frequent than page faults and thus must be handled more cheaply, we cannot afford an expensive software algorithm, as we can for page faults. As a result, many systems provide some support for randomly choosing an entry to replace. We'll examine replacement schemes in a little more detail in Section 5.8.

The Intrinsity FastMATH TLB

To see these ideas in a real processor, let's take a closer look at the TLB of the Intrinsity FastMATH. The memory system uses 4 KiB pages and just a 32-bit address space; thus, the virtual page number is 20 bits long. The physical address is the same size as the virtual address. The TLB contains 16 entries, it is fully associative, and it is shared between the instruction and data references. Each entry is 64 bits wide and contains a 20-bit tag (which is the virtual page number for that TLB entry), the corresponding physical page number (also 20 bits), a valid bit, a dirty bit, and other bookkeeping bits. Like most MIPS systems, it uses software to handle TLB misses.

Figure 5.31 shows the TLB and one of the caches, while Figure 5.32 shows the steps in processing a read or write request. When a TLB miss occurs, the hardware saves the page number of the reference in a special register and generates an

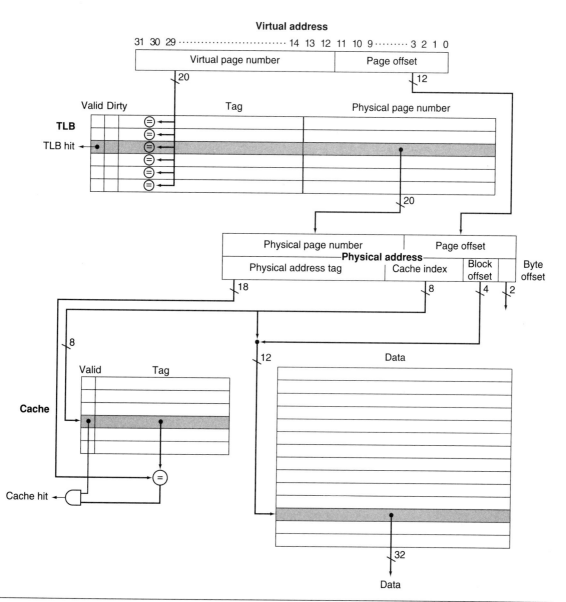

FIGURE 5.31 **The TLB and cache implement the process of going from a virtual address to a data item in the Intrinsity FastMATH.** This figure shows the organization of the TLB and the data cache, assuming a 4 KiB page size. Note that the address size for this computer is just 32 bits. This diagram focuses on a read; Figure 5.32 describes how to handle writes. Note that unlike Figure 5.12, the tag and data RAMs are split. By addressing the long but narrow data RAM with the cache index concatenated with the block offset, we select the desired word in the block without a 16:1 multiplexor. While the cache is direct mapped, the TLB is fully associative. Implementing a fully associative TLB requires that every TLB tag be compared against the virtual page number, since the entry of interest can be anywhere in the TLB. (See content addressable memories in the *Elaboration* on page 422.) If the valid bit of the matching entry is on, the access is a TLB hit, and bits from the physical page number together with bits from the page offset form the index that is used to access the cache.

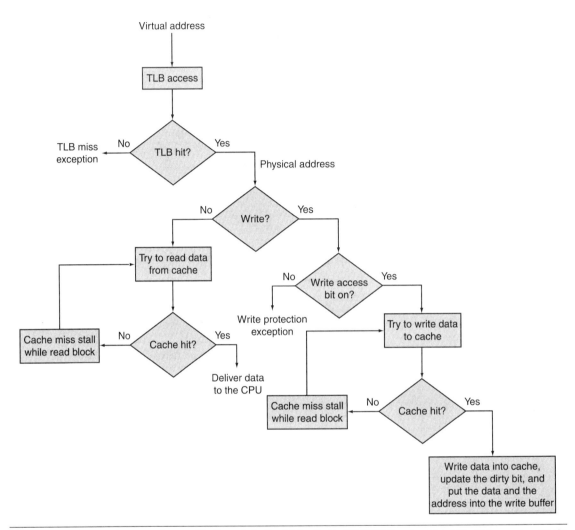

FIGURE 5.32 Processing a read or a write-through in the Intrinsity FastMATH TLB and cache. If the TLB generates a hit, the cache can be accessed with the resulting physical address. For a read, the cache generates a hit or miss and supplies the data or causes a stall while the data are brought from memory. If the operation is a write, a portion of the cache entry is overwritten for a hit and the data are sent to the write buffer if we assume write-through. A write miss is just like a read miss except that the block is modified after it is read from memory. Write-back requires writes to set a dirty bit for the cache block, and a write buffer is loaded with the whole block only on a read miss or write miss if the block to be replaced is dirty. Notice that a TLB hit and a cache hit are independent events, but a cache hit can only occur after a TLB hit occurs, which means that the data must be present in memory. The relationship between TLB misses and cache misses is examined further in the following example and the exercises at the end of this chapter. Note that the address size for this computer is just 32 bits.

exception. The exception invokes the operating system, which handles the miss in software. To find the physical address for the missing page, a TLB miss indexes the page table using the page number of the virtual address and the page table register, which indicates the starting address of the active process page table. Using a special set of system instructions that can update the TLB, the operating system places the physical address from the page table into the TLB. A TLB miss takes about 13 clock cycles, assuming the code and the page table entry are in the instruction cache and data cache, respectively. A true page fault occurs if the page table entry does not have a valid physical address. The hardware maintains an index that indicates the recommended entry to replace; it is chosen randomly.

There is an extra complication for write requests: namely, the write access bit in the TLB must be checked. This bit prevents the program from writing into pages for which it has only read access. If the program attempts a write and the write access bit is off, an exception is generated. The write access bit forms part of the protection mechanism, which we will discuss shortly.

Integrating Virtual Memory, TLBs, and Caches

Our virtual memory and cache systems work together as a hierarchy, so that data cannot be in the cache unless it is present in main memory. The operating system helps maintain this hierarchy by flushing the contents of any page from the cache when it decides to migrate that page to secondary memory. At the same time, the OS modifies the page tables and TLB, so that an attempt to access any data on the migrated page will generate a page fault.

Under the best of circumstances, a virtual address is translated by the TLB and sent to the cache where the appropriate data are found, retrieved, and sent back to the processor. In the worst case, a reference can miss in all three components of the memory hierarchy: the TLB, the page table, and the cache. The following example illustrates these interactions in more detail.

Overall Operation of a Memory Hierarchy

In a memory hierarchy like that of Figure 5.31, which includes a TLB and a cache organized as shown, a memory reference can encounter three different types of misses: a TLB miss, a page fault, and a cache miss. Consider all the combinations of these three events with one or more occurring (seven possibilities). For each possibility, state whether this event can actually occur and under what circumstances.

EXAMPLE

Figure 5.33 shows all combinations and whether each is possible in practice.

ANSWER

TLB	Page table	Cache	Possible? If so, under what circumstance?
Hit	Hit	Miss	Possible, although the page table is never really checked if TLB hits.
Miss	Hit	Hit	TLB misses, but entry found in page table; after retry, data is found in cache.
Miss	Hit	Miss	TLB misses, but entry found in page table; after retry, data misses in cache.
Miss	Miss	Miss	TLB misses and is followed by a page fault; after retry, data must miss in cache.
Hit	Miss	Miss	Impossible: cannot have a translation in TLB if page is not present in memory.
Hit	Miss	Hit	Impossible: cannot have a translation in TLB if page is not present in memory.
Miss	Miss	Hit	Impossible: data cannot be allowed in cache if the page is not in memory.

FIGURE 5.33   The possible combinations of events in the TLB, virtual memory system, and cache. Three of these combinations are impossible, and one is possible (TLB hit, page table hit, cache miss) but never detected.

PIPELINING

virtually addressed cache A cache that is accessed with a virtual address rather than a physical address.

aliasing A situation in which two addresses access the same object; it can occur in virtual memory when there are two virtual addresses for the same physical page.

physically addressed cache A cache that is addressed by a physical address.

Elaboration: Figure 5.33 assumes that all memory addresses are translated to physical addresses before the cache is accessed. In this organization, the cache is *physically indexed* and *physically tagged* (both the cache index and tag are physical, rather than virtual, addresses). In such a system, the amount of time to access memory, assuming a cache hit, must accommodate both a TLB access and a cache access; of course, these accesses can be **pipelined**.

Alternatively, the processor can index the cache with an address that is completely or partially virtual. This is called a **virtually addressed cache**, and it uses tags that are virtual addresses; hence, such a cache is *virtually indexed* and *virtually tagged*. In such caches, the address translation hardware (TLB) is unused during the normal cache access, since the cache is accessed with a virtual address that has not been translated to a physical address. This takes the TLB out of the critical path, reducing cache latency. When a cache miss occurs, however, the processor needs to translate the address to a physical address so that it can fetch the cache block from main memory.

When the cache is accessed with a virtual address and pages are shared between processes (which may access them with different virtual addresses), there is the possibility of **aliasing**. Aliasing occurs when the same object has two names—in this case, two virtual addresses for the same page. This ambiguity creates a problem, because a word on such a page may be cached in two different locations, each corresponding to distinct virtual addresses. This ambiguity would allow one program to write the data without the other program being aware that the data had changed. Completely virtually addressed caches either introduce design limitations on the cache and TLB to reduce aliases or require the operating system, and possibly the user, to take steps to ensure that aliases do not occur.

A common compromise between these two design points is caches that are virtually indexed—sometimes using just the page-offset portion of the address, which is really a physical address since it is not translated—but use physical tags. These designs, which are *virtually indexed but physically tagged*, attempt to achieve the performance advantages of virtually indexed caches with the architecturally simpler advantages of a physically addressed cache. For example, there is no alias problem in this case. Figure 5.31 assumed a 4 KiB page size, but it's really 16 KiB, so the Intrinsity FastMATH can use this trick. To pull it off, there must be careful coordination between the minimum page size, the cache size, and associativity. ARMv8 allows instruction caches to use either physical or virtual indexing, but they must be physically tagged. It requires data caches to *behave as though* physically tagged and indexed, but it does not mandate this implementation. For example, virtually indexed, physically tagged data caches could use additional logic to ensure that their behavior is consistent with the ARMv8 definition.

Implementing Protection with Virtual Memory

Perhaps the most important function of virtual memory today is to allow sharing of a single main memory by multiple processes, while providing memory protection among these processes and the operating system. The protection mechanism must ensure that although multiple processes are sharing the same main memory, one renegade process cannot write into the address space of another user process or into the operating system either intentionally or unintentionally. The write access bit in the TLB can protect a page from being written. Without this level of protection, computer viruses would be even more widespread.

To enable the operating system to implement protection in the virtual memory system, the hardware must provide at least the three basic capabilities summarized below. Note that the first two are the same requirements as needed for virtual machines (Section 5.6).

1. Support at least two modes that indicate whether the running process is a user process or an operating system process, variously called a supervisor process, a kernel process, or an *executive* process.

2. Provide a portion of the processor state that a user process can read but not write. This state includes the user/supervisor mode bit, which dictates whether the processor is in user or supervisor mode, the page table pointer, and the TLB. To write these elements, the operating system uses special instructions that are only available in supervisor mode.

3. Provide mechanisms whereby the processor can go from user mode to supervisor mode and vice versa. The first direction is typically accomplished by a system call exception, implemented as a special instruction (*SVC* in the ARMv8 instruction set) that transfers control to a dedicated location in supervisor code space. As with any other exception, the program counter from the point of the system call is saved in the *exception link register* (ELR), and the processor is placed in supervisor mode. To return to user mode from the exception, use the *exception return* (ERET) instruction, which resets to user mode and jumps to the address in ELR.

By using these mechanisms and storing the page tables in the operating system's address space, the operating system can change the page tables while preventing a user process from changing them, ensuring that a user process can access only the storage provided to it by the operating system.

Hardware/ Software Interface

supervisor mode Also called kernel mode. A mode indicating that a running process is an operating system process.

system call A special instruction that transfers control from user mode to a dedicated location in supervisor code space, invoking the exception mechanism in the process.

We also want to prevent a process from reading the data of another process. For example, we wouldn't want a student program to read the teacher's grades while they were in the processor's memory. Once we begin sharing main memory, we must provide the ability for a process to protect its data from both reading and writing by another process; otherwise, sharing the main memory will be a mixed blessing!

Remember that each process has its own virtual address space. Thus, if the operating system keeps the page tables organized so that the independent virtual pages map to disjoint physical pages, one process will not be able to access another's data. Of course, this also requires that a user process be unable to change the page table mapping. The operating system can assure safety if it prevents the user process from modifying its own page tables. However, the operating system must be able to modify the page tables. Placing the page tables in the protected address space of the operating system satisfies both requirements.

When processes want to share information in a limited way, the operating system must assist them, since accessing the information of another process requires changing the page table of the accessing process. The write access bit can be used to restrict the sharing to just read sharing, and, like the rest of the page table, this bit can be changed only by the operating system. To allow another process, say, P1, to read a page owned by process P2, P2 would ask the operating system to create a page table entry for a virtual page in P1's address space that points to the same physical page that P2 wants to share. The operating system could use the write protection bit to prevent P1 from writing the data, if that was P2's wish. Any bits that determine the access rights for a page must be included in both the page table and the TLB, because the page table is accessed only on a TLB *miss*.

Elaboration: When the operating system decides to change from running process P1 to running process P2 (called a **context switch** or *process switch*), it must ensure that P2 cannot get access to the page tables of P1 because that would compromise protection. If there is no TLB, it suffices to change the page table register to point to P2's page table (rather than to P1's); with a TLB, we must clear the TLB entries that belong to P1—both to protect the data of P1 and to force the TLB to load the entries for P2. If the process switch rate were high, this could be quite inefficient. For example, P2 might load only a few TLB entries before the operating system switched back to P1. Unfortunately, P1 would then find that all its TLB entries were gone and would have to pay TLB misses to reload them. This problem arises because the virtual addresses used by P1 and P2 are the same, and we must clear out the TLB to avoid confusing these addresses.

A common alternative is to extend the virtual address space by adding a *process identifier* or *task identifier*. The Intrinsity FastMATH has an 8-bit *address space ID* (ASID) field for this purpose. This small field identifies the currently running process; it is kept in a register loaded by the operating system when it switches processes. ARMv8 also offers ASID to reduce TLB flushes on context switches. The process identifier is concatenated to the tag portion of the TLB, so that a TLB hit occurs only if both the page number *and* the process identifier match. This combination eliminates the need to clear the TLB, except on rare occasions.

Similar problems can occur for a cache, since on a process switch, the cache will contain data from the running process. These problems arise in different ways for physically addressed and virtually addressed caches, and a variety of solutions, such as process identifiers, are used to ensure that a process gets its own data.

context switch
A changing of the internal state of the processor to allow a different process to use the processor that includes saving the state needed to return to the currently executing process.

Handling TLB Misses and Page Faults

Although the translation of virtual to physical addresses with a TLB is straightforward when we get a TLB hit, as we saw earlier, handling TLB misses and page faults is more complex. A TLB miss occurs when no entry in the TLB matches a virtual address. Recall that a TLB miss can indicate one of two possibilities:

1. The page is present in memory, and we need only create the missing TLB entry.

2. The page is not present in memory, and we need to transfer control to the operating system to deal with a page fault.

Handling a TLB miss or a page fault requires using the exception mechanism to interrupt the active process, transferring control to the operating system, and later resuming execution of the interrupted process. A page fault will be recognized sometime during the clock cycle used to access memory. To restart the instruction after the page fault is handled, the program counter of the instruction that caused the page fault must be saved. The *exception link register* (ELR) is used to hold this value.

In addition, a TLB miss or page fault exception must be asserted by the end of the same clock cycle that the memory access occurs, so that the next clock cycle will begin exception processing rather than continue normal instruction execution. If the page fault was not recognized in this clock cycle, a load instruction could overwrite a register, and this could be disastrous when we try to restart the instruction. For example, consider the instruction LDUR X1, [X1,#0]: the computer must be able to prevent the write pipeline stage from occurring; otherwise, it could not properly restart the instruction, since the contents of X1 would have been destroyed. A similar complication arises on stores. We must prevent the write into memory from actually completing when there is a page fault; this is usually done by deasserting the write control line to the memory.

Hardware/ Software Interface

Between the time we begin executing the exception handler in the operating system and the time that the operating system has saved all the state of the process, the operating system is particularly vulnerable. For instance, if another exception occurred when we were processing the first exception in the operating system, the control unit would overwrite the exception link register, making it impossible to return to the instruction that caused the page fault! We can avoid this disaster by providing the ability to disable and enable exceptions. When an exception first occurs, the processor sets a bit that disables all other exceptions; this could happen at the same time the processor sets the supervisor mode bit. The operating system will then save just enough state to allow it to recover if another exception occurs— namely, the *exception link register* (ELR) and the *exception syndrome register* (ESR), which as we saw in Chapter 4 records the reason for the exception. ELR and ESR in ARMv8 are two of the special control registers that help with exceptions, TLB misses, and page faults. The operating system can then re-enable exceptions. These steps make sure that exceptions will not cause the processor to lose any state and thereby be unable to restart execution of the interrupting instruction.

exception enable Also called interrupt enable. A signal or action that controls whether the process responds to an exception or not; necessary for preventing the occurrence of exceptions during intervals before the processor has safely saved the state needed to restart.

Once the operating system knows the virtual address that caused the page fault, it must complete three steps:

1. Look up the page table entry using the virtual address and find the location of the referenced page in secondary memory.

2. Choose a physical page to replace; if the chosen page is dirty, it must be written out to secondary memory before we can bring a new virtual page into this physical page.

3. Start a read to bring the referenced page from secondary memory into the chosen physical page.

Of course, this last step will take millions of processor clock cycles for disks (so will the second if the replaced page is dirty); accordingly, the operating system will usually select another process to execute in the processor until the disk access completes. Because the operating system has saved the state of the process, it can freely give control of the processor to another process.

When the read of the page from secondary memory is complete, the operating system can restore the state of the process that originally caused the page fault and execute the instruction that returns from the exception. This instruction will reset the processor from kernel to user mode, as well as restore the program counter. The user process then re-executes the instruction that faulted, accesses the requested page successfully, and continues execution.

Page fault exceptions for data accesses are difficult to implement properly in a processor because of a combination of three characteristics:

1. They occur in the middle of instructions, unlike instruction page faults.

2. The instruction cannot be completed before handling the exception.

3. After handling the exception, the instruction must be restarted as if nothing had occurred.

restartable instruction An instruction that can resume execution after an exception is resolved without the exception's affecting the result of the instruction.

Making instructions restartable, so that the exception can be handled and the instruction later continued, is relatively easy in an architecture like the ARMv8. Because each instruction writes only one data item and this write occurs at the end of the instruction cycle, we can simply prevent the instruction from completing (by not writing) and restart the instruction at the beginning.

Elaboration: For processors with more complex instructions that can touch many memory locations and write many data items, making instructions restartable is much harder. Processing one instruction may generate a number of page faults in the middle of the instruction. For example, x86 processors have block move instructions that touch thousands of data words. In such processors, instructions often cannot be restarted from the beginning, as we do for ARMv8 instructions. Instead, the instruction must be interrupted and later continued midstream in its execution. Resuming an instruction in the middle of its execution usually requires saving some special state, processing the exception, and restoring that special state. Making this work properly requires careful and detailed coordination between the exception-handling code in the operating system and the hardware.

Elaboration: Rather than pay an extra level of indirection on every memory access, the Virtual Memory Monitor (Section 5.6) maintains a *shadow page table* that maps directly from the guest virtual address space to the physical address space of the hardware. By detecting all modifications to the guest's page table, the VMM can ensure the shadow page table entries being used by the hardware for translations correspond to those of the guest OS environment, with the exception of the correct physical pages substituted for the real pages in the guest tables. Hence, the VMM must trap any attempt by the guest OS to change its page table or to access the page table pointer. This is commonly done by write protecting the guest page tables and trapping any access to the page table pointer by a guest OS. As noted above, the latter happens naturally if accessing the page table pointer is a privileged operation.

Elaboration: The final portion of the architecture to virtualize is I/O. This is by far the most difficult part of system virtualization because of the increasing number of I/O devices attached to the computer *and* the expanding diversity of I/O device types. Another difficulty is the sharing of a real device among multiple VMs, and yet another comes from supporting the myriad of device drivers that are required, especially if different guest OSes are supported on the same VM system. The VM illusion can be maintained by giving each VM generic versions of each type of I/O device driver, and then leaving it to the VMM to handle real I/O.

Elaboration: In addition to virtualizing the instruction set for a virtual machine, another challenge is virtualization of virtual memory, as each guest OS in every virtual machine manages its own set of page tables. To make this work, the VMM separates the notions of *real* and *physical memory* (which are often treated synonymously), and makes real memory a separate, intermediate level between virtual memory and physical memory. (Some use the terms *virtual memory, physical memory*, and *machine memory* to name the same three levels.) The guest OS maps virtual memory to real memory via its page tables, and the VMM page tables map the guest's real memory to physical memory. The virtual memory architecture is typically specified via page tables, as in IBM VM/370, the x86, and ARMv8.

Summary

Virtual memory is the name for the level of memory hierarchy that manages caching between the main memory and secondary memory. Virtual memory allows a single program to expand its address space beyond the limits of main memory. More importantly, virtual memory supports sharing of the main memory among multiple, simultaneously active processes, in a protected manner.

Managing the memory hierarchy between main memory and disk is challenging because of the high cost of page faults. Several techniques are used to reduce the miss rate:

1. Pages are made large to take advantage of spatial locality and to reduce the miss rate.

2. The mapping between virtual addresses and physical addresses, which is implemented with a page table, is made fully associative so that a virtual page can be placed anywhere in main memory.

3. The operating system uses techniques, such as LRU and a reference bit, to choose which pages to replace.

Writes to secondary memory are expensive, so virtual memory uses a write-back scheme and also tracks whether a page is unchanged (using a dirty bit) to avoid writing clean pages.

The virtual memory mechanism provides address translation from a virtual address used by the program to the physical address space used for accessing memory. This address translation allows protected sharing of the main memory and provides several additional benefits, such as simplifying memory allocation. Ensuring that processes are protected from each other requires that only the operating system can change the address translations, which is implemented by preventing user programs from altering the page tables. Controlled sharing of pages between processes can be implemented with the help of the operating system and access bits in the page table that indicate whether the user program has read or write access to a page.

If a processor had to access a page table resident in memory to translate every access, virtual memory would be too expensive, as caches would be pointless! Instead, a TLB acts as a cache for translations from the page table. Addresses are then translated from virtual to physical using the translations in the TLB.

Caches, virtual memory, and TLBs all rely on a common set of principles and policies. The next section discusses this common framework.

Understanding Program Performance

Although virtual memory was invented to enable a small memory to act as a large one, the performance difference between secondary memory and main memory means that if a program routinely accesses more virtual memory than it has physical memory, it will run very slowly. Such a program would be continuously swapping pages between main memory and secondary memory, called *thrashing*. Thrashing is a disaster if it occurs, but it is rare. If your program thrashes, the easiest solution is to run it on a computer with more memory or buy more memory for your computer. A more complex choice is to re-examine your algorithm and data structures to see if you can change the locality and thereby reduce the number of pages that your program uses simultaneously. This set of popular pages is informally called the *working set*.

A more common performance problem is TLB misses. Since a TLB might handle only 32–64 page entries at a time, a program could easily see a high TLB miss rate, as the processor may access less than a quarter mebibyte directly: 64×4 KiB = 0.25 MiB. For example, TLB misses are often a challenge for Radix Sort. To try to alleviate this problem, most computer architectures now offer support for larger page sizes. For instance, in addition to the minimum 4 KiB page, ARMv8 hardware supports 2 MiB and 1 GiB pages. Hence, if a program uses large page sizes, it can access more memory directly without TLB misses.

The practical challenge is getting the operating system to allow programs to select these larger page sizes. Once again, the more complex solution to reducing

TLB misses is to re-examine the algorithm and data structures to reduce the working set of pages; given the importance of memory accesses to performance and the frequency of TLB misses, some programs with large working sets have been redesigned with that goal.

Elaboration: ARMv8 supports the larger page sizes via the multi-level page table of Figure 5.29. In addition to pointing at the next level page table in levels 1 and 2, it allows a *block translation* to map the virtual address to a 1 GiB physical address (if the block translation is in level 1) or a 2 MiB physical address (if the block translation is in level 2). If the minimum page size is larger than 4 KiB, the block translations are also larger: 64 GiB and 32 MiB for 16 KiB pages and 4096 GiB and 512 MiB for 64 KiB pages.

Elaboration: ARMv8 has many options and optimizations that we do not have space to cover; the description of virtual memory takes 200 pages in the ARMv8 architecture manual. For example, while the minimum system has just two exception levels (*EL0* and *EL1*), to support virtual machines monitors there is an optional third exception level (*EL2*) and a fourth level (*EL3*) for security monitors. There are versions of the special ELR and ESR registers for each level. To get greater performance from the TLB, ARMv8 offers a hint that a single entry corresponds to 16 *contiguous* ranges that have the same permissions and attributes, thereby expanding the reach of that entry by a factor of 16. To offer different types of shared addresses for cores within a chip versus a cluster of chips, ARMv8 distinguishes *inner sharability* versus *outer sharability*, with the former always sharing the same domain as the latter.

Match the definitions in the right column to the terms in the left column.

Check Yourself

1. L1 cache	a. A cache for a cache
2. L2 cache	b. A cache for disks
3. Main memory	c. A cache for a main memory
4. TLB	d. A cache for page table entries

5.8 A Common Framework for Memory Hierarchy

By now, you've recognized that the different types of memory hierarchies have a great deal in common. Although many of the aspects of memory hierarchies differ quantitatively, many of the policies and features that determine how a hierarchy functions are similar qualitatively. Figure 5.34 shows how some of the quantitative

Feature	Typical values for L1 caches	Typical values for L2 caches	Typical values for paged memory	Typical values for a TLB
Total size in blocks	250–2000	2500–25,000	16,000–250,000	40–1024
Total size in kilobytes	16–64	125–2000	1,000,000–1,000,000,000	0.25–16
Block size in bytes	16–64	64–128	4000–64,000	4–32
Miss penalty in clocks	10–25	100–1000	10,000,000–100,000,000	10–1000
Miss rates (global for L2)	2%–5%	0.1%–2%	0.00001%–0.0001%	0.01%–2%

FIGURE 5.34   The key quantitative design parameters that characterize the major elements of memory hierarchy in a computer. These are typical values for these levels as of 2012. Although the range of values is wide, this is partially because many of the values that have shifted over time are related; for example, as caches become larger to overcome larger miss penalties, block sizes also grow. While not shown, server microprocessors today also have L3 caches, which can be 2 to 8 MiB and contain many more blocks than L2 caches. L3 caches lower the L2 miss penalty to 30 to 40 clock cycles.

characteristics of memory hierarchies can differ. In the rest of this section, we will discuss the common operational alternatives for memory hierarchies, and how these determine their behavior. We will examine these policies in a series of four questions that apply between any two levels of a memory hierarchy, although for simplicity, we will primarily use terminology for caches.

Question 1: Where Can a Block Be Placed?

We have seen that block placement in the upper level of the hierarchy can use a range of schemes, from direct mapped to set associative to fully associative. As mentioned above, this entire range of schemes can be thought of as variations on a set-associative scheme where the number of sets and the number of blocks per set varies:

Scheme name	Number of sets	Blocks per set
Direct mapped	Number of blocks in cache	1
Set associative	$\dfrac{\text{Number of blocks in the cache}}{\text{Associativity}}$	Associativity (typically 2–16)
Fully associative	1	Number of blocks in the cache

The advantage of increasing the degree of associativity is that it usually decreases the miss rate. The improvement in miss rate comes from reducing misses that compete for the same location. We will examine these in more detail shortly. First, let's look at how much improvement is gained. Figure 5.35 shows the miss rates for several cache sizes as associativity varies from direct mapped to eight-way set associative. The largest gains are obtained in going from direct mapped to two-way set associative, which yields between a 20% and 30% reduction in the miss rate. As cache sizes grow, the relative improvement from associativity increases only slightly; since the overall miss rate of a larger cache is lower, the opportunity for improving the miss rate decreases and the absolute improvement in the miss rate from associativity shrinks significantly. The potential disadvantages of associativity, as we mentioned earlier, are increased cost and slower access time.

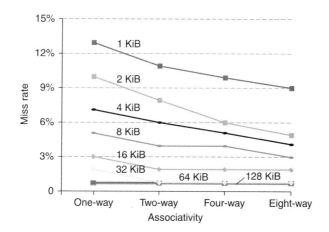

FIGURE 5.35 The data cache miss rates for each of eight cache sizes improve as the associativity increases. While the benefit of going from one-way (direct mapped) to two-way set associative is significant, the benefits of further associativity are smaller (e.g., 1–10% improvement going from two-way to four-way versus 20–30% improvement going from one-way to two-way). There is even less improvement in going from four-way to eight-way set associative, which, in turn, comes very close to the miss rates of a fully associative cache. Smaller caches obtain a significantly larger absolute benefit from associativity because the base miss rate of a small cache is larger. Figure 5.16 explains how these data were collected.

Question 2: How Is a Block Found?

The choice of how we locate a block depends on the block placement scheme, since that dictates the number of possible locations. We can summarize the schemes as follows:

Associativity	Location method	Comparisons required
Direct mapped	Index	1
Set associative	Index the set, search among elements	Degree of associativity
Full	Search all cache entries	Size of the cache
	Separate lookup table	0

The choice among direct-mapped, set-associative, or fully associative mapping in any memory hierarchy will depend on the cost of a miss versus the cost of implementing associativity, both in time and in extra hardware. Including the L2 cache on the chip enables much higher associativity, because the hit times are not as critical and the designer does not have to rely on standard SRAM chips as the building blocks. Fully associative caches are prohibitive except for small sizes, where the cost of the comparators is not overwhelming and where the absolute miss rate improvements are greatest.

In virtual memory systems, a separate mapping table—the page table—is kept to index the memory. In addition to the storage needed for the table, using an index table requires an extra memory access. The choice of full associativity for page placement and the extra table is motivated by these facts:

1. Full associativity is beneficial, since misses are very expensive.

2. Full associativity allows software to use sophisticated replacement schemes that are designed to reduce the miss rate.

3. The full map can be easily indexed with no extra hardware and no searching required.

Therefore, virtual memory systems almost always use fully associative placement.

Set-associative placement is often used for caches and TLBs, where the access combines indexing and the search of a small set. A few systems have used direct-mapped caches because of their advantage in access time and simplicity. The advantage in access time occurs because finding the requested block does not depend on a comparison. Such design choices depend on many details of the implementation, such as whether the cache is on-chip, the technology used for implementing the cache, and the critical role of cache access time in determining the processor cycle time.

Question 3: Which Block Should Be Replaced on a Cache Miss?

When a miss occurs in an associative cache, we must decide which block to replace. In a fully associative cache, all blocks are candidates for replacement. If the cache is set associative, we must choose among the blocks in the set. Of course, replacement is easy in a direct-mapped cache because there is only one candidate.

There are the two primary strategies for replacement in set-associative or fully associative caches:

■ *Random:* Candidate blocks are randomly selected, possibly using some hardware assistance.

■ *Least recently used* (LRU): The block replaced is the one that has been unused for the longest time.

In practice, LRU is too costly to implement for hierarchies with more than a small degree of associativity (two to four, typically), since tracking the usage information is expensive. Even for four-way set associativity, LRU is often approximated—for example, by keeping track of which pair of blocks is LRU (which requires 1 bit), and then tracking which block in each pair is LRU (which requires 1 bit per pair).

For larger associativity, either LRU is approximated or random replacement is used. In caches, the replacement algorithm is in hardware, which means that the scheme should be easy to implement. Random replacement is simple to build in hardware, and for a two-way set-associative cache, random replacement has a miss rate about 1.1 times higher than LRU replacement. As the caches become larger, the miss rate for both replacement strategies falls, and the absolute difference becomes small. In fact, random replacement can sometimes be better than the simple LRU approximations that are easily implemented in hardware.

In virtual memory, some form of LRU is always approximated, since even a tiny reduction in the miss rate can be important when the cost of a miss is enormous. Reference bits or equivalent functionality are often provided to make it easier for the operating system to track a set of less recently used pages. Because misses are so expensive and relatively infrequent, approximating this information primarily in software is acceptable.

Question 4: What Happens on a Write?

A key characteristic of any memory hierarchy is how it deals with writes. We have already seen the two basic options:

■ *Write-through:* The information is written to both the block in the cache and the block in the lower level of the memory hierarchy (main memory for a cache). The caches in Section 5.3 used this scheme.

■ *Write-back:* The information is written just to the block in the cache. The modified block is written to the lower level of the hierarchy only when it is replaced. Virtual memory systems always use write-back, for the reasons discussed in Section 5.7.

Both write-back and write-through have their advantages. The key advantages of write-back are the following:

■ Individual words can be written by the processor at the rate that the cache, rather than the memory, can accept them.

■ Multiple writes within a block require only one write to the lower level in the hierarchy.

■ When blocks are written back, the system can make effective use of a high-bandwidth transfer, since the entire block is written.

Write-through has these advantages:

■ Misses are simpler and cheaper because they never require a block to be written back to the lower level.

■ Write-through is easier to implement than write-back, although to be realistic, a write-through cache will still need to use a write buffer.

The BIG Picture

Caches, TLBs, and virtual memory may initially look very different, but they rely on the same two principles of locality, and they can be understood by their answers to four questions:

Question 1: Where can a block be placed?

Answer: One place (direct mapped), a few places (set associative), or any place (fully associative).

Question 2: How is a block found?

Answer: There are four methods: indexing (as in a direct-mapped cache), limited search (as in a set-associative cache), full search (as in a fully associative cache), and a separate lookup table (as in a page table).

Question 3: What block is replaced on a miss?

Answer: Typically, either the least recently used or a random block.

Question 4: How are writes handled?

Answer: Each level in the hierarchy can use either write-through or write-back.

three Cs model A cache model in which all cache misses are classified into one of three categories: compulsory misses, capacity misses, and conflict misses.

compulsory miss Also called cold-start miss. A cache miss caused by the first access to a block that has never been in the cache.

capacity miss A cache miss that occurs because the cache, even with full associativity, cannot contain all the blocks needed to satisfy the request.

conflict miss Also called collision miss. A cache miss that occurs in a set-associative or direct-mapped cache when multiple blocks compete for the same set and that are eliminated in a fully associative cache of the same size.

In virtual memory systems, only a write-back policy is practical because of the long latency of a write to the lower level of the hierarchy. The rate at which writes are generated by a processor generally exceeds the rate at which the memory system can process them, even allowing for physically and logically wider memories and burst modes for DRAM. Consequently, today lowest-level caches typically use write-back.

The Three Cs: An Intuitive Model for Understanding the Behavior of Memory Hierarchies

In this subsection, we look at a model that provides insight into the sources of misses in a memory hierarchy and how the misses will be affected by changes in the hierarchy. We will explain the ideas in terms of caches, although the ideas carry over directly to any other level in the hierarchy. In this model, all misses are classified into one of three categories (the three Cs):

■ Compulsory misses: These are cache misses caused by the first access to a block that has never been in the cache. These are also called cold-start misses.

■ Capacity misses: These are cache misses caused when the cache cannot contain all the blocks needed during execution of a program. Capacity misses occur when blocks are replaced and then later retrieved.

■ Conflict misses: These are cache misses that occur in set-associative or direct-mapped caches when multiple blocks compete for the same set. Conflict misses are those misses in a direct-mapped or set-associative cache that are eliminated in a fully associative cache of the same size. These cache misses are also called collision misses.

Figure 5.36 shows how the miss rate divides into the three sources. These sources of misses can be directly attacked by changing some aspect of the cache design. Since conflict misses arise straight from contention for the same cache block, increasing associativity reduces conflict misses. Associativity, however, may slow access time, leading to lower overall performance.

Capacity misses can easily be reduced by enlarging the cache; indeed, second-level caches have been growing steadily bigger for many years. Of course, when we make the cache larger, we must also be careful about increasing the access time, which could lead to lower overall performance. Thus, first-level caches have been growing slowly, if at all.

Because compulsory misses are generated by the first reference to a block, the primary way for the cache system to reduce the number of compulsory misses is to increase the block size. This will reduce the number of references required to touch each block of the program once, because the program will consist of fewer cache blocks. As mentioned above, increasing the block size too much can have a negative effect on performance because of the increase in the miss penalty.

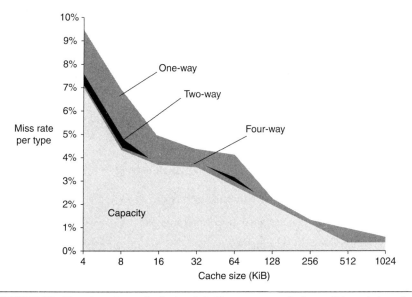

FIGURE 5.36 The miss rate can be broken into three sources of misses. This graph shows the total miss rate and its components for a range of cache sizes. These data are for the SPEC CPU2000 integer and floating-point benchmarks and are from the same source as the data in Figure 5.35. The compulsory miss component is 0.006% and cannot be seen in this graph. The next component is the capacity miss rate, which depends on cache size. The conflict portion, which depends both on associativity and on cache size, is shown for a range of associativities from one-way to eight-way. In each case, the labeled section corresponds to the increase in the miss rate that occurs when the associativity is changed from the next higher degree to the labeled degree of associativity. For example, the section labeled *two-way* indicates the additional misses arising when the cache has associativity of two rather than four. Thus, the difference in the miss rate incurred by a direct-mapped cache versus a fully associative cache of the same size is given by the sum of the sections marked *four-way, two-way,* and *one-way*. The difference between eight-way and four-way is so small that it is difficult to see on this graph.

Design change	Effect on miss rate	Possible negative performance effect
Increases cache size	Decreases capacity misses	May increase access time
Increases associativity	Decreases miss rate due to conflict misses	May increase access time
Increases block size	Decreases miss rate for a wide range of block sizes due to spatial locality	Increases miss penalty. Very large block could increase miss rate

FIGURE 5.37   Memory hierarchy design challenges.

 The **BIG** Picture

The challenge in designing memory hierarchies is that every change that potentially improves the miss rate can also negatively affect overall performance, as Figure 5.37 summarizes. This combination of positive and negative effects is what makes the design of a memory hierarchy interesting.

The decomposition of misses into the three Cs is a useful qualitative model. In real cache designs, many of the design choices interact, and changing one cache characteristic will often affect several components of the miss rate. Despite such shortcomings, this model is a useful way to gain insight into the performance of cache designs.

Check Yourself

Which of the following statements (if any) is generally true?

1. There is no way to reduce compulsory misses.

2. Fully associative caches have no conflict misses.

3. In reducing misses, associativity is more important than capacity.

 5.9

Using a Finite-State Machine to Control a Simple Cache

We can now build control for a cache, just as we implemented control for the single-cycle and pipelined datapaths in Chapter 4. This section starts with a definition of a simple cache and then a description of *finite-state machines* (FSMs). It finishes with the FSM of a controller for this simple cache. ▦ Section 5.12 goes into more depth, showing the cache and controller in a new hardware description language.

A Simple Cache

We're going to design a controller for a straightforward cache. Here are the key characteristics of the cache:

- Direct-mapped cache
- Write-back using write allocate
- Block size is four words (16 bytes or 128 bits)
- Cache size is 16 KiB, so it holds 1024 blocks
- 32-bit addresses
- The cache includes a valid bit and dirty bit per block

From Section 5.3, we can now calculate the fields of an address for the cache:

- Cache index is 10 bits
- Block offset is 4 bits
- Tag size is $32 - (10 + 4)$ or 18 bits

The signals between the processor to the cache are

- 1-bit Read or Write signal
- 1-bit Valid signal, saying whether there is a cache operation or not
- 32-bit address
- 32-bit data from processor to cache
- 32-bit data from cache to processor
- 1-bit Ready signal, saying the cache operation is complete

The interface between the memory and the cache has the same fields as between the processor and the cache, except that the data fields are now 128 bits wide. The extra memory width is generally found in microprocessors today, which deal with either 32-bit or 64-bit words in the processor while the DRAM controller is often 128 bits. Making the cache block match the width of the DRAM simplified the design. Here are the signals:

- 1-bit Read or Write signal
- 1-bit Valid signal, saying whether there is a memory operation or not
- 32-bit address
- 128-bit data from cache to memory
- 128-bit data from memory to cache
- 1-bit Ready signal, saying the memory operation is complete

Note that the interface to memory is not a fixed number of cycles. We assume a memory controller that will notify the cache via the Ready signal when the memory read or write is finished.

Before describing the cache controller, we need to review finite-state machines, which allow us to control an operation that can take multiple clock cycles.

Finite-State Machines

To design the control unit for the single-cycle datapath, we used truth tables that specified the setting of the control signals based on the instruction class. For a cache, the control is more complex because the operation can be a series of steps. The control for a cache must specify both the signals to be set in any step and the next step in the sequence.

The most common multistep control method is based on finite-state machines, which are usually represented graphically. A finite-state machine consists of a set of states and directions on how to change states. The directions are defined by a next-state function, which maps the current state and the inputs to a new state. When we use a finite-state machine for control, each state also specifies a set of outputs that are asserted when the machine is in that state. The implementation of a finite-state machine usually assumes that all outputs that are not explicitly asserted are deasserted. Similarly, the correct operation of the datapath depends on the fact that a signal that is not explicitly asserted is deasserted, rather than acting as a don't care.

Multiplexor controls are slightly different, since they select one of the inputs, whether they are 0 or 1. Thus, in the finite-state machine, we always specify the setting of all the multiplexor controls that we care about. When we implement the finite-state machine with logic, setting a control to 0 may be the default and therefore may not require any gates. A simple example of a finite-state machine appears in Appendix A, and if you are unfamiliar with the concept of a finite-state machine, you may want to examine Appendix A before proceeding.

A finite-state machine can be implemented with a temporary register that holds the current state and a block of combinational logic that determines both the data-path signals to be asserted and the next state. Figure 5.38 shows how such an implementation might look. Appendix C describes in detail how the finite-state machine is implemented using this structure. In Section A.3, the combinational control logic for a finite-state machine is implemented both with either a ROM (*read-only memory*) or a PLA (*programmable logic array*). (Also see Appendix A for a description of these logic elements.)

finite-state machine
A sequential logic function consisting of a set of inputs and outputs, a next-state function that maps the current state and the inputs to a new state, and an output function that maps the current state and possibly the inputs to a set of asserted outputs.

next-state function
A combinational function that, given the inputs and the current state, determines the next state of a finite-state machine.

Elaboration: Note that this simple design is called a *blocking* cache, in that the processor must wait until the cache has finished the request. Section 5.12 describes the alternative, which is called a *nonblocking* cache.

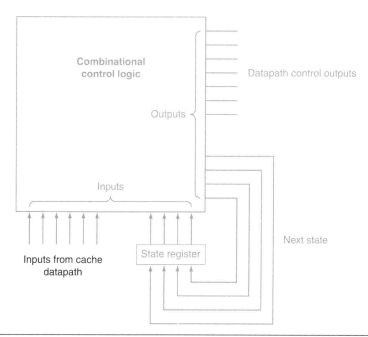

FIGURE 5.38 Finite-state machine controllers are typically implemented using a block of combinational logic and a register to hold the current state. The outputs of the combinational logic are the next-state number and the control signals to be asserted for the current state. The inputs to the combinational logic are the current state and any inputs used to determine the next state. Notice that in the finite-state machine used in this chapter, the outputs depend only on the current state, not on the inputs. We use color to indicate that these are control lines and logic versus data lines and logic. The *Elaboration* below explains this in more detail.

Elaboration: The style of finite-state machine in this book is called a Moore machine, after Edward Moore. Its identifying characteristic is that the output depends only on the current state. For a Moore machine, the box labeled combinational control logic can be split into two pieces. One piece has the control output and only the state input, while the other has just the next-state output.

An alternative style of machine is a Mealy machine, named after George Mealy. The Mealy machine allows both the input and the current state to be used to determine the output. Moore machines have potential implementation advantages in speed and size of the control unit. The speed advantages arise because the control outputs, which are needed early in the clock cycle, do not depend on the inputs, but only on the current state. In Appendix A, when the implementation of this finite-state machine is taken down to logic gates, the size advantage can be clearly seen. The potential disadvantage of a Moore machine is that it may require additional states. For example, in situations where there is a one-state difference between two sequences of states, the Mealy machine may unify the states by making the outputs depend on the inputs.

FSM for a Simple Cache Controller

Figure 5.39 shows the four states of our simple cache controller:

- *Idle:* This state waits for a valid read or write request from the processor, which moves the FSM to the Compare Tag state.

- *Compare Tag:* As the name suggests, this state tests to see if the requested read or write is a hit or a miss. The index portion of the address selects the tag to be compared. If the data in the cache block referred to by the index portion of the address are valid, and the tag portion of the address matches the tag, then it is a hit. Either the data are read from the selected word if it is a load or written to the selected word if it is a store. The Cache Ready signal is then set. If it is a write, the dirty bit is set to 1. Note that a write hit also sets the valid bit and the tag field; while it seems unnecessary, it is included because the tag is a single memory, so to change the dirty bit we likewise need to change the valid and tag fields. If it is a hit and the block is valid, the FSM returns to the idle state. A miss first updates the cache tag and then goes either to the Write-Back state, if the block at this location has dirty bit value of 1, or to the Allocate state if it is 0.

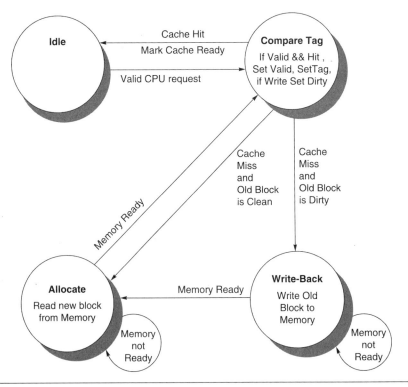

FIGURE 5.39   Four states of the simple controller.

- *Write-Back:* This state writes the 128-bit block to memory using the address composed from the tag and cache index. We remain in this state waiting for the Ready signal from memory. When the memory write is complete, the FSM goes to the Allocate state.

- *Allocate:* The new block is fetched from memory. We remain in this state waiting for the Ready signal from memory. When the memory read is complete, the FSM goes to the Compare Tag state. Although we could have gone to a new state to complete the operation instead of reusing the Compare Tag state, there is a good deal of overlap, including the update of the appropriate word in the block if the access was a write.

This simple model could easily be extended with more states to try to improve performance. For example, the Compare Tag state does both the compare and the read or write of the cache data in a single clock cycle. Often the compare and cache access are done in separate states to try to improve the clock cycle time. Another optimization would be to add a write buffer so that we could save the dirty block and then read the new block first so that the processor doesn't have to wait for two memory accesses on a dirty miss. The cache would next write the dirty block from the write buffer while the processor is operating on the requested data.

Section 5.12 goes into more detail about the FSM, showing the full controller in a hardware description language and a block diagram of this simple cache.

5.10 Parallelism and Memory Hierarchy: Cache Coherence

Given that a multicore multiprocessor means multiple processors on a single chip, these processors very likely share a common physical address space. Caching shared data introduces a new problem, because the view of memory held by two different processors is through their individual caches, which, without any additional precautions, could end up seeing two distinct values. Figure 5.40 illustrates the problem and shows how two different processors can have two different values for the same location. This difficulty is generally referred to as the *cache coherence problem*.

Informally, we could say that a memory system is coherent if any read of a data item returns the most recently written value of that data item. This definition, although intuitively appealing, is vague and simplistic; the reality is much more complex. This simple definition contains two different aspects of memory system behavior, both of which are critical to writing correct shared memory programs. The first aspect, called *coherence*, defines *what values* can be returned by a read. The second aspect, called *consistency*, determines *when* a written value will be returned by a read.

Let's look at coherence first. A memory system is coherent if

1. A read by a processor P to a location X that follows a write by P to X, with no writes of X by another processor occurring between the write and the read by P, always returns the value written by P. Thus, in Figure 5.40, if CPU A were to read X after time step 3, it should see the value 1.

2. A read by a processor to location X that follows a write by another processor to X returns the written value if the read and write are sufficiently separated in time and no other writes to X occur between the two accesses. Thus, in Figure 5.40, we need a mechanism so that the value 0 in the cache of CPU B is replaced by the value 1 after CPU A stores 1 into memory at address X in time step 3.

3. Writes to the same location are *serialized*; that is, two writes to the same location by any two processors are seen in the same order by all processors. For example, if CPU B stores 2 into memory at address X after time step 3, processors can never read the value at location X as 2 and then later read it as 1.

The first property simply preserves program order—we certainly expect this property to be true in uniprocessors, for instance. The second property defines the notion of what it means to have a coherent view of memory: if a processor could continuously read an old data value, we would clearly say that memory was incoherent.

The need for *write serialization* is more subtle, but equally important. Suppose we did not serialize writes, and processor P1 writes location X followed by P2 writing location X. Serializing the writes ensures that every processor will see the write done by P2 at some point. If we did not serialize the writes, it might be the

Time step	Event	Cache contents for CPU A	Cache contents for CPU B	Memory contents for location X
0				0
1	CPU A reads X	0		0
2	CPU B reads X	0	0	0
3	CPU A stores 1 into X	1	0	1

FIGURE 5.40 The cache coherence problem for a single memory location (X), read and written by two processors (A and B). We initially assume that neither cache contains the variable and that X has the value 0. We also assume a write-through cache; a write-back cache adds some additional but similar complications. After the value of X has been written by A, A's cache and the memory both contain the new value, but B's cache does not, and if B reads the value of X, it will receive 0!

case that some processor could see the write of P2 first and then see the write of P1, maintaining the value written by P1 indefinitely. The simplest way to avoid such difficulties is to ensure that all writes to the same location are seen in the identical order, which we call *write serialization*.

Basic Schemes for Enforcing Coherence

In a cache coherent multiprocessor, the caches provide both *migration* and *replication* of shared data items:

- *Migration:* A data item can be moved to a local cache and used there in a transparent fashion. Migration reduces both the latency to access a shared data item that is allocated remotely and the bandwidth demand on the shared memory.

- *Replication:* When shared data are being simultaneously read, the caches make a copy of the data item in the local cache. Replication reduces both latency of access and contention for a read shared data item.

Supporting migration and replication is critical to performance in accessing shared data, so many multiprocessors introduce a hardware protocol to maintain coherent caches. The protocols to maintain coherence for multiple processors are called *cache coherence protocols*. Key to implementing a cache coherence protocol is tracking the state of any sharing of a data block.

The most popular cache coherence protocol is *snooping*. Every cache that has a copy of the data from a block of physical memory also has a copy of the sharing status of the block, but no centralized state is kept. The caches are all accessible via some broadcast medium (a bus or network), and all cache controllers monitor or *snoop* on the medium to determine whether or not they have a copy of a block that is requested on a bus or switch access.

In the following section we explain snooping-based cache coherence as implemented with a shared bus, but any communication medium that broadcasts cache misses to all processors can be used to implement a snooping-based coherence scheme. This broadcasting to all caches makes snooping protocols simple to implement but also limits their scalability.

Snooping Protocols

One method of enforcing coherence is to ensure that a processor has exclusive access to a data item before it writes that item. This style of protocol is called a *write invalidate protocol* because it invalidates copies in other caches on a write. Exclusive access ensures that no other readable or writable copies of an item exist when the write occurs: all other cached copies of the item are invalidated.

Figure 5.41 shows an example of an invalidation protocol for a snooping bus with write-back caches in action. To see how this protocol ensures coherence, consider a write followed by a read by another processor: since the write requires

Processor activity	Bus activity	Contents of CPU A's cache	Contents of CPU B's cache	Contents of memory location X
				0
CPU A reads X	Cache miss for X	0		0
CPU B reads X	Cache miss for X	0	0	0
CPU A writes a 1 to X	Invalidation for X	1		0
CPU B reads X	Cache miss for X	1	1	1

FIGURE 5.41 An example of an invalidation protocol working on a snooping bus for a single cache block (X) with write-back caches. We assume that neither cache initially holds X and that the value of X in memory is 0. The CPU and memory contents show the value after the processor and bus activity have both completed. A blank indicates no activity or no copy cached. When the second miss by B occurs, CPU A responds with the value canceling the response from memory. In addition, both the contents of B's cache and the memory contents of X are updated. This update of memory, which occurs when a block becomes shared, simplifies the protocol, but it is possible to track the ownership and force the write-back only if the block is replaced. This requires the introduction of an additional state called "owner," which indicates that a block may be shared, but the owning processor is responsible for updating any other processors and memory when it changes the block or replaces it.

exclusive access, any copy held by the reading processor must be invalidated (hence the protocol name). Thus, when the read occurs, it misses in the cache, and the cache is forced to fetch a new copy of the data. For a write, we require that the writing processor have exclusive access, preventing any other processor from being able to write simultaneously. If two processors do attempt to write the same data at the same time, one of them wins the race, causing the other processor's copy to be invalidated. For the other processor to complete its write, it must obtain a new copy of the data, which must now contain the updated value. Therefore, this protocol also enforces write serialization.

Hardware/ Software Interface

false sharing When two unrelated shared variables are located in the same cache block and the full block is exchanged between processors even though the processors are accessing different variables.

One insight is that block size plays an important role in cache coherency. For example, take the case of snooping on a cache with a block size of eight words, with a single word alternatively written and read by two processors. Most protocols exchange full blocks between processors, thereby increasing coherency bandwidth demands.

Large blocks can also cause what is called false sharing: when two unrelated shared variables are located in the same cache block, the whole block is exchanged between processors even though the processors are accessing different variables. Programmers and compilers should lay out data carefully to avoid false sharing.

Elaboration: Although the three properties on page 478 are sufficient to ensure coherence, the question of when a written value is seen is also important. To see why, observe that we cannot require that a read of X in Figure 5.40 instantaneously sees the value written for X by some other processor. If, for example, a write of X on one processor precedes a read of X on another processor very shortly beforehand, it may be impossible to ensure that the read returns the value of the data written, since the written data may not even have left the processor at that point. The issue of exactly *when* a written value must be seen by a reader is defined by a *memory consistency model*.

We make the following two assumptions. First, a write does not complete (and allow the next write to occur) until all processors have seen the effect of that write. Second, the processor does not change the order of any write with respect to any other memory access. These two conditions mean that if a processor writes location X followed by location Y, any processor that sees the new value of Y must also see the new value of X. These restrictions allow the processor to reorder reads, but force the processor to finish a write in program order.

Elaboration: Since input can change memory behind the caches, and since output could need the latest value in a write back cache, there is also a cache coherency problem for I/O with the caches of a single processor as well as just between caches of multiple processors. The cache coherence problem for multiprocessors and I/O (see Chapter 6), although similar in origin, has different characteristics that affect the appropriate solution. Unlike I/O, where multiple data copies are a rare event—one to be avoided whenever possible—a program running on multiple processors will normally have copies of the same data in several caches.

Elaboration: In addition to the snooping cache coherence protocol where the status of shared blocks is distributed, a *directory-based* cache coherence protocol keeps the sharing status of a block of physical memory in just one location, called the *directory*. Directory-based coherence has slightly higher implementation overhead than snooping, but it can reduce traffic between caches and thus scale to larger processor counts.

Parallelism and Memory Hierarchy: Redundant Arrays of Inexpensive Disks

This online section describes how using many disks in conjunction can offer much higher throughput, which was the original inspiration of *Redundant Arrays of Inexpensive Disks* (RAID). The real popularity of RAID, however, was due more to the considerably greater dependability offered by including a modest number of redundant disks. The section explains the differences in performance, cost, and **dependability** between the RAID levels.

DEPENDABILITY

Advanced Material: Implementing Cache Controllers

This online section shows how to implement control for a cache, just as we implemented control for the single-cycle and pipelined datapaths in Chapter 4. This section starts with a description of finite-state machines and the implementation of a cache controller for a simple data cache, including a description of the cache controller in a hardware description language. It then goes into details of an example cache coherence protocol and the difficulties in implementing such a protocol.

Real Stuff: The ARM Cortex-A53 and Intel Core i7 Memory Hierarchies

In this section, we will look at the memory hierarchy of the same two microprocessors described in Chapter 4: the ARM Cortex-A53 and Intel Core i7. This section is in part based on Section 2.6 of *Computer Architecture: A Quantitative Approach*, 5th edition.

Figure 5.42 summarizes the address sizes and TLBs of the two processors. Note that the Cortex-A53 has two 10-entry fully associative micro-TLBs backed by a shared 512-entry four-way set associative main TLB with a 48-bit virtual address space and a 40-bit physical address space. The Core i7 has three TLBs with a 48-bit virtual address and a 44-bit physical address. Although the 64-bit registers of these processors could hold a larger virtual address, there was no software need for such a large space, and 48-bit virtual addresses shrinks both the page table memory footprint and the TLB hardware.

Figure 5.43 shows their caches. The Cortex-A53 has between one and four processors or cores while the Core i7 is fixed at four. Cortex-A53 has a 16 to 64 KiB, two-way L1 instruction cache (per core) and the Core i7 has a 32 KiB, four-way set associative, L1 instruction cache (per core). Both use 64 byte blocks. The Cortex-A53 increases the associativity to four-way for the data cache, other variables remain the same. Similarly, the Core i7 keeps everything the same except the associativity, which it increases to eight-way. The Core i7 provides a 256 KiB, eight-way set associative unified L2 cache (per core) with 64 byte blocks. In contrast, the Cortex-A53 provides a L2 cache that is shared between one and four cores. This cache is 16-way set associative with 64 byte blocks and between 128 KiB and 2 MiB in size. As the Core i7 is used for servers, it also offers an L3 cache shared by all the cores on the chip. Its size varies depending on the number of cores. With four cores, as in this case, the size is 8 MiB.

Characteristic	ARM Cortex-A53	Intel Core i7
Virtual address	48 bits	48 bits
Physical address	40 bits	44 bits
Page size	Variable: 4, 16, 64 KiB, 1, 2 MiB, 1 GiB	Variable: 4 KiB, 2/4 MiB
TLB organization	1 TLB for instructions and 1 TLB for data per core Both micro TLBs are fully associative, with 10 entries, round robin replacement 64-entry, four-way set-associative TLBs TLB misses handled in hardware	1 TLB for instructions and 1 TLB for data per core Both L1 TLBs are four-way set associative, LRU replacement L1 I-TLB has 128 entries for small pages, seven per thread for large pages L1 D-TLB has 64 entries for small pages, 32 for large pages The L2 TLB is four-way set associative, LRU replacement The L2 TLB has 512 entries TLB misses handled in hardware

FIGURE 5.42   Address translation and TLB hardware for the ARM Cortex-A53 and Intel Core i7 920. Both processors provide support for large pages, which are used for things like the operating system or mapping a frame buffer. The large-page scheme avoids using a large number of entries to map a single object that is always present.

PARALLELISM

A significant challenge facing cache designers is to support processors like the Cortex-A53 and the Core i7 that can execute more than one memory instruction per clock cycle. A popular technique is to break the cache into banks and allow multiple, independent, **parallel** accesses, provided the accesses are to different banks. The technique is similar to interleaved DRAM banks (see Section 5.2).

The Cortex-A53 and the Core i7 have additional optimizations that allow them to reduce the miss penalty. The first of these is the return of the requested word first on a miss. They also continue to execute instructions that access the data cache during a cache miss. Designers who are attempting to hide the cache miss latency commonly use this technique, called a nonblocking cache. They implement two flavors of nonblocking. *Hit under miss* allows additional cache hits during a miss, while *miss under miss* allows multiple outstanding cache misses. The aim of the first of these two is hiding some miss latency with other work, while the aim of the second is overlapping the latency of two different misses.

Overlapping a large fraction of miss times for multiple outstanding misses requires a high-bandwidth memory system capable of handling multiple misses in parallel. In a personal mobile device, the memory system below it can often pipeline, merge, reorder, or prioritize requests appropriately. Large servers and multiprocessors typically have memory systems capable of handling several outstanding misses in parallel.

nonblocking cache
A cache that allows the processor to make references to the cache while the cache is handling an earlier miss.

Characteristic	ARM Cortex-A53	Intel Core i7
L1 cache organization	Split instruction and data caches	Split instruction and data caches
L1 cache size	Configurable 16 to 64 KiB each for instructions/data	32 KiB each for instructions/data per core
L1 cache associativity	Two-way (I), four-way (D) set associative	Four-way (I), eight-way (D) set associative
L1 replacement	Random	Approximated LRU
L1 block size	64 bytes	64 bytes
L1 write policy	Write-back, variable allocation policies (default is Write-allocate)	Write-back, No-write-allocate
L1 hit time (load-use)	Two clock cycles	Four clock cycles, pipelined
L2 cache organization	Unified (instruction and data)	Unified (instruction and data) per core
L2 cache size	128 KiB to 2 MiB	256 KiB (0.25 MiB)
L2 cache associativity	16-way set associative	8-way set associative
L2 replacement	Approximated LRU	Approximated LRU
L2 block size	64 bytes	64 bytes
L2 write policy	Write-back, Write-allocate	Write-back, Write-allocate
L2 hit time	12 clock cycles	10 clock cycles
L3 cache organization	–	Unified (instruction and data)
L3 cache size	–	8 MiB, shared
L3 cache associativity	–	16-way set associative
L3 replacement	–	Approximated LRU
L3 block size	–	64 bytes
L3 write policy	–	Write-back, Write-allocate
L3 hit time	–	35 clock cycles

FIGURE 5.43   Caches in the ARM Cortex-A53 and Intel Core i7 920.

The Cortex-A53 and the Core i7 have prefetch mechanisms for data accesses. They look at a pattern of data misses and uses this information to try to predict the next address to start fetching the data before the miss occurs. Such techniques generally work best when accessing arrays in loops.

The sophisticated memory hierarchies of these chips and the large fraction of the dies dedicated to caches and TLBs show the significant design effort expended to try to close the gap between processor cycle times and memory latency.

Performance of the Cortex-A53 and Core i7 Memory Hierarchies

The memory hierarchy of the Cortex-A53 was measured using a 32 KiB two-way set associative L1 instruction cache, a 32 KiB four-way set associative L1 data cache, and a 1 MiB 16-way set associative L2 cache running the integer SPEC2006 benchmarks.

The Cortex-A53 instruction cache miss rates for these benchmarks are very small. Figure 5.44 shows the data cache results for the Cortex-A53, which have significant L1 and L2 miss rates. The L1 data cache miss rates go from 0.5% to 37.3%, with a mean of 6.4% and a median of 2.4%. The (global) L2 cache miss rates vary from 0.1% to 9.0%, with a mean of 1.3% and a median of 0.3%. The L1 miss penalty for a 1 GHz Cortex-A53 is 12 clock cycles, while the L2 miss penalty is 124 clock cycles. Using these miss penalties, Figure 5.45 shows the average miss penalty per data access. When these low miss rates are multiplied by their high miss penalties, you can see that they can represent a significant fraction of the CPI for 5 of the 12 SPEC2006 programs.

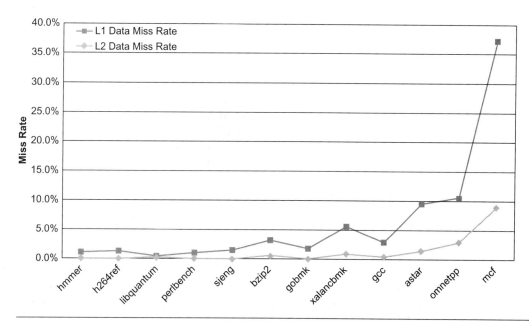

FIGURE 5.44 Data cache miss rates for ARM Cortex-A53 when running SPEC2006int. Applications with larger memory footprints tend to have higher miss rates in both L1 and L2. Note that the L2 rate is the global miss rate; that is, counting all references, including those that hit in L1. (See the *Elaboration* in Section 5.4.) mcf is known as a cache buster. Note that this figure is for the same systems and benchmarks as Figure 4.74 in Chapter 4.

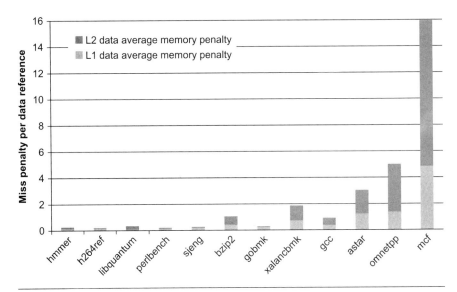

FIGURE 5.45   The average memory access penalty in clock cycles per data memory reference coming from L1 and L2 is shown for the ARM processor when running SPEC2006int. Although the miss rates for L1 are significantly higher, the L2 miss penalty, which is more than five times higher, means that the L2 misses can contribute significantly.

Figure 5.46 shows the miss rates for the caches of the Core i7 using the SPEC2006 benchmarks. The L1 instruction cache miss rate varies from 0.1% to 1.8%, averaging just over 0.4%. This rate is in keeping with other studies of instruction cache behavior for the SPECCPU2006 benchmarks, which show low instruction cache miss rates. With L1 data cache miss rates running 5% to 10%, and sometimes higher, the importance of the L2 and L3 caches should be obvious. Since the cost for a miss to memory is over 100 cycles, and the average data miss rate in L2 is 4%, L3 is obviously critical. Assuming about half the instructions is loads or stores, without L3 the L2 cache misses could add two cycles per instruction to the CPI! In comparison, the average L3 data miss rate of 1% is still significant but four times lower than the L2 miss rate and six times less than the L1 miss rate.

Elaboration: Because speculation may sometimes be wrong (see Chapter 4), there are references to the L1 data cache that do not correspond to loads or stores that eventually complete execution. The data in Figure 5.44 are measured against all data requests, including some that are cancelled. The miss rate when measured against only completed data accesses is 1.6 times higher (an average of 9.5% versus 5.9% for L1 Dcache misses).

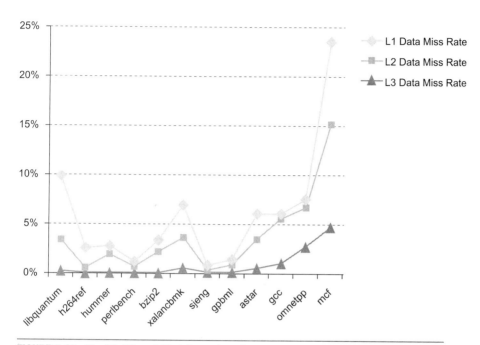

FIGURE 5.46 The L1, L2, and L3 data cache miss rates for the Intel Core i7 920 running the full integer SPECCPU2006 benchmarks.

5.14 Real Stuff: The Rest of the ARMv8 System and Special Instructions

Figure 5.48 lists the 63 remaining ARMv8 instructions in the special purpose and systems category. We'll describe the instructions in the figure top-down, from left to right.

The two "non-cache" or "no allocate" load pair (LDNP) and store pair (STNP) are intended for streaming through lots of data, so the data are unlikely to be used in the future; that is, no temporal locality. These instructions give hints to the memory hierarchy to *not* put the data read from or written to memory into caches, but to transfer data directly between main memory and processor registers. By specifying a pair of registers, large data transfers can proceed more quickly.

The barrier instructions provide synchronization barriers for instructions (ISB) and data (DSB and DMB). The latter two instructions are both barriers that affect data memory access ordering, and differ only in their strictness. The clear exclusive instruction (CLREX) tells the processor to give up exclusive access to a memory location that it requested earlier.

The eight CRC instructions are useful in calculating a cyclic redundancy checksum (CRC-32 or CRC-32C) on bytes, halfwords, words, or double words to help catch errors in large data sets (see the *Elaboration* on page 432 in Section 5.5). Similarly, the 15 Cryptographic instructions use the SIMD registers to accelerate the computation of Advanced Encryption Standard (AES) encryption, Galois/Counter Mode (GCM) encryption, and Secure Hash Algorithm (SHA) encryption.

The two move system register instructions either move data into the processor state registers from a general purpose register or an immediate (MSR) or move data from the processor state registers into a general purpose register (MRS). The 31 instructions above bring us to the halfway point of surveying the special and system instructions of ARMv8.

Starting at the top of the right column of Figure 5.48, the nine "unprivileged" loads and stores of different data widths let a processor at the EL1 interrupt level execute loads and stores that behave as if they were operating at the EL0 level, which means they can have protection faults. They act as normal loads and stores if executed at EL0 or levels higher than EL1.

The exception generation instructions include software breakpoint (BRK), halting software breakpoint (HLT), the supervisor call instructions (SVC), and two instructions similar to SVC except they go to higher exception levels: HVC goes to the hypervisor level (EL2) and SMC goes to the secure monitor level (EL3). Exception return (ERET), naturally enough, allows the program to return from an exception.

The four debugging instructions include three instructions that switch to a higher exception level (DCPS1, DCPS2, DCPS3) and one to restore to the previous processor element state (DRPS).

The six system management instructions include two general purpose ones (SYS and SYSL) and four for managing the memory hierarchy: instruction cache, data cache, address translation, and TLB invalidation.

Finally, the seven hint instructions provide a variety of architectural hints, including wait for interrupt, wait for event, send an event, yield, and NOP. These 32 instructions from the right column of Figure 5.48 bring the total to 63 special and system instructions in the full ARMv8 instruction set.

5.15   Going Faster: Cache Blocking and Matrix Multiply

Our next step in the continuing saga of improving performance of DGEMM by tailoring it to the underlying hardware is to add cache blocking to the subword parallelism and instruction level parallelism optimizations of Chapters 3 and 4. Figure 5.47 shows the blocked version of DGEMM from Figure 4.78. The changes are the same as was made earlier in going from unoptimized DGEMM in Figure 3.22 to blocked DGEMM in Figure 5.21 above. This time we take the unrolled version of DGEMM from Chapter 4 and invoke it many times on the submatrices of A, B, and C. Indeed, lines 28–34 and lines 7–8 in Figure 5.47 mirror lines 14–20 and lines 5–6 in Figure 5.21, except for incrementing the for loop in line 7 by the amount unrolled.

```
 1 #include <x86intrin.h>
 2 #define UNROLL (4)
 3 #define BLOCKSIZE 32
 4 void do_block (int n, int si, int sj, int sk,
 5                  double *A, double *B, double *C)
 6 {
 7   for ( int i = si; i < si+BLOCKSIZE; i+=UNROLL*4 )
 8     for ( int j = sj; j < sj+BLOCKSIZE; j++ ) {
 9       __m256d c[4];
10       for ( int x = 0; x < UNROLL; x++ )
11         c[x] = _mm256_load_pd(C+i+x*4+j*n);
12       /* c[x] = C[i][j] */
13       for( int k = sk; k < sk+BLOCKSIZE; k++ )
14       {
15         __m256d b = _mm256_broadcast_sd(B+k+j*n);
16       /* b = B[k][j] */
17         for (int x = 0; x < UNROLL; x++)
18           c[x] = _mm256_add_pd(c[x], /* c[x]+=A[i][k]*b */
19                 _mm256_mul_pd(_mm256_load_pd(A+n*k+x*4+i), b));
20       }
21
22
23       for ( int x = 0; x < UNROLL; x++ )
24         _mm256_store_pd(C+i+x*4+j*n, c[x]);
25         /* C[i][j] = c[x] */
26     }
27 }
28 void dgemm (int n, double* A, double* B, double* C)
29 {
30   for ( int sj = 0; sj < n; sj += BLOCKSIZE )
31     for ( int si = 0; si < n; si += BLOCKSIZE )
32       for ( int sk = 0; sk < n; sk += BLOCKSIZE )
33         do_block(n, si, sj, sk, A, B, C);
34 }
```

FIGURE 5.47 Optimized C version of DGEMM from Figure 4.78 using cache blocking. These changes are the same ones found in Figure 5.21. The assembly language produced by the compiler for the do_block function is nearly identical to Figure 4.79. Once again, there is no overhead to call the do_block because the compiler inlines the function call.

Unlike the earlier chapters, we do not show the resulting x86 code because the inner loop code is nearly identical to Figure 4.79, as the blocking does not affect the computation, just the order that it accesses data in memory. What does change is the bookkeeping integer instructions to implement the loops. It expands from 14 instructions before the inner loop and eight after the loop for Figure 4.78 to 40 and 28 instructions respectively for the bookkeeping code generated

Type	Mnemonic	Instruction	Type	Mnemonic	Instruction
Non-cache	LDNP	Load Non-temporal Pair	Unprivileged	LDTR	Load Unprivileged register
	STNP	Store Non-temporal Pair		LDTRB	Load Unprivileged byte
Barrier	CLREX	Clear exclusive monitor		LDTRSB	Load Unprivileged signed byte
	DSB	Data synchronization barrier		LDTRH	Load Unprivileged halfword
	DMB	Data memory barrier		LDTRSH	Load Unprivileged signed halfword
	ISB	Instruction synchronization barrier		LDTRSW	Load Unprivileged signed word
CRC	CRC32B	CRC-32 sum from byte		STTR	Store Unprivileged register
	CRC32H	CRC-32 sum from halfword		STTRB	Store Unprivileged byte
	CRC32W	CRC-32 sum from word		STTRH	Store Unprivileged halfword
	CRC32X	CRC-32 sum from doubleword	Exception	BRK	Software breakpoint instruction
	CRC32CB	CRC-32C sum from byte		HLT	Halting software breakpoint instruction
	CRC32CH	CRC-32C sum from halfword		HVC	Generate exception targeting Exception level 2
	CRC32CW	CRC-32C sum from word		SMC	Generate exception targeting Exception level 3
	CRC32CX	CRC-32C sum from doubleword		**SVC**	Generate exception targeting Exception level 1
Crypto	AESD	AES single round decryption		**ERET**	Exception return using current ELR and SPSR
	AESE	AES single round encryption	Debug	DCPS1	Debug switch to Exception level 1
	AESIMC	AES inverse mix columns		DCPS2	Debug switch to Exception level 2
	AESMC	AES mix columns		DCPS3	Debug switch to Exception level 3
	PMULL	Polynomial multiply long		DRPS	Debug restore PE state
	SHA1C	SHA1 hash update (choose)	System	SYS	System instruction
	SHA1H	SHA1 fixed rotate		SYSL	System instruction with result
	SHA1M	SHA1 hash update (majority)		IC	Instruction cache maintenance
	SHA1P	SHA1 hash update (parity)		DC	Data cache maintenance
	SHA1SU0	SHA1 schedule update 0		AT	Address translation
	SHA1SU1	SHA1 schedule update 1		TLBI	TLB Invalidate
	SHA256H	SHA256 hash update (part 1)	Hint	NOP	No operation
	SHA256H2	SHA256 hash update (part 2)		YIELD	Yield hint
	SHA256SU0	SHA256 schedule update 0		WFE	Wait for event
	SHA256SU1	SHA256 schedule update 1		WFI	Wait for interrupt
Sys Reg	MRS	Move system register to general-purpose register		SEV	Send event
				SEVL	Send event local
	MSR	Move general-purpose register or immediate to system register		HINT	Unallocated hint

FIGURE 5.48   The list of assembly language instructions for the systems and special operations in the full ARMv8 instruction set. Bold means the instruction is also in LEGv8.

for Figure 5.47. Nevertheless, the extra instructions executed pale in comparison to the performance improvement of reducing cache misses. Figure 5.49 compares unoptimized to optimized for subword parallelism, instruction level parallelism, and caches. Blocking improves performance over unrolled AVX code by factors of 2 to 2.5 for the larger matrices. When we compare unoptimized code to the code with all three optimizations, the performance improvement is factors of 8 to 15, with the largest increase for the largest matrix.

Elaboration: As mentioned in the *Elaboration* in Section 3.9, these results are with Turbo mode turned off. As in Chapters 3 and 4, when we turn it on, we improve all the results by the temporary increase in the clock rate of $3.3/2.6 = 1.27$. Turbo mode works particularly well in this case because it is using only a single core of an eight-core chip. However, if we want to run fast we should use all cores, which we'll see in Chapter 6.

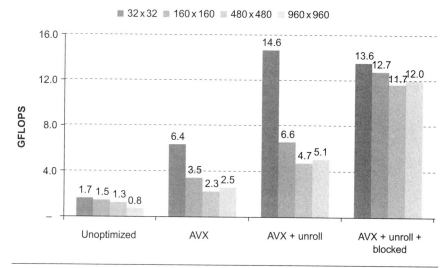

FIGURE 5.49 Performance of four versions of DGEMM from matrix dimensions 32 × 32 to 960 × 960. The fully optimized code for the largest matrix is almost 15 times as fast the unoptimized version in Figure 3.22 in Chapter 3.

5.16 Fallacies and Pitfalls

As one of the most naturally quantitative aspects of computer architecture, the memory hierarchy would seem to be less vulnerable to fallacies and pitfalls. Not only have there been many fallacies propagated and pitfalls encountered, but some have led to major negative outcomes. We start with a pitfall that often traps students in exercises and exams.

Pitfall: Ignoring memory system behavior when writing programs or when generating code in a compiler.

This could be rewritten as a fallacy: "Programmers can ignore memory hierarchies in writing code." The evaluation of sort in Figure 5.19 and of cache blocking in Section 5.14 demonstrate that programmers can easily double performance if they factor the behavior of the memory system into the design of their algorithms.

Pitfall: Forgetting to account for byte addressing or the cache block size in simulating a cache.

When simulating a cache (by hand or by computer), we need to make sure we account for the effect of byte addressing and multiword blocks in determining into which cache block a given address maps. For example, if we have a 32-byte direct-mapped cache with a block size of 4 bytes, the byte address 36 maps into block 1 of the cache, since byte address 36 is block address 9 and (9 modulo 8) = 1. On the other hand, if address 36 is a word address, then it maps into block (36 mod 8) = 4. Make sure the problem clearly states the base of the address.

In like fashion, we must account for the block size. Suppose we have a cache with 256 bytes and a block size of 32 bytes. Into which block does the byte address 300 fall? If we break the address 300 into fields, we can see the answer:

63	62	61	. . .	. . .	. . .	11	10	9	8	7	6	5	4	3	2	1	0
0	0	0	. . .	. . .	. . .	0	0	0	1	0	0	1	0	1	1	0	0

Cache block number (bits 7–5), Block offset (bits 4–0)

Block address

Byte address 300 is block address

$$\left\lceil \frac{300}{32} \right\rceil = 9$$

The number of blocks in the cache is

$$\left\lceil \frac{256}{32} \right\rceil = 8$$

Block number 9 falls into cache block number (9 modulo 8) = 1.

This mistake catches many people, including the authors (in earlier drafts) and instructors who forget whether they intended the addresses to be in doublewords, words, bytes, or block numbers. Remember this pitfall when you tackle the exercises.

Pitfall: Having less set associativity for a shared cache than the number of cores or threads sharing that cache.

Without extra care, a **parallel** program running on 2^n processors or threads can easily allocate data structures to addresses that would map to the same set of a shared L2 cache. If the cache is at least 2^n-way associative, then these accidental conflicts are hidden by the hardware from the program. If not, programmers could face apparently mysterious performance bugs—actually due to L2 conflict misses—when migrating from, say, a 16-core design to 32-core design if both use 16-way associative L2 caches.

PARALLELISM

Pitfall: Using average memory access time to evaluate the memory hierarchy of an out-of-order processor.

If a processor stalls during a cache miss, then you can separately calculate the memory-stall time and the processor execution time, and hence evaluate the memory hierarchy independently using average memory access time (see page 413).

If the processor continues to execute instructions, and may even sustain more cache misses during a cache miss, then the only accurate assessment of the memory hierarchy is to simulate the out-of-order processor along with the memory hierarchy.

Pitfall: Extending an address space by adding segments on top of an unsegmented address space.

During the 1970s, many programs grew so large that not all the code and data could be addressed with just a 16-bit address. Computers were then revised to offer 32-bit addresses, either through an unsegmented 32-bit address space (also called a *flat address space*) or by adding 16 bits of segment to the existing 16-bit address. From a marketing point of view, adding segments that were programmer-visible and that forced the programmer and compiler to decompose programs into segments could solve the addressing problem. Unfortunately, there is trouble any time a programming language wants an address that is larger than one segment, such as indices for large arrays, unrestricted pointers, or reference parameters. Moreover, adding segments can turn every address into two words—one for the segment number and one for the segment offset—causing problems in the use of addresses in registers.

Fallacy: Disk failure rates in the field match their specifications.

Two recent studies evaluated large collections of disks to check the relationship between results in the field compared to specifications. One study was of almost 100,000 disks that had quoted MTTF of 1,000,000 to 1,500,000 hours, or AFR of 0.6% to 0.8%. They found AFRs of 2% to 4% to be common, often three to five times higher than the specified rates [Schroeder and Gibson, 2007]. A second study of more than 100,000 disks at Google, which had a quoted AFR of about 1.5%, saw failure rates of 1.7% for drives in their first year rise to 8.6% for drives in their third year, or about five to six times the declared rate [Pinheiro, Weber, and Barroso, 2007].

Fallacy: Operating systems are the best place to schedule disk accesses.

As mentioned in Section 5.2, higher-level disk interfaces offer logical block addresses to the host operating system. Given this high-level abstraction, the best an OS can do to try to help performance is to sort the logical block addresses into increasing order. However, since the disk knows the actual mapping of the logical addresses onto the physical geometry of sectors, tracks, and surfaces, it can reduce the rotational and seek latencies by rescheduling.

For example, suppose the workload is four reads [Anderson, 2003]:

Operation	Starting LBA	Length
Read	724	8
Read	100	16
Read	9987	1
Read	26	128

The host might reorder the four reads into logical block order:

Operation	Starting LBA	Length
Read	26	128
Read	100	16
Read	724	8
Read	9987	1

Depending on the relative location of the data on the disk, reordering could make it worse, as Figure 5.50 shows. The disk-scheduled reads complete in three-quarters of a disk revolution, but the OS-scheduled reads take three revolutions.

Pitfall: Implementing a virtual machine monitor on an instruction set architecture that wasn't designed to be virtualizable.

Many architects in the 1970s and 1980s weren't careful to make sure that all instructions reading or writing information related to hardware resource information were privileged. This *laissez-faire* attitude causes problems for VMMs for all of these architectures, including the x86, which we use here as an example.

Figure 5.51 describes the 18 instructions that cause problems for virtualization [Robin and Irvine, 2000]. The two broad classes are instructions that

- Read control registers in user mode that reveals that the guest operating system is running in a virtual machine (such as POPF, mentioned earlier)

- Check protection as required by the segmented architecture but assume that the operating system is running at the highest privilege level

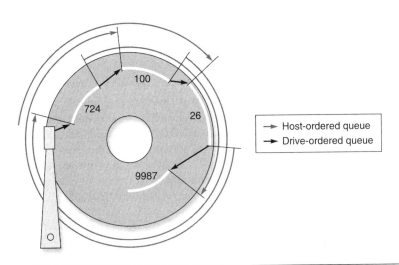

FIGURE 5.50   Example showing OS versus disk schedule accesses, labeled host-ordered versus drive-ordered. The former takes three revolutions to complete the four reads, while the latter completes them in just three-fourths of a revolution. *From Anderson [2003].*

Problem category	Problem x86 instructions
Access sensitive registers without trapping when running in user mode	Store global descriptor table register (SGDT)
	Store local descriptor table register (SLDT)
	Store interrupt descriptor table register (SIDT)
	Store machine status word (SMSW)
	Push flags (PUSHF, PUSHFD)
	Pop flags (POPF, POPFD)
When accessing virtual memory mechanisms in user mode, instructions fail the x86 protection checks	Load access rights from segment descriptor (LAR)
	Load segment limit from segment descriptor (LSL)
	Verify if segment descriptor is readable (VERR)
	Verify if segment descriptor is writable (VERW)
	Pop to segment register (POP CS, POP SS, . . .)
	Push segment register (PUSH CS, PUSH SS, . . .)
	Far call to different privilege level (CALL)
	Far return to different privilege level (RET)
	Far jump to different privilege level (JMP)
	Software interrupt (INT)
	Store segment selector register (STR)
	Move to/from segment registers (MOVE)

FIGURE 5.51 Summary of 18 x86 instructions that cause problems for virtualization [Robin and Irvine, 2000]. The first five instructions in the top group allow a program in user mode to read a control register, such as descriptor table registers, without causing a trap. The pop flags instruction modifies a control register with sensitive information but fails silently when in user mode. The protection checking of the segmented architecture of the x86 is the downfall of the bottom group, as each of these instructions checks the privilege level implicitly as part of instruction execution when reading a control register. The checking assumes that the OS must be at the highest privilege level, which is not the case for guest VMs. Only the Move to segment register tries to modify control state, and protection checking foils it as well.

To simplify implementations of VMMs on the x86, both AMD and Intel have proposed extensions to the architecture via a new mode. Intel's VT-x provides a new execution mode for running VMs, an architected definition of the VM state, instructions to swap VMs rapidly, and a large set of parameters to select the circumstances where a VMM must be invoked. Altogether, VT-x adds 11 new instructions for the x86. AMD's Pacifica makes similar proposals.

An alternative to modifying the hardware is to make small changes to the operating system to avoid using the troublesome pieces of the architecture. This technique is called *paravirtualization*, and the open source Xen VMM is a good example. The Xen VMM provides a guest OS with a virtual machine abstraction that uses only the easy-to-virtualize parts of the physical x86 hardware on which the VMM runs.

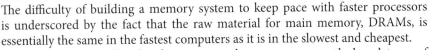

5.17 Concluding Remarks

The difficulty of building a memory system to keep pace with faster processors is underscored by the fact that the raw material for main memory, DRAMs, is essentially the same in the fastest computers as it is in the slowest and cheapest.

It is the principle of locality that gives us a chance to overcome the long latency of memory access—and the soundness of this strategy is demonstrated at all levels of the **memory hierarchy**. Although these levels of the hierarchy look quite different in quantitative terms, they follow similar strategies in their operation and exploit the same properties of locality.

HIERARCHY

Multilevel caches make it possible to use more cache optimizations more easily for two reasons. First, the design parameters of a lower-level cache are different from a first-level cache. For example, because a lower-level cache will be much larger, it is possible to use bigger block sizes. Second, a lower-level cache is not constantly being used by the processor, as a first-level cache is. This allows us to consider having the lower-level cache do something when it is idle that may be useful in preventing future misses.

Another trend is to seek software help. Efficiently managing the memory hierarchy using a variety of program transformations and hardware facilities is a major focus of compiler enhancements. Two different ideas are being explored. One idea is to reorganize the program to enhance its spatial and temporal locality. This approach focuses on loop-oriented programs that use sizable arrays as the major data structure; large linear algebra problems are a typical example, such as DGEMM. By restructuring the loops that access the arrays, substantially improved locality—and, therefore, cache performance—can be obtained.

PREDICTION

prefetching A technique in which data blocks needed in the future are brought into the cache early by using special instructions that specify the address of the block.

Another approach is *prefetching*. In prefetching, a block of data is brought into the cache before it is actually referenced. Many microprocessors use hardware prefetching to try to *predict* accesses that may be difficult for software to notice.

A third approach is special cache-aware instructions that optimize memory transfer. For example, the microprocessors in Section 6.10 in Chapter 6 use an optimization that does not fetch the contents of a block from memory on a write miss because the program is going to write the full block. This optimization significantly reduces memory traffic for one kernel.

As we will see in Chapter 6, memory systems are a central design issue for parallel processors. The growing significance of the memory hierarchy in determining system performance means that this important area will continue to be a focus for both designers and researchers for some years to come.

5.18 Historical Perspective and Further Reading

This section, which appears online, gives an overview of memory technologies, from mercury delay lines to DRAM, the invention of the memory hierarchy, protection mechanisms, and virtual machines, and concludes with a brief history of operating systems, including CTSS, MULTICS, UNIX, BSD UNIX, MS-DOS, Windows, and Linux.

5.19 Exercises

Assume memory is byte addressable and words are 64 bits, unless specified otherwise.

5.1 In this exercise we look at memory locality properties of matrix computation. The following code is written in C, where elements within the same row are stored contiguously. Assume each word is a 64-bit integer.

```
for (I=0; I<8; I++)
  for (J=0; J<8000; J++)
    A[I][J]=B[I][0]+A[J][I];
```

5.1.1 [5] <§5.1> How many 64-bit integers can be stored in a 16-byte cache block?

5.1.2 [5] <§5.1> Which variable references exhibit temporal locality?

5.1.3 [5] <§5.1> Which variable references exhibit spatial locality?

Locality is affected by both the reference order and data layout. The same computation can also be written below in Matlab, which differs from C in that it stores matrix elements within the same column contiguously in memory.

```
for I=1:8
  for J=1:8000
    A(I,J)=B(I,0)+A(J,I);
  end
end
```

5.1.4 [5] <§5.1> Which variable references exhibit temporal locality?

5.1.5 [5] <§5.1> Which variable references exhibit spatial locality?

5.1.6 [15] <§5.1> How many 16-byte cache blocks are needed to store all 64-bit matrix elements being referenced using Matlab's matrix storage? How many using C's matrix storage? (Assume each row contains more than one element.)

5.2 Caches are important to providing a high-performance memory hierarchy to processors. Below is a list of 64-bit memory address references, given as word addresses.

```
0x03, 0xb4, 0x2b, 0x02, 0xbf, 0x58, 0xbe, 0x0e, 0xb5,
0x2c, 0xba, 0xfd
```

5.2.1 [10] <§5.3> For each of these references, identify the binary word address, the tag, and the index given a direct-mapped cache with 16 one-word blocks. Also list whether each reference is a hit or a miss, assuming the cache is initially empty.

5.2.2 [10] <§5.3> For each of these references, identify the binary word address, the tag, the index, and the offset given a direct-mapped cache with two-word blocks and a total size of eight blocks. Also list if each reference is a hit or a miss, assuming the cache is initially empty.

5.2.3 [20] <§§5.3, 5.4> You are asked to optimize a cache design for the given references. There are three direct-mapped cache designs possible, all with a total of eight words of data:

- C1 has 1-word blocks,
- C2 has 2-word blocks, and
- C3 has 4-word blocks.

5.3 By convention, a cache is named according to the amount of data it contains (i.e., a 4 KiB cache can hold 4 KiB of data); however, caches also require SRAM to store metadata such as tags and valid bits. For this exercise, you will examine how a cache's configuration affects the total amount of SRAM needed to implement it as well as the performance of the cache. For all parts, assume that the caches are byte addressable, and that addresses and words are 64 bits.

5.3.1 [10] <§5.3> Calculate the total number of bits required to implement a 32 KiB cache with two-word blocks.

5.3.2 [10] <§5.3> Calculate the total number of bits required to implement a 64 KiB cache with 16-word blocks. How much bigger is this cache than the 32 KiB cache described in Exercise 5.3.1? (Notice that, by changing the block size, we doubled the amount of data without doubling the total size of the cache.)

5.3.3 [5] <§5.3> Explain why this 64 KiB cache, despite its larger data size, might provide slower performance than the first cache.

5.3.4 [10] <§§5.3, 5.4> Generate a series of read requests that have a lower miss rate on a 32 KiB two-way set associative cache than on the cache described in Exercise 5.3.1.

5.4 [15] <§5.3> Section 5.3 shows the typical method to index a direct-mapped cache, specifically (Block address) modulo (Number of blocks in the cache). Assuming a 64-bit address and 1024 blocks in the cache, consider a different indexing function, specifically (Block address[63:54] XOR Block address[53:44]). Is it possible to use this to index a direct-mapped cache? If so, explain why and discuss any changes that might need to be made to the cache. If it is not possible, explain why.

5.5 For a direct-mapped cache design with a 64-bit address, the following bits of the address are used to access the cache.

Tag	Index	Offset
63–10	9–5	4–0

5.5.1 [5] <§5.3> What is the cache block size (in words)?

5.5.2 [5] <§5.3> How many blocks does the cache have?

5.5.3 [5] <§5.3> What is the ratio between total bits required for such a cache implementation over the data storage bits?

Beginning from power on, the following byte-addressed cache references are recorded.

Address											
Hex	00	04	10	84	E8	A0	400	1E	8C	C1C	B4	884
Dec	0	4	16	132	232	160	1024	30	140	3100	180	2180

5.5.4 [20] <§5.3> For each reference, list (1) its tag, index, and offset, (2) whether it is a hit or a miss, and (3) which bytes were replaced (if any).

5.5.5 [5] <§5.3> What is the hit ratio?

5.5.6 [5] <§5.3> List the final state of the cache, with each valid entry represented as a record of <index, tag, data>. For example,

```
<0, 3, Mem[0xC00]-Mem[0xC1F]>
```

5.6 Recall that we have two write policies and two write allocation policies, and their combinations can be implemented either in L1 or L2 cache. Assume the following choices for L1 and L2 caches:

L1	L2
Write through, non-write allocate	Write back, write allocate

5.6.1 [5] <§§5.3, 5.8> Buffers are employed between different levels of memory hierarchy to reduce access latency. For this given configuration, list the possible buffers needed between L1 and L2 caches, as well as L2 cache and memory.

5.6.2 [20] <§§5.3, 5.8> Describe the procedure of handling an L1 write-miss, considering the components involved and the possibility of replacing a dirty block.

5.6.3 [20] <§§5.3, 5.8> For a multilevel exclusive cache configuration (a block can only reside in one of the L1 and L2 caches), describe the procedures of handling an L1 write-miss and an L1 read-miss, considering the components involved and the possibility of replacing a dirty block.

5.7 Consider the following program and cache behaviors.

Data Reads per 1000 Instructions	Data Writes per 1000 Instructions	Instruction Cache Miss Rate	Data Cache Miss Rate	Block Size (bytes)
250	100	0.30%	2%	64

5.7.1 [10] <§§5.3, 5.8> Suppose a CPU with a write-through, write-allocate cache achieves a CPI of 2. What are the read and write bandwidths (measured by bytes per cycle) between RAM and the cache? (Assume each miss generates a request for one block.)

5.7.2 [10] <§§5.3, 5.8> For a write-back, write-allocate cache, assuming 30% of replaced data cache blocks are dirty, what are the read and write bandwidths needed for a CPI of 2?

5.8 Media applications that play audio or video files are part of a class of workloads called "streaming" workloads (i.e., they bring in large amounts of data but do not reuse much of it). Consider a video streaming workload that accesses a 512 KiB working set sequentially with the following word address stream:

```
0, 1, 2, 3, 4, 5, 6, 7, 8, 9 . . .
```

5.8.1 [10] <§§5.4, 5.8> Assume a 64 KiB direct-mapped cache with a 32-byte block. What is the miss rate for the address stream above? How is this miss rate sensitive to the size of the cache or the working set? How would you categorize the misses this workload is experiencing, based on the 3C model?

5.8.2 [5] <§§5.1, 5.8> Re-compute the miss rate when the cache block size is 16 bytes, 64 bytes, and 128 bytes. What kind of locality is this workload exploiting?

5.8.3 [10] <§5.13> "Prefetching" is a technique that leverages predictable address patterns to speculatively bring in additional cache blocks when a particular cache block is accessed. One example of prefetching is a stream buffer that prefetches sequentially adjacent cache blocks into a separate buffer when a particular cache block is brought in. If the data are found in the prefetch buffer, it is considered as a hit, moved into the cache, and the next cache block is prefetched. Assume a two-entry stream buffer; and, assume that the cache latency is such that a cache block can be loaded before the computation on the previous cache block is completed. What is the miss rate for the address stream above?

5.9 Cache block size (B) can affect both miss rate and miss latency. Assuming a machine with a base CPI of 1, and an average of 1.35 references (both instruction and data) per instruction, find the block size that minimizes the total miss latency given the following miss rates for various block sizes.

8: 4%	16: 3%	32: 2%	64: 1.5%	128: 1%

5.9.1 [10] <§5.3> What is the optimal block size for a miss latency of $20 \times B$ cycles?

5.9.2 [10] <§5.3> What is the optimal block size for a miss latency of $24 + B$ cycles?

5.9.3 [10] <§5.3> For constant miss latency, what is the optimal block size?

5.10 In this exercise, we will look at the different ways capacity affects overall performance. In general, cache access time is proportional to capacity. Assume that main memory accesses take 70 ns and that 36% of all instructions access data memory. The following table shows data for L1 caches attached to each of two processors, P1 and P2.

	L1 Size	L1 Miss Rate	L1 Hit Time
P1	2 KiB	8.0%	0.66 ns
P2	4 KiB	6.0%	0.90 ns

5.10.1 [5] <§5.4> Assuming that the L1 hit time determines the cycle times for P1 and P2, what are their respective clock rates?

5.10.2 [10] <§5.4> What is the Average Memory Access Time for P1 and P2 (in cycles)?

5.10.3 [5] <§5.4> Assuming a base CPI of 1.0 without any memory stalls, what is the total CPI for P1 and P2? Which processor is faster? (When we say a "base CPI of 1.0", we mean that instructions complete in one cycle, unless either the instruction access or the data access causes a cache miss.)

For the next three problems, we will consider the addition of an L2 cache to P1 (to presumably make up for its limited L1 cache capacity). Use the L1 cache capacities and hit times from the previous table when solving these problems. The L2 miss rate indicated is its local miss rate.

L2 Size	L2 Miss Rate	L2 Hit Time
1 MiB	95%	5.62 ns

5.10.4 [10] <§5.4> What is the AMAT for P1 with the addition of an L2 cache? Is the AMAT better or worse with the L2 cache?

5.10.5 [5] <§5.4> Assuming a base CPI of 1.0 without any memory stalls, what is the total CPI for P1 with the addition of an L2 cache?

5.10.6 [10] <§5.4> What would the L2 miss rate need to be in order for P1 with an L2 cache to be faster than P1 without an L2 cache?

5.10.7 [15] <§5.4> What would the L2 miss rate need to be in order for P1 with an L2 cache to be faster than P2 without an L2 cache?

5.11 This exercise examines the effect of different cache designs, specifically comparing associative caches to the direct-mapped caches from Section 5.4. For these exercises, refer to the sequence of word address shown below.

    0x03, 0xb4, 0x2b, 0x02, 0xbe, 0x58, 0xbf, 0x0e, 0x1f,
    0xb5, 0xbf, 0xba, 0x2e, 0xce

5.11.1 [10] <§5.4> Sketch the organization of a three-way set associative cache with two-word blocks and a total size of 48 words. Your sketch should have a style similar to Figure 5.18, but clearly show the width of the tag and data fields.

5.11.2 [10] <§5.4> Trace the behavior of the cache from Exercise 5.11.1. Assume a true LRU replacement policy. For each reference, identify

- the binary word address,

- the tag,

- the index,

- the offset

- whether the reference is a hit or a miss, and

- which tags are in each way of the cache after the reference has been handled.

5.11.3 [5] <§5.4> Sketch the organization of a fully associative cache with one-word blocks and a total size of eight words. Your sketch should have a style similar to Figure 5.18, but clearly show the width of the tag and data fields.

5.11.4 [10] <§5.4> Trace the behavior of the cache from Exercise 5.11.3. Assume a true LRU replacement policy. For each reference, identify

- the binary word address,

- the tag,

- the index,

- the offset

- whether the reference is a hit or a miss, and

■ the contents of the cache after each reference has been handled.

5.11.5 [5] <§5.4> Sketch the organization of a fully associative cache with two-word blocks and a total size of eight words. Your sketch should have a style similar to Figure 5.18, but clearly show the width of the tag and data fields.

5.11.6 [10] <§5.4> Trace the behavior of the cache from Exercise 5.11.5. Assume an LRU replacement policy. For each reference, identify

■ the binary word address,

■ the tag,

■ the index,

■ the offset

■ whether the reference is a hit or a miss, and

■ the contents of the cache after each reference has been handled.

5.11.7 [10] <§5.4> Repeat Exercise 5.11.6 using MRU (*most recently used*) replacement.

5.11.8 [15] <§5.4> Repeat Exercise 5.11.6 using the optimal replacement policy (i.e., the one that gives the lowest miss rate).

5.12 Multilevel caching is an important technique to overcome the limited amount of space that a first-level cache can provide while still maintaining its speed. Consider a processor with the following parameters:

Base CPI, No Memory Stalls	Processor Speed	Main Memory Access Time	First-Level Cache Miss Rate per Instruction**	Second-Level Cache, Direct-Mapped Speed	Miss Rate with Second-Level Cache, Direct-Mapped	Second-Level Cache, Eight-Way Set Associative Speed	Miss Rate with Second-Level Cache, Eight-Way Set Associative
1.5	2 GHz	100 ns	7%	12 cycles	3.5%	28 cycles	1.5%

**First Level Cache miss rate is per instruction. Assume the total number of L1 cache misses (instruction and data combined) is equal to 7% of the number of instructions.

5.12.1 [10] <§5.4> Calculate the CPI for the processor in the table using: 1) only a first-level cache, 2) a second-level direct-mapped cache, and 3) a second-level eight-way set associative cache. How do these numbers change if main memory access time doubles? (Give each change as both an absolute CPI and a percent change.) Notice the extent to which an L2 cache can hide the effects of a slow memory.

5.12.2 [10] <§5.4> It is possible to have an even greater cache hierarchy than two levels? Given the processor above with a second-level, direct-mapped cache, a designer wants to add a third-level cache that takes 50 cycles to access and will have a 13% miss rate. Would this provide better performance? In general, what are the advantages and disadvantages of adding a third-level cache?

5.12.3 [20] <§5.4> In older processors, such as the Intel Pentium or Alpha 21264, the second level of cache was external (located on a different chip) from the main processor and the first-level cache. While this allowed for large second-level caches, the latency to access the cache was much higher, and the bandwidth was typically lower because the second-level cache ran at a lower frequency. Assume a 512 KiB off-chip second-level cache has a miss rate of 4%. If each additional 512 KiB of cache lowered miss rates by 0.7%, and the cache had a total access time of 50 cycles, how big would the cache have to be to match the performance of the second-level direct-mapped cache listed above?

5.13 *Mean time between failures* (MTBF), *mean time to replacement* (MTTR), and *mean time to failure* (MTTF) are useful metrics for evaluating the reliability and availability of a storage resource. Explore these concepts by answering the questions about a device with the following metrics:

MTTF	MTTR
3 Years	1 Day

5.13.1 [5] <§5.5> Calculate the MTBF for such a device.

5.13.2 [5] <§5.5> Calculate the availability for such a device.

5.13.3 [5] <§5.5> What happens to availability as the MTTR approaches 0? Is this a realistic situation?

5.13.4 [5] <§5.5> What happens to availability as the MTTR gets very high, i.e., a device is difficult to repair? Does this imply the device has low availability?

5.14 This exercise examines the *single error correcting, double error detecting* (SEC/DED) Hamming code.

5.14.1 [5] <§5.5> What is the minimum number of parity bits required to protect a 128-bit word using the SEC/DED code?

5.14.2 [5] <§5.5> Section 5.5 states that modern server memory modules (DIMMs) employ SEC/DED ECC to protect each 64 bits with 8 parity bits. Compute the cost/performance ratio of this code to the code from Exercise 5.14.1. In this case, cost is the relative number of parity bits needed while performance is the relative number of errors that can be corrected. Which is better?

5.14.3 [5] <§5.5> Consider a SEC code that protects 8 bit words with 4 parity bits. If we read the value 0x375, is there an error? If so, correct the error.

5.15 For a high-performance system such as a B-tree index for a database, the page size is determined mainly by the data size and disk performance. Assume that, on average, a B-tree index page is 70% full with fix-sized entries. The utility of a page is its B-tree depth, calculated as $\log_2$(entries). The following table shows that for 16-byte entries, and a 10-year-old disk with a 10 ms latency and 10 MB/s transfer rate, the optimal page size is 16 K.

Page Size (KiB)	Page Utility or B-Tree Depth (Number of Disk Accesses Saved)	Index Page Access Cost (ms)	Utility/Cost
2	6.49 (or $\log_2$(2048/16 × 0.7))	10.2	0.64
4	7.49	10.4	0.72
8	8.49	10.8	0.79
16	9.49	11.6	0.82
32	10.49	13.2	0.79
64	11.49	16.4	0.70
128	12.49	22.8	0.55
256	13.49	35.6	0.38

5.15.1 [10] <§5.7> What is the best page size if entries now become 128 bytes?

5.15.2 [10] <§5.7> Based on Exercise 5.15.1, what is the best page size if pages are half full?

5.15.3 [20] <§5.7> Based on Exercise 5.15.2, what is the best page size if using a modern disk with a 3 ms latency and 100 MB/s transfer rate? Explain why future servers are likely to have larger pages.

Keeping "frequently used" (or "hot") pages in DRAM can save disk accesses, but how do we determine the exact meaning of "frequently used" for a given system? Data engineers use the cost ratio between DRAM and disk access to quantify the reuse time threshold for hot pages. The cost of a disk access is $Disk/accesses_per_sec, while the cost to keep a page in DRAM is $DRAM_MiB/page_size. The typical DRAM and disk costs and typical database page sizes at several time points are listed below:

Year	DRAM Cost ($/MiB)	Page Size (KiB)	Disk Cost ($/disk)	Disk Access Rate (access/sec)
1987	5000	1	15,000	15
1997	15	8	2000	64
2007	0.05	64	80	83

5.15.4 [20] <§5.7> What other factors can be changed to keep using the same page size (thus avoiding software rewrite)? Discuss their likeliness with current technology and cost trends.

5.16 As described in Section 5.7, virtual memory uses a page table to track the mapping of virtual addresses to physical addresses. This exercise shows how this table must be updated as addresses are accessed. The following data constitute a stream of virtual byte addresses as seen on a system. Assume 4 KiB pages, a four-entry fully associative TLB, and true LRU replacement. If pages must be brought in from disk, increment the next largest page number.

Decimal	4669	2227	13916	34587	48870	12608	49225
hex	0x123d	0x08b3	0x365c	0x871b	0xbee6	0x3140	0xc049

TLB

Valid	Tag	Physical Page Number	Time Since Last Access
1	0xb	12	4
1	0x7	4	1
1	0x3	6	3
0	0x4	9	7

Page table

Index	Valid	Physical Page or in Disk
0	1	5
1	0	Disk
2	0	Disk
3	1	6
4	1	9
5	1	11
6	0	Disk
7	1	4
8	0	Disk
9	0	Disk
a	1	3
b	1	12

5.16.1 [10] <§5.7> For each access shown above, list

■ whether the access is a hit or miss in the TLB,

■ whether the access is a hit or miss in the page table,

■ whether the access is a page fault,

■ the updated state of the TLB.

5.16.2 [15] <§5.7> Repeat Exercise 5.16.1, but this time use 16 KiB pages instead of 4 KiB pages. What would be some of the advantages of having a larger page size? What are some of the disadvantages?

5.16.3 [15] <§5.7> Repeat Exercise 5.16.1, but this time use 4 KiB pages and a two-way set associative TLB.

5.16.4 [15] <§5.7> Repeat Exercise 5.16.1, but this time use 4 KiB pages and a direct mapped TLB.

5.16.5 [10] <§§5.4, 5.7> Discuss why a CPU must have a TLB for high performance. How would virtual memory accesses be handled if there were no TLB?

5.17 There are several parameters that affect the overall size of the page table. Listed below are key page table parameters.

Virtual Address Size	Page Size	Page Table Entry Size
32 bits	8 KiB	4 bytes

5.17.1 [5] <§5.7> Given the parameters shown above, calculate the maximum possible page table size for a system running five processes.

5.17.2 [10] <§5.7> Given the parameters shown above, calculate the total page table size for a system running five applications that each utilize half of the virtual memory available, given a two-level page table approach with up to 256 entries at the 1st level. Assume each entry of the main page table is 6 bytes. Calculate the minimum and maximum amount of memory required for this page table.

5.17.3 [10] <§5.7> A cache designer wants to increase the size of a 4 KiB virtually indexed, physically tagged cache. Given the page size shown above, is it possible to make a 16 KiB direct-mapped cache, assuming two 64-bit words per block? How would the designer increase the data size of the cache?

5.18 In this exercise, we will examine space/time optimizations for page tables. The following list provides parameters of a virtual memory system.

Virtual Address (bits)	Physical DRAM Installed	Page Size	PTE Size (byte)
43	16 GiB	4 KiB	4

5.18.1 [10] <§5.7> For a single-level page table, how many *page table entries* (PTEs) are needed? How much physical memory is needed for storing the page table?

5.18.2 [10] <§5.7> Using a multi-level page table can reduce the physical memory consumption of page tables by only keeping active PTEs in physical memory. How many levels of page tables will be needed if the segment tables (the upper-level page tables) are allowed to be of unlimited size? How many memory references are needed for address translation if missing in TLB?

5.18.3 [10] <§5.7> Suppose the segments are limited to the 4 KiB page size (so that they can be paged). Is 4 bytes large enough for all page table entries (including those in the segment tables?

5.18.4 [10] <§5.7> How many levels of page tables are needed if the segments are limited to the 4 KiB page size?

5.18.5 [15] <§5.7> An inverted page table can be used to further optimize space and time. How many PTEs are needed to store the page table? Assuming a hash table implementation, what are the common case and worst case numbers of memory references needed for servicing a TLB miss?

5.19 The following table shows the contents of a four-entry TLB.

Entry-ID	Valid	VA Page	Modified	Protection	PA Page
1	1	140	1	RW	30
2	0	40	0	RX	34
3	1	200	1	RO	32
4	1	280	0	RW	31

5.19.1 [5] <§5.7> Under what scenarios would entry 3's valid bit be set to zero?

5.19.2 [5] <§5.7> What happens when an instruction writes to VA page 30? When would a software managed TLB be faster than a hardware managed TLB?

5.19.3 [5] <§5.7> What happens when an instruction writes to VA page 200?

5.20 In this exercise, we will examine how replacement policies affect miss rate. Assume a two-way set associative cache with four one-word blocks. Consider the following word address sequence: 0, 1, 2, 3, 4, 2, 3, 4, 5, 6, 7, 0, 1, 2, 3, 4, 5, 6, 7, 0.

Consider the following address sequence: 0, 2, 4, 8, 10, 12, 14, 16, 0

5.20.1 [5] <§§5.4, 5.8> Assuming an LRU replacement policy, which accesses are hits?

5.20.2 [5] <§§5.4, 5.8> Assuming an MRU (*most recently used*) replacement policy, which accesses are hits?

5.20.3 [5] <§§5.4, 5.8> Simulate a random replacement policy by flipping a coin. For example, "heads" means to evict the first block in a set and "tails" means to evict the second block in a set. How many hits does this address sequence exhibit?

5.20.4 [10] <§§5.4, 5.8> Describe an optimal replacement policy for this sequence. Which accesses are hits using this policy?

5.20.5 [10] <§§5.4, 5.8> Describe why it is difficult to implement a cache replacement policy that is optimal for all address sequences.

5.20.6 [10] <§§5.4, 5.8> Assume you could make a decision upon each memory reference whether or not you want the requested address to be cached. What effect could this have on miss rate?

5.21 One of the biggest impediments to widespread use of virtual machines is the performance overhead incurred by running a virtual machine. Listed below are various performance parameters and application behavior.

Base CPI	Priviliged O/S Accesses per 10,000 Instructions	Overhead to Trap to the Guest O/S	Overhead to Trap to VMM	I/O Access per 10,000 Instructions	I/O Access Time (Includes Time to Trap to Guest O/S)
1.5	120	15 cycles	175 cycles	30	1100 cycles

5.21.1 [10] <§5.6> Calculate the CPI for the system listed above assuming that there are no accesses to I/O. What is the CPI if the VMM overhead doubles? If it is cut in half? If a virtual machine software company wishes to limit the performance degradation to 10%, what is the longest possible penalty to trap to the VMM?

5.21.2 [15] <§5.6> I/O accesses often have a large effect on overall system performance. Calculate the CPI of a machine using the performance characteristics above, assuming a non-virtualized system. Calculate the CPI again, this time using a virtualized system. How do these CPIs change if the system has half the I/O accesses?

5.22 [15] <§§5.6, 5.7> Compare and contrast the ideas of virtual memory and virtual machines. How do the goals of each compare? What are the pros and cons of each? List a few cases where virtual memory is desired, and a few cases where virtual machines are desired.

5.23 [10] <§5.6> Section 5.6 discusses virtualization under the assumption that the virtualized system is running the same ISA as the underlying hardware. However, one possible use of virtualization is to emulate non-native ISAs. An example of this is QEMU, which emulates a variety of ISAs such as MIPS, SPARC, and PowerPC. What are some of the difficulties involved in this kind of virtualization? Is it possible for an emulated system to run faster than on its native ISA?

5.24 In this exercise, we will explore the control unit for a cache controller for a processor with a write buffer. Use the finite state machine found in Figure 5.39 as a starting point for designing your own finite state machines. Assume that the cache controller is for the simple direct-mapped cache described on page 476 (Figure 5.39 in Section 5.9), but you will add a write buffer with a capacity of one block.

Recall that the purpose of a write buffer is to serve as temporary storage so that the processor doesn't have to wait for two memory accesses on a dirty miss. Rather than writing back the dirty block before reading the new block, it buffers the dirty block and immediately begins reading the new block. The dirty block can then be written to main memory while the processor is working.

5.24.1 [10] <§§5.8, 5.9> What should happen if the processor issues a request that *hits* in the cache while a block is being written back to main memory from the write buffer?

5.24.2 [10] <§§5.8, 5.9> What should happen if the processor issues a request that *misses* in the cache while a block is being written back to main memory from the write buffer?

5.24.3 [30] <§§5.8, 5.9> Design a finite state machine to enable the use of a write buffer.

5.25 Cache coherence concerns the views of multiple processors on a given cache block. The following data show two processors and their read/write operations on two different words of a cache block X (initially X[0] = X[1] = 0).

P1	P2
X[0] ++; X[1] = 3;	X[0] = 5; X[1] +=2;

5.25.1 [15] <§5.10> List the possible values of the given cache block for a correct cache coherence protocol implementation. List at least one more possible value of the block if the protocol doesn't ensure cache coherency.

5.25.2 [15] <§5.10> For a snooping protocol, list a valid operation sequence on each processor/cache to finish the above read/write operations.

5.25.3 [10] <§5.10> What are the best-case and worst-case numbers of cache misses needed to execute the listed read/write instructions?

Memory consistency concerns the views of multiple data items. The following data show two processors and their read/write operations on different cache blocks (A and B initially 0).

P1	P2
A = 1; B = 2; A+=2; B++;	C = B; D = A;

5.25.4 [15] <§5.10> List the possible values of C and D for all implementations that ensure both consistency assumptions on page 481.

5.25.5 [15] <§5.10> List at least one more possible pair of values for C and D if such assumptions are not maintained.

5.25.6 [15] <§§5.3, 5.10> For various combinations of write policies and write allocation policies, which combinations make the protocol implementation simpler?

5.26 Chip multiprocessors (CMPs) have multiple cores and their caches on a single chip. CMP on-chip L2 cache design has interesting trade-offs. The following table shows the miss rates and hit latencies for two benchmarks with private vs. shared L2 cache designs. Assume the L1 cache has a 3% miss rate and a 1-cycle access time.

	Private	Shared
Benchmark A miss rate	10%	4%
Benchmark B miss rate	2%	1%

Assume the following hit latencies:

Private Cache	Shared Cache	Memory
5	20	180

5.26.1 [15] <§5.13> Which cache design is better for each of these benchmarks? Use data to support your conclusion.

5.26.2 [15] <§5.13> Off-chip bandwidth becomes the bottleneck as the number of CMP cores increases. How does this bottleneck affect private and shared cache systems differently? Choose the best design if the latency of the first off-chip link doubles.

5.26.3 [10] <§5.13> Discuss the pros and cons of shared vs. private L2 caches for both single-threaded, multi-threaded, and multiprogrammed workloads, and reconsider them if having on-chip L3 caches.

5.26.4 [10] <§5.13> Would a non-blocking L2 cache produce more improvement on a CMP with a shared L2 cache or a private L2 cache? Why?

5.26.5 [10] <§5.13> Assume new generations of processors double the number of cores every 18 months. To maintain the same level of per-core performance, how much more off-chip memory bandwidth is needed for a processor released in three years?

5.26.6 [15] <§5.13> Consider the entire memory hierarchy. What kinds of optimizations can improve the number of concurrent misses?

5.27 In this exercise we show the definition of a web server log and examine code optimizations to improve log processing speed. The data structure for the log is defined as follows:

```
struct entry {
int srcIP; // remote IP address
char URL[128]; // request URL (e.g., "GET index.html")
long long refTime; // reference time
int status; // connection status
char browser[64]; // client browser name
} log [NUM_ENTRIES];
```

Assume the following processing function for the log:

```
topK_sourceIP (int hour);
```

This function determines the most frequently observed source IPs during the given hour.

5.27.1 [5] <§5.15> Which fields in a log entry will be accessed for the given log processing function? Assuming 64-byte cache blocks and no prefetching, how many cache misses per entry does the given function incur on average?

5.27.2 [5] <§5.15> How can you reorganize the data structure to improve cache utilization and access locality?

5.27.3 [10] <§5.15> Give an example of another log processing function that would prefer a different data structure layout. If both functions are important, how would you rewrite the program to improve the overall performance? Supplement the discussion with code snippet and data.

5.28 For the problems below, use data from "Cache Performance for SPEC CPU2000 Benchmarks" (http://www.cs.wisc.edu/multifacet/misc/spec2000cache-data/) for the pairs of benchmarks shown in the following table.

a.	Mesa/gcc
b.	mcf/swim

5.28.1 [10] <§5.15> For 64 KiB data caches with varying set associativities, what are the miss rates broken down by miss types (cold, capacity, and conflict misses) for each benchmark?

5.28.2 [10] <§5.15> Select the set associativity to be used by a 64 KiB L1 data cache shared by both benchmarks. If the L1 cache has to be directly mapped, select the set associativity for the 1 MiB L2 cache.

5.28.3 [20] <§5.15> Give an example in the miss rate table where higher set associativity actually increases miss rate. Construct a cache configuration and reference stream to demonstrate this.

5.29 To support multiple virtual machines, two levels of memory virtualization are needed. Each virtual machine still controls the mapping of *virtual address* (VA) to *physical address* (PA), while the hypervisor maps the *physical address* (PA) of each virtual machine to the actual *machine address* (MA). To accelerate such mappings, a software approach called "shadow paging" duplicates each virtual machine's page tables in the hypervisor, and intercepts VA to PA mapping changes to keep both copies consistent. To remove the complexity of shadow page tables, a hardware approach called *nested page table* (NPT) explicitly supports two classes of page tables (VA⇒PA and PA⇒MA) and can walk such tables purely in hardware.

Consider the following sequence of operations: (1) Create process; (2) TLB miss; (3) page fault; (4) context switch;

5.29.1 [10] <§§5.6, 5.7> What would happen for the given operation sequence for shadow page table and nested page table, respectively?

5.29.2 [10] <§§5.6, 5.7> Assuming an x86-based four-level page table in both guest and nested page table, how many memory references are needed to service a TLB miss for native vs. nested page table?

5.29.3 [15] <§§5.6, 5.7> Among TLB miss rate, TLB miss latency, page fault rate, and page fault handler latency, which metrics are more important for shadow page table? Which are important for nested page table?

Assume the following parameters for a shadow paging system.

TLB Misses per 1000 Instructions	NPT TLB Miss Latency	Page Faults per 1000 Instructions	Shadowing Page Fault Overhead
0.2	200 cycles	0.001	30,000 cycles

5.29.4 [10] <§5.6> For a benchmark with native execution CPI of 1, what are the CPI numbers if using shadow page tables vs. NPT (assuming only page table virtualization overhead)?

5.29.5 [10] <§5.6> What techniques can be used to reduce page table shadowing induced overhead?

5.29.6 [10] <§5.6> What techniques can be used to reduce NPT induced overhead?

§5.1, page 391: 1 and 4. (3 is false because the cost of the memory hierarchy varies per computer, but in 2016 the highest cost is usually the DRAM.)
§5.3, page 412: 1 and 4: A lower miss penalty can enable smaller blocks, since you don't have that much latency to amortize, yet higher memory bandwidth usually leads to larger blocks, since the miss penalty is only slightly larger.
§5.4, page 431: 1.
§5.7, page 465: 1-a, 2-c, 3-b, 4-d.
§5.8, page 472: 2. (Both large block sizes and prefetching may reduce compulsory misses, so 1 is false.)
§5.11, page 5.11-8: All are true.

Answers to Check Yourself

6

Parallel Processors from Client to Cloud

"I swing big, with everything I've got. I hit big or I miss big. I like to live as big as I can."

Babe Ruth
American baseball player

Multiprocessor or Cluster Organization

Computer

Computer

Network

Computer

Computer

6.1   Introduction

multiprocessor
A computer system with at least two processors. This computer is in contrast to a uniprocessor, which has one, and is increasingly hard to find today.

PARALLELISM

task-level parallelism or **process-level parallelism** Utilizing multiple processors by running independent programs simultaneously.

parallel processing program A single program that runs on multiple processors simultaneously.

cluster A set of computers connected over a local area network that function as a single large multiprocessor.

Computer architects have long sought the "The City of Gold" (El Dorado) of computer design: to create powerful computers simply by connecting many existing smaller ones. This golden vision is the fountainhead of multiprocessors. Ideally, customers order as many processors as they can afford and receive a commensurate amount of performance. Thus, multiprocessor software must be designed to work with a variable number of processors. As mentioned in Chapter 1, energy has become the overriding issue for both microprocessors and datacenters. Replacing large inefficient processors with many smaller, efficient processors can deliver better performance per joule both in the large and in the small, if software can efficiently use them. Therefore, improved energy efficiency joins scalable performance in the case for multiprocessors.

Since multiprocessor software should scale, some designs support operation in the presence of broken hardware; that is, if a single processor fails in a multiprocessor with *n* processors, these systems would continue to provide service with *n* – 1 processors. Hence, multiprocessors can also improve availability (see Chapter 5).

High performance can mean greater throughput for independent tasks, called task-level parallelism or process-level parallelism. These tasks are independent single-threaded applications, and they are an important and popular use of multiple processors. This approach contrasts with running a single job on multiple processors. We use the term parallel processing program to refer to a single program that runs on multiple processors simultaneously.

There have long been scientific problems that have needed much faster computers, and this class of problems has been used to justify many novel parallel computers over the decades. Some of these problems can be handled simply today, using a cluster composed of microprocessors housed in many independent servers (see Section 6.7). In addition, clusters can serve equally demanding applications outside the sciences, such as search engines, Web servers, email servers, and databases.

As described in Chapter 1, multiprocessors have been shoved into the spotlight because the energy problem means that future increases in performance will primarily come from explicit hardware parallelism rather than much higher clock rates or vastly improved CPI. As we said in Chapter 1, they are called multicore microprocessors instead of multiprocessor microprocessors, presumably to avoid redundancy in naming. Hence, processors are often called *cores* in a multicore chip. The number of cores is expected to increase with **Moore's Law**. These multicores are almost always Shared Memory Processors (SMPs), as they usually share a single physical address space. We'll see SMPs more in Section 6.5.

The state of technology today means that programmers who care about performance must become parallel programmers, for sequential code now means slow code.

The tall challenge facing the industry is to create hardware and software that will make it easy to write correct parallel processing programs that will execute efficiently in performance and energy as the number of cores per chip scales.

This abrupt shift in microprocessor design caught many off guard, so there is a great deal of confusion about the terminology and what it means. Figure 6.1 tries to clarify the terms serial, parallel, sequential, and concurrent. The columns of this figure represent the software, which is either inherently sequential or concurrent. The rows of the figure represent the hardware, which is either serial or parallel. For example, the programmers of compilers think of them as sequential programs: the steps include parsing, code generation, optimization, and so on. In contrast, the programmers of operating systems normally think of them as concurrent programs: cooperating processes handling I/O events due to independent jobs running on a computer.

The point of these two axes of Figure 6.1 is that concurrent software can run on serial hardware, such as operating systems for the Intel Pentium 4 uniprocessor, or on parallel hardware, such as an OS on the more recent Intel Core i7. The same is true for sequential software. For example, the MATLAB programmer writes a matrix multiply thinking about it sequentially, but it could run serially on the Pentium 4 or in parallel on the Intel Core i7.

You might guess that the only challenge of the parallel revolution is figuring out how to make naturally sequential software have high performance on parallel hardware, but it is also to make concurrent programs have high performance on multiprocessors as the number of processors increases. With this distinction made, in the rest of this chapter we will use *parallel processing program* or *parallel software* to mean either sequential or concurrent software running on parallel hardware. The next section of this chapter describes why it is hard to create efficient parallel processing programs.

multicore microprocessor A microprocessor containing multiple processors ("cores") in a single integrated circuit. Virtually all microprocessors today in desktops and servers are multicore.

shared memory multiprocessor (SMP) A parallel processor with a single physical address space.

MOORE'S LAW

Software		
	Sequential	**Concurrent**
Hardware Serial	Matrix Multiply written in MatLab running on an Intel Pentium 4	Windows Vista Operating System running on an Intel Pentium 4
Parallel	Matrix Multiply written in MATLAB running on an Intel Core i7	Windows Vista Operating System running on an Intel Core i7

FIGURE 6.1 Hardware/software categorization and examples of application perspective on concurrency versus hardware perspective on parallelism.

Before proceeding further down the path to parallelism, don't forget our initial incursions from the earlier chapters:

- Chapter 2, Section 2.11: Parallelism and Instructions: Synchronization

- Chapter 3, Section 3.6: Parallelism and Computer Arithmetic: Subword Parallelism

- Chapter 4, Section 4.10: Parallelism via Instructions

- Chapter 5, Section 5.10: Parallelism and Memory Hierarchy: Cache Coherence

Check Yourself

True or false: To benefit from a multiprocessor, an application must be concurrent.

6.2   The Difficulty of Creating Parallel Processing Programs

The difficulty with parallelism is not the hardware; it is that too few important application programs have been rewritten to complete tasks sooner on multiprocessors. It is difficult to write software that uses multiple processors to complete one task faster, and the problem gets worse as the number of processors increases.

Why has this been so? Why have parallel processing programs been so much harder to develop than sequential programs?

The first reason is that you *must* get better performance or better energy efficiency from a parallel processing program on a multiprocessor; otherwise, you would just use a sequential program on a uniprocessor, as sequential programming is simpler. In fact, uniprocessor design techniques, such as superscalar and out-of-order execution, take advantage of instruction-level parallelism (see Chapter 4), normally without the involvement of the programmer. Such innovations reduced the demand for rewriting programs for multiprocessors, since programmers could do nothing and yet their sequential programs would run faster on new computers.

Why is it difficult to write parallel processing programs that are fast, especially as the number of processors increases? In Chapter 1, we used the analogy of eight reporters trying to write a single story in hopes of doing the work eight times faster. To succeed, the task must be broken into eight equal-sized pieces, because otherwise some reporters would be idle while waiting for the ones with larger pieces to finish. Another speed-up obstacle could be that the reporters would spend too much time communicating with each other instead of writing their pieces of the story. For both this analogy and parallel programming, the challenges include scheduling, partitioning the work into parallel pieces, balancing the load evenly between the workers, time to synchronize, and overhead for communication between the parties. The challenge is stiffer with the more reporters for a newspaper story and with the more processors for parallel programming.

Our discussion in Chapter 1 reveals another obstacle, namely Amdahl's Law. It reminds us that even small parts of a program must be parallelized if the program is to make good use of many cores.

Speed-up Challenge

EXAMPLE

Suppose you want to achieve a speed-up of 90 times faster with 100 processors. What percentage of the original computation can be sequential?

Amdahl's Law (Chapter 1) says

ANSWER

Execution time after improvement =

$$\frac{\text{Execution time affected by improvement}}{\text{Amount of improvement}} + \text{Execution time unaffected}$$

We can reformulate Amdahl's Law in terms of speed-up versus the initial execution time:

$$\text{Speed-up} = \frac{\text{Execution time before}}{(\text{Execution time before} - \text{Execution time affected}) + \dfrac{\text{Execution time affected}}{\text{Amount of improvement}}}$$

This formula is usually rewritten assuming that the execution time before is 1 for some unit of time, and the execution time affected by improvement is considered the fraction of the original execution time:

$$\text{Speed-up} = \frac{1}{(1 - \text{Fraction time affected}) + \dfrac{\text{Fraction time affected}}{\text{Amount of improvement}}}$$

Substituting 90 for speed-up and 100 for the amount of improvement into the formula above:

$$90 = \frac{1}{(1 - \text{Fraction time affected}) + \dfrac{\text{Fraction time affected}}{100}}$$

Then simplifying the formula and solving for fraction time affected:

$$90 \times (1 - 0.99 \times \text{Fraction time affected}) = 1$$
$$90 - (90 \times 0.99 \times \text{Fraction time affected}) = 1$$
$$90 - 1 = 90 \times 0.99 \times \text{Fraction time affected}$$
$$\text{Fraction time affected} = 89/89.1 = 0.999$$

Thus, to achieve a speed-up of 90 from 100 processors, the sequential percentage can only be 0.1%.

However, there are applications with plenty of parallelism, as we shall see next.

EXAMPLE

Speed-up Challenge: Bigger Problem

Suppose you want to perform two sums: one is a sum of 10 scalar variables, and one is a matrix sum of a pair of two-dimensional arrays, with dimensions 10 by 10. For now let's assume only the matrix sum is parallelizable; we'll see soon how to parallelize scalar sums. What speed-up do you get with 10 versus 40 processors? Next, calculate the speed-ups assuming the matrices grow to 20 by 20.

ANSWER

If we assume performance is a function of the time for an addition, t, then there are 10 additions that do not benefit from parallel processors and 100 additions that do. If the time for a single processor is 110 t, the execution time for 10 processors is

$$\text{Execution time after improvement} =$$
$$\frac{\text{Execution time affected by improvement}}{\text{Amount of improvement}} + \text{Execution time unaffected}$$

$$\text{Execution time after improvement} = \frac{100t}{10} + 10t = 20t$$

so the speed-up with 10 processors is $110t/20t = 5.5$. The execution time for 40 processors is

$$\text{Execution time after improvement} = \frac{100t}{40} + 10t = 12.5t$$

so the speed-up with 40 processors is $110t/12.5t = 8.8$. Thus, for this problem size, we get about 55% of the potential speed-up with 10 processors, but only 22% with 40.

Look what happens when we increase the matrix. The sequential program now takes $10t+400t = 410t$. The execution time for 10 processors is

$$\text{Execution time after improvement} = \frac{400t}{10} + 10t = 50t$$

so the speed-up with 10 processors is $410t/50t = 8.2$. The execution time for 40 processors is

$$\text{Execution time after improvement} = \frac{400t}{40} + 10t = 20t$$

so the speed-up with 40 processors is $410t/20t = 20.5$. Thus, for this larger problem size, we get 82% of the potential speed-up with 10 processors and 51% with 40.

These examples show that getting good speed-up on a multiprocessor while keeping the problem size fixed is harder than getting good speed-up by increasing the size of the problem. This insight allows us to introduce two terms that describe ways to scale up.

Strong scaling means measuring speed-up while keeping the problem size fixed. **Weak scaling** means that the problem size grows proportionally to the increase in the number of processors. Let's assume that the size of the problem, M, is the working set in main memory, and we have P processors. Then the memory per processor for strong scaling is approximately M/P, and for weak scaling, it is about M.

Note that the **memory hierarchy** can interfere with the conventional wisdom about weak scaling being easier than strong scaling. For example, if the weakly scaled dataset no longer fits in the last level cache of a multicore microprocessor, the resulting performance could be much worse than by using strong scaling.

Depending on the application, you can argue for either scaling approach. For example, the TPC-C debit-credit database benchmark requires that you scale up the number of customer accounts in proportion to the higher transactions per minute. The argument is that it's nonsensical to think that a given customer base is suddenly going to start using ATMs 100 times a day just because the bank gets a faster computer. Instead, if you're going to demonstrate a system that can perform 100 times the numbers of transactions per minute, you should run the experiment with 100 times as many customers. Bigger problems often need more data, which is an argument for weak scaling.

This final example shows the importance of load balancing.

strong scaling Speed-up achieved on a multiprocessor without increasing the size of the problem.

weak scaling Speed-up achieved on a multiprocessor while increasing the size of the problem proportionally to the increase in the number of processors.

HIERARCHY

Speed-up Challenge: Balancing Load

To achieve the speed-up of 20.5 on the previous larger problem with 40 processors, we assumed the load was perfectly balanced. That is, each of the 40

EXAMPLE

processors had 2.5% of the work to do. Instead, show the impact on speed-up if one processor's load is higher than all the rest. Calculate at twice the load (5%) and five times the load (12.5%) for that hardest working processor. How well utilized are the rest of the processors?

ANSWER

If one processor has 5% of the parallel load, then it must do 5% × 400 or 20 additions, and the other 39 will share the remaining 380. Since they are operating simultaneously, we can just calculate the execution time as a maximum

$$\text{Execution time after improvement} = \text{Max}\left(\frac{380t}{39}, \frac{20t}{1}\right) + 10t = 30t$$

The speed-up drops from 20.5 to 410t/30t = 14. The remaining 39 processors are utilized less than half the time: while waiting 20t for the hardest working processor to finish, they only compute for 380t/39 = 9.7t.

If one processor has 12.5% of the load, it must perform 50 additions. The formula is:

$$\text{Execution time after improvement} = \text{Max}\left(\frac{350t}{39}, \frac{50t}{1}\right) + 10t = 60t$$

The speed-up drops even further to 410t/60t = 7. The rest of the processors are utilized less than 20% of the time (9t/50t). This example demonstrates the importance of balancing load, for just a single processor with twice the load of the others cuts speed-up by a third, and five times the load on just one processor reduces speed-up by almost a factor of three.

Now that we better understand the goals and challenges of parallel processing, we give an overview of the rest of the chapter. Section 6.3 describes a much older classification scheme than in Figure 6.1. In addition, it describes two styles of instruction set architectures that support running of sequential applications on parallel hardware, namely *SIMD* and *vector*. Section 6.4 then describes *multithreading*, a term often confused with multiprocessing, in part because it relies upon similar concurrency in programs. Section 6.5 describes the first the two alternatives of a fundamental parallel hardware characteristic, which is whether or not all the processors in the systems rely upon a single physical address space. As mentioned above, the two popular versions of these alternatives are called *shared memory multiprocessors* (SMPs) and *clusters*, and this section covers the former. Section 6.6 describes a relatively new style of computer from the graphics hardware community, called a *graphics-processing unit* (GPU) that also assumes a single physical address. (**Appendix B** describes GPUs in even more detail.) Section 6.7 describes clusters, a popular example of a computer with multiple physical address spaces. Section 6.8 shows typical topologies used to connect many processors together, either server nodes in a cluster or cores in a microprocessor.   Section 6.9 describes the hardware and software for communicating between nodes in a cluster using Ethernet. It shows how to optimize its performance using custom software and hardware. We next discuss the difficulty of finding parallel benchmarks in Section 6.10. This section also includes a simple, yet insightful performance model that helps in the design of applications as well as architectures. We use this model as well as parallel benchmarks in Section 6.11 to compare a multicore computer to a GPU. Section 6.12 divulges the final and largest step in our journey of accelerating matrix multiply. For matrices that don't fit in the cache, parallel processing uses 16 cores to improve performance by a factor of 14. We close with fallacies and pitfalls and our conclusions for parallelism.

In the next section, we introduce acronyms that you probably have already seen to identify different types of parallel computers.

Check Yourself

True or false: Strong scaling is not bound by Amdahl's Law.

6.3 SISD, MIMD, SIMD, SPMD, and Vector

One categorization of parallel hardware proposed in the 1960s is still used today. It was based on the number of instruction streams and the number of data streams. Figure 6.2 shows the categories. Thus, a conventional uniprocessor has a single instruction stream and single data stream, and a conventional multiprocessor has multiple instruction streams and multiple data streams. These two categories are abbreviated SISD and MIMD, respectively.

While it is possible to write separate programs that run on different processors on a MIMD computer and yet work together for a grander, coordinated goal, programmers normally write a single program that runs on all processors of a MIMD computer, relying on conditional statements when different processors should execute distinct sections of code. This style is called Single Program Multiple Data (SPMD), but it is just the normal way to program a MIMD computer.

The closest we can come to multiple instruction streams and single data stream (MISD) processor might be a "stream processor" that would perform a series of computations on a single data stream in a pipelined fashion: parse the input from the network, decrypt the data, decompress it, search for match, and so on. The inverse of MISD is much more popular. SIMD computers operate on vectors of

SISD or Single Instruction stream, Single Data stream. A uniprocessor.

MIMD or Multiple Instruction streams, Multiple Data streams. A multiprocessor.

SPMD Single Program, Multiple Data streams. The conventional MIMD programming model, where a single program runs across all processors.

SIMD or Single Instruction stream, Multiple Data streams. The same instruction is applied to many data streams, as in a vector processor.

		Data Streams	
		Single	Multiple
Instruction Streams	Single	SISD: Intel Pentium 4	SIMD: SSE instructions of x86
	Multiple	MISD: No examples today	MIMD: Intel Core i7

FIGURE 6.2 Hardware categorization and examples based on number of instruction streams and data streams: SISD, SIMD, MISD, and MIMD.

data. For example, a single SIMD instruction might add 64 numbers by sending 64 data streams to 64 ALUs to form 64 sums within a single clock cycle. The subword parallel instructions that we saw in Sections 3.6 and 3.7 are another example of SIMD; indeed, the middle letter of Intel's SSE acronym stands for SIMD.

The virtues of SIMD are that all the parallel execution units are synchronized, and they all respond to a single instruction that emanates from a single *program counter* (PC). From a programmer's perspective, this is close to the already familiar SISD. Although every unit will be executing the same instruction, each execution unit has its own address registers, and so each unit can have different data addresses. Thus, in terms of Figure 6.1, a sequential application might be compiled to run on serial hardware organized as a SISD or in parallel hardware that was organized as a SIMD.

The original motivation behind SIMD was to amortize the cost of the control unit over dozens of execution units. Another advantage is the reduced instruction bandwidth and space—SIMD needs only one copy of the code that is being simultaneously executed, while message-passing MIMDs may need a copy in every processor and shared memory MIMD will need multiple instruction caches.

SIMD works best when dealing with arrays in `for` loops. Hence, for parallelism to work in SIMD, there must be a great deal of identically structured data, which is called data-level parallelism. SIMD is at its weakest in `case` or `switch` statements, where each execution unit must perform a different operation on its data, depending on what data it has. Execution units with the wrong data must be disabled so that units with proper data may continue. If there are n cases, in these situations, SIMD processors essentially run at $1/n$th of peak performance.

The so-called array processors that inspired the SIMD category have faded into history (see ⊞ Section 6.15 online), but two current interpretations of SIMD remain active today.

data-level parallelism Parallelism achieved by performing the same operation on independent data.

SIMD in x86: Multimedia Extensions

As described in Chapter 3, subword parallelism for narrow integer data was the original inspiration of the *Multimedia Extension* (MMX) instructions of the x86 in 1996. As **Moore's Law** continued, more instructions were added, leading first to *Streaming SIMD Extensions* (SSE) and now *Advanced Vector Extensions* (AVX). AVX supports the simultaneous execution of four 64-bit floating-point numbers. The width of the operation and the registers is encoded in the opcode of these multimedia instructions. As the data width of the registers and operations grew, the number of opcodes for multimedia instructions exploded, and now there are hundreds of SSE and AVX instructions (see Chapter 3).

MOORE'S LAW

Vector

An older and, as we shall see, more elegant interpretation of SIMD is called a *vector architecture*, which has been closely identified with computers designed by Seymour Cray starting in the 1970s. It is also a great match to problems with lots of data-level parallelism. Rather than having 64 ALUs perform 64 additions simultaneously, like the old array processors, the vector architectures pipelined the ALU to get good performance at lower cost. The basic philosophy of vector architecture is to collect data elements from memory, put them in order into a large set of registers, operate on them sequentially in registers using **pipelined execution units**, and then write the results back to memory. A key feature of vector architectures is therefore a set of vector registers. Thus, a vector architecture might have 32 vector registers, each with 64 64-bit elements.

PIPELINING

Comparing Vector to Conventional Code

EXAMPLE

Suppose we extend the LEGv8 instruction set architecture with vector instructions and vector registers. Vector operations use the same names as LEGv8 operations, but with the letter "V" appended. For example, FADDDV adds two double-precision vectors. Let's also expand the 32 V registers used for SIMD instructions from two 64-bit elements to sixty-four 64-bit elements. The vector instructions take as their input either a pair of vector (V) registers (FADDDVS) or a vector register and a scalar register (FADDDVS). In the latter case, the value in the scalar register is used as the input for all operations—the operation FADDDVS will add the contents of a scalar register to each element in a vector register. The names LDURDV and STURDV denote vector load and vector store, and they load or store an entire vector of double-precision data. One operand is the vector register to be loaded or stored; the other operand, which is a LEGv8 general-purpose register, is the starting address of the vector in memory.

Given this short description, show the conventional LEGv8 code versus the vector LEGv8 code for

$$Y = a \times X + Y$$

where X and Y are vectors of 64 double precision floating-point numbers, initially resident in memory, and a is a scalar double precision variable. (This example is the so-called DAXPY loop that forms the inner loop of the Linpack benchmark; DAXPY stands for double precision $a \times X$ plus Y.) Assume that the starting addresses of X and Y are in X19 and X20, respectively.

Here is the conventional LEGv8 code for DAXPY:

ANSWER

```
        LDURD   D0,[X28,a]    //load scalar a
        ADDI    X0,X19,512    //upper bound of what to load
loop:   LDURD   D2,[X19,#0]   //load x(i)
        FMULD   D2,D2,D0      //a x x(i)
        LDURD   D4,[X20,#0]   //load y(i)
        FADDD   D4,D4,D2      //a x x(i) + y(i)
        STURD   D4,[X20,#0]   //store into y(i)
        ADDI    X19,X19,#8    //increment index to x
        ADDI    X20,X20,#8    //increment index to y
        CMPB    X0,X19        //compute bound
        B.NE    loop          //check if done
```

Here is the vector LEGv8 code for DAXPY:

```
LDURD    D0,[X28,a]      //load scalar a
LDURDV   V1,[X19,#0]     //load vector x
FMULDVS  V2,V1,D0        //vector-scalar multiply
LDURDV   V3,[X20,#0]     //load vector y
FADDDV   V4,V2,V3        //add y to product
STURDV   V4,[X20,#0]     //store the result
```

There are some interesting comparisons between the two code segments in this example. The most dramatic is that the vector processor greatly reduces the dynamic instruction bandwidth, executing only six instructions versus almost 600 for the traditional LEGv8 architecture. This reduction occurs both because the vector operations work on 64 elements at a time and because the overhead instructions that constitute nearly half the loop on LEGv8 are not present in the vector code. As you might expect, this reduction in instructions fetched and executed saves energy.

Another important difference is the frequency of **pipeline** hazards (Chapter 4). In the straightforward LEGv8 code, every FADDD must wait for a FMULD, every STURD must wait for the FADDD and every FADDD *and* FMULD must wait on LDURD. On the vector processor, each vector instruction will only stall for the first element in each vector, and then subsequent elements will flow smoothly down the pipeline. Thus, pipeline stalls are required only once per vector *operation*, rather than once per vector *element*. In this example, the pipeline stall frequency on LEGv8 will be about 64 times higher than it is on the vector version of LEGv8. The pipeline stalls can be reduced on LEGv8 by using loop unrolling (see Chapter 4). However, the large difference in instruction bandwidth cannot be reduced.

Since the vector elements are independent, they can be operated on in parallel, much like subword parallelism for the Intel x86 AVX instructions. All modern vector computers have vector functional units with multiple parallel pipelines (called *vector lanes*; see Figures 6.2 and 6.3) that can produce two or more results per clock cycle.

PIPELINING

Elaboration: The loop in the example above exactly matched the vector length. When loops are shorter, vector architectures use a register that reduces the length of vector operations. When loops are larger, we add bookkeeping code to iterate full-length vector operations and to handle the leftovers. This latter process is called *strip mining*.

Vector versus Scalar

Vector instructions have several important properties compared to conventional instruction set architectures, which are called *scalar architectures* in this context:

- A single vector instruction specifies a great deal of work—it is equivalent to executing an entire loop. The instruction fetch and decode bandwidth needed is dramatically reduced.

- By using a vector instruction, the compiler or programmer indicates that the computation of each result in the vector is independent of the computation of other results in the same vector, so hardware does not have to check for data hazards within a vector instruction.

- Vector architectures and compilers have a reputation for making it much easier than when using MIMD multiprocessors to write efficient applications when they contain data-level parallelism.

- Hardware need only check for data hazards between two vector instructions once per vector operand, not once for every element within the vectors. Reduced checking can save energy as well as time.

- Vector instructions that access memory have a known access pattern. If the vector's elements are all adjacent, then fetching the vector from a set of heavily interleaved memory banks works very well. Thus, the cost of the latency to main memory is seen only once for the entire vector, rather than once for each word of the vector.

- Because a complete loop is replaced by a vector instruction whose behavior is predetermined, control hazards that would normally arise from the loop branch are nonexistent.

- The savings in instruction bandwidth and hazard checking plus the efficient use of memory bandwidth give vector architectures advantages in power and energy versus scalar architectures.

For these reasons, vector operations can be made faster than a sequence of scalar operations on the same number of data items, and designers are motivated to include vector units if the application domain can often use them.

Vector versus Multimedia Extensions

Like multimedia extensions found in the x86 AVX instructions, a vector instruction specifies multiple operations. However, multimedia extensions typically denote a few operations while vector specifies dozens of operations. Unlike multimedia extensions, the number of elements in a vector operation is not in the opcode but in a separate register. This distinction means different versions of the vector architecture can be implemented with a different number of elements just by changing the contents of that register and hence retain binary compatibility. In contrast, a new large set of opcodes is added each time the "vector" length changes in the multimedia extension architecture of the x86: MMX, SSE, SSE2, AVX, AVX2, ...

Also, unlike multimedia extensions, the data transfers need not be contiguous. Vectors support both strided accesses, where the hardware loads every *n*th data element in memory, and indexed accesses, where hardware finds the addresses of the items to be loaded into a vector register. Indexed accesses are also called *gather-scatter*, in that indexed loads gather elements from main memory into contiguous vector elements, and indexed stores scatter vector elements across main memory.

Like multimedia extensions, vector architectures easily capture the flexibility in data widths, so it is easy to make a vector operation work on 32 64-bit data elements or 64 32-bit data elements or 128 16-bit data elements or 256 8-bit data elements. The parallel semantics of a vector instruction allows an implementation to execute these operations using a deeply **pipelined** functional unit, an array of parallel functional units, or a combination of parallel and pipelined functional units. Figure 6.3 illustrates how to improve vector performance by using parallel pipelines to execute a vector add instruction.

PIPELINING

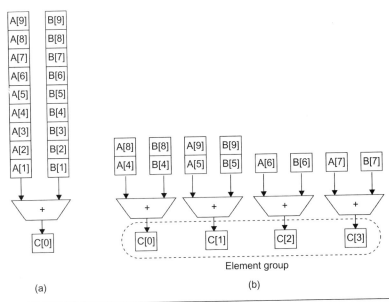

FIGURE 6.3   Using multiple functional units to improve the performance of a single vector add instruction, C = A + B. The vector processor (a) on the left has a single add pipeline and can complete one addition per cycle. The vector processor (b) on the right has four add pipelines or lanes and can complete four additions per cycle. The elements within a single vector add instruction are interleaved across the four lanes.

vector lane One or more vector functional units and a portion of the vector register file. Inspired by lanes on highways that increase traffic speed, multiple lanes execute vector operations simultaneously.

PARALLELISM

Vector arithmetic instructions usually only allow element N of one vector register to take part in operations with element N from other vector registers. This dramatically simplifies the construction of a highly parallel vector unit, which can be structured as multiple parallel **vector lanes**. As with a traffic highway, we can increase the peak throughput of a vector unit by adding more lanes. Figure 6.4 shows the structure of a four-lane vector unit. Thus, going to four lanes from one lane reduces the number of clocks per vector instruction by roughly a factor of four. For multiple lanes to be advantageous, both the applications and the architecture must support long vectors. Otherwise, they will execute so quickly that you'll run out of instructions, requiring instruction level **parallel** techniques like those in Chapter 4 to supply enough vector instructions.

Generally, vector architectures are a very efficient way to execute data parallel processing programs; they are better matches to compiler technology than multimedia extensions; and they are easier to evolve over time than the multimedia extensions to the x86 architecture.

Given these classic categories, we next see how to exploit parallel streams of instructions to improve the performance of a *single* processor, which we will reuse with multiple processors.

Check Yourself

True or false: As exemplified in the x86, multimedia extensions can be thought of as a vector architecture with short vectors that support only contiguous vector data transfers.

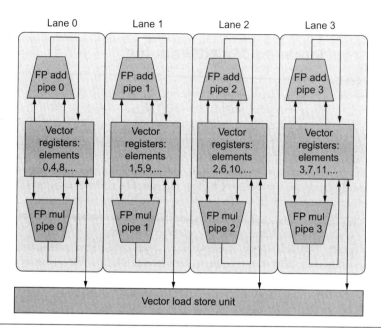

Lane 0   Lane 1   Lane 2   Lane 3

FIGURE 6.4 Structure of a vector unit containing four lanes. The vector-register storage is divided across the lanes, with each lane holding every fourth element of each vector register. The figure shows three vector functional units: an FP add, an FP multiply, and a load-store unit. Each of the vector arithmetic units contains four execution pipelines, one per lane, which acts in concert to complete a single vector instruction. Note how each section of the vector-register file only needs to provide enough read and write ports (see Chapter 4) for functional units local to its lane.

Elaboration: Given the advantages of vector, why aren't they more popular outside high-performance computing? There were concerns about the larger state for vector registers increasing context switch time and the difficulty of handling page faults in vector loads and stores, and SIMD instructions achieved some of the benefits of vector instructions. In addition, as long as advances in instruction-level parallelism could deliver on the performance promise of **Moore's Law**, there was little reason to take the chance of changing architecture styles.

MOORE'S LAW

Elaboration: Another advantage of vector and multimedia extensions is that it is relatively easy to extend a scalar instruction set architecture with these instructions to improve performance of data parallel operations.

Elaboration: The Haswell-generation x86 processors from Intel support AVX2, which has a gather operation but not a scatter operation.

hardware multithreading Increasing utilization of a processor by switching to another thread when one thread is stalled.

thread A thread includes the program counter, the register state, and the stack. It is a lightweight process; whereas threads commonly share a single address space, processes don't.

process A process includes one or more threads, the address space, and the operating system state. Hence, a process switch usually invokes the operating system, but not a thread switch.

fine-grained multithreading A version of hardware multithreading that implies switching between threads after every instruction.

coarse-grained multithreading A version of hardware multithreading that implies switching between threads only after significant events, such as a last-level cache miss.

PIPELINING

6.4 Hardware Multithreading

A related concept to MIMD, especially from the programmer's perspective, is hardware multithreading. While MIMD relies on multiple processes or threads to try to keep many processors busy, hardware multithreading allows multiple threads to share the functional units of a *single* processor in an overlapping fashion to try to utilize the hardware resources efficiently. To permit this sharing, the processor must duplicate the independent state of each thread. For example, each thread would have a separate copy of the register file and the program counter. The memory itself can be shared through the virtual memory mechanisms, which already support multi-programming. In addition, the hardware must support the ability to change to a different thread relatively quickly. In particular, a thread switch should be much more efficient than a process switch, which typically requires hundreds to thousands of processor cycles while a thread switch can be instantaneous.

There are two main approaches to hardware multithreading. Fine-grained multithreading switches between threads on each instruction, resulting in interleaved execution of multiple threads. This interleaving is often done in a round-robin fashion, skipping any threads that are stalled at that clock cycle. To make fine-grained multithreading practical, the processor must be able to switch threads on every clock cycle. One advantage of fine-grained multithreading is that it can hide the throughput losses that arise from both short and long stalls, since instructions from other threads can be executed when one thread stalls. The primary disadvantage of fine-grained multithreading is that it slows down the execution of the individual threads, since a thread that is ready to execute without stalls will be delayed by instructions from other threads.

Coarse-grained multithreading was invented as an alternative to fine-grained multithreading. Coarse-grained multithreading switches threads only on expensive stalls, such as last-level cache misses. This change relieves the need to have thread switching be extremely fast and is much less likely to slow down the execution of an individual thread, since instructions from other threads will only be issued when a thread encounters a costly stall. Coarse-grained multithreading suffers, however, from a major drawback: it is limited in its ability to overcome throughput losses, especially from shorter stalls. This limitation arises from the **pipeline** start-up costs of coarse-grained multithreading. Because a processor with coarse-grained multithreading issues instructions from a single thread, when a stall occurs, the pipeline must be emptied or frozen. The new thread that begins executing after the stall must fill the pipeline before instructions are able to complete. Due to this start-up overhead, coarse-grained multithreading is much more useful for reducing the penalty of high-cost stalls, where pipeline refill is negligible compared to the stall time.

Simultaneous multithreading (SMT) is a variation on hardware multithreading that uses the resources of a multiple-issue, dynamically scheduled **pipelined** processor to exploit thread-level parallelism at the same time it exploits instruction-level parallelism (see Chapter 4). The key insight that motivates SMT is that multiple-issue processors often have more functional unit parallelism available than most single threads can effectively use. Furthermore, with register renaming and dynamic scheduling (see Chapter 4), multiple instructions from independent threads can be issued without regard to the dependences among them; the resolution of the dependences can be handled by the dynamic scheduling capability.

Since SMT relies on the existing dynamic mechanisms, it does not switch resources every cycle. Instead, SMT is *always* executing instructions from multiple threads, leaving it up to the hardware to associate instruction slots and renamed registers with their proper threads.

Figure 6.5 conceptually illustrates the differences in a processor's ability to exploit superscalar resources for the following processor configurations. The top portion shows

PIPELINING

simultaneous multithreading (SMT) A version of multithreading that lowers the cost of multithreading by utilizing the resources needed for multiple issue, dynamically scheduled microarchitecture.

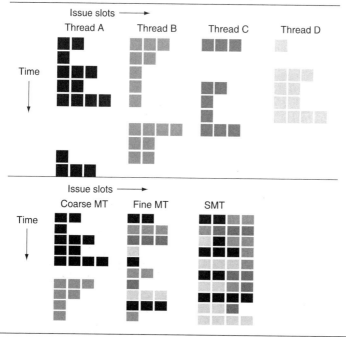

FIGURE 6.5 How four threads use the issue slots of a superscalar processor in different approaches. The four threads at the top show how each would execute running alone on a standard superscalar processor without multithreading support. The three examples at the bottom show how they would execute running together in three multithreading options. The horizontal dimension represents the instruction issue capability in each clock cycle. The vertical dimension represents a sequence of clock cycles. An empty (white) box indicates that the corresponding issue slot is unused in that clock cycle. The shades of gray and color correspond to four different threads in the multithreading processors. The additional pipeline start-up effects for coarse multithreading, which are not illustrated in this figure, would lead to further loss in throughput for coarse multithreading.

how four threads would execute independently on a superscalar with no multithreading support. The bottom portion shows how the four threads could be combined to execute on the processor more efficiently using three multithreading options:

- A superscalar with coarse-grained multithreading
- A superscalar with fine-grained multithreading
- A superscalar with simultaneous multithreading

PARALLELISM

In the superscalar without hardware multithreading support, the use of issue slots is limited by a lack of **instruction-level parallelism**. In addition, a major stall, such as an instruction cache miss, can leave the entire processor idle.

In the coarse-grained multithreaded superscalar, the long stalls are partially hidden by switching to another thread that uses the resources of the processor. Although this reduces the number of completely idle clock cycles, the pipeline start-up overhead still leads to idle cycles, and limitations to ILP mean all issue slots will not be used. In the fine-grained case, the interleaving of threads mostly eliminates idle clock cycles. Because only a single thread issues instructions in a given clock cycle, however, limitations in instruction-level parallelism still lead to idle slots within some clock cycles.

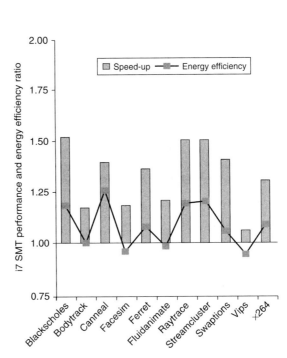

FIGURE 6.6 The speed-up from using multithreading on one core on an i7 processor averages 1.31 for the PARSEC benchmarks (see 🌐 Section 6.9) and the energy efficiency improvement is 1.07. These data were collected and analyzed by Esmaeilzadeh et al. [2011].

In the SMT case, thread-level parallelism and instruction-level parallelism are both exploited, with multiple threads using the issue slots in a single clock cycle. Ideally, the issue slot usage is limited by imbalances in the resource needs and resource availability over multiple threads. In practice, other factors can restrict how many slots are used. Although Figure 6.5 greatly simplifies the real operation of these processors, it does illustrate the potential performance advantages of multithreading in general and SMT in particular.

Figure 6.6 plots the performance and energy benefits of multithreading on a single processor of the Intel Core i7 960, which has hardware support for two threads. The average speed-up is 1.31, which is not bad given the modest extra resources for hardware multithreading. The average improvement in energy efficiency is 1.07, which is excellent. In general, you'd be happy with a performance speed-up being energy neutral.

Now that we have seen how multiple threads can utilize the resources of a single processor more effectively, we next show how to use them to exploit multiple processors.

1. True or false: Both multithreading and multicore rely on parallelism to get more efficiency from a chip.

2. True or false: *Simultaneous multithreading* (SMT) uses threads to improve resource utilization of a dynamically scheduled, out-of-order processor.

Check Yourself

6.5 Multicore and Other Shared Memory Multiprocessors

While hardware multithreading improved the efficiency of processors at modest cost, the big challenge of the last decade has been to deliver on the performance potential of **Moore's Law** by efficiently programming the increasing number of processors per chip.

Given the difficulty of rewriting old programs to run well on parallel hardware, a natural question is: what can computer designers do to simplify the task? One answer was to provide a single physical address space that all processors can share, so that programs need not concern themselves with where their data are, merely that programs may be executed in parallel. In this approach, all variables of a program can be made available at any time to any processor. The alternative is to have a separate address space per processor that requires that sharing must be explicit; we'll describe this option in the Section 6.7. When the physical address space is common then the hardware typically provides cache coherence to give a consistent view of the shared memory (see Section 5.8).

As mentioned above, a *shared memory multiprocessor* (SMP) is one that offers the programmer a *single physical address space* across all processors—which is

MOORE'S LAW

uniform memory access (UMA) A multiprocessor in which latency to any word in main memory is about the same no matter which processor requests the access.

nonuniform memory access (NUMA) A type of single address space multiprocessor in which some memory accesses are much faster than others depending on which processor asks for which word.

synchronization The process of coordinating the behavior of two or more processes, which may be running on different processors.

lock A synchronization device that allows access to data to only one processor at a time.

nearly always the case for multicore chips—although a more accurate term would have been shared-*address* multiprocessor. Processors communicate through shared variables in memory, with all processors capable of accessing any memory location via loads and stores. Figure 6.7 shows the classic organization of an SMP. Note that such systems can still run independent jobs in their own virtual address spaces, even if they all share a physical address space.

Single address space multiprocessors come in two styles. In the first style, the latency to a word in memory does not depend on which processor asks for it. Such machines are called uniform memory access (UMA) multiprocessors. In the second style, some memory accesses are much faster than others, depending on which processor asks for which word, typically because main memory is divided and attached to different microprocessors or to different memory controllers on the same chip. Such machines are called nonuniform memory access (NUMA) multiprocessors. As you might expect, the programming challenges are harder for a NUMA multiprocessor than for a UMA multiprocessor, but NUMA machines can scale to larger sizes, and NUMAs can have lower latency to nearby memory.

As processors operating in parallel will normally share data, they also need to coordinate when operating on shared data; otherwise, one processor could start working on data before another is finished with it. This coordination is called synchronization, which we saw in Chapter 2. When sharing is supported with a single address space, there must be a separate mechanism for synchronization. One approach uses a lock for a shared variable. Only one processor at a time can acquire the lock, and other processors interested in shared data must wait until the original processor unlocks the variable. Section 2.11 of Chapter 2 describes the instructions for locking in the ARMv8 instruction set.

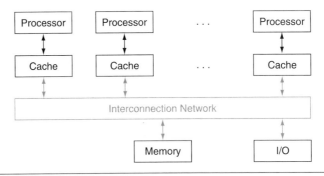

FIGURE 6.7 Classic organization of a shared memory multiprocessor.

A Simple Parallel Processing Program for a Shared Address Space

Suppose we want to sum 64,000 numbers on a shared memory multiprocessor computer with uniform memory access time. Let's assume we have 64 processors.

The first step is to ensure a balanced load per processor, so we split the set of numbers into subsets of the same size. We do not allocate the subsets to a different memory space, since there is a single memory space for this machine; we just give different starting addresses to each processor. Pn is the number that identifies the processor, between 0 and 63. All processors start the program by running a loop that sums their subset of numbers:

```
sum[Pn] = 0;
for (i = 1000*Pn; i < 1000*(Pn+1); i += 1)
  sum[Pn] += A[i]; /*sum the assigned areas*/
```

(Note the C code i += 1 is just a shorter way to say i = i + 1.)

The next step is to add these 64 partial sums. This step is called a **reduction**, where we divide to conquer. Half of the processors add pairs of partial sums, and then a quarter add pairs of the new partial sums, and so on until we have the single, final sum. Figure 6.8 illustrates the hierarchical nature of this reduction.

In this example, the two processors must synchronize before the "consumer" processor tries to read the result from the memory location written by the "producer" processor; otherwise, the consumer may read the old value of

reduction A function that processes a data structure and returns a single value.

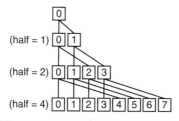

FIGURE 6.8 The last four levels of a reduction that sums results from each processor, from bottom to top. For all processors whose number i is less than half, add the sum produced by processor number (i + half) to its sum.

the data. We want each processor to have its own version of the loop counter variable i, so we must indicate that it is a "private" variable. Here is the code (half is private also):

```
half = 64; /*64 processors in multiprocessor*/
do
    synch(); /*wait for partial sum completion*/
    if (half%2 != 0 && Pn == 0)
        sum[0] += sum[half-1];
        /*Conditional sum needed when half is
        odd; Processor0 gets missing element */
        half = half/2; /*dividing line on who sums */
        if (Pn < half) sum[Pn] += sum[Pn+half];
while (half > 1); /*exit with final sum in Sum[0] */
```

Hardware/ Software Interface

OpenMP An API for shared memory multiprocessing in C, C++, or Fortran that runs on UNIX and Microsoft platforms. It includes compiler directives, a library, and runtime directives.

Given the long-term interest in parallel programming, there have been hundreds of attempts to build parallel programming systems. A limited but popular example is OpenMP. It is just an *Application Programmer Interface* (API) along with a set of compiler directives, environment variables, and runtime library routines that can extend standard programming languages. It offers a portable, scalable, and simple programming model for shared memory multiprocessors. Its primary goal is to parallelize loops and perform reductions.

Most C compilers already have support for OpenMP. The command to use the OpenMP API with the UNIX C compiler is just:

```
cc -fopenmp foo.c
```

OpenMP extends C using *pragmas*, which are just commands to the C macro preprocessor like #define and #include. To set the number of processors we want to use to be 64, as we wanted in the example above, we just use the command

```
#define P 64 /* define a constant that we'll use a few times */
#pragma omp parallel num_threads(P)
```

That is, the runtime libraries should use 64 parallel threads.

To turn the sequential for loop into a parallel for loop that divides the work equally between all the threads that we told it to use, we just write (assuming sum is initialized to 0)

```
#pragma omp parallel for
for (Pn = 0; Pn < P; Pn += 1)
    for (i = 0; 1000*Pn; i < 1000*(Pn+1); i += 1)
        sum[Pn] += A[i]; /*sum the assigned areas*/
```

To perform the reduction, we can use another command that tells OpenMP what the reduction operator is and what variable you need to use to place the result of the reduction.

```
#pragma omp parallel for reduction(+ : FinalSum)
for (i = 0; i < P; i += 1)
        FinalSum += sum[i]; /* Reduce to a single number */
```

Note that it is now up to the OpenMP library to find efficient code to sum 64 numbers efficiently using 64 processors.

While OpenMP makes it easy to write simple parallel code, it is not very helpful with debugging, so many programmers use more sophisticated parallel programming systems than OpenMP, just as many programmers today use more productive languages than C.

Given this tour of classic MIMD hardware and software, our next path is a more exotic tour of a type of MIMD architecture with a different heritage and thus a very different perspective on the parallel programming challenge.

True or false: Shared memory multiprocessors cannot take advantage of task-level parallelism.

Check Yourself

Elaboration: Some writers repurposed the acronym SMP to mean *symmetric multiprocessor*, to indicate that the latency from processor to memory was about the same for all processors. This shift was done to contrast them from large-scale NUMA multiprocessors, as both classes used a single address space. As clusters proved much more popular than large-scale NUMA multiprocessors, in this book we restore SMP to its original meaning, and use it to contrast against those that use multiple address spaces, such as clusters.

Elaboration: An alternative to sharing the physical address space would be to have separate physical address spaces but share a common virtual address space, leaving it up to the operating system to handle communication. This approach has been tried, but it has too high an overhead to offer a practical shared memory abstraction to the performance-oriented programmer.

MOORE'S LAW

6.6   Introduction to Graphics Processing Units

The original justification for adding SIMD instructions to existing architectures was that many microprocessors were connected to graphics displays in PCs and workstations, so an increasing fraction of processing time was used for graphics. As **Moore's Law** increased the number of transistors available to microprocessors, it therefore made sense to improve graphics processing.

A major driving force for improving graphics processing was the computer game industry, both on PCs and in dedicated game consoles such as the Sony PlayStation. The rapidly growing game market encouraged many companies to make increasing investments in developing faster graphics hardware, and this positive feedback loop led graphics processing to improve at a quicker rate than general-purpose processing in mainstream microprocessors.

Given that the graphics and game community had different goals than the microprocessor development community, it evolved its own style of processing and terminology. As the graphics processors increased in power, they earned the name *Graphics Processing Units* or *GPUs* to distinguish themselves from CPUs.

For a few hundred dollars, anyone can buy a GPU today with hundreds of parallel floating-point units, which makes high-performance computing more accessible. The interest in GPU computing blossomed when this potential was combined with a programming language that made GPUs easier to program. Hence, many programmers of scientific and multimedia applications today are pondering whether to use GPUs or CPUs.

(This section concentrates on using GPUs for computing. To see how GPU computing combines with the traditional role of graphics acceleration, see 🌐 Appendix B.)

Here are some of the key characteristics as to how GPUs vary from CPUs:

- GPUs are accelerators that supplement a CPU, so they do not need to be able to perform all the tasks of a CPU. This role allows them to dedicate all their resources to graphics. It's fine for GPUs to perform some tasks poorly or not at all, given that in a system with both a CPU and a GPU, the CPU can do them if needed.

- The GPU problem sizes are typically hundreds of megabytes to gigabytes, but not hundreds of gigabytes to terabytes.

These differences led to different styles of architecture:

- Perhaps the biggest difference is that GPUs do not rely on multilevel caches to overcome the long latency to memory, as do CPUs. Instead, GPUs rely on hardware multithreading (Section 6.4) to hide the latency to memory. That is, between the time of a memory request and the time that data arrive, the GPU executes hundreds or thousands of threads that are independent of that request.

- The GPU memory is thus oriented toward bandwidth rather than latency. There are even special graphics DRAM chips for GPUs that are wider and have higher bandwidth than DRAM chips for CPUs. In addition, GPU memories have traditionally had smaller main memories than conventional microprocessors. In 2013, GPUs typically have 4 to 6 GiB or less, while CPUs have 32 to 256 GiB. Finally, keep in mind that for general-purpose computation, you must include the time to transfer the data between CPU memory and GPU memory, since the GPU is a coprocessor.

- Given the reliance on many threads to deliver good memory bandwidth, GPUs can accommodate many parallel processors (MIMD) as well as many threads. Hence, each GPU processor is more highly multithreaded than a typical CPU, plus they have more processors.

Although GPUs were designed for a narrower set of applications, some programmers wondered if they could specify their applications in a form that would let them tap the high potential performance of GPUs. After tiring of trying to specify their problems using the graphics APIs and languages, they developed C-inspired programming languages to allow them to write programs directly for the GPUs. An example is NVIDIA's CUDA (*Compute Unified Device Architecture*), which enables the programmer to write C programs to execute on GPUs, albeit with some restrictions. ⊞ Appendix B gives examples of CUDA code. (OpenCL is a multi-company initiative to develop a portable programming language that provides many of the benefits of CUDA.)

NVIDIA decided that the unifying theme of all these forms of parallelism is the *CUDA Thread*. Using this lowest level of parallelism as the programming primitive, the compiler and the hardware can gang thousands of CUDA threads together to utilize the various styles of parallelism within a GPU: multithreading, MIMD, SIMD, and instruction-level parallelism. These threads are blocked together and executed in groups of 32 at a time. A multithreaded processor inside a GPU executes these blocks of threads, and a GPU consists of 8 to 32 of these multithreaded processors.

Hardware/ Software Interface

An Introduction to the NVIDIA GPU Architecture

We use NVIDIA systems as our example as they are representative of GPU architectures. Specifically, we follow the terminology of the CUDA parallel programming language and use the Fermi architecture as the example.

Like vector architectures, GPUs work well only with data-level parallel problems. Both styles have gather-scatter data transfers, and GPU processors have even more registers than do vector processors. Unlike most vector architectures, GPUs also rely on hardware multithreading within a single multithreaded SIMD processor to hide memory latency (see Section 6.4).

A multithreaded SIMD processor is similar to a vector processor, but the former has many parallel functional units instead of just a few that are deeply pipelined, as does the latter.

As mentioned above, a GPU contains a collection of multithreaded SIMD processors; that is, a GPU is a MIMD composed of multithreaded SIMD processors. For example, NVIDIA has four implementations of the Fermi architecture at different price points with 7, 11, 14, or 15 multithreaded SIMD processors. To provide transparent scalability across models of GPUs with differing number of multithreaded SIMD processors, the Thread Block Scheduler hardware assigns blocks of threads to multithreaded SIMD processors. Figure 6.9 shows a simplified block diagram of a multithreaded SIMD processor.

Dropping down one more level of detail, the machine object that the hardware creates, manages, schedules, and executes is a *thread of SIMD instructions*, which we will also call a *SIMD thread*. It is a traditional thread, but it contains exclusively SIMD instructions. These SIMD threads have their own program counters, and they run on a multithreaded SIMD processor. The *SIMD Thread Scheduler* includes a controller that lets it know which threads of SIMD instructions are ready to run, and then it sends them off to a dispatch unit to be run on the multithreaded

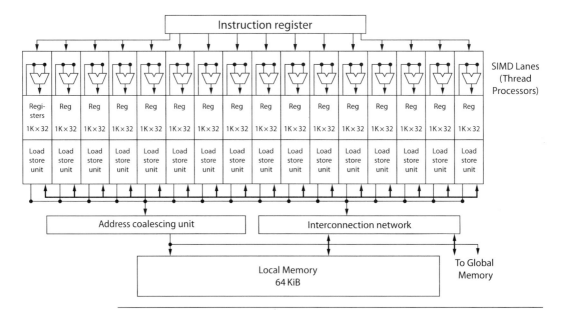

FIGURE 6.9   Simplified block diagram of the datapath of a multithreaded SIMD Processor.
It has 16 SIMD lanes. The SIMD Thread Scheduler has many independent SIMD threads that it chooses from to run on this processor.

SIMD processor. It is identical to a hardware thread scheduler in a traditional multithreaded processor (see Section 6.4), except that it is scheduling threads of SIMD instructions. Thus, GPU hardware has two levels of hardware schedulers:

1. The *Thread Block Scheduler* that assigns blocks of threads to multithreaded SIMD processors, and

2. The SIMD Thread Scheduler *within* a SIMD processor, which schedules when SIMD threads should run.

The SIMD instructions of these threads are 32 wide, so each thread of SIMD instructions would compute 32 of the elements of the computation. Since the thread consists of SIMD instructions, the SIMD processor must have parallel functional units to perform the operation. We call them *SIMD Lanes*, and they are quite similar to the Vector Lanes in Section 6.3.

Elaboration: The number of lanes per SIMD processor varies across GPU generations. With Fermi, each 32-wide thread of SIMD instructions is mapped to 16 SIMD lanes, so each SIMD instruction in a thread of SIMD instructions takes two clock cycles to complete. Each thread of SIMD instructions is executed in lock step. Staying with the analogy of a SIMD processor as a vector processor, you could say that it has 16 lanes, and the vector length would be 32. This wide but shallow nature is why we use the term SIMD processor instead of vector processor, as it is more intuitive.

Since by definition the threads of SIMD instructions are independent, the SIMD Thread Scheduler can pick whatever thread of SIMD instructions is ready, and need not stick with the next SIMD instruction in the sequence within a single thread. Thus, using the terminology of Section 6.4, it uses fine-grained multithreading.

To hold these memory elements, a Fermi SIMD processor has an impressive 32,768 32-bit registers. Just like a vector processor, these registers are divided logically across the vector lanes or, in this case, SIMD lanes. Each SIMD thread is limited to no more than 64 registers, so you might think of a SIMD thread as having up to 64 vector registers, with each vector register having 32 elements and each element being 32 bits wide.

Since Fermi has 16 SIMD lanes, each contains 2048 registers. Each CUDA thread gets one element of each of the vector registers. Note that a CUDA thread is just a vertical cut of a thread of SIMD instructions, corresponding to one element executed by one SIMD lane. Beware that CUDA threads are very different from POSIX threads; you can't make arbitrary system calls or synchronize arbitrarily in a CUDA thread.

NVIDIA GPU Memory Structures

Figure 6.10 shows the memory structures of an NVIDIA GPU. We call the on-chip memory that is local to each multithreaded SIMD processor *Local Memory*. It is shared by the SIMD lanes within a multithreaded SIMD processor, but this memory is not shared between multithreaded SIMD processors. We call the off-chip DRAM shared by the whole GPU and all thread blocks *GPU Memory*.

Rather than rely on large caches to contain the entire working sets of an application, GPUs traditionally use smaller streaming caches and rely on extensive multithreading of threads of SIMD instructions to hide the long latency to DRAM,

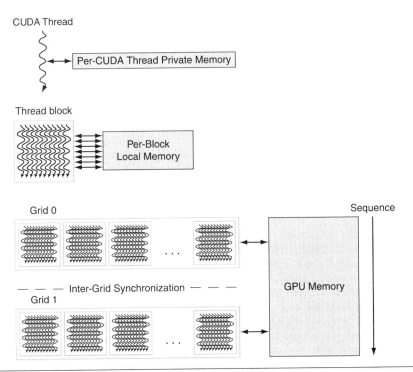

FIGURE 6.10 GPU Memory structures. GPU Memory is shared by the vectorized loops. All threads of SIMD instructions within a thread block share Local Memory.

since their working sets can be hundreds of megabytes. Thus, they will not fit in the last-level cache of a multicore microprocessor. Given the use of hardware multithreading to hide DRAM latency, the chip area used for caches in system processors is spent instead on computing resources and on the large number of registers to hold the state of the many threads of SIMD instructions.

Elaboration: While hiding memory latency is the underlying philosophy, note that the latest GPUs and vector processors have added caches. For example, the recent Fermi architecture has added caches, but they are thought of as either bandwidth filters to reduce demands on GPU Memory or as accelerators for the few variables whose latency cannot be hidden by multithreading. Local memory for stack frames, function calls, and register spilling is a good match to caches, since latency matters when calling a function. Caches can also save energy, since on-chip cache accesses take much less energy than accesses to multiple, external DRAM chips.

Putting GPUs into Perspective

At a high level, multicore computers with SIMD instruction extensions do share similarities with GPUs. Figure 6.11 summarizes the similarities and differences. Both are MIMDs whose processors use multiple SIMD lanes, although GPUs have more processors and many more lanes. Both use hardware multithreading to improve processor utilization, although GPUs have hardware support for many more threads. Both use caches, although GPUs use smaller streaming caches and multicore computers use large multilevel caches that try to contain whole working sets completely. Both use a 64-bit address space, although the physical main memory is much smaller in GPUs. While GPUs support memory protection at the page level, they do not yet support demand paging.

SIMD processors are also similar to vector processors. The multiple SIMD processors in GPUs act as independent MIMD cores, just as many vector computers have multiple vector processors. This view would consider the Fermi GTX 580 as a 16-core machine with hardware support for multithreading, where each core has 16 lanes. The biggest difference is multithreading, which is fundamental to GPUs and missing from most vector processors.

GPUs and CPUs do not go back in computer architecture genealogy to a shared ancestor; there is no Missing Link that explains both. As a result of this uncommon heritage, GPUs have not used the terms common in the computer architecture community, which has led to confusion about what GPUs are and how they work. To help resolve the confusion, Figure 6.12 (from left to right) lists the more descriptive term used in this section, the closest term from mainstream computing, the official NVIDIA GPU term in case you are interested, and then a short description of the term. This "GPU Rosetta Stone" may help relate this section and ideas to more conventional GPU descriptions, such as those found in ⬚ Appendix B.

While GPUs are moving toward mainstream computing, they can't abandon their responsibility to continue to excel at graphics. Thus, the design of GPUs may make more sense when architects ask, given the hardware invested to do graphics

Feature	Multicore with SIMD	GPU
SIMD processors	4 to 8	8 to 16
SIMD lanes/processor	2 to 4	8 to 16
Multithreading hardware support for SIMD threads	2 to 4	16 to 32
Largest cache size	8 MiB	0.75 MiB
Size of memory address	64-bit	64-bit
Size of main memory	8 GiB to 256 GiB	4 GiB to 6 GiB
Memory protection at level of page	Yes	Yes
Demand paging	Yes	No
Cache coherent	Yes	No

FIGURE 6.11 Similarities and differences between multicore with Multimedia SIMD extensions and recent GPUs.

Type	More descriptive name	Closest old term outside of GPUs	Official CUDA/ NVIDIA GPU term	Book definition
Program abstractions	Vectorizable Loop	Vectorizable Loop	Grid	A vectorizable loop, executed on the GPU, made up of one or more Thread Blocks (bodies of vectorized loop) that can execute in parallel.
	Body of Vectorized Loop	Body of a (Strip-Mined) Vectorized Loop	Thread Block	A vectorized loop executed on a multithreaded SIMD Processor, made up of one or more threads of SIMD instructions. They can communicate via Local Memory.
	Sequence of SIMD Lane Operations	One iteration of a Scalar Loop	CUDA Thread	A vertical cut of a thread of SIMD instructions corresponding to one element executed by one SIMD Lane. Result is stored depending on mask and predicate register.
Machine object	A Thread of SIMD Instructions	Thread of Vector Instructions	Warp	A traditional thread, but it contains just SIMD instructions that are executed on a multithreaded SIMD Processor. Results stored depending on a per-element mask.
	SIMD Instruction	Vector Instruction	PTX Instruction	A single SIMD instruction executed across SIMD Lanes.
Processing hardware	Multithreaded SIMD Processor	(Multithreaded) Vector Processor	Streaming Multiprocessor	A multithreaded SIMD Processor executes threads of SIMD instructions, independent of other SIMD Processors.
	Thread Block Scheduler	Scalar Processor	Giga Thread Engine	Assigns multiple Thread Blocks (bodies of vectorized loop) to multithreaded SIMD Processors.
	SIMD Thread Scheduler	Thread scheduler in a Multithreaded CPU	Warp Scheduler	Hardware unit that schedules and issues threads of SIMD instructions when they are ready to execute; includes a scoreboard to track SIMD Thread execution.
	SIMD Lane	Vector lane	Thread Processor	A SIMD Lane executes the operations in a thread of SIMD instructions on a single element. Results stored depending on mask.
Memory hardware	GPU Memory	Main Memory	Global Memory	DRAM memory accessible by all multithreaded SIMD Processors in a GPU.
	Local Memory	Local Memory	Shared Memory	Fast local SRAM for one multithreaded SIMD Processor, unavailable to other SIMD Processors.
	SIMD Lane Registers	Vector Lane Registers	Thread Processor Registers	Registers in a single SIMD Lane allocated across a full thread block (body of vectorized loop).

FIGURE 6.12 Quick guide to GPU terms. We use the first column for hardware terms. Four groups cluster these 12 terms. From top to bottom: Program Abstractions, Machine Objects, Processing Hardware, and Memory Hardware.

well, how can we supplement it to improve the performance of a wider range of applications?

Having covered two different styles of MIMD that have a shared address space, we next introduce parallel processors where each processor has its own private address space, which makes it considerably easier to build much larger systems. The Internet services that you use every day depend on these large-scale systems.

Elaboration: While the GPU was introduced as having a separate memory from the CPU, both AMD and Intel have announced "fused" products that combine GPUs and CPUs to share a single memory. The challenge will be to maintain the high bandwidth memory in a fused architecture that has been a foundation of GPUs.

True or false: GPUs rely on graphics DRAM chips to reduce memory latency and thereby increase performance on graphics applications.

Check Yourself

6.7 Clusters, Warehouse Scale Computers, and Other Message-Passing Multiprocessors

The alternative approach to sharing an address space is for the processors to each have their own private physical address space. Figure 6.13 shows the classic organization of a multiprocessor with multiple private address spaces. This alternative multiprocessor must communicate via explicit message passing, which traditionally is the name of such style of computers. Provided the system has routines to send and receive messages, coordination is built in with message passing, since one processor knows when a message is sent, and the receiving processor knows when a message arrives. If the sender needs confirmation that the message has arrived, the receiving processor can then send an acknowledgment message back to the sender.

There have been several attempts to build large-scale computers based on high-performance message-passing networks, and they do offer better absolute

message passing Communicating between multiple processors by explicitly sending and receiving information.

send message routine A routine used by a processor in machines with private memories to pass a message to another processor.

receive message routine A routine used by a processor in machines with private memories to accept a message from another processor.

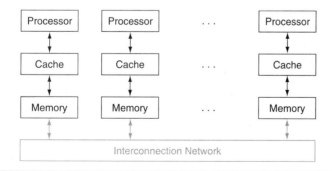

FIGURE 6.13 Classic organization of a multiprocessor with multiple private address spaces, traditionally called a message-passing multiprocessor. Note that unlike the SMP in Figure 6.7, the interconnection network is not between the caches and memory but is instead between processor-memory nodes.

communication performance than clusters built using local area networks. Indeed, many supercomputers today use custom networks. The problem is that they are much more expensive than local area networks like Ethernet. Few applications today outside of high-performance computing can justify the higher communication performance, given the much higher costs.

Hardware/ Software Interface

Computers that rely on message passing for communication rather than cache coherent shared memory are much easier for hardware designers to build (see Section 5.8). There is an advantage for programmers as well, in that communication is explicit, which means there are fewer performance surprises than with the implicit communication in cache-coherent shared memory computers. The downside for programmers is that it's harder to port a sequential program to a message-passing computer, since every communication must be identified in advance or the program doesn't work. Cache-coherent shared memory allows the hardware to figure out what data need to be communicated, which makes porting easier. There are differences of opinion as to which is the shortest path to high performance, given the pros and cons of implicit communication, but there is no confusion in the marketplace today. Multicore microprocessors use shared physical memory and nodes of a cluster communicate with each other using message passing.

Some concurrent applications run well on parallel hardware, independent of whether it offers shared addresses or message passing. In particular, task-level parallelism and applications with little communication—like Web search, mail servers, and file servers—do not require shared addressing to run well. As a result, **clusters** have become the most widespread example today of the message-passing parallel computer. Given the separate memories, each node of a cluster runs a distinct copy of the operating system. In contrast, the cores inside a microprocessor are connected using a high-speed network inside the chip, and a multichip shared-memory system uses the memory interconnect for communication. The memory interconnect has higher bandwidth and lower latency, allowing much better communication performance for shared memory multiprocessors.

clusters Collections of computers connected via I/O over standard network switches to form a message-passing multiprocessor.

The weakness of separate memories for user memory from a parallel programming perspective turns into a strength in system dependability (see Section 5.5). Since a cluster consists of independent computers connected through a local area network, it is much easier to replace a computer without bringing down the system in a cluster than in a shared memory multiprocessor. Fundamentally, the shared address means that it is difficult to isolate a processor and replace it without heroic work by the operating system and in the physical design of the server. It is also easy for clusters to scale down gracefully when a server fails, thereby improving **dependability**. Since the cluster software is a layer that runs on top of the local operating systems running on each computer, it is much easier to disconnect and replace a broken computer.

D E P E N D A B I L I T Y

Given that clusters are constructed from whole computers and independent, scalable networks, this isolation also makes it easier to expand the system without bringing down the application that runs on top of the cluster.

Their lower cost, higher availability, and rapid, incremental expandability make clusters attractive to service Internet providers, despite their poorer communication performance when compared to large-scale shared-memory multiprocessors. The search engines that hundreds of millions of us use every day depend upon this technology. Amazon, Facebook, Google, Microsoft, and others all have multiple datacenters each with clusters of tens of thousands of servers. Clearly, the use of multiple processors in Internet service companies has been hugely successful.

Warehouse-Scale Computers

Internet services, such as those described above, necessitated the construction of new buildings to house, power, and cool 100,000 servers. Although they may be classified as just large clusters, their architecture and operation are more sophisticated. They act as one giant computer and cost on the order of $150M for the building, the electrical and cooling infrastructure, the servers, and the networking equipment that connects and houses 50,000 to 100,000 servers. We consider them a new class of computer, called *Warehouse-Scale Computers* (WSC).

Anyone can build a fast CPU. The trick is to build a fast system.

Seymour Cray, considered the father of the supercomputer.

Hardware/ Software Interface

The most popular framework for batch processing in a WSC is MapReduce [Dean, 2008] and its open-source twin Hadoop. Inspired by the Lisp functions of the same name, Map first applies a programmer-supplied function to each logical input record. Map runs on thousands of servers to produce an intermediate result of key-value pairs. Reduce collects the output of those distributed tasks and collapses them using another programmer-defined function. With appropriate software support, both are highly parallel yet easy to understand and to use. Within 30 minutes, a novice programmer can run a MapReduce task on thousands of servers.

For example, one MapReduce program calculates the number of occurrences of every English word in a large collection of documents. Below is a simplified version of that program, which shows only the inner loop and assumes just one occurrence of all English words found in a document:

```
map(String key, String value):
    // key: document name
    // value: document contents
    for each word w in value:
    EmitIntermediate(w, "1"); // Produce list of all words
reduce(String key, Iterator values):
// key: a word
// values: a list of counts
    int result = 0;
    for each v in values:
    result += ParseInt(v); // get integer from key-value pair
    Emit(AsString(result));
```

The function `EmitIntermediate` used in the Map function emits each word in the document and the value one. Then the Reduce function sums all the values per word for each document using `ParseInt()` to get the number of occurrences per word in all documents. The MapReduce runtime environment schedules map tasks and reduce tasks to the servers of a WSC.

At this extreme scale, which requires innovation in power distribution, cooling, monitoring, and operations, the WSC is a modern descendant of the 1970s supercomputers—making Seymour Cray the godfather of today's WSC architects. His extreme computers handled computations that could be done nowhere else, but were so expensive that only a few companies could afford them. This time the target is providing information technology for the world instead of high-performance computing for scientists and engineers. Hence, WSCs surely play a more important societal role today than Cray's supercomputers did in the past.

While they share some common goals with servers, WSCs have three major distinctions:

software as a service (SaaS) Rather than selling software that is installed and run on customers' own computers, software is run at a remote site and made available over the Internet typically via a Web interface to customers. SaaS customers are charged based on use versus on ownership.

1. *Ample, easy parallelism*: A concern for a server architect is whether the applications in the targeted marketplace have enough parallelism to justify the amount of parallel hardware and whether the cost is too high for sufficient communication hardware to exploit this parallelism. A WSC architect has no such concern. First, batch applications like MapReduce benefit from the large number of independent data sets that need independent processing, such as billions of Web pages from a Web crawl. Second, interactive Internet service applications, also known as Software as a Service (SaaS), can benefit from millions of independent users of interactive Internet services. Reads and writes are rarely dependent in SaaS, so SaaS rarely needs to synchronize. For example, search uses a read-only index and email is normally reading and writing independent information. We call this type of easy parallelism *Request-Level Parallelism*, as many independent efforts can proceed in parallel naturally with little need for communication or synchronization.

2. *Operational Costs Count*: Traditionally, server architects design their systems for peak performance within a cost budget and worry about energy only to make sure they don't exceed the cooling capacity of their enclosure. They usually ignored operational costs of a server, assuming that they pale in comparison to purchase costs. WSCs have longer lifetimes—the building and electrical and cooling infrastructure are often amortized over 10 or more years—so the operational costs add up: energy, power distribution, and cooling represent more than 30% of the costs of a WSC over 10 years.

3. *Scale and the Opportunities/Problems Associated with Scale*: To construct a single WSC, you must purchase 100,000 servers along with the supporting infrastructure, which means volume discounts. Hence, WSCs are so massive

PARALLELISM

internally that you get economy of scale even if there are few WSCs. These economies of scale led to *cloud computing*, as the lower per unit costs of a WSC meant that cloud companies could rent servers at a profitable rate and still be below what it costs outsiders to do it themselves. The flip side of the economic opportunity of scale is the need to cope with the failure frequency of scale. Even if a server had a Mean Time To Failure of an amazing 25 years (200,000 hours), the WSC architect would need to design for five server failures every day. Section 5.15 mentioned annualized disk failure rate (AFR) was measured at Google at 2% to 4%. If there were four disks per server and their annual failure rate was 2%, the WSC architect should expect to see one disk fail every *hour*. Thus, fault tolerance is even more important for the WSC architect than for the server architect.

The economies of scale uncovered by WSC have realized the long dreamed of goal of computing as a utility. Cloud computing means anyone anywhere with good ideas, a business model, and a credit card can tap thousands of servers to deliver their vision almost instantly around the world. Of course, there are important obstacles that could limit the growth of cloud computing—such as security, privacy, standards, and the rate of growth of Internet bandwidth—but we foresee them being addressed so that WSCs and cloud computing can flourish.

To put the growth rate of cloud computing into perspective, in 2012 Amazon Web Services announced that it adds enough new server capacity *every day* to support all of Amazon's global infrastructure as of 2003, when Amazon was a $5.2Bn annual revenue enterprise with 6000 employees.

Now that we understand the importance of message-passing multiprocessors, especially for cloud computing, we next cover ways to connect the nodes of a WSC together. Thanks to **Moore's Law** and the increasing number of cores per chip, we now need networks inside a chip as well, so these topologies are important in the small as well as in the large.

MOORE'S LAW

Elaboration: The MapReduce framework shuffles and sorts the key-value pairs at the end of the Map phase to produce groups that all share the same key. These groups are next passed to the Reduce phase.

Elaboration: Another form of large-scale computing is *grid computing*, where the computers are spread across large areas, and then the programs that run across them must communicate via long haul networks. The most popular and unique form of grid computing was pioneered by the SETI@home project. As millions of PCs are idle at any one time doing nothing useful, they could be harvested and put to good use if someone developed software that could run on those computers and then gave each PC an independent piece of the problem to work on. The first example was the *Search for ExtraTerrestrial Intelligence* (SETI), which was launched at UC Berkeley in 1999. Over 5 million computer users in more than 200 countries have signed up for SETI@home, with more than 50% outside the US. By the end of 2011, the average performance of the SETI@home grid was 3.5 PetaFLOPS.

1. True or false: Like SMPs, message-passing computers rely on locks for synchronization.

2. True or false: Clusters have separate memories and thus need many copies of the operating system.

6.8 Introduction to Multiprocessor Network Topologies

Multicore chips require on-chip networks to connect cores together, and clusters require local area networks to connect servers together. This section reviews the pros and cons of different interconnection network topologies.

Network costs include the number of switches, the number of links on a switch to connect to the network, the width (number of bits) per link, and length of the links when the network is mapped into silicon. For example, some cores or servers may be adjacent and others may be on the other side of the chip or the other side of the datacenter. Network performance is multifaceted as well. It includes the latency on an unloaded network to send and receive a message, the throughput in terms of the maximum number of messages that can be transmitted in a given time period, delays caused by contention for a portion of the network, and variable performance depending on the pattern of communication. Another obligation of the network may be fault tolerance, since systems may be required to operate in the presence of broken components. Finally, in this era of energy-limited systems, the energy efficiency of different organizations may trump other concerns.

Networks are normally drawn as graphs, with each edge of the graph representing a link of the communication network. In the figures in this section, the processor-memory node is shown as a black square and the switch is shown as a colored circle. We assume here that all links are *bidirectional*; that is, information can flow in either direction. All networks consist of *switches* whose links go to processor-memory nodes and to other switches. The first network connects a sequence of nodes together:

This topology is called a *ring*. Since some nodes are not directly connected, some messages will have to hop along intermediate nodes until they arrive at the final destination.

Unlike a bus—a shared set of wires that allows broadcasting to all connected devices—a ring is capable of many simultaneous transfers.

Because there are numerous topologies to choose from, performance metrics are needed to distinguish these designs. Two are popular. The first is *total network bandwidth*, which is the bandwidth of each link multiplied by the number of links. This represents the peak bandwidth. For the ring network above, with P processors, the total network bandwidth would be P times the bandwidth of one link; the total network bandwidth of a bus is just the bandwidth of that bus.

To balance this best bandwidth case, we include another metric that is closer to the worst case: the *bisection bandwidth*. This metric is calculated by dividing the machine into two halves. Then you sum the bandwidth of the links that cross that imaginary dividing line. The bisection bandwidth of a ring is two times the link bandwidth. It is one times the link bandwidth for the bus. If a single link is as fast as the bus, the ring is only twice as fast as a bus in the worst case, but it is P times faster in the best case.

Since some network topologies are not symmetric, the question arises of where to draw the imaginary line when bisecting the machine. Bisection bandwidth is a worst-case metric, so the answer is to choose the division that yields the most pessimistic network performance. Stated alternatively, calculate all possible bisection bandwidths and pick the smallest. We take this pessimistic view because parallel programs are often limited by the weakest link in the communication chain.

At the other extreme from a ring is a *fully connected network*, where every processor has a bidirectional link to every other processor. For fully connected networks, the total network bandwidth is $P \times (P-1)/2$, and the bisection bandwidth is $(P/2)^2$.

The tremendous improvement in performance of fully connected networks is offset by the tremendous increase in cost. This consequence inspires engineers to invent new topologies that are between the cost of rings and the performance of fully connected networks. The evaluation of success depends in large part on the nature of the communication in the workload of parallel programs run on the computer.

The number of different topologies that have been discussed in publications would be difficult to count, but only a few have been used in commercial parallel processors. Figure 6.14 illustrates two of the popular topologies.

An alternative to placing a processor at every node in a network is to leave only the switch at some of these nodes. The switches are smaller than processor-memory-switch nodes, and thus may be packed more densely, thereby lessening distance and increasing performance. Such networks are frequently called *multistage networks* to reflect the multiple steps that a message may travel. Types of multistage networks are as numerous as single-stage networks; Figure 6.15 illustrates two of the popular multistage organizations. A *fully connected* or *crossbar network* allows any node to communicate with any other node in one pass through the network. An *Omega network* uses less hardware than the crossbar network ($2n \log_2 n$ versus n^2 switches), but contention can occur between messages, depending on the pattern

network bandwidth Informally, the peak transfer rate of a network; can refer to the speed of a single link or the collective transfer rate of all links in the network.

bisection bandwidth The bandwidth between two equal parts of a multiprocessor. This measure is for a worst-case split of the multiprocessor.

fully connected network A network that connects processor-memory nodes by supplying a dedicated communication link between every node.

multistage network A network that supplies a small switch at each node.

crossbar network A network that allows any node to communicate with any other node in one pass through the network.

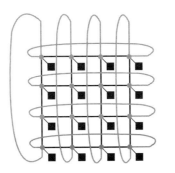

a. 2-D grid or mesh of 16 nodes   b. *n*-cube tree of 8 nodes ($8 = 2^3$ so $n = 3$)

FIGURE 6.14   Network topologies that have appeared in commercial parallel processors. The colored circles represent switches and the black squares represent processor-memory nodes. Even though a switch has many links, generally only one goes to the processor. The Boolean *n*-cube topology is an *n*-dimensional interconnect with 2^n nodes, requiring *n* links per switch (plus one for the processor) and thus *n* nearest-neighbor nodes. Frequently, these basic topologies have been supplemented with extra arcs to improve performance and reliability.

of communication. For example, the Omega network in Figure 6.15 cannot send a message from P_0 to P_6 at the same time that it sends a message from P_1 to P_4.

Implementing Network Topologies

This simple analysis of all the networks in this section ignores important practical considerations in the construction of a network. The distance of each link affects the cost of communicating at a high clock rate—generally, the longer the distance, the more expensive it is to run at a high clock rate. Shorter distances also make it easier to assign more wires to the link, as the power to drive many wires is less if the wires are short. Shorter wires are also cheaper than longer wires. Another practical limitation is that the three-dimensional drawings must be mapped onto chips that are essentially two-dimensional media. The final concern is energy. Energy concerns may force multicore chips to rely on simple grid topologies, for example. The bottom line is that topologies that appear elegant when sketched on the blackboard may be impractical when constructed in silicon or in a datacenter.

Now that we understand the importance of clusters and have seen topologies that we can follow to connect them together, we next look at the hardware and software of the interface of the network to the processor.

Check Yourself

True or false: For a ring with P nodes, the ratio of the total network bandwidth to the bisection bandwidth is P/2.

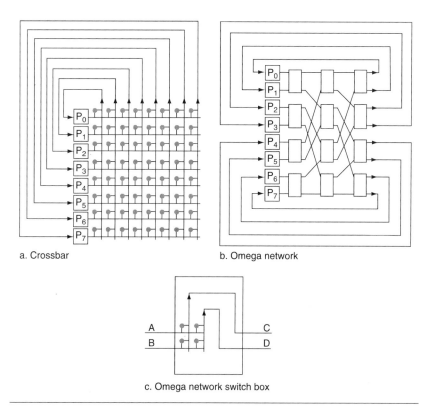

a. Crossbar   b. Omega network

c. Omega network switch box

FIGURE 6.15 Popular multistage network topologies for eight nodes. The switches in these drawings are simpler than in earlier drawings because the links are unidirectional; data come in at the left and exit out the right link. The switch box in c can pass A to C and B to D or B to C and A to D. The crossbar uses n^2 switches, where n is the number of processors, while the Omega network uses $2n \log_2 n$ of the large switch boxes, each of which is logically composed of four of the smaller switches. In this case, the crossbar uses 64 switches versus 12 switch boxes, or 48 switches, in the Omega network. The crossbar, however, can support any combination of messages between processors, while the Omega network cannot.

Communicating to the Outside World: Cluster Networking

This online section describes the networking hardware and software used to connect the nodes of a cluster together. The example is 10 gigabit/second Ethernet connected to the computer using *Peripheral Component Interconnect Express* (PCIe). It shows both software and hardware optimizations how to improve network performance, including zero copy messaging, user space communication, using polling instead of I/O interrupts, and hardware calculation of checksums. While the example is networking, the techniques in this section apply to storage controllers and other I/O devices as well.

After covering the performance of network at a low level of detail in this online section, the next section shows how to benchmark multiprocessors of all kinds with much higher-level programs.

6.10 Multiprocessor Benchmarks and Performance Models

As we saw in Chapter 1, benchmarking systems is always a sensitive topic, because it is a highly visible way to try to determine which system is better. The results affect not only the sales of commercial systems, but also the reputation of the designers of those systems. Hence, all participants want to win the competition, but they also want to be sure that if someone else wins, they deserve it because they have a genuinely better system. This desire leads to rules to ensure that the benchmark results are not simply engineering tricks for that benchmark, but are instead advances that improve performance of real applications.

To avoid possible tricks, a typical rule is that you can't change the benchmark. The source code and data sets are fixed, and there is a single proper answer. Any deviation from those rules makes the results invalid.

Many multiprocessor benchmarks follow these traditions. A common exception is to be able to increase the size of the problem so that you can run the benchmark on systems with a widely different number of processors. That is, many benchmarks allow weak scaling rather than require strong scaling, even though you must take care when comparing results for programs running different problem sizes.

Figure 6.16 gives a summary of several parallel benchmarks, also described below:

- *Linpack* is a collection of linear algebra routines, and the routines for performing Gaussian elimination constitute what is known as the Linpack benchmark. The DGEMM routine in the example on page 223 represents a small fraction of the source code of the Linpack benchmark, but it accounts for most of the execution time for the benchmark. It allows weak scaling, letting the user pick any size problem. Moreover, it allows the user to rewrite Linpack in almost any form and in any language, as long as it computes the proper result and performs the same number of floating point operations for a given problem size. Twice a year, the 500 computers with the fastest Linpack performance are published at www.top500.org. The first on this list is considered by the press to be the world's fastest computer.

- *SPECrate* is a throughput metric based on the SPEC CPU benchmarks, such as SPEC CPU 2006 (see Chapter 1). Rather than report performance of the individual programs, SPECrate runs many copies of the program simultaneously. Thus, it measures task-level parallelism, as there is no communication between the tasks. You can run as many copies of the programs as you want, so this is again a form of weak scaling.

Benchmark	Scaling?	Reprogram?	Description
Linpack	Weak	Yes	Dense matrix linear algebra [Dongarra, 1979]
SPECrate	Weak	No	Independent job parallelism [Henning, 2007]
Stanford Parallel Applications for Shared Memory SPLASH 2 [Woo et al., 1995]	Strong (although offers two problem sizes)	No	Complex 1D FFT Blocked LU Decomposition Blocked Sparse Cholesky Factorization Integer Radix Sort Barnes-Hut Adaptive Fast Multipole Ocean Simulation Hierarchical Radiosity Ray Tracer Volume Renderer Water Simulation with Spatial Data Structure Water Simulation without Spatial Data Structure
NAS Parallel Benchmarks [Bailey et al., 1991]	Weak	Yes (C or Fortran only)	EP: embarrassingly parallel MG: simplified multigrid CG: unstructured grid for a conjugate gradient method FT: 3-D partial differential equation solution using FFTs IS: large integer sort
PARSEC Benchmark Suite [Bienia et al., 2008]	Weak	No	Blackscholes—Option pricing with Black-Scholes PDE Bodytrack—Body tracking of a person Canneal—Simulated cache-aware annealing to optimize routing Dedup—Next-generation compression with data deduplication Facesim—Simulates the motions of a human face Ferret—Content similarity search server Fluidanimate—Fluid dynamics for animation with SPH method Freqmine—Frequent itemset mining Streamcluster—Online clustering of an input stream Swaptions—Pricing of a portfolio of swaptions Vips—Image processing x264—H.264 video encoding
Berkeley Design Patterns [Asanovic et al., 2006]	Strong or Weak	Yes	Finite-State Machine Combinational Logic Graph Traversal Structured Grid Dense Matrix Sparse Matrix Spectral Methods (FFT) Dynamic Programming N-Body MapReduce Backtrack/Branch and Bound Graphical Model Inference Unstructured Grid

FIGURE 6.16   Examples of parallel benchmarks.

■ *SPLASH* and *SPLASH 2* (Stanford Parallel Applications for Shared Memory) were efforts by researchers at Stanford University in the 1990s to put together a parallel benchmark suite similar in goals to the SPEC CPU benchmark suite. It includes both kernels and applications, including many from the high-performance computing community. This benchmark requires strong scaling, although it comes with two data sets.

- The *NAS (NASA Advanced Supercomputing) parallel benchmarks* were another attempt from the 1990s to benchmark multiprocessors. Taken from computational fluid dynamics, they consist of five kernels. They allow weak scaling by defining a few data sets. Like Linpack, these benchmarks can be rewritten, but the rules require that the programming language can only be C or Fortran.

- The recent *PARSEC (Princeton Application Repository for Shared Memory Computers) benchmark suite* consists of multithreaded programs that use Pthreads (POSIX threads) and OpenMP (*Open MultiProcessing*; see Section 6.5). They focus on emerging computational domains and consist of nine applications and three kernels. Eight rely on data parallelism, three rely on pipelined parallelism, and one on unstructured parallelism.

- On the cloud front, the goal of the *Yahoo! Cloud Serving Benchmark* (YCSB) is to compare performance of cloud data services. It offers a framework that makes it easy for a client to benchmark new data services, using Cassandra and HBase as representative examples [Cooper, 2010].

- Researchers at the University of California at Berkeley have advocated one approach. They identified 13 design patterns that they claim will be part of applications of the future. Frameworks or kernels implement these design patterns. Examples are sparse matrices, structured grids, finite-state machines, map reduce, and graph traversal. By keeping the definitions at a high level, they hope to encourage innovations at any level of the system. Thus, the system with the fastest sparse matrix solver is welcome to use any data structure, algorithm, and programming language, in addition to novel architectures and compilers.

The downside of such traditional restrictions to benchmarks is that innovation is chiefly limited to the architecture and compiler. Better data structures, algorithms, programming languages, and so on often cannot be used, since that would give a misleading result. The system could win because of, say, the algorithm, and not because of the hardware or the compiler.

While these guidelines are understandable when the foundations of computing are relatively stable—as they were in the 1990s and the first half of this decade—they are undesirable during a programming revolution. For this revolution to succeed, we need to encourage innovation at all levels.

Performance Models

A topic related to benchmarks is performance models. As we have seen with the increasing architectural diversity in this chapter—multithreading, SIMD, GPUs—it would be especially helpful if we had a simple model that offered insights into the performance of different architectures. It need not be perfect, just insightful.

The 3Cs for cache performance from Chapter 5 is an example performance model. It is not a perfect performance model, since it ignores potentially important

Pthreads A UNIX API for creating and manipulating threads. It is structured as a library.

factors like block size, block allocation policy, and block replacement policy. Moreover, it has quirks. For example, a miss can be ascribed due to capacity in one design, and to a conflict miss in another cache of the same size. Yet 3Cs model has been popular for 25 years, because it offers insight into the behavior of programs, helping both architects and programmers improve their creations based on insights from that model.

To find such a model for parallel computers, let's start with small kernels, like those from the 13 Berkeley design patterns in Figure 6.16. While there are versions with different data types for these kernels, floating point is popular in several implementations. Hence, peak floating-point performance is a limit on the speed of such kernels on a given computer. For multicore chips, peak floating-point performance is the collective peak performance of all the cores on the chip. If there were multiple microprocessors in the system, you would multiply the peak per chip by the total number of chips.

The demands on the memory system can be estimated by dividing this peak floating-point performance by the average number of floating-point operations per byte accessed:

$$\frac{\text{Floating-Point Operations/Sec}}{\text{Floating-Point Operations/Byte}} = \text{Bytes/Sec}$$

The ratio of floating-point operations per byte of memory accessed is called the **arithmetic intensity**. It can be calculated by taking the total number of floating-point operations for a program divided by the total number of data bytes transferred to main memory during program execution. Figure 6.17 shows the arithmetic intensity of several of the Berkeley design patterns from Figure 6.16.

arithmetic intensity
The ratio of floating-point operations in a program to the number of data bytes accessed by a program from main memory.

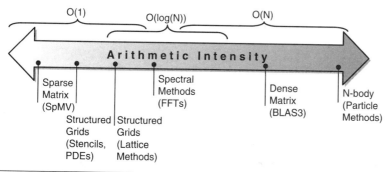

FIGURE 6.17   Arithmetic intensity, specified as the number of floating-point operations to run the program divided by the number of bytes accessed in main memory [Williams, Waterman, and Patterson 2009]. Some kernels have an arithmetic intensity that scales with problem size, such as Dense Matrix, but there are many kernels with arithmetic intensities independent of problem size. For kernels in this former case, weak scaling can lead to different results, since it puts much less demand on the memory system.

The Roofline Model

This simple model ties floating-point performance, arithmetic intensity, and memory performance together in a two-dimensional graph [Williams, Waterman, and Patterson 2009]. Peak floating-point performance can be found using the hardware specifications mentioned above. The working sets of the kernels we consider here do not fit in on-chip caches, so peak memory performance may be defined by the memory system behind the caches. One way to find the peak memory performance is the Stream benchmark. (See the *Elaboration* on page 395 in Chapter 5.)

Figure 6.18 shows the model, which is done once for a computer, not for each kernel. The vertical Y-axis is achievable floating-point performance from 0.5 to 64.0 GFLOPs/second. The horizontal X-axis is arithmetic intensity, varying from 1/8 FLOPs/DRAM byte accessed to 16 FLOPs/DRAM byte accessed. Note that the graph is a log-log scale.

For a given kernel, we can find a point on the X-axis based on its arithmetic intensity. If we draw a vertical line through that point, the performance of the kernel on that computer must lie somewhere along that line. We can plot a horizontal line showing peak floating-point performance of the computer. Obviously, the actual floating-point performance can be no higher than the horizontal line, since that is a hardware limit.

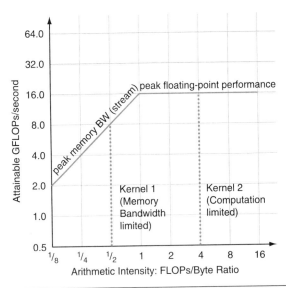

FIGURE 6.18   Roofline Model [Williams, Waterman, and Patterson 2009]. This example has a peak floating-point performance of 16 GFLOPS/sec and a peak memory bandwidth of 16 GB/sec from the Stream benchmark. (Since Stream is actually four measurements, this line is the average of the four.) The dotted vertical line in color on the left represents Kernel 1, which has an arithmetic intensity of 0.5 FLOPs/byte. It is limited by memory bandwidth to no more than 8 GFLOPS/sec on this Opteron X2. The dotted vertical line to the right represents Kernel 2, which has an arithmetic intensity of 4 FLOPs/byte. It is limited only computationally to 16 GFLOPS/s. (These data are based on the AMD Opteron X2 (Revision F) using dual cores running at 2 GHz in a dual socket system.)

How could we plot the peak memory performance, which is measured in bytes/second? Since the X-axis is FLOPs/byte and the Y-axis FLOPs/second, bytes/second is just a diagonal line at a 45-degree angle in this figure. Hence, we can plot a third line that gives the maximum floating-point performance that the memory system of that computer can support for a given arithmetic intensity. We can express the limits as a formula to plot the line in the graph in Figure 6.18:

$$\text{Attainable GFLOPs/sec} = \text{Min (Peak Memory BW} \times \text{Arithmetic Intensity, Peak Floating-Point Performance)}$$

The horizontal and diagonal lines give this simple model its name and indicate its value. The "roofline" sets an upper bound on performance of a kernel depending on its arithmetic intensity. Given a roofline of a computer, you can apply it repeatedly, since it doesn't vary by kernel.

If we think of arithmetic intensity as a pole that hits the roof, either it hits the slanted part of the roof, which means performance is ultimately limited by memory bandwidth, or it hits the flat part of the roof, which means performance is computationally limited. In Figure 6.18, kernel 1 is an example of the former, and kernel 2 is an example of the latter.

Note that the "ridge point," where the diagonal and horizontal roofs meet, offers an interesting insight into the computer. If it is far to the right, then only kernels with very high arithmetic intensity can achieve the maximum performance of that computer. If it is far to the left, then almost any kernel can potentially hit the maximum performance.

Comparing Two Generations of Opterons

The AMD Opteron X4 (Barcelona) with four cores is the successor to the Opteron X2 with two cores. To simplify board design, they use the same socket. Hence, they have the same DRAM channels and thus the same peak memory bandwidth. In addition to doubling the number of cores, the Opteron X4 also has twice the peak floating-point performance per core: Opteron X4 cores can issue two floating-point SSE2 instructions per clock cycle, while Opteron X2 cores issue at most one. As the two systems we're comparing have similar clock rates—2.2 GHz for Opteron X2 versus 2.3 GHz for Opteron X4—the Opteron X4 has about four times the peak floating-point performance of the Opteron X2 with the same DRAM bandwidth. The Opteron X4 also has a 2MiB L3 cache, which is not found in the Opteron X2.

In Figure 6.19 the roofline models for both systems are compared. As we would expect, the ridge point moves to the right, from 1 in the Opteron X2 to 5 in the Opteron X4. Hence, to see a performance gain in the next generation, kernels need an arithmetic intensity higher than 1, or their working sets must fit in the caches of the Opteron X4.

The roofline model gives an upper bound to performance. Suppose your program is far below that bound. What optimizations should you perform, and in what order?

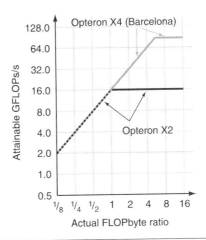

FIGURE 6.19 Roofline models of two generations of Opterons. The Opteron X2 roofline, which is the same as in Figure 6.18, is in black, and the Opteron X4 roofline is in color. The bigger ridge point of Opteron X4 means that kernels that were computationally bound on the Opteron X2 could be memory-performance bound on the Opteron X4.

To reduce computational bottlenecks, the following two optimizations can help almost any kernel:

PARALLELISM

1. *Floating-point operation mix.* Peak floating-point performance for a computer typically requires an equal number of nearly simultaneous additions and multiplications. That balance is necessary either because the computer supports a fused multiply-add instruction (see the *Elaboration* on page 228 in Chapter 3) or because the floating-point unit has an equal number of floating-point adders and floating-point multipliers. The best performance also requires that a significant fraction of the instruction mix is floating-point operations and not integer instructions.

2. *Improve **instruction-level parallelism** and apply SIMD.* For modern architectures, the highest performance comes when fetching, executing, and committing three to four instructions per clock cycle (see Section 4.10). The goal for this step is to improve the code from the compiler to increase ILP. One way is by unrolling loops, as we saw in Section 4.12. For the x86 architectures, a single AVX instruction can operate on four double precision operands, so they should be used whenever possible (see Sections 3.7 and 3.8).

To reduce memory bottlenecks, the following two optimizations can help:

PREDICTION

1. *Software prefetching.* Usually the highest performance requires keeping many memory operations in flight, which is easier to do by performing **predicting** accesses via software prefetch instructions rather than waiting until the data are required by the computation.

2. *Memory affinity.* Microprocessors today include a memory controller on the same chip with the microprocessor, which improves performance of the **memory hierarchy.** If the system has multiple chips, this means that some addresses go to the DRAM that is local to one chip, and the rest require accesses over the chip interconnect to access the DRAM that is local to another chip. This split results in non-uniform memory accesses, which we described in Section 6.5. Accessing memory through another chip lowers performance. This second optimization tries to allocate data and the threads tasked to operate on that data to the same memory-processor pair, so that the processors rarely have to access the memory of the other chips.

HIERARCHY

The roofline model can help decide which of these two optimizations to perform and the order in which to perform them. We can think of each of these optimizations as a "ceiling" below the appropriate roofline, meaning that you cannot break through a ceiling without performing the associated optimization.

The computational roofline can be found from the manuals, and the memory roofline can be found from running the Stream benchmark. The computational ceilings, such as floating-point balance, can also come from the manuals for that computer. A memory ceiling, such as memory affinity, requires running experiments on each computer to determine the gap between them. The good news is that this process only need be done once per computer, for once someone characterizes a computer's ceilings, everyone can use the results to prioritize their optimizations for that computer.

Figure 6.20 adds ceilings to the roofline model in Figure 6.18, showing the computational ceilings in the top graph and the memory bandwidth ceilings on the bottom graph. Although the higher ceilings are not labeled with both optimizations, they are implied in this figure; to break through the highest ceiling, you need to have already broken through all the ones below.

The width of the gap between the ceiling and the next higher limit is the reward for trying that optimization. Thus, Figure 6.20 suggests that optimization 2, which improves ILP, has a large benefit for improving computation on that computer, and optimization 4, which improves memory affinity, has a large benefit for improving memory bandwidth on that computer.

Figure 6.21 combines the ceilings of Figure 6.20 into a single graph. The arithmetic intensity of a kernel determines the optimization region, which in turn suggests which optimizations to try. Note that the computational optimizations and the memory bandwidth optimizations overlap for much of the arithmetic intensity. Three regions are shaded differently in Figure 6.21 to indicate the diferent optimization strategies. For example, Kernel 2 falls in the trapezoid on the right, which suggests working only on the computational optimizations. Kernel 1 falls in the parallelogram in the middle, which suggests trying both types of optimizations. Moreover, it suggests starting with optimizations 2 and 4. Note that the Kernel 1 vertical lines fall below the foating-point imbalance optimization, so optimization 1 may be unnecessary. If a kernel fell in the gray triangle on the lower left, it would suggest trying just memory optimizations.

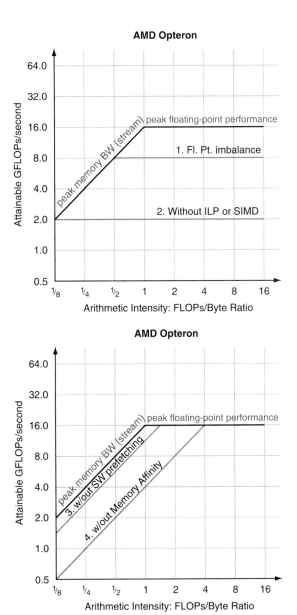

FIGURE 6.20 Roofline model with ceilings. The top graph shows the computational "ceilings" of 8 GFLOPs/sec if the floating-point operation mix is imbalanced and 2 GFLOPs/sec if the optimizations to increase ILP and SIMD are also missing. The bottom graph shows the memory bandwidth ceilings of 11 GB/sec without software prefetching and 4.8 GB/sec if memory affinity optimizations are also missing.

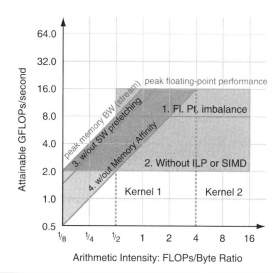

FIGURE 6.21   Roofline model with ceilings, overlapping areas shaded, and the two kernels from Figure 6.18. Kernels whose arithmetic intensity land in the trapezoid on the right should focus on computation optimizations, and kernels whose arithmetic intensity land in the gray triangle in the lower left should focus on memory bandwidth optimizations. Tose that land in the parallelogram in the middle need to worry about both. As Kernel 1 falls in the parallelogram in the middle, try optimizing ILP and SIMD, memory afnity, and sofware prefetching. Kernel 2 falls in the trapezoid on the right, so try optimizing ILP and SIMD and the balance of floating-point operations.

Thus far, we have been assuming that the arithmetic intensity is fixed, but that is not really the case. First, there are kernels where the arithmetic intensity increases with problem size, such as for Dense Matrix and N-body problems (see Figure 6.17). Indeed, this can be a reason that programmers have more success with weak scaling than with strong scaling. Second, the effectiveness of the **memory hierarchy** affects the number of accesses that go to memory, so optimizations that improve cache performance also improve arithmetic intensity. One example is improving temporal locality by unrolling loops and then grouping together statements with similar addresses. Many computers have special cache instructions that allocate data in a cache but do not first fill the data from memory at that address, since it will soon be over-written. Both these optimizations reduce memory traffic, thereby moving the arithmetic intensity pole to the right by a factor of, say, 1.5. This shift right could put the kernel in a different optimization region.

HIERARCHY

While the examples above show how to help programmers improve performance, architects can also use the model to decide where they should optimize hardware to improve the performance of the kernels that they think will be important.

The next section uses the roofline model to demonstrate the performance difference between a multicore microprocessor and a GPU and to see whether these differences reflect performance of real programs.

Elaboration: The ceilings are ordered so that lower ceilings are easier to optimize. Clearly, a programmer can optimize in any order, but following this sequence reduces the chances of wasting effort on an optimization that has no benefit due to other constraints. Like the 3Cs model, as long as the roofline model delivers on insights, a model can have assumptions that may prove optimistic. For example, roofline assumes the load is balanced between all processors.

Elaboration: An alternative to the Stream benchmark is to use the raw DRAM bandwidth as the roofline. While the raw bandwidth definitely is a hard upper bound, actual memory performance is often so far from that boundary that it's not that useful. That is, no program can go close to that bound. The downside to using Stream is that very careful programming may exceed the Stream results, so the memory roofline may not be as hard a limit as the computational roofline. We stick with Stream because few programmers will be able to deliver more memory bandwidth than Stream discovers.

Elaboration: Although the roofline model shown is for multicore processors, it clearly would work for a uniprocessor as well.

Check Yourself True or false: The main drawback with conventional approaches to benchmarks for parallel computers is that the rules that ensure fairness also slow software innovation.

6.11 Real Stuff: Benchmarking and Rooflines of the Intel Core i7 960 and the NVIDIA Tesla GPU

A group of Intel researchers published a paper [Lee et al., 2010] comparing a quad-core Intel Core i7 960 with multimedia SIMD extensions to the previous generation GPU, the NVIDIA Tesla GTX 280. Figure 6.22 lists the characteristics of the two systems. Both products were purchased in Fall 2009. The Core i7 is in Intel's 45-nanometer semiconductor technology while the GPU is in TSMC's 65-nanometer technology. Although it might have been fairer to have a comparison by a neutral party or by both interested parties, the purpose of this section is *not* to determine how much faster one product is than another, but to try to understand the relative value of features of these two contrasting architecture styles.

The rooflines of the Core i7 960 and GTX 280 in Figure 6.23 illustrate the differences in the computers. Not only does the GTX 280 have much higher memory bandwidth and double-precision floating-point performance, but also its double-precision ridge point is considerably to the left. The double-precision ridge point is 0.6 for the GTX 280 versus 3.1 for the Core i7. As mentioned above, it is much easier to hit peak computational performance the further the ridge point of

	Core I7-960	GTX 280	GTX 480	Ratio 280/i7	Ratio 480/i7
Number of processing elements (cores or SMs)	4	30	15	7.5	3.8
Clock frequency (GHz)	3.2	1.3	1.4	0.41	0.44
Die size	263	576	520	2.2	2.0
Technology	Intel 45 nm	TSMC 65 nm	TSMC 40 nm	1.6	1.0
Power (chip, not module)	130	130	167	1.0	1.3
Transistors	700 M	1400 M	3030 M	2.0	4.4
Memory brandwidth (GBytes/sec)	32	141	177	4.4	5.5
Single-precision SIMD width	4	8	32	2.0	8.0
Double-precision SIMD width	2	1	16	0.5	8.0
Peak single-precision scalar FLOPS (GFLOPs/sec)	26	117	63	4.6	2.5
Peak single-precision SIMD FLOPS (GFLOPs/sec)	102	311 to 933	515 or 1344	3.0–9.1	6.6–13.1
(SP 1 add or multiply)	N.A.	(311)	(515)	(3.0)	(6.6)
(SP 1 instruction fused multiply-adds)	N.A.	(622)	(1344)	(6.1)	(13.1)
(Rare SP dual issue fused multiply-add and multiply)	N.A.	(933)	N.A.	(9.1)	–
Peak double-precision SIMD FLOPS (GFLOPs/sec)	51	78	515	1.5	10.1

FIGURE 6.22 Intel Core i7-960, NVIDIA GTX 280, and GTX 480 specifications. The rightmost columns show the ratios of the Tesla GTX 280 and the Fermi GTX 480 to Core i7. Although the case study is between the Tesla 280 and i7, we include the Fermi 480 to show its relationship to the Tesla 280 since it is described in this chapter. Note that these memory bandwidths are higher than in Figure 6.23 because these are DRAM pin bandwidths and those in Figure 6.23 are at the processors as measured by a benchmark program. (From Table 2 in Lee et al. [2010].)

the roofline is to the left. For single-precision performance, the ridge point moves far to the right for both computers, so it's considerably harder to hit the roof of single-precision performance. Note that the arithmetic intensity of the kernel is based on the bytes that go to main memory, not the bytes that go to cache memory. Thus, as mentioned above, caching can change the arithmetic intensity of a kernel on a particular computer, if most references really go to the cache. Note also that this bandwidth is for unit-stride accesses in both architectures. Real gather-scatter addresses can be slower on the GTX 280 and on the Core i7, as we shall see.

The researchers selected the benchmark programs by analyzing the computational and memory characteristics of four recently proposed benchmark suites and then "formulated the set of *throughput computing kernels* that capture these characteristics." Figure 6.24 shows the performance results, with larger numbers meaning faster. The Rooflines help explain the relative performance in this case study.

Given that the raw performance specifications of the GTX 280 vary from 2.5× slower (clock rate) to 7.5× faster (cores per chip) while the performance varies

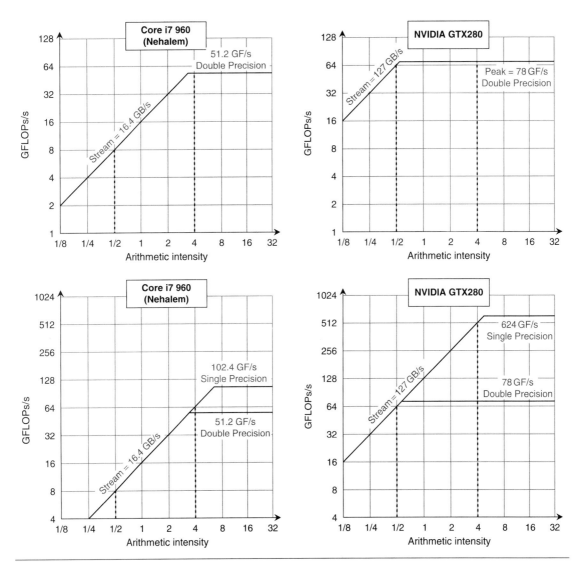

FIGURE 6.23 Roofline model [Williams, Waterman, and Patterson 2009]. These rooflines show double-precision floating-point performance in the top row and single-precision performance in the bottom row. (The DP FP performance ceiling is also in the bottom row to give perspective.) The Core i7 960 on the left has a peak DP FP performance of 51.2 GFLOPs/sec, a SP FP peak of 102.4 GFLOPs/sec, and a peak memory bandwidth of 16.4 GBytes/sec. The NVIDIA GTX 280 has a DP FP peak of 78 GFLOPs/sec, SP FP peak of 624 GFLOPs/sec, and 127 GBytes/sec of memory bandwidth. The dashed vertical line on the left represents an arithmetic intensity of 0.5 FLOP/byte. It is limited by memory bandwidth to no more than 8 DP GFLOPs/sec or 8 SP GFLOPs/sec on the Core i7. The dashed vertical line to the right has an arithmetic intensity of 4 FLOP/byte. It is limited only computationally to 51.2 DP GFLOPs/sec and 102.4 SP GFLOPs/sec on the Core i7 and 78 DP GFLOPs/sec and 624 SP GFLOPs/sec on the GTX 280. To hit the highest computation rate on the Core i7 you need to use all four cores and SSE instructions with an equal number of multiplies and adds. For the GTX 280, you need to use fused multiply-add instructions on all multithreaded SIMD processors.

Kernel	Units	Core i7-960	GTX 280	GTX 280/ i7-960
SGEMM	GFLOPs/sec	94	364	3.9
MC	Billion paths/sec	0.8	1.4	1.8
Conv	Million pixels/sec	1250	3500	2.8
FFT	GFLOPs/sec	71.4	213	3.0
SAXPY	GBytes/sec	16.8	88.8	5.3
LBM	Million lookups/sec	85	426	5.0
Solv	Frames/sec	103	52	0.5
SpMV	GFLOPs/sec	4.9	9.1	1.9
GJK	Frames/sec	67	1020	15.2
Sort	Million elements/sec	250	198	0.8
RC	Frames/sec	5	8.1	1.6
Search	Million queries/sec	50	90	1.8
Hist	Million pixels/sec	1517	2583	1.7
Bilat	Million pixels/sec	83	475	5.7

FIGURE 6.24 Raw and relative performance measured for the two platforms. In this study, SAXPY is just used as a measure of memory bandwidth, so the right unit is GBytes/sec and not GFLOP/sec. (Based on Table 3 in [Lee et al., 2010].)

from 2.0× slower (Solv) to 15.2× faster (GJK), the Intel researchers decided to find the reasons for the differences:

■ *Memory bandwidth.* The GPU has 4.4× the memory bandwidth, which helps explain why LBM and SAXPY run 5.0 and 5.3× faster; their working sets are hundreds of megabytes and hence don't fit into the Core i7 cache. (So as to access memory intensively, they purposely did not use cache blocking as in Chapter 5.) Hence, the slope of the rooflines explains their performance. SpMV also has a large working set, but it only runs 1.9× faster because the double-precision floating point of the GTX 280 is only 1.5× as fast as the Core i7.

■ *Compute bandwidth.* Five of the remaining kernels are compute bound: SGEMM, Conv, FFT, MC, and Bilat. The GTX is faster by 3.9, 2.8, 3.0, 1.8, and 5.7×, respectively. The first three of these use single-precision floating-point arithmetic, and GTX 280 single precision is 3 to 6× faster. MC uses double precision, which explains why it's only 1.8× faster since DP performance is only 1.5× faster. Bilat uses transcendental functions, which the GTX 280 supports directly. The Core i7 spends two-thirds of its time calculating transcendental functions for Bilat, so the GTX 280 is 5.7× faster. This observation helps point out the value of hardware support for operations that occur in your workload: double-precision floating point and perhaps even transcendentals.

- *Cache benefits. Ray casting* (RC) is only 1.6× faster on the GTX because cache blocking with the Core i7 caches prevents it from becoming memory bandwidth bound (see Sections 5.4 and 5.14), as it is on GPUs. Cache blocking can help Search, too. If the index trees are small so that they fit in the cache, the Core i7 is twice as fast. Larger index trees make them memory bandwidth bound. Overall, the GTX 280 runs search 1.8× faster. Cache blocking also helps Sort. While most programmers wouldn't run Sort on a SIMD processor, it can be written with a 1-bit Sort primitive called *split*. However, the split algorithm executes many more instructions than a scalar sort does. As a result, the Core i7 runs 1.25× as fast as the GTX 280. Note that caches also help other kernels on the Core i7, since cache blocking allows SGEMM, FFT, and SpMV to become compute bound. This observation re-emphasizes the importance of cache blocking optimizations in Chapter 5.

- *Gather-Scatter.* The multimedia SIMD extensions are of little help if the data are scattered throughout main memory; optimal performance comes only when accesses to data are aligned on 16-byte boundaries. Thus, GJK gets little benefit from SIMD on the Core i7. As mentioned above, GPUs offer gather-scatter addressing that is found in a vector architecture but omitted from most SIMD extensions. The memory controller even batches accesses to the same DRAM page together (see Section 5.2). This combination means the GTX 280 runs GJK a startling 15.2× as fast as the Core i7, which is larger than any single physical parameter in Figure 6.22. This observation reinforces the importance of gather-scatter to vector and GPU architectures that is missing from SIMD extensions.

- *Synchronization.* The performance of synchronization is limited by atomic updates, which are responsible for 28% of the total runtime on the Core i7 despite its having a hardware fetch-and-increment instruction. Thus, Hist is only 1.7× faster on the GTX 280. Solv solves a batch of independent constraints in a small amount of computation followed by barrier synchronization. The Core i7 benefits from the atomic instructions and a memory consistency model that ensures the right results even if not all previous accesses to memory hierarchy have completed. Without the memory consistency model, the GTX 280 version launches some batches from the system processor, which leads to the GTX 280 running 0.5× as fast as the Core i7. This observation points out how synchronization performance can be important for some data parallel problems.

It is striking how often weaknesses in the Tesla GTX 280 that were uncovered by kernels selected by Intel researchers were already being addressed in the successor architecture to Tesla: Fermi has faster double-precision floating-point performance, faster atomic operations, and caches. It was also interesting that the gather-scatter support of vector architectures that predate the SIMD instructions by decades was so important to the effective usefulness of these SIMD extensions, which some had predicted before the comparison. The Intel researchers noted that six of the 14 kernels would exploit SIMD better with more efficient gather-scatter support on the Core i7. This study certainly establishes the importance of cache blocking as well.

Now that we have seen a wide range of results of benchmarking different multiprocessors, let's return to our DGEMM example to see in detail how much we have to change the C code to exploit multiple processors.

6.12 Going Faster: Multiple Processors and Matrix Multiply

This section is the final and largest step in our incremental performance journey of adapting DGEMM to the underlying hardware of the Intel Core i7 (Sandy Bridge). Each Core i7 has eight cores, and the computer we have been using has two Core i7s. Thus, we have 16 cores on which to run DGEMM.

Figure 6.25 shows the OpenMP version of DGEMM that utilizes those cores. Note that line 30 is the *single* line added to Figure 5.48 to make this code run on multiple processors: an OpenMP pragma that tells the compiler to use multiple threads in the outermost loop. It tells the computer to spread the work of the outermost loop across all the threads.

Figure 6.26 plots a classic multiprocessor speed-up graph, showing the performance improvement versus a single thread as the number of threads increase. This graph makes it easy to see the challenges of strong scaling versus weak scaling. When everything fits in the first-level data cache, as is the case for 32 × 32 matrices, adding threads actually hurts performance. The 16-threaded version of DGEMM is almost half as fast as the single-threaded version in this case. In contrast, the two largest matrices get a 14 × speedup from 16 threads, and hence the classic two "up and to the right" lines in Figure 6.26.

Figure 6.27 shows the absolute performance increase as we increase the number of threads from one to 16. DGEMM now operates at 174 GLOPS for 960 × 960 matrices. As our unoptimized C version of DGEMM in Figure 3.22 ran this code at just 0.8 GFLOPS, the optimizations in Chapters 3 to 6 that tailor the code to the underlying hardware result in a speed-up of over 200 times!

Next up is our warnings of the fallacies and pitfalls of multiprocessing. The computer architecture graveyard is filled with parallel processing projects that have ignored them.

Elaboration: These results are with Turbo mode turned off. We are using a dual chip system in this system, so not surprisingly, we can get the full Turbo speed-up (3.3/2.6 = 1.27) with either one thread (only one core on one of the chips) or two threads (one core per chip). As we increase the number of threads and hence the number of active cores, the benefit of Turbo mode decreases, as there is less of the power budget to spend on the active cores. For four threads the average Turbo speed-up is 1.23, for eight it is 1.13, and for 16 it is 1.11.

```
1 #include <x86intrin.h>
2 #define UNROLL (4)
3 #define BLOCKSIZE 32
4 void do_block (int n, int si, int sj, int sk,
5                double *A, double *B, double *C)
6 {
7   for ( int i = si; i < si+BLOCKSIZE; i+=UNROLL*4 )
8     for ( int j = sj; j < sj+BLOCKSIZE; j++ ) {
9        __m256d c[4];
10       for ( int x = 0; x < UNROLL; x++ )
11         c[x] = _mm256_load_pd(C+i+x*4+j*n);
12     /* c[x] = C[i][j] */
13       for( int k = sk; k < sk+BLOCKSIZE; k++ )
14       {
15          __m256d b = _mm256_broadcast_sd(B+k+j*n);
16     /* b = B[k][j] */
17         for (int x = 0; x < UNROLL; x++)
18           c[x] = _mm256_add_pd(c[x], /* c[x]+=A[i][k]*b */
19                  _mm256_mul_pd(_mm256_load_pd(A+n*k+x*4+i), b));
20       }
21
22       for ( int x = 0; x < UNROLL; x++ )
23         _mm256_store_pd(C+i+x*4+j*n, c[x]);
24         /* C[i][j] = c[x] */
25     }
26 }
27
28 void dgemm (int n, double* A, double* B, double* C)
29 {
30 #pragma omp parallel for
31   for ( int sj = 0; sj < n; sj += BLOCKSIZE )
32     for ( int si = 0; si < n; si += BLOCKSIZE )
33       for ( int sk = 0; sk < n; sk += BLOCKSIZE )
34         do_block(n, si, sj, sk, A, B, C);
35 }
```

FIGURE 6.25   OpenMP version of DGEMM from Figure 5.48. Line 30 is the only OpenMP code, making the outermost for loop operate in parallel. This line is the only difference from Figure 5.48.

Elaboration: Although the Sandy Bridge supports two hardware threads per core, we do not get more performance from 32 threads. The reason is that a single AVX hardware is shared between the two threads multiplexed onto one core, so assigning two threads per core actually hurts performance due to the multiplexing overhead.

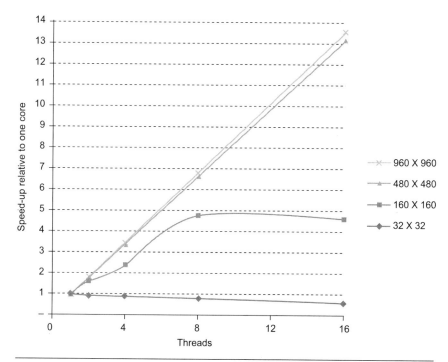

FIGURE 6.26 Performance improvements relative to a single thread as the number of threads increase. The most honest way to present such graphs is to make performance relative to the best version of a single processor program, which we did. This plot is relative to the performance of the code in Figure 5.48 *without* including OpenMP pragmas.

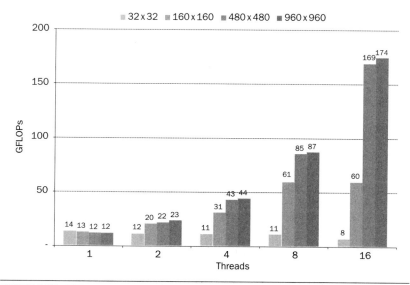

FIGURE 6.27 DGEMM performance versus the number of threads for four matrix sizes. The performance improvement compared unoptimized code in Figure 3.22 for the 960 × 960 matrix with 16 threads is an astounding 212 times faster!

6.13 Fallacies and Pitfalls

For over a decade prophets have voiced the contention that the organization of a single computer has reached its limits and that truly significant advances can be made only by interconnection of a multiplicity of computers in such a manner as to permit cooperative solution. …Demonstration is made of the continued validity of the single processor approach …

Gene Amdahl, "Validity of the single processor approach to achieving large scale computing capabilities," Spring Joint Computer Conference, 1967

The many assaults on parallel processing have uncovered numerous fallacies and pitfalls. We cover four here.

Fallacy: Amdahl's Law doesn't apply to parallel computers.

In 1987, the head of a research organization claimed that a multiprocessor machine had broken Amdahl's Law. To try to understand the basis of the media reports, let's see the quote that gave us Amdahl's Law [1967, p. 483]:

A fairly obvious conclusion which can be drawn at this point is that the effort expended on achieving high parallel processing rates is wasted unless it is accompanied by achievements in sequential processing rates of very nearly the same magnitude.

This statement must still be true; the neglected portion of the program must limit performance. One interpretation of the law leads to the following lemma: portions of every program must be sequential, so there must be an economic upper bound to the number of processors—say, 100. By showing linear speed-up with 1000 processors, this lemma is disproved; hence the claim that Amdahl's Law was broken.

The approach of the researchers was just to use weak scaling: rather than going 1000 times faster on the same data set, they computed 1000 times more work in comparable time. For their algorithm, the sequential portion of the program was constant, independent of the size of the input, and the rest was fully parallel—hence, linear speed-up with 1000 processors.

Amdahl's Law obviously applies to parallel processors. What this research does point out is that one of the main uses of faster computers is to run larger problems. Just be sure that users really care about those problems versus being a justification to buying an expensive computer by finding a problem that simply keeps lots of processors busy.

Fallacy: Peak performance tracks observed performance.

The supercomputer industry once used this metric in marketing, and the fallacy is exacerbated with parallel machines. Not only are marketers using the nearly unattainable peak performance of a uniprocessor node, but also they are then multiplying it by the total number of processors, assuming perfect speed-up! Amdahl's Law suggests how difficult it is to reach either peak; multiplying the two together multiplies the sins. The roofline model helps put peak performance in perspective.

Pitfall: Not developing the software to take advantage of, or optimize for, a multiprocessor architecture.

There is a long history of parallel software lagging behind parallel hardware, possibly because the software problems are much harder. We give one example to show the subtlety of the issues, but there are many examples we could choose!

One frequently encountered problem occurs when software designed for a uniprocessor is adapted to a multiprocessor environment. For example, the Silicon Graphics operating system originally protected the page table with a single lock, assuming that page allocation is infrequent. In a uniprocessor, this does not represent a performance problem. In a multiprocessor, it can become a major performance bottleneck for some programs. Consider a program that uses a large number of pages that are initialized at start-up, which UNIX does for statically allocated pages. Suppose the program is parallelized so that multiple processes allocate the pages. Because page allocation requires the use of the page table, which is locked whenever it is in use, even an OS kernel that allows multiple threads in the OS will be serialized if the processes all try to allocate their pages at once (which is exactly what we might expect at initialization time!).

This page table serialization eliminates parallelism in initialization and has a significant impact on overall parallel performance. This performance bottleneck persists even for task-level parallelism. For example, suppose we split the parallel processing program apart into separate jobs and run them, one job per processor, so that there is no sharing between the jobs. (This is exactly what one user did, since he reasonably believed that the performance problem was due to unintended sharing or interference in his application.) Unfortunately, the lock still serializes all the jobs—so even the independent job performance is poor.

This pitfall indicates the kind of subtle but significant performance bugs that can arise when software runs on multiprocessors. Like many other key software components, the OS algorithms and data structures must be rethought in a multiprocessor context. Placing locks on smaller portions of the page table effectively eliminated the problem.

Fallacy: You can get good vector performance without providing memory bandwidth.

As we saw in the Roofline model, memory bandwidth is quite important to all architectures. DAXPY requires 1.5 memory references per floating-point operation, and this ratio is typical of many scientific codes. Even if the floating-point operations took no time, a Cray-1 could not increase the DAXPY performance of the vector sequence used, since it was memory limited. The Cray-1 performance on Linpack jumped when the compiler used blocking to change the computation so that values could be kept in the vector registers. This approach lowered the number of memory references per FLOP and improved the performance by nearly a factor of two! Thus, the memory bandwidth on the Cray-1 became sufficient for a loop that formerly required more bandwidth, which is just what the Roofline model would predict.

6.14   Concluding Remarks

The dream of building computers by simply aggregating processors has been around since the earliest days of computing. Progress in building and using effective and efficient parallel processors, however, has been slow. This rate of progress has been limited by difficult software problems as well as by a long process of evolving the architecture of multiprocessors to enhance usability and improve efficiency. We have discussed many of the software challenges in this chapter, including the difficulty of writing programs that obtain good speed-up due to Amdahl's Law. The wide variety of different architectural approaches and the limited success and short life of many of the parallel architectures of the past have compounded the software difficulties. We discuss the history of the development of these multiprocessors in online 🌐 Section 6.15. To go into even greater depth on topics in this chapter, see Chapter 4 of *Computer Architecture: A Quantitative Approach, Fifth Edition* for more on GPUs and comparisons between GPUs and CPUs and Chapter 6 for more on WSCs.

As we said in Chapter 1, despite this long and checkered past, the information technology industry has now tied its future to parallel computing. Although it is easy to make the case that this effort will fail like many in the past, there are reasons to be hopeful:

- Clearly, *software as a service* (SaaS) is growing in importance, and clusters have proven to be a very successful way to deliver such services. By providing redundancy at a higher level, including geographically distributed datacenters, such services have delivered 24 × 7 × 365 availability for customers around the world.

- We believe that Warehouse-Scale Computers are changing the goals and principles of server design, just as the needs of mobile clients are changing the goals and principles of microprocessor design. Both are revolutionizing the software industry as well. Performance per dollar and performance per joule drive both mobile client hardware and the WSC hardware, and parallelism is the key to delivering on those sets of goals.

- SIMD and vector operations are a good match to multimedia applications, which are playing a larger role in the post-PC era. They share the advantage of being easier for the programmer than classic parallel MIMD programming and being more energy-efficient than MIMD. To put into perspective the importance of SIMD versus MIMD, Figure 6.28 plots the number of cores for MIMD versus the number of 32-bit and 64-bit operations per clock cycle in SIMD mode for x86 computers over time. For x86 computers, we expect to see two additional cores per chip about every 2 years and the SIMD width to double about every 4 years. Given these assumptions, over the next decade the potential speed-up from SIMD parallelism is twice that of

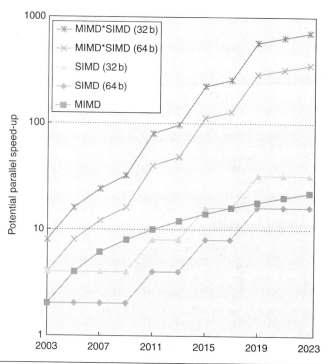

FIGURE 6.28   Potential speed-up via parallelism from MIMD, SIMD, and both MIMD and SIMD over time for x86 computers. This figure assumes that two cores per chip for MIMD will be added every 2 years and the number of operations for SIMD will double every 4 years.

MIMD parallelism. Given the effectiveness of SIMD for multimedia and its increasing importance in the post-PC era, that emphasis may be appropriate. Hence, it's as least as important to understand SIMD parallelism as MIMD parallelism, even though the latter has received much more attention.

■ The use of parallel processing in domains such as scientific and engineering computation is popular. This application domain has an almost limitless thirst for more computation. It also has many applications that have lots of natural concurrency. Once again, clusters dominate this application area. For example, using the 2012 Top 500 report, clusters are responsible for more than 80% of the 500 fastest Linpack results.

■ All desktop and server microprocessor manufacturers are building multiprocessors to achieve higher performance, so, unlike in the past, there is no easy path to higher performance for sequential applications. As we said earlier, sequential programs are now slow programs. Hence, programmers who need higher performance *must* parallelize their codes or write new parallel processing programs.

■ In the past, microprocessors and multiprocessors were subject to different definitions of success. When scaling uniprocessor performance, microprocessor architects were happy if single thread performance went up by the square root of the increased silicon area. Thus, they were pleased with sublinear performance in terms of resources. Multiprocessor success used to be defined as *linear* speed-up as a function of the number of processors, assuming that the cost of purchase or cost of administration of *n* processors was *n* times as much as one processor. Now that parallelism is happening on-chip via multicore, we can use the traditional microprocessor metric of being successful with sublinear performance improvement.

■ The success of just-in-time runtime compilation and autotuning makes it feasible to think of software adapting itself to take advantage of the increasing number of cores per chip, which provides flexibility that is not available when limited to static compilers.

■ Unlike in the past, the open source movement has become a critical portion of the software industry. This movement is a meritocracy, where better engineering solutions can win the mind share of the developers over legacy concerns. It also embraces innovation, inviting change to old software and welcoming new languages and software products. Such an open culture could be extremely helpful during this time of rapid change.

To motivate readers to embrace this revolution, we demonstrated the potential of parallelism concretely for matrix multiply on the Intel Core i7 (Sandy Bridge) in the Going Faster sections of Chapters 3 to 6:

■ Data-level parallelism in Chapter 3 improved performance by a factor of 3.85 by executing four 64-bit floating-point operations in parallel using the 256-bit operands of the AVX instructions, demonstrating the value of SIMD.

■ Instruction-level parallelism in Chapter 4 pushed performance up by another factor of 2.3 by unrolling loops four times to give the out-of-order execution hardware more instructions to schedule.

■ Cache optimizations in Chapter 5 improved performance of matrices that didn't fit into the L1 data cache by another factor of 2.0 to 2.5 by using cache blocking to reduce cache misses.

■ Thread-level parallelism in this chapter improved performance of matrices that don't fit into a single L1 data cache by another factor of 4 to 14 by utilizing all 16 cores of our multicore chips, demonstrating the value of MIMD. We did this by adding a single line using an OpenMP pragma.

Using the ideas in this book and tailoring the software to this computer added 24 lines of code to DGEMM. For the matrix sizes of 32×32, 160×160, 480×480, and 960×960, the overall performance speed-up from these ideas realized in those two-dozen lines of code is factors of 8, 39, 129, and 212!

This parallel revolution in the hardware/software interface is perhaps the greatest challenge facing the field in the last 60 years. You can also think of it as an outstanding opportunity, as our Going Faster sections demonstrate. This revolution will provide many new research and business prospects inside and outside the IT field, and the companies that dominate the multicore era may not be the same ones that dominated the uniprocessor era. After understanding the underlying hardware trends and learning to adapt software to them, perhaps you will be one of the innovators who will seize the opportunities that are certain to appear in the uncertain times ahead. We look forward to benefiting from your inventions!

Historical Perspective and Further Reading

This section online gives the rich and often disastrous history of multiprocessors over the last 50 years.

References

B. F. Cooper, A. Silberstein, E. Tam, R. Ramakrishnan, R. Sears. Benchmarking cloud serving systems with YCSB, In: Proceedings of the 1st ACM Symposium on Cloud computing, June 10–11, 2010, Indianapolis, Indiana, USA, doi:10.1145/1807128.1807152.

G. Regnier, S. Makineni, R. Illikkal, R. Iyer, D. Minturn, R. Huggahalli, D. Newell, L. Cline, and A. Foong. TCP onloading for data center servers. IEEE Computer, 37(11):48–58, 2004.

6.16 Exercises

6.1 First, write down a list of your daily activities that you typically do on a weekday. For instance, you might get out of bed, take a shower, get dressed, eat breakfast, dry your hair, brush your teeth. Make sure to break down your list so you have a minimum of 10 activities.

6.1.1 [5] <§6.2> Now consider which of these activities is already exploiting some form of parallelism (e.g., brushing multiple teeth at the same time, versus one at a time, carrying one book at a time to school, versus loading them all into your backpack and then carry them "in parallel"). For each of your activities, discuss if they are already working in parallel, but if not, why they are not.

6.1.2 [5] <§6.2> Next, consider which of the activities could be carried out concurrently (e.g., eating breakfast and listening to the news). For each of your activities, describe which other activity could be paired with this activity.

6.1.3 [5] <§6.2> For Exercise 6.1.2, what could we change about current systems (e.g., showers, clothes, TVs, cars) so that we could perform more tasks in parallel?

6.1.4 [5] <§6.2> Estimate how much shorter time it would take to carry out these activities if you tried to carry out as many tasks in parallel as possible.

6.2 You are trying to bake three blueberry pound cakes. Cake ingredients are as follows:

1 cup butter, softened
1 cup sugar
4 large eggs
1 teaspoon vanilla extract
1/2 teaspoon salt
1/4 teaspoon nutmeg
1 1/2 cups flour
1 cup blueberries

The recipe for a single cake is as follows:

Step 1: Preheat oven to 325°F (160°C). Grease and flour your cake pan.

Step 2: In large bowl, beat together with a mixer butter and sugar at medium speed until light and fluffy. Add eggs, vanilla, salt and nutmeg. Beat until thoroughly blended. Reduce mixer speed to low and add flour, 1/2 cup at a time, beating just until blended.

Step 3: Gently fold in blueberries. Spread evenly in prepared baking pan. Bake for 60 minutes.

6.2.1 [5] <§6.2> Your job is to cook three cakes as efficiently as possible. Assuming that you only have one oven large enough to hold one cake, one large bowl, one cake pan, and one mixer, come up with a schedule to make three cakes as quickly as possible. Identify the bottlenecks in completing this task.

6.2.2 [5] <§6.2> Assume now that you have three bowls, three cake pans and three mixers. How much faster is the process now that you have additional resources?

6.2.3 [5] <§6.2> Assume now that you have two friends that will help you cook, and that you have a large oven that can accommodate all three cakes. How will this change the schedule you arrived at in Exercise 6.2.1 above?

6.2.4 [5] <§6.2> Compare the cake-making task to computing three iterations of a loop on a parallel computer. Identify data-level parallelism and task-level parallelism in the cake-making loop.

6.3 Many computer applications involve searching through a set of data and sorting the data. A number of efficient searching and sorting algorithms have been devised in order to reduce the runtime of these tedious tasks. In this problem we will consider how best to parallelize these tasks.

6.3.1 [10] <§6.2> Consider the following binary search algorithm (a classic divide and conquer algorithm) that searches for a value X in a sorted N-element array A and returns the index of matched entry:

```
BinarySearch(A[0..N−1], X) {
    low = 0
    high = N −1
    while (low <= high) {
        mid = (low + high) / 2
        if (A[mid] >X)
            high = mid −1
        else if (A[mid] <X)
            low = mid + 1
        else
            return mid // found
    }
    return −1 // not found
}
```

Assume that you have Y cores on a multi-core processor to run BinarySearch. Assuming that Y is much smaller than N, express the speed-up factor you might expect to obtain for values of Y and N. Plot these on a graph.

6.3.2 [5] <§6.2> Next, assume that Y is equal to N. How would this affect your conclusions in your previous answer? If you were tasked with obtaining the best speed-up factor possible (i.e., strong scaling), explain how you might change this code to obtain it.

6.4 Consider the following piece of C code:

```
for (j=2;j<=1000;j++)
    D[j] = D[j−1]+D[j−2];
```

The ARMv8 code corresponding to the above fragment is:

```
            MOV     X10 #8000
            ADD     X2, X0, X10
            ADDI    X1, X0, #16
    LOOP:   LDUR    D0, [X1, #-16]
            LDUR    D2, [X1, #-8]
            FADDD   D4, D0, D2
```

```
STUR    D4, [X1, #0]
ADDI    X1, X1, #8
CMP     X1, X2
B.LE    LOOP
```

The latency of an instruction is the number of cycles that must come between that instruction and an instruction using the result. Assume floating point instructions have the following associated latencies (in cycles):

FADD	LDUR	STUR
4	6	1

6.4.1 [10] <§6.2> How many cycles does it take to execute this code?

6.4.2 [10] <§6.2> Re-order the code to reduce stalls. Now, how many cycles does it take to execute this code? (Hint: You can remove additional stalls by changing the offset on the STUR instruction.)

6.4.3 [10] <§6.2> When an instruction in a later iteration of a loop depends upon a data value produced in an earlier iteration of the same loop, we say that there is a *loop-carried dependence* between iterations of the loop. Identify the loop-carried dependences in the above code. Identify the dependent program variable and assembly-level registers. You can ignore the loop induction variable j.

6.4.4 [15] <§6.2> Rewrite the code by using registers to carry the data between iterations of the loop (as opposed to storing and re-loading the data from main memory). Show where this code stalls and calculate the number of cycles required to execute.

6.4.5 [10] <§6.2> Loop unrolling was described in Chapter 4. Unroll and optimize the loop above so that each unrolled loop handles three iterations of the original loop. Show where this code stalls and calculate the number of cycles required to execute.

6.4.6 [10] <§6.2> The unrolling from Exercise 6.4.5. works nicely because we happen to want a multiple of three iterations. What happens if the number of iterations is not known at compile time? How can we efficiently handle a number of iterations that isn't a multiple of the number of iterations per unrolled loop?

6.4.7 [15] <§6.2> Consider running this code on a two-node distributed memory message passing system. Assume that we are going to use message passing as described in Section 6.7, where we introduce a new operation send (x, y) that sends to node x the value y, and an operation receive( ) that waits for the value being sent to it. Assume that send operations take one cycle to issue (i.e., later instructions on the same node can proceed on the next cycle), but take several cycles to be received on the receiving node. Receive instructions stall execution on the node where they are executed until they receive a message. Can you use such a system to speed up the code for this exercise? If so, what is the maximum latency for receiving information that can be tolerated? If not, why not?

6.5 Consider the following recursive mergesort algorithm (another classic divide and conquer algorithm). Mergesort was first described by John Von Neumann in 1945. The basic idea is to divide an unsorted list x of m elements into two sublists of about half the size of the original list. Repeat this operation on each sublist, and continue until we have lists of size 1 in length. Then starting with sublists of length 1, "merge" the two sublists into a single sorted list.

```
Mergesort(m)
    var list left, right, result
    if length(m) ≤ 1
        return m
    else
        var middle = length(m) / 2
        for each x in m up to middle
            add x to left
        for each x in m after middle
            add x to right
        left = Mergesort(left)
        right = Mergesort(right)
        result = Merge(left, right)
        return result
```

The merge step is carried out by the following code:

```
Merge(left,right)
    var list result
    while length(left) >0 and length(right) > 0
        if first(left) ≤ first(right)
            append first(left) to result
            left = rest(left)
        else
            append first(right) to result
            right = rest(right)
    if length(left) >0
        append rest(left) to result
    if length(right) >0
        append rest(right) to result
    return result
```

6.5.1 [10] <§6.2> Assume that you have Y cores on a multicore processor to run Mergesort. Assuming that Y is much smaller than length (m), express the speed-up factor you might expect to obtain for values of Y and length (m). Plot these on a graph.

6.5.2 [10] <§6.2> Next, assume that Y is equal to length (m). How would this affect your conclusions in your previous answer? If you were tasked with obtaining the best speed-up factor possible (i.e., strong scaling), explain how you might change this code to obtain it.

6.6 Matrix multiplication plays an important role in a number of applications. Two matrices can only be multiplied if the number of columns of the first matrix is equal to the number of rows in the second.

Let's assume we have an $m \times n$ matrix A and we want to multiply it by an $n \times p$ matrix B. We can express their product as an $m \times p$ matrix denoted by AB (or $A \cdot B$). If we assign $C = AB$, and $c_{i,j}$ denotes the entry in C at position (i, j), then for each element i and j with $1 \leq i \leq m$ and $1 \leq j \leq p$ $c_{i,j} = \sum_{k=1}^{n} a_{i,k} \times b_{k,j}$. Now we want to see if we can parallelize the computation of C. Assume that matrices are laid out in memory sequentially as follows: $a_{1,1}, a_{2,1}, a_{3,1}, a_{4,1}, \ldots$, etc.

6.6.1 [10] <§6.5> Assume that we are going to compute C on both a single-core shared-memory machine and a four-core shared-memory machine. Compute the speed-up we would expect to obtain on the four-core machine, ignoring any memory issues.

6.6.2 [10] <§6.5> Repeat Exercise 6.6.1, assuming that updates to C incur a cache miss due to false sharing when consecutive elements are in a row (i.e., index i) are updated.

6.6.3 [10] <§6.5> How would you fix the false sharing issue that can occur?

6.7 Consider the following portions of two different programs running at the same time on four processors in a *symmetric multicore processor* (SMP). Assume that before this code is run, both x and y are 0.

Core 1: x = 2;

Core 2: y = 2;

Core 3: w = x + y + 1;

Core 4: z = x + y;

6.7.1 [10] <§6.5> What are all the possible resulting values of w, x, y, and z? For each possible outcome, explain how we might arrive at those values. You will need to examine all possible interleavings of instructions.

6.7.2 [5] <§6.5> How could you make the execution more deterministic so that only one set of values is possible?

6.8 The dining philosopher's problem is a classic problem of synchronization and concurrency. The general problem is stated as philosophers sitting at a round table doing one of two things: eating or thinking. When they are eating, they are not thinking, and when they are thinking, they are not eating. There is a bowl of pasta in the center. A fork is placed in between each philosopher. The result is that each philosopher has one fork to her left and one fork to her right. Given the nature of eating pasta, the philosopher needs two forks to eat, and can only use the forks on her immediate left and right. The philosophers do not speak to one another.

6.8.1 [10] <§6.7> Describe the scenario where none of philosophers ever eats (i.e., starvation). What is the sequence of events that happen that lead up to this problem?

6.8.2 [10] <§6.7> Describe how we can solve this problem by introducing the concept of a priority. Can we guarantee that we will treat all the philosophers fairly? Explain.

Now assume we hire a waiter who is in charge of assigning forks to philosophers. Nobody can pick up a fork until the waiter says they can. The waiter has global knowledge of all forks. Further, if we impose the policy that philosophers will always request to pick up their left fork before requesting to pick up their right fork, then we can guarantee to avoid deadlock.

6.8.3 [10] <§6.7> We can implement requests to the waiter as either a queue of requests or as a periodic retry of a request. With a queue, requests are handled in the order they are received. The problem with using the queue is that we may not always be able to service the philosopher whose request is at the head of the queue (due to the unavailability of resources). Describe a scenario with five philosophers where a queue is provided, but service is not granted even though there are forks available for another philosopher (whose request is deeper in the queue) to eat.

6.8.4 [10] <§6.7> If we implement requests to the waiter by periodically repeating our request until the resources become available, will this solve the problem described in Exercise 6.8.3? Explain.

6.9 Consider the following three CPU organizations:

CPU SS: A two-core superscalar microprocessor that provides out-of-order issue capabilities on two *function units* (FUs). Only a single thread can run on each core at a time.

CPU MT: A fine-grained multithreaded processor that allows instructions from two threads to be run concurrently (i.e., there are two functional units), though only instructions from a single thread can be issued on any cycle.

CPU SMT: An SMT processor that allows instructions from two threads to be run concurrently (i.e., there are two functional units), and instructions from either or both threads can be issued to run on any cycle.

Assume we have two threads X and Y to run on these CPUs that include the following operations:

Thread X	Thread Y
A1 – takes three cycles to execute	B1 – take two cycles to execute
A2 – no dependences	B2 – conflicts for a functional unit with B1
A3 – conflicts for a functional unit with A1	B3 – depends on the result of B2
A4 – depends on the result of A3	B4 – no dependences and takes two cycles to execute

Assume all instructions take a single cycle to execute unless noted otherwise or they encounter a hazard.

6.9.1 [10] <§6.4> Assume that you have one SS CPU. How many cycles will it take to execute these two threads? How many issue slots are wasted due to hazards?

6.9.2 [10] <§6.4> Now assume you have two SS CPUs. How many cycles will it take to execute these two threads? How many issue slots are wasted due to hazards?

6.9.3 [10] <§6.4> Assume that you have one MT CPU. How many cycles will it take to execute these two threads? How many issue slots are wasted due to hazards?

6.9.4 [10] <§6.4> Assume you have one SMT CPU. How many cycles will it take to execute the two threads? How many issue slots are wasted due to hazards?

6.10 Virtualization software is being aggressively deployed to reduce the costs of managing today's high-performance servers. Companies like VMWare, Microsoft and IBM have all developed a range of virtualization products. The general concept, described in Chapter 5, is that a hypervisor layer can be introduced between the hardware and the operating system to allow multiple operating systems to share the same physical hardware. The hypervisor layer is then responsible for allocating CPU and memory resources, as well as handling services typically handled by the operating system (e.g., I/O).

Virtualization provides an abstract view of the underlying hardware to the hosted operating system and application software. This will require us to rethink how multi-core and multiprocessor systems will be designed in the future to support the sharing of CPUs and memories by a number of operating systems concurrently.

6.10.1 [30] <§6.4> Select two hypervisors on the market today, and compare and contrast how they virtualize and manage the underlying hardware (CPUs and memory).

6.10.2 [15] <§6.4> Discuss what changes may be necessary in future multi-core CPU platforms in order to better match the resource demands placed on these systems. For instance, can multithreading play an effective role in alleviating the competition for computing resources?

6.11 We would like to execute the loop below as efficiently as possible. We have two different machines, a MIMD machine and a SIMD machine.

```
for (i=0; i<2000; i++)
  for (j=0; j<3000; j++)
      X_array[i][j] = Y_array[j][i] + 200;
```

6.11.1 [10] <§6.3> For a four CPU MIMD machine, show the sequence of LEGv8 instructions that you would execute on each CPU. What is the speed-up for this MIMD machine?

6.11.2 [20] <§6.3> For an eight-wide SIMD machine (i.e., eight parallel SIMD functional units), write an assembly program in using your own SIMD extensions to LEGv8 to execute the loop. Compare the number of instructions executed on the SIMD machine to the MIMD machine.

6.12 A systolic array is an example of an MISD machine. A systolic array is a pipeline network or "wavefront" of data processing elements. Each of these elements does not need a program counter since execution is triggered by the arrival of data. Clocked systolic arrays compute in "lock-step" with each processor undertaking alternate compute and communication phases.

6.12.1 [10] <§6.3> Consider proposed implementations of a systolic array (you can find these on the Internet or in technical publications). Then attempt to program the loop provided in Exercise 6.11 using this MISD model. Discuss any difficulties you encounter.

6.12.2 [10] <§6.3> Discuss the similarities and differences between an MISD and SIMD machines. Answer this question in terms of data-level parallelism.

6.13 Assume we want to execute the DAXP loop shown on page 526 in ARMv8t assembly on the NVIDIA 8800 GTX GPU described in this chapter. In this problem, we will assume that all math operations are performed on single-precision floating-point numbers (we will rename the loop SAXPY). Assume that instructions take the following number of cycles to execute.

Loads	Stores	Add.S	Mult.S
5	2	3	4

6.13.1 [20] <§6.6> Describe how you will constructs warps for the SAXPY loop to exploit the eight cores provided in a single multiprocessor.

6.14 Download the CUDA Toolkit and SDK from https://developer.nvidia.com/cuda-toolkit. Make sure to use the "emurelease" (Emulation Mode) version of the code. (You will not need actual NVIDIA hardware for this assignment.) Build the example programs provided in the SDK, and confirm that they run on the emulator.

6.14.1 [90] <§6.6> Using the "template" SDK sample as a starting point, write a CUDA program to perform the following vector operations:

1) $a - b$ (vector-vector subtraction)

2) $a \cdot b$ (vector dot product)

The dot product of two vectors $a = [a_1, a_2, ..., a_n]$ and $b = [b_1, b_2, ..., b_n]$ is defined as:

$$a \cdot b = \sum_{i=1}^{n} a_i b_i = a_1 b_1 + a_2 b_2 + \cdots + a_n b_n$$

Submit code for each program that demonstrates each operation and verifies the correctness of the results.

6.14.2 [90] <§6.6> If you have GPU hardware available, complete a performance analysis on your program, examining the computation time for the GPU and a CPU version of your program for a range of vector sizes. Explain any results you see.

6.15 AMD has recently announced integrating a graphics processing unit with their x86 cores into a single package (though with different clocks for each of the cores). This is an example of a heterogeneous multiprocessor system. One of the key design points is to allow for fast data communication between the CPU and the GPU. Before AMD's Fusion architecture, communications were needed between discrete CPU and GPU chips. Presently, the plan is to use multiple (at least 16) PCI express channels to facilitate intercommunication.

6.15.1 [25] <§6.6> Compare the bandwidth and latency associated with these two interconnect technologies.

6.16 Refer to Figure 6.14b, which shows an n-cube interconnect topology of order 3 that interconnects eight nodes. One attractive feature of an n-cube interconnection network topology is its ability to sustain broken links and still provide connectivity.

6.16.1 [10] <§6.8> Develop an equation that computes how many links in the n-cube (where n is the order of the cube) can fail and we can still guarantee an unbroken link will exist to connect any node in the n-cube.

6.16.2 [10] <§6.8> Compare the resiliency to failure of n-cube to a fully connected interconnection network. Plot a comparison of reliability as a function of the added number of links for the two topologies.

6.17 Benchmarking is a field of study that involves identifying representative workloads to run on specific computing platforms in order to be able to objectively compare performance of one system to another. In this exercise we will compare two classes of benchmarks: the Whetstone CPU benchmark and the PARSEC Benchmark suite. Select one program from PARSEC. All programs should be freely available on the Internet. Consider running multiple copies of Whetstone versus running the PARSEC Benchmark on any of the systems described in Section 6.11.

6.17.1 [60] <§6.10> What is inherently different between these two classes of workload when run on these multi-core systems?

6.17.2 [60] <§6.10> In terms of the Roofline Model, how dependent will the results you obtain when running these benchmarks be on the amount of sharing and synchronization present in the workload used?

6.18 When performing computations on sparse matrices, latency in the memory hierarchy becomes much more of a factor. Sparse matrices lack the spatial locality in the datastream typically found in matrix operations. As a result, new matrix representations have been proposed.

One of the earliest sparse matrix representations is the Yale Sparse Matrix Format. It stores an initial sparse $m \times n$ matrix, M in row form using three one-dimensional arrays. Let R be the number of nonzero entries in M. We construct an array A of length R that contains all nonzero entries of M (in left-to-right top-to-bottom order). We also construct a second array IA of length $m+1$ (i.e., one entry per row, plus one). $IA(i)$ contains the index in A of the first nonzero element of row i. Row i of the original matrix extends from $A(IA(i))$ to $A(IA(i+1)-1)$. The third array, JA, contains the column index of each element of A, so it also is of length R.

6.18.1 [15] <§6.10> Consider the sparse matrix X below and write C code that would store this code in Yale Sparse Matrix Format.

```
Row 1 [1, 2, 0, 0, 0, 0]
Row 2 [0, 0, 1, 1, 0, 0]
Row 3 [0, 0, 0, 0, 9, 0]
Row 4 [2, 0, 0, 0, 0, 2]
Row 5 [0, 0, 3, 3, 0, 7]
Row 6 [1, 3, 0, 0, 0, 1]
```

6.18.2 [10] <§6.10> In terms of storage space, assuming that each element in matrix X is single-precision floating point, compute the amount of storage used to store the matrix above in Yale Sparse Matrix Format.

6.18.3 [15] <§6.10> Perform matrix multiplication of matrix X by matrix Y shown below.

```
[2, 4, 1, 99, 7, 2]
```

Put this computation in a loop, and time its execution. Make sure to increase the number of times this loop is executed to get good resolution in your timing measurement. Compare the runtime of using a naïve representation of the matrix, and the Yale Sparse Matrix Format.

6.18.4 [15] <§6.10> Can you find a more efficient sparse matrix representation (in terms of space and computational overhead)?

6.19 In future systems, we expect to see heterogeneous computing platforms constructed out of heterogeneous CPUs. We have begun to see some appear in the embedded processing market in systems that contain both floating-point DSPs and microcontroller CPUs in a multichip module package.

Assume that you have three classes of CPU:

CPU A—A moderate-speed multi-core CPU (with a floating-point unit) that can execute multiple instructions per cycle.

CPU B—A fast single-core integer CPU (i.e., no floating-point unit) that can execute a single instruction per cycle.

CPU C—A slow vector CPU (with floating-point capability) that can execute multiple copies of the same instruction per cycle.

Assume that our processors run at the following frequencies:

CPU A	CPU B	CPU C
1 GHz	3 GHz	250 MHz

CPU A can execute two instructions per cycle, CPU B can execute one instruction per cycle, and CPU C can execute eight instructions (through the same instruction) per cycle. Assume all operations can complete execution in a single cycle of latency without any hazards.

All three CPUs have the ability to perform integer arithmetic, though CPU B cannot perform floating point arithmetic. CPU A and B have an instruction set similar to a LEGv8 processor. CPU C can only perform floating point add and subtract operations, as well as memory loads and stores. Assume all CPUs have access to shared memory and that synchronization has zero cost.

The task at hand is to compare two matrices X and Y that each contain 1024×1024 floating-point elements. The output should be a count of the number of indices where the value in X was larger or equal to the value in Y.

6.19.1 [10] <§6.11> Describe how you would partition the problem on the three different CPUs to obtain the best performance.

6.19.2 [10] <§6.11> What kind of instruction would you add to the vector CPU C to obtain better performance?

6.20 This question looks at the amount of queuing that is occurring in the system given a maximum transaction processing rate, and the latency observed on average for a transaction. The latency includes both the service time (which is computed by the maximum rate) and the queue time.

Assume a quad-core computer system can process database queries at a steady state maximum rate of rate requests per second. Also assume that each transaction takes, on average, lat ms to process. For each of the pairs in the table, answer the following questions:

Average Transaction Latency	Maximum transaction processing rate
1 ms	5000/sec
2 ms	5000/sec
1 ms	10,000/sec
2 ms	10,000/sec

For each of the pairs in the table, answer the following questions:

6.20.1 [10] <§6.11> On average, how many requests are being processed at any given instant?

6.20.2 [10] <§6.11> If we move to an eight-core system, ideally, what will happen to the system throughput (i.e., how many queries/second will the computer process)?

6.20.3 [10] <§6.11> Discuss why we rarely obtain this kind of speed-up by simply increasing the number of cores.

§6.1, page 518: False. Task-level parallelism can help sequential applications and sequential applications can be made to run on parallel hardware, although it is more challenging.

§6.2, page 523: False. *Weak* scaling can compensate for a serial portion of the program that would otherwise limit scalability, but not so for strong scaling.

§6.3, page 528: True, but they are missing useful vector features like gather-scatter and vector length registers that improve the efficiency of vector architectures. (As an elaboration in this section mentions, the AVX2 SIMD extensions offers indexed loads via a gather operation but *not* scatter for indexed stores. The Haswell generation x86 microprocessor is the first to support AVX2.)

§6.4, page 533: 1. True. 2. True.

§6.5, page 537: False. Since the shared address is a *physical* address, multiple tasks each in their own *virtual* address spaces can run well on a shared memory multiprocessor.

§6.6, page 545: False. Graphics DRAM chips are prized for their higher bandwidth.

§6.7, page 550: 1. False. Sending and receiving a message is an implicit synchronization, as well as a way to share data. 2. True.

§6.8, page 552: True.

§6.9, page 6.9-10: 1. polling with loads and stores. 2. interrupts with DMA.

§6.10, page 564: True. We likely need innovation at all levels of the hardware and software stack for parallel computing to succeed.

Answers to Check Yourself

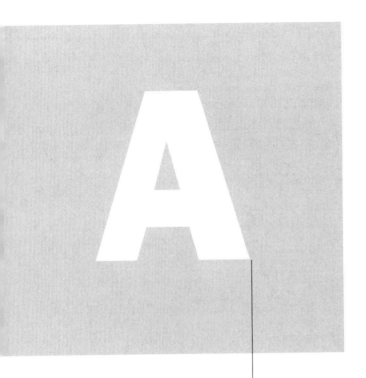

APPENDIX

The Basics of Logic Design

I always loved that word, Boolean.

Claude Shannon
IEEE Spectrum, April 1992
(Shannon's master's thesis showed
that the algebra invented by George
Boole in the 1800s could represent the
workings of electrical switches.)

Introduction

This appendix provides a brief discussion of the basics of logic design. It does not replace a course in logic design, nor will it enable you to design significant working logic systems. If you have little or no exposure to logic design, however, this appendix will provide sufficient background to understand all the material in this book. In addition, if you are looking to understand some of the motivation behind how computers are implemented, this material will serve as a useful introduction. If your curiosity is aroused but not sated by this appendix, the references at the end provide several additional sources of information.

Section A.2 introduces the basic building blocks of logic, namely, *gates*. Section A.3 uses these building blocks to construct simple *combinational* logic systems, which contain no memory. If you have had some exposure to logic or digital systems, you will probably be familiar with the material in these first two sections. Section A.5 shows how to use the concepts of Sections A.2 and A.3 to design an ALU for the LEGv8 processor. Section A.6 shows how to make a fast adder, and may be safely skipped if you are not interested in this topic. Section A.7 is a short introduction to the topic of clocking, which is necessary to discuss how memory elements work. Section A.8 introduces memory elements, and Section A.9 extends it to focus on random access memories; it describes both the characteristics that are important to understanding how they are used, as discussed in Chapter 4, and the background that motivates many of the aspects of memory hierarchy design discussed in Chapter 5. Section A.10 describes the design and use of finite-state machines, which are sequential logic blocks. If you intend to read 🖦 Appendix C, you should thoroughly understand the material in Sections A.2 through A.10. If you intend to read only the material on control in Chapter 4, you can skim the appendices; however, you should have some familiarity with all the material except Section A.11. Section A.11 is intended for those who want a deeper understanding of clocking methodologies and timing. It explains the basics of how edge-triggered clocking works, introduces another clocking scheme, and briefly describes the problem of synchronizing asynchronous inputs.

Throughout this appendix, where it is appropriate, we also include segments to demonstrate how logic can be represented in Verilog, which we introduce in Section A.4. A more extensive and complete Verilog tutorial is available online on the Companion Web site for this book.

A.2   Gates, Truth Tables, and Logic Equations

The electronics inside a modern computer are *digital*. Digital electronics operate with only two voltage levels of interest: a high voltage and a low voltage. All other voltage values are temporary and occur while transitioning between the values. (As we discuss later in this section, a possible pitfall in digital design is sampling a signal when it not clearly either high or low.) The fact that computers are digital is also a key reason they use binary numbers, since a binary system matches the underlying abstraction inherent in the electronics. In various logic families, the values and relationships between the two voltage values differ. Thus, rather than refer to the voltage levels, we talk about signals that are (logically) true, or 1, or are asserted; or signals that are (logically) false, or 0, or are deasserted. The values 0 and 1 are called *complements* or *inverses* of one another.

Logic blocks are categorized as one of two types, depending on whether they contain memory. Blocks without memory are called *combinational*; the output of a combinational block depends only on the current input. In blocks with memory, the outputs can depend on both the inputs and the value stored in memory, which is called the *state* of the logic block. In this section and the next, we will focus

asserted signal A signal that is (logically) true, or 1.

deasserted signal A signal that is (logically) false, or 0.

only on combinational logic. After introducing different memory elements in Section A.8, we will describe how sequential logic, which is logic including state, is designed.

Truth Tables

Because a combinational logic block contains no memory, it can be completely specified by defining the values of the outputs for each possible set of input values. Such a description is normally given as a *truth table*. For a logic block with n inputs, there are 2^n entries in the truth table, since there are that many possible combinations of input values. Each entry specifies the value of all the outputs for that particular input combination.

combinational logic
A logic system whose blocks do not contain memory and hence compute the same output given the same input.

sequential logic
A group of logic elements that contain memory and hence whose value depends on the inputs as well as the current contents of the memory.

Truth Tables

Consider a logic function with three inputs, A, B, and C, and three outputs, D, E, and F. The function is defined as follows: D is true if at least one input is true, E is true if exactly two inputs are true, and F is true only if all three inputs are true. Show the truth table for this function.

EXAMPLE

The truth table will contain $2^3 = 8$ entries. Here it is:

ANSWER

Inputs			Outputs		
A	B	C	D	E	F
0	0	0	0	0	0
0	0	1	1	0	0
0	1	0	1	0	0
0	1	1	1	1	0
1	0	0	1	0	0
1	0	1	1	1	0
1	1	0	1	1	0
1	1	1	1	0	1

Truth tables can completely describe any combinational logic function; however, they grow in size quickly and may not be easy to understand. Sometimes we want to construct a logic function that will be 0 for many input combinations, and we use a shorthand of specifying only the truth table entries for the nonzero outputs. This approach is used in Chapter 4 and ▦ Appendix C.

Boolean Algebra

Another approach is to express the logic function with logic equations. This is done with the use of *Boolean algebra* (named after Boole, a 19th-century mathematician). In Boolean algebra, all the variables have the values 0 or 1 and, in typical formulations, there are three operators:

- The OR operator is written as +, as in $A + B$. The result of an OR operator is 1 if either of the variables is 1. The OR operation is also called a *logical sum*, since its result is 1 if either operand is 1.

- The AND operator is written as $\cdot$, as in $A \cdot B$. The result of an AND operator is 1 only if both inputs are 1. The AND operator is also called *logical product*, since its result is 1 only if both operands are 1.

- The unary operator NOT is written as $\overline{A}$. The result of a NOT operator is 1 only if the input is 0. Applying the operator NOT to a logical value results in an inversion or negation of the value (i.e., if the input is 0 the output is 1, and vice versa).

There are several laws of Boolean algebra that are helpful in manipulating logic equations.

- Identity law: $A + 0 = A$ and $A \cdot 1 = A$
- Zero and One laws: $A + 1 = 1$ and $A \cdot 0 = 0$
- Inverse laws: $A + \overline{A} = 1$ and $A \cdot \overline{A} = 0$
- Commutative laws: $A + B = B + A$ and $A \cdot B = B \cdot A$
- Associative laws: $A + (B + C) = (A + B) + C$ and $A \cdot (B \cdot C) = (A \cdot B) \cdot C$
- Distributive laws: $A \cdot (B + C) = (A \cdot B) + (A \cdot C)$ and
  $A + (B \cdot C) = (A + B) \cdot (A + C)$

In addition, there are two other useful theorems, called DeMorgan's laws, that are discussed in more depth in the exercises.

Any set of logic functions can be written as a series of equations with an output on the left-hand side of each equation and a formula consisting of variables and the three operators above on the right-hand side.

Logic Equations

Show the logic equations for the logic functions, D, E, and F, described in the previous example.

Here's the equation for D:

$$D = A + B + C$$

F is equally simple:

$$F = A \cdot B \cdot C$$

E is a little tricky. Think of it in two parts: what must be true for E to be true (two of the three inputs must be true), and what cannot be true (all three cannot be true). Thus we can write E as

$$E = ((A \cdot B) + (A \cdot C) + (B \cdot C)) \cdot (\overline{A \cdot B \cdot C})$$

We can also derive E by realizing that E is true only if exactly two of the inputs are true. Then we can write E as an OR of the three possible terms that have two true inputs and one false input:

$$E = (A \cdot B \cdot \overline{C}) + (A \cdot C \cdot \overline{B}) + (B \cdot C \cdot \overline{A})$$

Proving that these two expressions are equivalent is explored in the exercises.

In Verilog, we describe combinational logic whenever possible using the assign statement, which is described beginning on page A-23. We can write a definition for E using the Verilog exclusive-OR operator as `assign E = (A ^ B ^ C) *  (A + B + C) * (A * B * C)`, which is yet another way to describe this function. D and F have even simpler representations, which are just like the corresponding C code: `D = A | B | C` and `F = A & B & C`.

Gates

gate A device that implements basic logic functions, such as AND or OR.

Logic blocks are built from gates that implement basic logic functions. For example, an AND gate implements the AND function, and an OR gate implements the OR function. Since both AND and OR are commutative and associative, an AND or an OR gate can have multiple inputs, with the output equal to the AND or OR of all the inputs. The logical function NOT is implemented with an inverter that always has a single input. The standard representation of these three logic building blocks is shown in Figure A.2.1.

Rather than draw inverters explicitly, a common practice is to add "bubbles" to the inputs or outputs of a gate to cause the logic value on that input line or output line to be inverted. For example, Figure A.2.2 shows the logic diagram for the function $\overline{A} + B$, using explicit inverters on the left and bubbled inputs and outputs on the right.

Any logical function can be constructed using AND gates, OR gates, and inversion; several of the exercises give you the opportunity to try implementing some common logic functions with gates. In the next section, we'll see how an implementation of any logic function can be constructed using this knowledge.

In fact, all logic functions can be constructed with only a single gate type, if that gate is inverting. The two common inverting gates are called NOR and NAND and correspond to inverted OR and AND gates, respectively. NOR and NAND gates are called *universal*, since any logic function can be built using this one gate type. The exercises explore this concept further.

NOR gate An inverted OR gate.

NAND gate An inverted AND gate.

Check Yourself

Are the following two logical expressions equivalent? If not, find a setting of the variables to show they are not:

- $(A \cdot B \cdot \overline{C}) + (A \cdot C \cdot \overline{B}) + (B \cdot C \cdot \overline{A})$
- $B \cdot (A \cdot \overline{C} + C \cdot \overline{A})$

FIGURE A.2.1   Standard drawing for an AND gate, OR gate, and an inverter, shown from left to right. The signals to the left of each symbol are the inputs, while the output appears on the right. The AND and OR gates both have two inputs. Inverters have a single input.

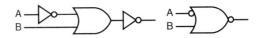

FIGURE A.2.2   Logic gate implementation of $\overline{A} + B$ using explicit inverts on the left and bubbled inputs and outputs on the right. This logic function can be simplified to $A \cdot \overline{B}$ or in Verilog, A & ~ B.

A.3 Combinational Logic

In this section, we look at a couple of larger logic building blocks that we use heavily, and we discuss the design of structured logic that can be automatically implemented from a logic equation or truth table by a translation program. Last, we discuss the notion of an array of logic blocks.

Decoders

One logic block that we will use in building larger components is a decoder. The most common type of decoder has an n-bit input and 2^n outputs, where only one output is asserted for each input combination. This decoder translates the n-bit input into a signal that corresponds to the binary value of the n-bit input. The outputs are thus usually numbered, say, Out0, Out1, ..., Out$2^n -1$. If the value of the input is i, then Out$i$ will be true and all other outputs will be false. Figure A.3.1 shows a 3-bit decoder and the truth table. This decoder is called a *3-to-8 decoder* since there are three inputs and eight (2^3) outputs. There is also a logic element called an *encoder* that performs the inverse function of a decoder, taking 2^n inputs and producing an n-bit output.

decoder A logic block that has an n-bit input and 2^n outputs, where only one output is asserted for each input combination.

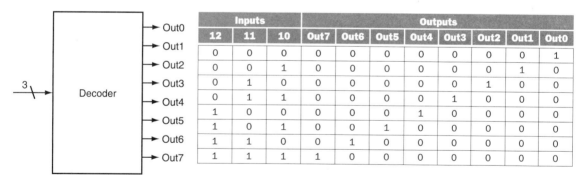

	Inputs						Outputs			
I2	I1	I0	Out7	Out6	Out5	Out4	Out3	Out2	Out1	Out0
0	0	0	0	0	0	0	0	0	0	1
0	0	1	0	0	0	0	0	0	1	0
0	1	0	0	0	0	0	0	1	0	0
0	1	1	0	0	0	0	1	0	0	0
1	0	0	0	0	0	1	0	0	0	0
1	0	1	0	0	1	0	0	0	0	0
1	1	0	0	1	0	0	0	0	0	0
1	1	1	1	0	0	0	0	0	0	0

a. A 3-bit decoder

b. The truth table for a 3-bit decoder

FIGURE A.3.1 A 3-bit decoder has three inputs, called I2, I1, and I0, and $2^3 = 8$ outputs, called Out0 to Out7. Only the output corresponding to the binary value of the input is true, as shown in the truth table. The label 3 on the input to the decoder says that the input signal is 3 bits wide.

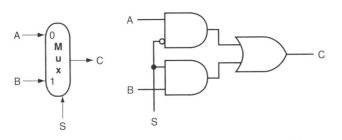

FIGURE A.3.2   A two-input multiplexor on the left and its implementation with gates on the right. The multiplexor has two data inputs (*A* and *B*), which are labeled *0* and *1*, and one selector input (*S*), as well as an output *C*. Implementing multiplexors in Verilog requires a little more work, especially when they are wider than two inputs. We show how to do this beginning on page A-23.

Multiplexors

selector value Also called **control value**. The control signal that is used to select one of the input values of a multiplexor as the output of the multiplexor.

One basic logic function that we use quite often in Chapter 4 is the *multiplexor*. A multiplexor might more properly be called a *selector*, since its output is one of the inputs that is selected by a control. Consider the two-input multiplexor. The left side of Figure A.3.2 shows this multiplexor has three inputs: two data values and a selector (or control) value. The selector value determines which of the inputs becomes the output. We can represent the logic function computed by a two-input multiplexor, shown in gate form on the right side of Figure A.3.2, as $C = (A \cdot \overline{S}) + (B \cdot S)$.

Multiplexors can be created with an arbitrary number of data inputs. When there are only two inputs, the selector is a single signal that selects one of the inputs if it is true (1) and the other if it is false (0). If there are n data inputs, there will need to be $\lceil \log_2 n \rceil$ selector inputs. In this case, the multiplexor basically consists of three parts:

1. A decoder that generates n signals, each indicating a different input value

2. An array of n AND gates, each combining one of the inputs with a signal from the decoder

3. A single large OR gate that incorporates the outputs of the AND gates

To associate the inputs with selector values, we often label the data inputs numerically (i.e., 0, 1, 2, 3, ..., $n-1$) and interpret the data selector inputs as a binary number. Sometimes, we make use of a multiplexor with undecoded selector signals.

Multiplexors are easily represented combinationally in Verilog by using *if* expressions. For larger multiplexors, *case* statements are more convenient, but care must be taken to synthesize combinational logic.

Two-Level Logic and PLAs

As pointed out in the previous section, any logic function can be implemented with only AND, OR, and NOT functions. In fact, a much stronger result is true. Any logic function can be written in a canonical form, where every input is either a true or complemented variable and there are only two levels of gates—one being AND and the other OR—with a possible inversion on the final output. Such a representation is called a *two-level representation*, and there are two forms, called sum of products and *product of sums*. A sum-of-products representation is a logical sum (OR) of products (terms using the AND operator); a product of sums is just the opposite. In our earlier example, we had two equations for the output E:

$$E = ((A \cdot B) + (A \cdot C) + (B \cdot C)) \cdot \overline{(A \cdot B \cdot C)}$$

and

$$E = (A \cdot B \cdot \overline{C}) + (A \cdot C \cdot \overline{B}) \cdot (B \cdot C \cdot \overline{A})$$

This second equation is in a sum-of-products form: it has two levels of logic and the only inversions are on individual variables. The first equation has three levels of logic.

sum of products A form of logical representation that employs a logical sum (OR) of products (terms joined using the AND operator).

Elaboration: We can also write E as a product of sums:

$$E = \overline{(\overline{A} + \overline{B} + C) \cdot (\overline{A} + \overline{C} + B) \cdot (\overline{B} + C + A)}$$

To derive this form, you need to use *DeMorgan's theorems*, which are discussed in the exercises.

In this text, we use the sum-of-products form. It is easy to see that any logic function can be represented as a sum of products by constructing such a representation from the truth table for the function. Each truth table entry for which the function is true corresponds to a product term. The product term consists of a logical product of all the inputs or the complements of the inputs, depending on whether the entry in the truth table has a 0 or 1 corresponding to this variable. The logic function is the logical sum of the product terms where the function is true. This is more easily seen with an example.

EXAMPLE

Sum of Products

Show the sum-of-products representation for the following truth table for D.

Inputs			Outputs
A	**B**	**C**	**D**
0	0	0	0
0	0	1	1
0	1	0	1
0	1	1	0
1	0	0	1
1	0	1	0
1	1	0	0
1	1	1	1

ANSWER

There are four product terms, since the function is true (1) for four different input combinations. These are:

$$\overline{A} \cdot \overline{B} \cdot C$$
$$\overline{A} \cdot B \cdot C$$
$$A \cdot \overline{B} \cdot \overline{C}$$
$$A \cdot B \cdot C$$

Thus, we can write the function for D as the sum of these terms:

$$D = (\overline{A} \cdot \overline{B} \cdot C)(\overline{A} \cdot B \cdot \overline{C})(A \cdot \overline{B} \cdot \overline{C})(A \cdot B \cdot C)$$

Note that only those truth table entries for which the function is true generate terms in the equation.

programmable logic array (PLA) A structured-logic element composed of a set of inputs and corresponding input complements and two stages of logic: the first generates product terms of the inputs and input complements, and the second generates sum terms of the product terms. Hence, PLAs implement logic functions as a sum of products.

minterms Also called **product terms**. A set of logic inputs joined by conjunction (AND operations); the product terms form the first logic stage of the *programmable logic array* (PLA).

We can use this relationship between a truth table and a two-level representation to generate a gate-level implementation of any set of logic functions. A set of logic functions corresponds to a truth table with multiple output columns, as we saw in the example on page A-5. Each output column represents a different logic function, which may be directly constructed from the truth table.

The sum-of-products representation corresponds to a common structured-logic implementation called a **programmable logic array (PLA)**. A PLA has a set of inputs and corresponding input complements (which can be implemented with a set of inverters), and two stages of logic. The first stage is an array of AND gates that form a set of **product terms** (sometimes called **minterms**); each product term can consist of any of the inputs or their complements. The second stage is an array of OR gates, each of which forms a logical sum of any number of the product terms. Figure A.3.3 shows the basic form of a PLA.

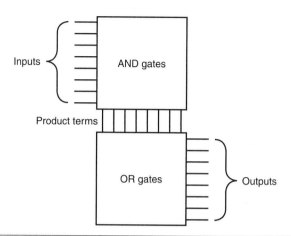

FIGURE A.3.3   The basic form of a PLA consists of an array of AND gates followed by an array of OR gates. Each entry in the AND gate array is a product term consisting of any number of inputs or inverted inputs. Each entry in the OR gate array is a sum term consisting of any number of these product terms.

A PLA can directly implement the truth table of a set of logic functions with multiple inputs and outputs. Since each entry where the output is true requires a product term, there will be a corresponding row in the PLA. Each output corresponds to a potential row of OR gates in the second stage. The number of OR gates corresponds to the number of truth table entries for which the output is true. The total size of a PLA, such as that shown in Figure A.3.3, is equal to the sum of the size of the AND gate array (called the *AND plane*) and the size of the OR gate array (called the *OR plane*). Looking at Figure A.3.3, we can see that the size of the AND gate array is equal to the number of inputs times the number of different product terms, and the size of the OR gate array is the number of outputs times the number of product terms.

A PLA has two characteristics that help make it an efficient way to implement a set of logic functions. First, only the truth table entries that produce a true value for at least one output have any logic gates associated with them. Second, each different product term will have only one entry in the PLA, even if the product term is used in multiple outputs. Let's look at an example.

PLAs

Consider the set of logic functions defined in the example on page A-5. Show a PLA implementation of this example for *D*, *E*, and *F*.

EXAMPLE

ANSWER

Here is the truth table we constructed earlier:

Inputs			Outputs		
A	B	C	D	E	F
0	0	0	0	0	0
0	0	1	1	0	0
0	1	0	1	0	0
0	1	1	1	1	0
1	0	0	1	0	0
1	0	1	1	1	0
1	1	0	1	1	0
1	1	1	1	0	1

Since there are seven unique product terms with at least one true value in the output section, there will be seven columns in the AND plane. The number of rows in the AND plane is three (since there are three inputs), and there are also three rows in the OR plane (since there are three outputs). Figure A.3.4 shows the resulting PLA, with the product terms corresponding to the truth table entries from top to bottom.

Rather than drawing all the gates, as we do in Figure A.3.4, designers often show just the position of AND gates and OR gates. Dots are used on the intersection of a product term signal line and an input line or an output line when a corresponding AND gate or OR gate is required. Figure A.3.5 shows how the PLA of Figure A.3.4 would look when drawn in this way. The contents of a PLA are fixed when the PLA is created, although there are also forms of PLA-like structures, called *PALs*, that can be programmed electronically when a designer is ready to use them.

ROMs

Another form of structured logic that can be used to implement a set of logic functions is a read-only memory (ROM). A ROM is called a memory because it has a set of locations that can be read; however, the contents of these locations are fixed, usually at the time the ROM is manufactured. There are also programmable ROMs (PROMs) that can be programmed electronically, when a designer knows their contents. There are also erasable PROMs; these devices require a slow erasure process using ultraviolet light, and thus are used as read-only memories, except during the design and debugging process.

A ROM has a set of input address lines and a set of outputs. The number of addressable entries in the ROM determines the number of address lines: if the

read-only memory (ROM) A memory whose contents are designated at creation time, after which the contents can only be read. ROM is used as structured logic to implement a set of logic functions by using the terms in the logic functions as address inputs and the outputs as bits in each word of the memory.

programmable ROM (PROM) A form of read-only memory that can be programmed when a designer knows its contents.

ROM contains 2^m addressable entries, called the *height*, then there are m input lines. The number of bits in each addressable entry is equal to the number of output bits and is sometimes called the *width* of the ROM. The total number of bits in the ROM is equal to the height times the width. The height and width are sometimes collectively referred to as the *shape* of the ROM.

Inputs

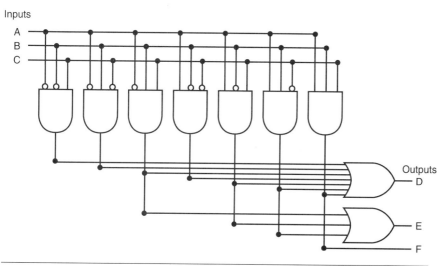

FIGURE A.3.4   The PLA for implementing the logic function described in the example.

A ROM can encode a collection of logic functions directly from the truth table. For example, if there are n functions with m inputs, we need a ROM with m address lines (and 2^m entries), with each entry being n bits wide. The entries in the input portion of the truth table represent the addresses of the entries in the ROM, while the contents of the output portion of the truth table constitute the contents of the ROM. If the truth table is organized so that the sequence of entries in the input portion constitutes a sequence of binary numbers (as have all the truth tables we have shown so far), then the output portion gives the ROM contents in order as well. In the example starting on page A-13, there were three inputs and three outputs. This leads to a ROM with $2^3 = 8$ entries, each 3 bits wide. The contents of those entries in increasing order by address are directly given by the output portion of the truth table that appears on page A-14.

ROMs and PLAs are closely related. A ROM is fully decoded: it contains a full output word for every possible input combination. A PLA is only partially decoded. This means that a ROM will always contain more entries. For the earlier truth table on page A-14, the ROM contains entries for all eight possible inputs, whereas the PLA contains only the seven active product terms. As the number of inputs grows,

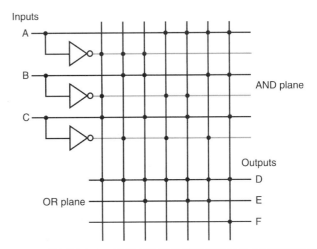

FIGURE A.3.5   A PLA drawn using dots to indicate the components of the product terms and sum terms in the array. Rather than use inverters on the gates, usually all the inputs are run the width of the AND plane in both true and complement forms. A dot in the AND plane indicates that the input, or its inverse, occurs in the product term. A dot in the OR plane indicates that the corresponding product term appears in the corresponding output.

the number of entries in the ROM grows exponentially. In contrast, for most real logic functions, the number of product terms grows much more slowly (see the examples in ⊞ **Appendix C**). This difference makes PLAs generally more efficient for implementing combinational logic functions. ROMs have the advantage of being able to implement any logic function with the matching number of inputs and outputs. This advantage makes it easier to change the ROM contents if the logic function changes, since the size of the ROM need not change.

In addition to ROMs and PLAs, modern logic synthesis systems will also translate small blocks of combinational logic into a collection of gates that can be placed and wired automatically. Although some small collections of gates are usually not area-efficient, for small logic functions they have less overhead than the rigid structure of a ROM and PLA and so are preferred.

For designing logic outside of a custom or semicustom integrated circuit, a common choice is a field programming device; we describe these devices in Section A.12.

Don't Cares

Often in implementing some combinational logic, there are situations where we do not care what the value of some output is, either because another output is true or because a subset of the input combinations determines the values of the outputs. Such situations are referred to as *don't cares*. Don't cares are important because they make it easier to optimize the implementation of a logic function.

There are two types of don't cares: output don't cares and input don't cares, both of which can be represented in a truth table. *Output don't cares* arise when we don't care about the value of an output for some input combination. They appear as Xs in the output portion of a truth table. When an output is a don't care for some input combination, the designer or logic optimization program is free to make the output true or false for that input combination. *Input don't cares* arise when an output depends on only some of the inputs, and they are also shown as Xs, though in the input portion of the truth table.

Don't Cares

Consider a logic function with inputs A, B, and C defined as follows:

- If A or C is true, then output D is true, whatever the value of B.

- If A or B is true, then output E is true, whatever the value of C.

- Output F is true if exactly one of the inputs is true, although we don't care about the value of F, whenever D and E are both true.

Show the full truth table for this function and the truth table using don't cares. How many product terms are required in a PLA for each of these?

EXAMPLE

Here's the full truth table, without don't cares:

ANSWER

Inputs			Outputs		
A	B	C	D	E	F
0	0	0	0	0	0
0	0	1	1	0	1
0	1	0	0	1	1
0	1	1	1	1	0
1	0	0	1	1	1
1	0	1	1	1	0
1	1	0	1	1	0
1	1	1	1	1	0

This requires seven product terms without optimization. The truth table written with output don't cares looks like this:

Inputs			Outputs		
A	B	C	D	E	F
0	0	0	0	0	0
0	0	1	1	0	1
0	1	0	0	1	1
0	1	1	1	1	X
1	0	0	1	1	X
1	0	1	1	1	X
1	1	0	1	1	X
1	1	1	1	1	X

If we also use the input don't cares, this truth table can be further simplified to yield the following:

Inputs			Outputs		
A	B	C	D	E	F
0	0	0	0	0	0
0	0	1	1	0	1
0	1	0	0	1	1
X	1	1	1	1	X
1	X	X	1	1	X

This simplified truth table requires a PLA with four minterms, or it can be implemented in discrete gates with one two-input AND gate and three OR gates (two with three inputs and one with two inputs). This compares to the original truth table that had seven minterms and would have required four AND gates.

Logic minimization is critical to achieving efficient implementations. One tool useful for hand minimization of random logic is *Karnaugh maps*. Karnaugh maps represent the truth table graphically, so that product terms that may be combined are easily seen. Nevertheless, hand optimization of significant logic functions using Karnaugh maps is impractical, both because of the size of the maps and their complexity. Fortunately, the process of logic minimization is highly mechanical and can be performed by design tools. In the process of minimization, the tools take advantage of the don't cares, so specifying them is important. The textbook references at the end of this appendix provide further discussion on logic minimization, Karnaugh maps, and the theory behind such minimization algorithms.

Arrays of Logic Elements

Many of the combinational operations to be performed on data have to be done to an entire word (64 bits) of data. Thus we often want to build an array of logic elements, which we can represent simply by showing that a given operation will happen to an entire collection of inputs. Inside a machine, much of the time we want to select between a pair of *buses*. A **bus** is a collection of data lines that is treated together as a single logical signal. (The term *bus* is also used to indicate a shared collection of lines with multiple sources and uses.)

For example, in the LEGv8 instruction set, the result of an instruction that is written into a register can come from one of two sources. A multiplexor is used to choose which of the two buses (each 64 bits wide) will be written into the Result register. The 1-bit multiplexor, which we showed earlier, will need to be replicated 64 times.

We indicate that a signal is a bus rather than a single 1-bit line by showing it with a thicker line in a figure. Most buses are 64 bits wide; those that are not are explicitly labeled with their width. When we show a logic unit whose inputs and outputs are buses, this means that the unit must be replicated a sufficient number of times to accommodate the width of the input. Figure A.3.6 shows how we draw a multiplexor that selects between a pair of 64-bit buses and how this expands in terms of 1-bit-wide multiplexors. Sometimes we need to construct an array of logic elements where the inputs for some elements in the array are outputs from earlier elements. For example, this is how a multibit-wide ALU is constructed. In such cases, we must explicitly show how to create wider arrays, since the individual elements of the array are no longer independent, as they are in the case of a 64-bit-wide multiplexor.

bus In logic design, a collection of data lines that is treated together as a single logical signal; also, a shared collection of lines with multiple sources and uses.

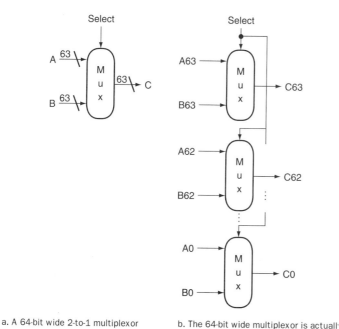

a. A 64-bit wide 2-to-1 multiplexor

b. The 64-bit wide multiplexor is actually an array of 64 1-bit multiplexors

FIGURE A.3.6 A multiplexor is arrayed 64 times to perform a selection between two 64-bit inputs. Note that there is still only one data selection signal used for all 64 1-bit multiplexors.

Check
Yourself Parity is a function in which the output depends on the number of 1s in the input. For an even parity function, the output is 1 if the input has an even number of ones. Suppose a ROM is used to implement an even parity function with a 4-bit input. Which of A, B, C, or D represents the contents of the ROM?

Address	A	B	C	D
0	0	1	0	1
1	0	1	1	0
2	0	1	0	1
3	0	1	1	0
4	0	1	0	1
5	0	1	1	0
6	0	1	0	1
7	0	1	1	0
8	1	0	0	1
9	1	0	1	0
10	1	0	0	1
11	1	0	1	0
12	1	0	0	1
13	1	0	1	0
14	1	0	0	1
15	1	0	1	0

A.4   Using a Hardware Description Language

hardware description language
A programming language for describing hardware, used for generating simulations of a hardware design and also as input to synthesis tools that can generate actual hardware.

Verilog One of the two most common hardware description languages.

VHDL One of the two most common hardware description languages.

Today most digital design of processors and related hardware systems is done using a hardware description language. Such a language serves two purposes. First, it provides an abstract description of the hardware to simulate and debug the design. Second, with the use of logic synthesis and hardware compilation tools, this description can be compiled into the hardware implementation.

In this section, we introduce the hardware description language Verilog and show how it can be used for combinational design. In the rest of the appendix, we expand the use of Verilog to include design of sequential logic. In the optional sections of Chapter 4 that appear online, we use Verilog to describe processor implementations. In the optional section from Chapter 5 that appears online, we use system Verilog to describe cache controller implementations. System Verilog adds structures and some other useful features to Verilog.

Verilog is one of the two primary hardware description languages; the other is VHDL. Verilog is somewhat more heavily used in industry and is based on C, as opposed to VHDL, which is based on Ada. The reader generally familiar with C will find the basics of Verilog, which we use in this appendix, easy to follow.

Readers already familiar with VHDL should find the concepts simple, provided they have been exposed to the syntax of C.

Verilog can specify both a behavioral and a structural definition of a digital system. A behavioral specification describes how a digital system functionally operates. A structural specification describes the detailed organization of a digital system, usually using a hierarchical description. A structural specification can be used to describe a hardware system in terms of a hierarchy of basic elements such as gates and switches. Thus, we could use Verilog to describe the exact contents of the truth tables and datapath of the last section.

With the arrival of hardware synthesis tools, most designers now use Verilog or VHDL to structurally describe only the datapath, relying on logic synthesis to generate the control from a behavioral description. In addition, most CAD systems provide extensive libraries of standardized parts, such as ALUs, multiplexors, register files, memories, and programmable logic blocks, as well as basic gates.

Obtaining an acceptable result using libraries and logic synthesis requires that the specification be written with an eye toward the eventual synthesis and the desired outcome. For our simple designs, this primarily means making clear what we expect to be implemented in combinational logic and what we expect to require in sequential logic. In most of the examples we use in this section and the remainder of this appendix, we have written the Verilog with the eventual synthesis in mind.

behavioral specification Describes how a digital system operates functionally.

structural specification Describes how a digital system is organized in terms of a hierarchical connection of elements.

hardware synthesis tools Computer-aided design software that can generate a gate-level design based on behavioral descriptions of a digital system.

Datatypes and Operators in Verilog

There are two primary datatypes in Verilog:

1. A wire specifies a combinational signal.

2. A reg (register) holds a value, which can vary with time. A reg need not necessarily correspond to an actual register in an implementation, although it often will.

wire In Verilog, specifies a combinational signal.

reg In Verilog, a register.

A register or wire, named X, that is 64 bits wide is declared as an array: `reg [63:0] X` or `wire [63:0] X`, which also sets the index of 0 to designate the least significant bit of the register. Because we often want to access a subfield of a register or wire, we can refer to a contiguous set of bits of a register or wire with the notation [starting bit: ending bit], where both indices must be constant values.

An array of registers is used for a structure like a register file or memory. Thus, the declaration

```
reg [63:0] registerfile[0:31]
```

specifies a variable registerfile that is equivalent to a LEGv8 registerfile, where register 0 is the first. When accessing an array, we can refer to a single element, as in C, using the notation registerfile[regnum].

The possible values for a register or wire in Verilog are

- 0 or 1, representing logical false or true

- X, representing unknown, the initial value given to all registers and to any wire not connected to something

- Z, representing the high-impedance state for tristate gates, which we will not discuss in this appendix

Constant values can be specified as decimal numbers as well as binary, octal, or hexadecimal. We often want to say exactly how large a constant field is in bits. This is done by prefixing the value with a decimal number specifying its size in bits. For example:

- 4'b0100 specifies a 4-bit binary constant with the value 4, as does 4'd4.

- −8'h4 specifies an 8-bit constant with the value −4 (in two's complement representation)

Values can also be concatenated by placing them within { } separated by commas. The notation {x{bitfield}} replicates bitfield x times. For example:

- {32{2'b01}} creates a 64-bit value with the pattern 0101 ... 01.

- {A[31:16],B[15:0]} creates a value whose upper 16 bits come from A and whose lower 16 bits come from B.

Verilog provides the full set of unary and binary operators from C, including the arithmetic operators (+, −, *, /), the logical operators (&, |, ~), the comparison operators (= =, ! =, >, <, < =, > =), the shift operators (<<, >>), and C's conditional operator (?, which is used in the form condition ? expr1 :expr2 and returns expr1 if the condition is true and expr2 if it is false). Verilog adds a set of unary logic reduction operators (&, |, ^) that yield a single bit by applying the logical operator to all the bits of an operand. For example, &A returns the value obtained by ANDing all the bits of A together, and ^A returns the reduction obtained by using exclusive OR on all the bits of A.

Check Yourself

Which of the following define exactly the same value?

1. 8'bimoooo

2. 8'hF0

3. 8'd240

4. {{4{1'b1}},{4{1'b0}}}

5. {4'b1,4'b0)

Structure of a Verilog Program

A Verilog program is structured as a set of modules, which may represent anything from a collection of logic gates to a complete system. Modules are similar to classes in C++, although not nearly as powerful. A module specifies its input and output ports, which describe the incoming and outgoing connections of a module. A module may also declare additional variables. The body of a module consists of:

- `initial` constructs, which can initialize `reg` variables

- Continuous assignments, which define only combinational logic

- `always` constructs, which can define either sequential or combinational logic

- Instances of other modules, which are used to implement the module being defined

Representing Complex Combinational Logic in Verilog

A continuous assignment, which is indicated with the keyword `assign`, acts like a combinational logic function: the output is continuously assigned the value, and a change in the input values is reflected immediately in the output value. Wires may only be assigned values with continuous assignments. Using continuous assignments, we can define a module that implements a half-adder, as Figure A.4.1 shows.

Assign statements are one sure way to write Verilog that generates combinational logic. For more complex structures, however, assign statements may be awkward or tedious to use. It is also possible to use the `always` block of a module to describe a combinational logic element, although care must be taken. Using an `always` block allows the inclusion of Verilog control constructs, such as *if-then-else, case* statements, *for* statements, and *repeat* statements, to be used. These statements are similar to those in C with small changes.

An `always` block specifies an optional list of signals on which the block is sensitive (in a list starting with @). The `always` block is re-evaluated if any of the

```
module half_adder (A,B,Sum,Carry);
    input A,B; //two 1-bit inputs
    output Sum, Carry; //two 1-bit outputs
    assign Sum = A ^ B; //sum is A xor B
    assign Carry = A & B; //Carry is A and B
endmodule
```

FIGURE A.4.1   A Verilog module that defines a half-adder using continuous assignments.

sensitivity list The list of
signals that specifies when
an always block should
be re-evaluated.

listed signals changes value; if the list is omitted, the always block is constantly re-evaluated. When an always block is specifying combinational logic, the sensitivity list should include all the input signals. If there are multiple Verilog statements to be executed in an always block, they are surrounded by the keywords begin and end, which take the place of the { and } in C. An always block thus looks like this:

```
always @(list of signals that cause reevaluation) begin
    Verilog statements including assignments and other
    control statements end
```

Reg variables may only be assigned inside an always block, using a procedural assignment statement (as distinguished from continuous assignment we saw earlier). There are, however, two different types of procedural assignments. The assignment operator = executes as it does in C; the right-hand side is evaluated, and the left-hand side is assigned the value. Furthermore, it executes like the normal C assignment statement: that is, it is completed before the next statement is executed. Hence, the assignment operator = has the name blocking assignment. This blocking can be useful in the generation of sequential logic, and we will return to it shortly. The other form of assignment (nonblocking) is indicated by <=. In nonblocking assignment, all right-hand sides of the assignments in an always group are evaluated and the assignments are done simultaneously. As a first example of combinational logic implemented using an always block, Figure A.4.2 shows the implementation of a 4-to-1 multiplexor, which uses a case construct to make it easy to write. The case construct looks like a C switch statement. Figure A.4.3 shows a definition of a LEGv8 ALU, which also uses a case statement.

Since only reg variables may be assigned inside always blocks, when we want to describe combinational logic using an always block, care must be taken to ensure that the reg does not synthesize into a register. A variety of pitfalls are described in the elaboration below.

blocking assignment In
Verilog, an assignment
that completes before
the execution of the next
statement.

**nonblocking
assignment** An
assignment that continues
after evaluating the right-
hand side, assigning the
left-hand side the value
only after all right-hand
sides are evaluated.

Elaboration: Continuous assignment statements always yield combinational logic, but other Verilog structures, even when in always blocks, can yield unexpected results during logic synthesis. The most common problem is creating sequential logic by implying the existence of a latch or register, which results in an implementation that is both slower and more costly than perhaps intended. To ensure that the logic that you intend to be combinational is synthesized that way, make sure you do the following:

1. Place all combinational logic in a continuous assignment or an always block.

2. Make sure that all the signals used as inputs appear in the sensitivity list of an always block.

3. Ensure that every path through an always block assigns a value to the exact same set of bits.

The last of these is the easiest to overlook; read through the example in Figure A.5.15 to convince yourself that this property is adhered to.

```
module Mult4to1 (In1,In2,In3,In4,Sel,Out);
   input [31:0] In1, In2, In3, In4; /four 32-bit inputs
   input [1:0] Sel; //selector signal
   output reg [31:0] Out;// 32-bit output
   always @(In1, In2, In3, In4, Sel)
   case (Sel) //a 4->1 multiplexor
      0: Out <= In1;
      1: Out <= In2;
      2: Out <= In3;
      default: Out <= In4;
   endcase
endmodule
```

FIGURE A.4.2 A Verilog definition of a 4-to-1 multiplexor with 32-bit inputs, using a case **statement.** The case statement acts like a C switch statement, except that in Verilog only the code associated with the selected case is executed (as if each case state had a break at the end) and there is no fall-through to the next statement.

```
module MIPSALU (ALUctl, A, B, ALUOut, Zero);
   input [3:0] ALUctl;
   input [31:0] A,B;
   output reg [31:0] ALUOut;
   output Zero;
   assign Zero = (ALUOut==0); //Zero is true if ALUOut is 0; goes anywhere
   always @(ALUctl, A, B) //reevaluate if these change
      case (ALUctl)
         0: ALUOut <= A & B;
         1: ALUOut <= A | B;
         2: ALUOut <= A + B;
         6: ALUOut <= A - B;
         7: ALUOut <= A < B ? 1:0;
         12: ALUOut <= ~(A | B); // result is nor
         default: ALUOut <= 0; //default to 0, should not happen;
      endcase
endmodule
```

FIGURE A.4.3 A Verilog behavioral definition of a LEGv8 ALU. This could be synthesized using a module library containing basic arithmetic and logical operations.

Check Yourself Assuming all values are initially zero, what are the values of A and B after executing this Verilog code inside an `always` block?

```
C = 1;
A <= C;
B = C;
```

A.5 Constructing a Basic Arithmetic Logic Unit

ALU n. [Arthritic Logic Unit or (rare) Arithmetic Logic Unit] A random-number generator supplied as standard with all computer systems.

Stan Kelly-Bootle, *The Devil's DP Dictionary,* 1981

The arithmetic logic unit (ALU) is the brawn of the computer, the device that performs the arithmetic operations like addition and subtraction or logical operations like AND and OR. This section constructs an ALU from four hardware building blocks (AND and OR gates, inverters, and multiplexors) and illustrates how combinational logic works. In the next section, we will see how addition can be sped up through more clever designs.

Because the LEGv8 word is 64 bits wide, we need a 64-bit-wide ALU. Let's assume that we will connect 64 1-bit ALUs to create the desired ALU. We'll therefore start by constructing a 1-bit ALU.

A 1-Bit ALU

The logical operations are easiest, because they map directly onto the hardware components in Figure A.2.1.

The 1-bit logical unit for AND and OR looks like Figure A.5.1. The multiplexor on the right then selects *a* AND *b* or *a* OR *b*, depending on whether the value of *Operation* is 0 or 1. The line that controls the multiplexor is shown in color to distinguish it from the lines containing data. Notice that we have renamed the control and output lines of the multiplexor to give them names that reflect the function of the ALU.

The next function to include is addition. An adder must have two inputs for the operands and a single-bit output for the sum. There must be a second output to pass on the carry, called *CarryOut*. Since the CarryOut from the neighbor adder must be included as an input, we need a third input. This input is called *CarryIn*. Figure A.5.2 shows the inputs and the outputs of a 1-bit adder. Since we know what addition is supposed to do, we can specify the outputs of this "black box" based on its inputs, as Figure A.5.3 demonstrates.

We can express the output functions CarryOut and Sum as logical equations, and these equations can in turn be implemented with logic gates. Let's do CarryOut. Figure A.5.4 shows the values of the inputs when CarryOut is a 1.

We can turn this truth table into a logical equation:

$$CarryOut = (b \cdot CarryIn) + (a \cdot CarryIn) + (a \cdot b) + (a \cdot b \cdot CarryIn)$$

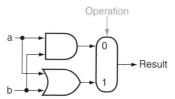

FIGURE A.5.1 The 1-bit logical unit for AND and OR.

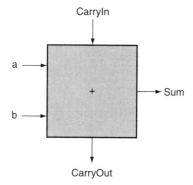

FIGURE A.5.2 A 1-bit adder. This adder is called a full adder; it is also called a (3,2) adder because it has three inputs and two outputs. An adder with only the a and b inputs is called a (2,2) adder or half-adder.

Inputs			Outputs		
a	b	CarryIn	CarryOut	Sum	Comments
0	0	0	0	0	$0 + 0 + 0 = 00_{two}$
0	0	1	0	1	$0 + 0 + 1 = 01_{two}$
0	1	0	0	1	$0 + 1 + 0 = 01_{two}$
0	1	1	1	0	$0 + 1 + 1 = 10_{two}$
1	0	0	0	1	$1 + 0 + 0 = 01_{two}$
1	0	1	1	0	$1 + 0 + 1 = 10_{two}$
1	1	0	1	0	$1 + 1 + 0 = 10_{two}$
1	1	1	1	1	$1 + 1 + 1 = 11_{two}$

FIGURE A.5.3 Input and output specification for a 1-bit adder.

If $a \cdot b \cdot CarryIn$ is true, then all of the other three terms must also be true, so we can leave out this last term corresponding to the fourth line of the table. We can thus simplify the equation to

$$CarryOut = (b \cdot CarryIn) + (a \cdot CarryIn) + (a \cdot b)$$

Figure A.5.5 shows that the hardware within the adder black box for CarryOut consists of three AND gates and one OR gate. The three AND gates correspond exactly to the three parenthesized terms of the formula above for CarryOut, and the OR gate sums the three terms.

Inputs		
a	**b**	**CarryIn**
0	1	1
1	0	1
1	1	0
1	1	1

FIGURE A.5.4   Values of the inputs when CarryOut is a 1.

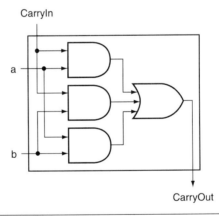

FIGURE A.5.5   Adder hardware for the CarryOut signal. The rest of the adder hardware is the logic for the Sum output given in the equation on this page.

The Sum bit is set when exactly one input is 1 or when all three inputs are 1. The Sum results in a complex Boolean equation (recall that $\bar{a}$ means NOT a):

$$Sum = (a \cdot \bar{b} \cdot \overline{CarryIn}) + (\bar{a} \cdot b \cdot \overline{CarryIn}) + (\bar{a} \cdot \bar{b} \cdot CarryIn) + (a \cdot b \cdot CarryIn)$$

The drawing of the logic for the Sum bit in the adder black box is left as an exercise for the reader.

Figure A.5.6 shows a 1-bit ALU derived by combining the adder with the earlier components. Sometimes designers also want the ALU to perform a few more simple operations, such as generating 0. The easiest way to add an operation is to expand the multiplexor controlled by the Operation line and, for this example, to connect 0 directly to the new input of that expanded multiplexor.

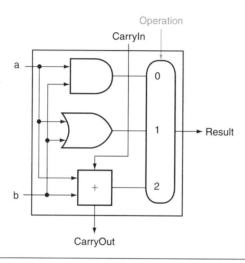

FIGURE A.5.6   A 1-bit ALU that performs AND, OR, and addition (see Figure A.5.5).

A 64-Bit ALU

Now that we have completed the 1-bit ALU, the full 64-bit ALU is created by connecting adjacent "black boxes." Using xi to mean the ith bit of x, Figure A.5.7 shows a 64-bit ALU. Just as a single stone can cause ripples to radiate to the shores of a quiet lake, a single carry out of the least significant bit (Result0) can ripple all the way through the adder, causing a carry out of the most signifcant bit (Result63). Hence, the adder created by directly linking the carries of 1-bit adders is called a *ripple carry* adder. We'll see a faster way to connect the 1-bit adders starting on page A-38.

Subtraction is the same as adding the negative version of an operand, and this is how adders perform subtraction. Recall that the shortcut for negating a two's complement number is to invert each bit (sometimes called the *one's complement*) and then add 1. To invert each bit, we simply add a 2:1 multiplexor that chooses between b and b̄, as Figure A.5.8 shows.

Suppose we connect 64 of these 1-bit ALUs, as we did in Figure A.5.7. The added multiplexor gives the option of b or its inverted value, depending on Binvert, but

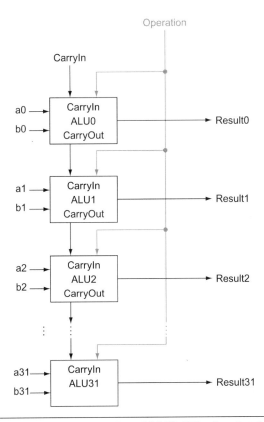

FIGURE A.5.7   A 64-bit ALU constructed from 64 1-bit ALUs. CarryOut of the less significant bit is connected to the CarryIn of the more significant bit. This organization is called ripple carry.

this is only one step in negating a two's complement number. Notice that the least significant bit still has a CarryIn signal, even though it's unnecessary for addition. What happens if we set this CarryIn to 1 instead of 0? The adder will then calculate $a + b + 1$. By selecting the inverted version of b, we get exactly what we want:

$$a + \bar{b} + 1 = a + (\bar{b} + 1) = a + (-b) = a - b$$

The simplicity of the hardware design of a two's complement adder helps explain why two's complement representation has become the universal standard for integer computer arithmetic.

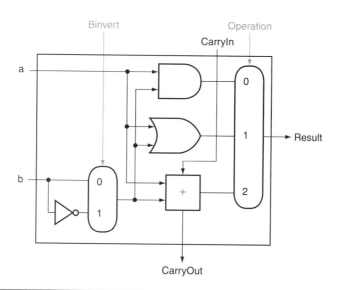

FIGURE A.5.8   A 1-bit ALU that performs AND, OR, and addition on a and b or a and $\bar{b}$. By selecting $\bar{b}$ (Binvert = 1) and setting CarryIn to 1 in the least significant bit of the ALU, we get two's comple- ment subtraction of b from a instead of addition of b to a.

A LEGv8 ALU also needs a NOR function. Instead of adding a separate gate for NOR, we can reuse much of the hardware already in the ALU, like we did for subtract. The insight comes from the following truth about NOR:

$$\overline{(a + b)} = \bar{a} \cdot \bar{b}$$

That is, NOT (a OR b) is equivalent to NOT a AND NOT b. This fact is called DeMorgan's theorem and is explored in the exercises in more depth.

Since we have AND and NOT b, we only need to add NOT a to the ALU. Figure A.5.9 shows that change.

Tailoring the 64-Bit ALU to LEGv8

These four operations—add, subtract, AND, OR—are found in the ALU of almost every computer, and the operations of most LEGv8 instructions can be performed by this ALU. But the design of the ALU is incomplete.

One instruction that still needs support is the compare and branch zero (CBZ). Recall that the operation the ALU just passes the register input value. For the ALU to perform CBZ, we first need to expand the three-input multiplexor in Figure A.5.8 to add an input for the CBZ result. We call that new input Pass and use it only for CBZ.

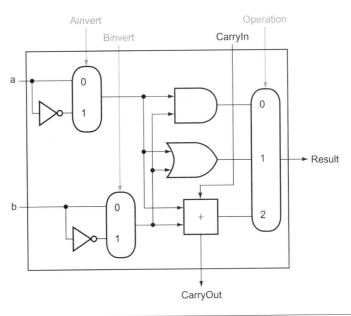

FIGURE A.5.9 A 1-bit ALU that performs AND, OR, and addition on a and b or ā and b̄. By selecting ā (Ainvert = 1) and b̄ (Binvert = 1), we get a NOR b instead of a AND b.

The top drawing of Figure A.5.10 shows the new 1-bit ALU with the expanded multiplexor. Figure A.5.11 shows the 64-bit ALU.

To further tailor the ALU to the LEGv8 instruction set, we must support conditional branch instructions. These instructions branch either if two registers are equal or if they are unequal. The easiest way to test equality with the ALU is to subtract b from a and then test to see if the result is 0, since

$$(a - b = 0) \Rightarrow a = b$$

Thus, if we add hardware to test if the result is 0, we can test for equality. The simplest way is to OR all the outputs together and then send that signal through an inverter:

$$\text{Zero} = \overline{(\text{Result63} + \text{Result62} + \cdots + \text{Result2} + \text{Result1} + \text{Result0})}$$

Figure A.5.12 shows the revised 64-bit ALU. We can think of the combination of the 1-bit Ainvert line, the 1-bit Binvert line, and the 2-bit Operation lines as 4-bit control lines for the ALU, telling it to perform add, subtract, AND, OR, or set on less than. Figure A.5.13 shows the ALU control lines and the corresponding ALU operation.

Finally, now that we have seen what is inside a 64-bit ALU, we will use the universal symbol for a complete ALU, as shown in Figure A.5.14.

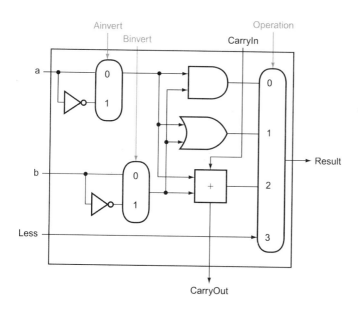

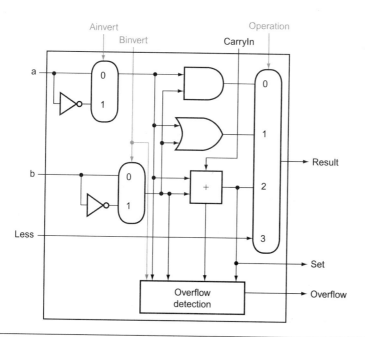

FIGURE A.5.10 (Top) A 1-bit ALU that performs AND, OR, and addition on a and b or b̄, and (bottom) a 1-bit ALU for the most significant bit. The top drawing includes a direct input that is connected to perform the set on less than operation (see Figure A.5.11); the bottom has a direct output from the adder for the less than comparison called Set. (See Exercise A.24 at the end of this appendix to see how to calculate overflow with fewer inputs.)

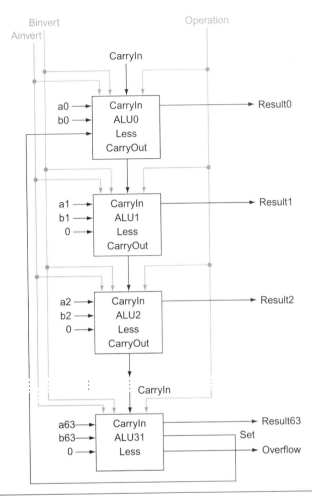

FIGURE A.5.11   A 64-bit ALU constructed from the 63 copies of the 1-bit ALU in the top of Figure A.5.10 and one 1-bit ALU in the bottom of that figure. The Less inputs are connected to 0 except for the least significant bit, which is connected to the Set output of the most significant bit. If the ALU performs a − b and we select the input 3 in the multiplexor in Figure A.5.10, then Result = 0 … 001 if a < b, and Result = 0 … 000 otherwise.

Defining the LEGv8 ALU in Verilog

Figure A.5.15 shows how a combinational LEGv8 ALU might be specified in Verilog; such a specification would probably be compiled using a standard parts library that provided an adder, which could be instantiated. For completeness, we show the ALU control for LEGv8 in Figure A.5.16, which is used in Chapter 4, where we build a Verilog version of the LEGv8 datapath.

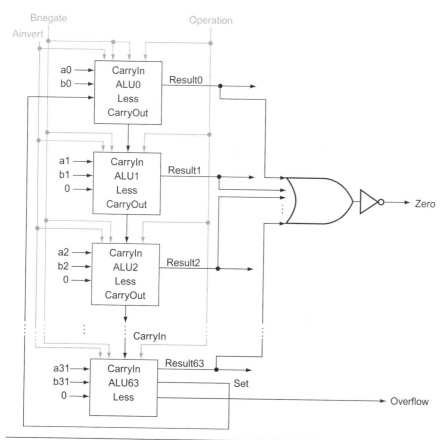

FIGURE A.5.12 The final 64-bit ALU. This adds a Zero detector to Figure A.5.11.

ALU control lines	Function
0000	AND
0001	OR
0010	add
0110	subtract
0111	set on less than
1100	NOR

FIGURE A.5.13 The values of the three ALU control lines, Bnegate, and Operation, and the corresponding ALU operations.

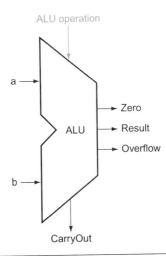

FIGURE A.5.14   The symbol commonly used to represent an ALU, as shown in Figure A.5.12. This symbol is also used to represent an adder, so it is normally labeled either with ALU or Adder.

```
module MIPSALU (ALUctl, A, B, ALUOut, Zero);
    input [3:0] ALUctl;
    input [63:0] A,B;
    output reg [63:0] ALUOut;
    output Zero;
    assign Zero = (ALUOut==0); //Zero is true if ALUOut is 0
    always @(ALUctl, A, B) begin //reevaluate if these change
        case (ALUctl)
            0: ALUOut <= A & B;
            1: ALUOut <= A | B;
            2: ALUOut <= A + B;
            6: ALUOut <= A - B;
            7: ALUOut <= A < B ? 1 : 0;
            12: ALUOut <= ~(A | B); // result is nor
            default: ALUOut <= 0;
        endcase
    end
endmodule
```

FIGURE A.5.15   A Verilog behavioral definition of a LEGv8 ALU.

```
module ALUControl (ALUOp, FuncCode, ALUCtl);
    input [1:0] ALUOp;
    input [5:0] FuncCode;
    output [3:0] reg ALUCtl;

    always case (FuncCode)
    32: ALUOp<=2; // add
    34: ALUOp<=6; //subtract
    36: ALUOP<=0; // and
    37: ALUOp<=1; // or
    39: ALUOp<=12; // nor
    42: ALUOp<=7; // slt
    default: ALUOp<=15; // should not happen
    endcase
endmodule
```

FIGURE A.5.16 The LEGv8 ALU control: a simple piece of combinational control logic.

The next question is, "How quickly can this ALU add two 64-bit operands?" We can determine the a and b inputs, but the CarryIn input depends on the operation in the adjacent 1-bit adder. If we trace all the way through the chain of dependencies, we connect the most significant bit to the least significant bit, so the most significant bit of the sum must wait for the *sequential* evaluation of all 64 1-bit adders. This sequential chain reaction is too slow to be used in time-critical hardware. The next section explores how to speed-up addition. This topic is not crucial to understanding the rest of the appendix and may be skipped.

Suppose you wanted to add the operation NOT (a AND b), called NAND. How could the ALU change to support it?

Check Yourself

1. <u>No change.</u> You can calculate NAND quickly using the current ALU since $\overline{(a \cdot b)} = \overline{a} + \overline{b}$ and we already have NOT a, NOT b, and OR.

2. You must expand the big multiplexor to add another input, and then add new logic to calculate NAND.

A.6 Faster Addition: Carry Lookahead

The key to speeding up addition is determining the carry in to the high-order bits sooner. There are a variety of schemes to anticipate the carry so that the worst-case scenario is a function of the $\log_2$ of the number of bits in the adder. These anticipatory signals are faster because they go through fewer gates in sequence, but it takes many more gates to anticipate the proper carry.

A key to understanding fast-carry schemes is to remember that, unlike software, hardware executes in parallel whenever inputs change.

Fast Carry Using "Infinite" Hardware

As we mentioned earlier, any equation can be represented in two levels of logic. Since the only external inputs are the two operands and the CarryIn to the least significant bit of the adder, in theory we could calculate the CarryIn values to all the remaining bits of the adder in just two levels of logic.

For example, the CarryIn for bit 2 of the adder is exactly the CarryOut of bit 1, so the formula is

$$CarryIn2 = (b1 \cdot CarryIn1) + (a1 \cdot CarryIn1) + (a1 \cdot b1)$$

Similarly, CarryIn1 is defined as

$$CarryIn1 = (b0 \cdot CarryIn0) + (a0 \cdot CarryIn0) + (a0 \cdot b0)$$

Using the shorter and more traditional abbreviation of ci for CarryIni, we can rewrite the formulas as

$$c2 = (b1 \cdot c1) + (a1 \cdot c1) + (a1 \cdot b1)$$
$$c1 = (b0 \cdot c0) + (a0 \cdot c0) + (a0 \cdot b0)$$

Substituting the definition of c1 for the first equation results in this formula:

$$c2 = (a1 \cdot a0 \cdot b0) + (a1 \cdot a0 \cdot c0) \cdot (a1 \cdot b0 \cdot c0)$$
$$+(b1 \cdot a0 \cdot b0) + (b1 \cdot a0 \cdot c0) + (b1 \cdot b0 \cdot c0) + (a1 \cdot b1)$$

You can imagine how the equation expands as we get to higher bits in the adder; it grows rapidly with the number of bits. This complexity is reflected in the cost of the hardware for fast carry, making this simple scheme prohibitively expensive for wide adders.

Fast Carry Using the First Level of Abstraction: Propagate and Generate

Most fast-carry schemes limit the complexity of the equations to simplify the hardware, while still making substantial speed improvements over ripple carry. One such scheme is a *carry-lookahead adder*. In Chapter 1, we said computer systems cope with complexity by using levels of abstraction. A carry-lookahead adder relies on levels of abstraction in its implementation.

Let's factor our original equation as a first step:

$$c_i + 1 = (b_i \cdot c_i) + (a_i \cdot c_i) + (a_i \cdot b_i)$$
$$= (a_i \cdot b_i) + (a_i + b_i) \cdot c_i$$

If we were to rewrite the equation for c_2 using this formula, we would see some repeated patterns:

$$c_2 = (a_1 \cdot b_1) + (a_1 \cdot b_1) \cdot ((a_0 \cdot b_0) + (a_0 + b_0) \cdot c_0)$$

Note the repeated appearance of $(a_i \cdot b_i)$ and $(a_i + b_i)$ in the formula above. These two important factors are traditionally called *generate* (g_i) and *propagate* (p_i):

$$g_i = a_i \cdot b_i$$
$$p_i = a_i + b_i$$

Using them to define $c_i + 1$, we get

$$c_i + 1 = g_i + p_i \cdot c_i$$

To see where the signals get their names, suppose g_i is 1. Then

$$c_i + 1 = g_i + p_i \cdot c_i = 1 + p_i \cdot c_i = 1$$

That is, the adder *generates* a CarryOut ($c_i + 1$) independent of the value of CarryIn (c_i). Now suppose that g_i is 0 and p_i is 1. Then

$$c_i + 1 = g_i + p_i \cdot c_i = 0 + 1 \cdot c_i = c_i$$

That is, the adder *propagates* CarryIn to a CarryOut. Putting the two together, CarryIn$_i$ + 1 is a 1 if either g_i is 1 or both p_i is 1 and CarryIn$_i$ is 1.

As an analogy, imagine a row of dominoes set on edge. The end domino can be tipped over by pushing one far away, provided there are no gaps between the two. Similarly, a carry out can be made true by a generate far away, provided all the propagates between them are true.

Relying on the definitions of propagate and generate as our first level of abstraction, we can express the CarryIn signals more economically. Let's show it for 4 bits:

$$c_1 = g_0 + (p_0 \cdot c_0)$$
$$c_2 = g_1 + (p_1 \cdot g_0) + (p_1 \cdot p_0 \cdot c_0)$$
$$c_3 = g_2 + (p_2 \cdot g_1) + (p_2 \cdot p_1 \cdot g_0) + (p_2 \cdot p_1 \cdot p_0 \cdot c_0)$$
$$c_4 = g_3 + (p_3 \cdot g_2) + (p_3 \cdot p_2 \cdot g_1) + (p_3 \cdot p_2 \cdot p_1 \cdot g_0)$$
$$+ (p_3 \cdot p_2 \cdot p_1 \cdot p_0 \cdot c_0)$$

These equations just represent common sense: CarryIni is a 1 if some earlier adder generates a carry and all intermediary adders propagate a carry. Figure A.6.1 uses plumbing to try to explain carry lookahead.

Even this simplified form leads to large equations and, hence, considerable logic even for a 16-bit adder. Let's try moving to two levels of abstraction.

Fast Carry Using the Second Level of Abstraction

First, we consider this 4-bit adder with its carry-lookahead logic as a single building block. If we connect them in ripple carry fashion to form a 16-bit adder, the add will be faster than the original with a little more hardware.

To go faster, we'll need carry lookahead at a higher level. To perform carry lookahead for 4-bit adders, we need to propagate and generate signals at this higher level. Here they are for the four 4-bit adder blocks:

$$P0 = p3 \cdot p2 \cdot p1 \cdot p0$$
$$P1 = p7 \cdot p6 \cdot p5 \cdot p4$$
$$P2 = p11 \cdot p10 \cdot p9 \cdot p8$$
$$P3 = p15 \cdot p14 \cdot p13 \cdot p12$$

That is, the "super" propagate signal for the 4-bit abstraction (Pi) is true only if each of the bits in the group will propagate a carry.

For the "super" generate signal (Gi), we care only if there is a carry out of the most significant bit of the 4-bit group. This obviously occurs if generate is true for that most significant bit; it also occurs if an earlier generate is true *and* all the intermediate propagates, including that of the most significant bit, are also true:

$$G0 = g3 + (p3 \cdot g2) + (p3 \cdot p2 \cdot g1) + (p3 \cdot p2 \cdot p1 \cdot g0)$$
$$G1 = g7 + (p7 \cdot g6) + (p7 \cdot p6 \cdot g5) + (p7 \cdot p6 \cdot p5 \cdot g4)$$
$$G2 = g11 + (p11 \cdot g10) + (p11 \cdot p10 \cdot g9) + (p11 \cdot p10 \cdot p9 \cdot g8)$$
$$G3 = g15 + (p15 \cdot g14) + (p15 \cdot p14 \cdot g13) + (p15 \cdot p14 \cdot p13 \cdot g12)$$

Figure A.6.2 updates our plumbing analogy to show P0 and G0.

Then the equations at this higher level of abstraction for the carry in for each 4-bit group of the 16-bit adder (C1, C2, C3, C4 in Figure A.6.3) are very similar to the carry out equations for each bit of the 4-bit adder (c1, c2, c3, c4) on page A-40:

$$C1 = G0 + (P0 \cdot c0)$$
$$C2 = G1 + (P1 \cdot G0) + (P1 \cdot P0 \cdot c0)$$
$$C3 = G2 + (P2 \cdot G1) + (P2 \cdot P1 \cdot G0) + (P2 \cdot P1 \cdot P0 \cdot c0)$$
$$C4 = G3 + (P3 \cdot G2) + (P3 \cdot P2 \cdot G1) + (P3 \cdot P2 \cdot P1 \cdot G0)$$
$$+ (P3 \cdot P2 \cdot P1 \cdot P0 \cdot c0)$$

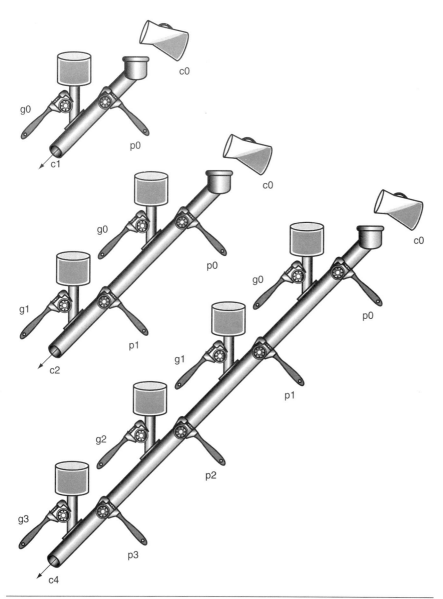

FIGURE A.6.1 A plumbing analogy for carry lookahead for 1 bit, 2 bits, and 4 bits using water pipes and valves. The wrenches are turned to open and close valves. Water is shown in color. The output of the pipe ($ci + 1$) will be full if either the nearest generate value (gi) is turned on or if the i propagate value (pi) is on and there is water further upstream, either from an earlier generate or a propagate with water behind it. CarryIn ($c0$) can result in a carry out without the help of any generates, but with the help of *all* propagates.

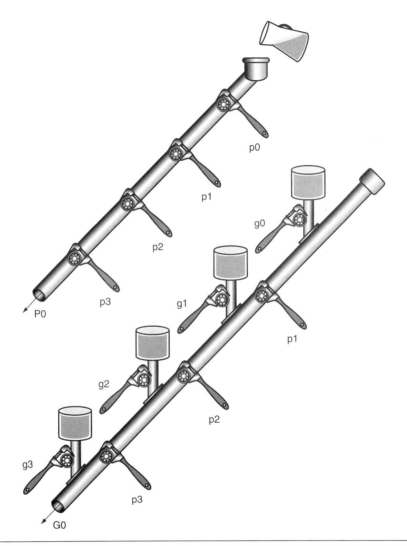

FIGURE A.6.2   A plumbing analogy for the next-level carry-lookahead signals P0 and G0.
P0 is open only if all four propagates (pi) are open, while water flows in G0 only if at least one generate (gi) is open and all the propagates downstream from that generate are open.

Figure A.6.3 shows 4-bit adders connected with such a carry-lookahead unit. The exercises explore the speed differences between these carry schemes, different notations for multibit propagate and generate signals, and the design of a 64-bit adder.

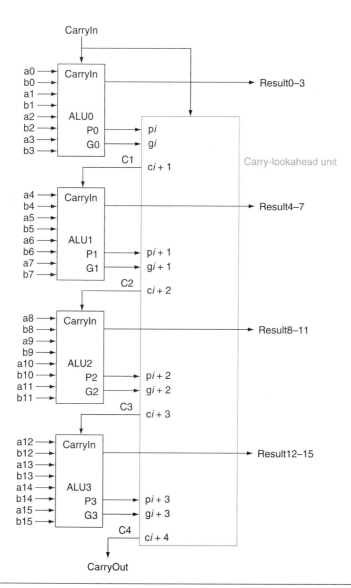

FIGURE A.6.3   Four 4-bit ALUs using carry lookahead to form a 16-bit adder. Note that the carries come from the carry-lookahead unit, not from the 4-bit ALUs.

Both Levels of the Propagate and Generate

EXAMPLE

Determine the gi, pi, Pi, and Gi values of these two 16-bit numbers:

```
a:        0001  1010  0011  0011₂ ₒ
b:        1110  0101  1110  1011₂ ₒ
```

Also, what is CarryOut15 (C4)?

ANSWER

Aligning the bits makes it easy to see the values of generate gi ($ai \cdot bi$) and propagate pi ($ai + bi$):

```
a:        0001  1010  0011  0011
b:        1110  0101  1110  1011
g i :     0000  0000  0010  0011
p i :     1111  1111  1111  1011
```

where the bits are numbered 15 to 0 from left to right. Next, the "super" propagates (P3, P2, P1, P0) are simply the AND of the lower-level propagates:

$$P3 = 1 \cdot 1 \cdot 1 \cdot 1 = 1$$
$$P2 = 1 \cdot 1 \cdot 1 \cdot 1 = 1$$
$$P1 = 1 \cdot 1 \cdot 1 \cdot 1 = 1$$
$$P0 = 1 \cdot 0 \cdot 1 \cdot 1 = 0$$

The "super" generates are more complex, so use the following equations:

$$G0 = g3 + (p3 \cdot g2) + (p3 \cdot p2 \cdot g1) + (p3 \cdot p2 \cdot p1 \cdot g0)$$
$$= 0 + (1 \cdot 0) + (1 \cdot 0 \cdot 1) + (1 \cdot 0 \cdot 1 \cdot 1) = 0 + 0 + 0 + 0 = 0$$
$$G1 = g7 + (p7 \cdot g6) + (p7 \cdot p6 \cdot g5) + (p7 \cdot p6 \cdot p5 \cdot g4)$$
$$= 0 + (1 \cdot 0) + (1 \cdot 1 \cdot 1) + (1 \cdot 1 \cdot 1 \cdot 0) = 0 + 0 + 1 + 0 = 1$$
$$G2 = g11 + (p11 \cdot g10) + (p11 \cdot p10 \cdot g9) + (p11 \cdot p10 \cdot p9 \cdot g8)$$
$$= 0 + (1 \cdot 0) + (1 \cdot 1 \cdot 0) + (1 \cdot 1 \cdot 1 \cdot 0) = 0 + 0 + 0 + 0 = 0$$
$$G3 = g15 + (p15 \cdot g14) + (p15 \cdot p14 \cdot g13) + (p15 \cdot p14 \cdot p13 \cdot g12)$$
$$= 0 + (1 \cdot 0) + (1 \cdot 1 \cdot 0) + (1 \cdot 1 \cdot 1 \cdot 0) = 0 + 0 + 0 + 0 = 0$$

Finally, CarryOut15 is

$$C4 = G3 + (P3 \cdot G2) + (P3 \cdot P2 \cdot G1) + (P3 \cdot P2 \cdot P1 \cdot G0)$$
$$+ (P3 \cdot P2 \cdot P1 \cdot P0 \cdot c0)$$
$$= 0 + (1 \cdot 0) + (1 \cdot 1 \cdot 1) + (1 \cdot 1 \cdot 1 \cdot 0) + (1 \cdot 1 \cdot 1 \cdot 0 \cdot 0)$$
$$= 0 + 0 + 1 + 0 + 0 = 1$$

Hence, there *is* a carry out when adding these two 16-bit numbers.

The reason carry lookahead can make carries faster is that all logic begins evaluating the moment the clock cycle begins, and the result will not change once the output of each gate stops changing. By taking the shortcut of going through fewer gates to send the carry in signal, the output of the gates will stop changing sooner, and hence the time for the adder can be less.

To appreciate the importance of carry lookahead, we need to calculate the relative performance between it and ripple carry adders.

Speed of Ripple Carry versus Carry Lookahead

One simple way to model time for logic is to assume each AND or OR gate takes the same time for a signal to pass through it. Time is estimated by simply counting the number of gates along the path through a piece of logic. Compare the number of *gate delays* for paths of two 16-bit adders, one using ripple carry and one using two-level carry lookahead.

EXAMPLE

Figure A.5.5 on page A-28 shows that the carry out signal takes two gate delays per bit. Then the number of gate delays between a carry in to the least significant bit and the carry out of the most significant is $32 \times 2 = 64$.

ANSWER

For carry lookahead, the carry out of the most significant bit is just C4, defined in the example. It takes two levels of logic to specify C4 in terms of Pi and Gi (the OR of several AND terms). Pi is specified in one level of logic (AND) using pi, and Gi is specified in two levels using pi and gi, so the worst case for this next level of abstraction is two levels of logic. pi and gi are each one level of logic, defined in terms of ai and bi. If we assume one gate delay for each level of logic in these equations, the worst case is $2 + 2 + 1 = 5$ gate delays.

Hence, for the path from carry in to carry out, the 16-bit addition by a carry-lookahead adder is six times faster, using this very simple estimate of hardware speed.

Summary

Carry lookahead offers a faster path than waiting for the carries to ripple through all 32 1-bit adders. This faster path is paved by two signals, generate and propagate.

The former creates a carry regardless of the carry input, and the latter passes a carry along. Carry lookahead also gives another example of how abstraction is important in computer design to cope with complexity.

Check Yourself

Using the simple estimate of hardware speed above with gate delays, what is the relative performance of a ripple carry 8-bit add versus a 64-bit add using carry-lookahead logic?

1. A 64-bit carry-lookahead adder is three times faster: 8-bit adds are 16 gate delays and 64-bit adds are seven gate delays.

2. They are about the same speed, since 64-bit adds need more levels of logic in the 16-bit adder.

3. Eight-bit adds are faster than 64 bits, even with carry lookahead.

Elaboration: We have now accounted for all but one of the arithmetic and logical operations for the core LEGv8 instruction set: the ALU in Figure A.5.14 omits support of shift instructions. It would be possible to widen the ALU multiplexor to include a left shift by 1 bit or a right shift by 1 bit. But hardware designers have created a circuit called a *barrel shifter*, which can shift from 1 to 63 bits in no more time than it takes to add two 64-bit numbers, so shifting is normally done outside the ALU.

Elaboration: The logic equation for the Sum output of the full adder on page A-28 can be expressed more simply by using a more powerful gate than AND and OR. An *exclusive OR* gate is true if the two operands disagree; that is,

$$x \neq y \Rightarrow 1 \text{ and } x == y \Rightarrow 0$$

In some technologies, exclusive OR is more efficient than two levels of AND and OR gates. Using the symbol $\oplus$ to represent exclusive OR, here is the new equation:

$$\text{Sum} = a \oplus b \oplus \text{CarryIn}$$

Also, we have drawn the ALU the traditional way, using gates. Computers are designed today in CMOS transistors, which are basically switches. CMOS ALU and barrel shifters take advantage of these switches and have many fewer multiplexors than shown in our designs, but the design principles are similar.

Elaboration: Using lowercase and uppercase to distinguish the hierarchy of generate and propagate symbols breaks down when you have more than two levels. An alternate notation that scales is $g_{i..j}$ and $p_{i..j}$ for the generate and propagate signals for bits i to j. Thus, $g_{1..1}$ is generated for bit 1, $g_{4..1}$ is for bits 4 to 1, and $g_{16..1}$ is for bits 16 to 1.

A.7 Clocks

Before we discuss memory elements and sequential logic, it is useful to discuss briefly the topic of clocks. This short section introduces the topic and is similar to the discussion found in Section 4.2. More details on clocking and timing methodologies are presented in Section A.11.

Clocks are needed in sequential logic to decide when an element that contains state should be updated. A clock is simply a free-running signal with a fixed *cycle time*; the *clock frequency* is simply the inverse of the cycle time. As shown in Figure A.7.1, the *clock cycle time* or *clock period* is divided into two portions: when the clock is high and when the clock is low. In this text, we use only edge-triggered clocking. This means that all state changes occur on a clock edge. We use an edge-triggered methodology because it is simpler to explain. Depending on the technology, it may or may not be the best choice for a clocking methodology.

edge-triggered clocking A clocking scheme in which all state changes occur on a clock edge.

clocking methodology The approach used to determine when data are valid and stable relative to the clock.

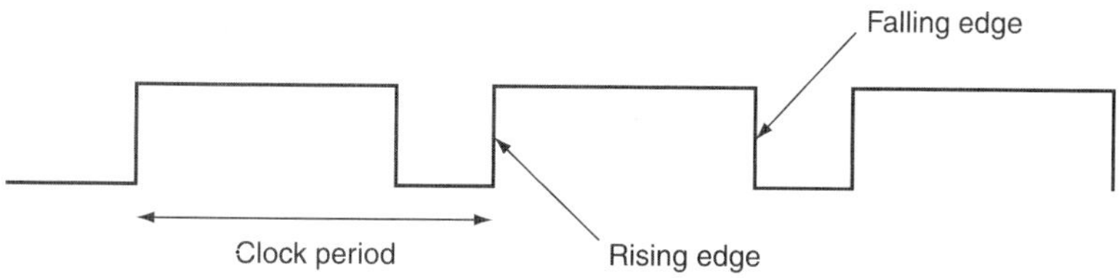

FIGURE A.7.1   A clock signal oscillates between high and low values. The clock period is the time for one full cycle. In an edge-triggered design, either the rising or falling edge of the clock is active and causes state to be changed.

In an edge-triggered methodology, either the rising edge or the falling edge of the clock is *active* and causes state changes to occur. As we will see in the next section, the state elements in an edge-triggered design are implemented so that the contents of the state elements only change on the active clock edge. The choice of which edge is active is influenced by the implementation technology and does not affect the concepts involved in designing the logic.

The clock edge acts as a sampling signal, causing the value of the data input to a state element to be sampled and stored in the state element. Using an edge trigger means that the sampling process is essentially instantaneous, eliminating problems that could occur if signals were sampled at slightly different times.

The major constraint in a clocked system, also called a synchronous system, is that the signals that are written into state elements must be *valid* when the active

state element A memory element.

synchronous system A memory system that employs clocks and where data signals are read only when the clock indicates that the signal values are stable.

clock edge occurs. A signal is valid if it is stable (i.e., not changing), and the value will not change again until the inputs change. Since combinational circuits cannot have feedback, if the inputs to a combinational logic unit are not changed, the outputs will eventually become valid.

Figure A.7.2 shows the relationship among the state elements and the combinational logic blocks in a synchronous, sequential logic design. The state elements, whose outputs change only after the clock edge, provide valid inputs to the combinational logic block. To ensure that the values written into the state elements on the active clock edge are valid, the clock must have a long enough period so that all the signals in the combinational logic block stabilize, and then the clock edge samples those values for storage in the state elements. This constraint sets a lower bound on the length of the clock period, which must be long enough for all state element inputs to be valid.

In the rest of this appendix, as well as in Chapter 4, we usually omit the clock signal, since we are assuming that all state elements are updated on the same clock edge. Some state elements will be written on every clock edge, while others will be written only under certain conditions (such as a register being updated). In such cases, we will have an explicit write signal for that state element. The write signal must still be gated with the clock so that the update occurs only on the clock edge if the write signal is active. We will see how this is done and used in the next section.

One other advantage of an edge-triggered methodology is that it is possible to have a state element that is used as both an input and output to the same combinational logic block, as shown in Figure A.7.3. In practice, care must be taken to prevent races in such situations and to ensure that the clock period is long enough; this topic is discussed further in Section A.11.

Now that we have discussed how clocking is used to update state elements, we can discuss how to construct the state elements.

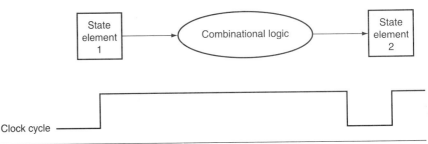

FIGURE A.7.2   The inputs to a combinational logic block come from a state element, and the outputs are written into a state element. The clock edge determines when the contents of the state elements are updated.

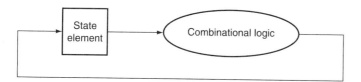

FIGURE A.7.3 An edge-triggered methodology allows a state element to be read and written in the same clock cycle without creating a race that could lead to undetermined data values. Of course, the clock cycle must still be long enough so that the input values are stable when the active clock edge occurs.

Elaboration Occasionally, designers find it useful to have a small number of state elements that change on the opposite clock edge from the majority of the state elements. Doing so requires extreme care, because such an approach has effects on both the inputs and the outputs of the state element. Why then would designers ever do this? Consider the case where the amount of combinational logic before and after a state element is small enough so that each could operate in one-half clock cycle, rather than the more usual full clock cycle. Then the state element can be written on the clock edge corresponding to a half clock cycle, since the inputs and outputs will both be usable after one-half clock cycle. One common place where this technique is used is in **register files**, where simply reading or writing the register file can often be done in half the normal clock cycle. Chapter 4 makes use of this idea to reduce the pipelining overhead.

register file A state element that consists of a set of registers that can be read and written by supplying a register number to be accessed.

Memory Elements: Flip-Flops, Latches, and Registers

In this section and the next, we discuss the basic principles behind memory elements, starting with flip-flops and latches, moving on to register files, and finishing with memories. All memory elements store state: the output from any memory element depends both on the inputs and on the value that has been stored inside the memory element. Thus all logic blocks containing a memory element contain state and are sequential.

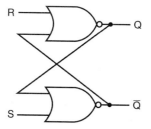

FIGURE A.8.1 A pair of cross-coupled NOR gates can store an internal value. The value stored on the output Q is recycled by inverting it to obtain $\overline{Q}$ and then inverting $\overline{Q}$ to obtain Q. If either R or $\overline{Q}$ is asserted, Q will be deasserted and vice versa.

The simplest type of memory elements are *unclocked*; that is, they do not have any clock input. Although we only use clocked memory elements in this text, an unclocked latch is the simplest memory element, so let's look at this circuit first. Figure A.8.1 shows an *S-R latch* (set-reset latch), built from a pair of NOR gates (OR gates with inverted outputs). The outputs Q and $\bar{Q}$ represent the value of the stored state and its complement. When neither S nor R are asserted, the cross-coupled NOR gates act as inverters and store the previous values of Q and $\bar{Q}$.

For example, if the output, Q, is true, then the bottom inverter produces a false output (which is $\bar{Q}$), which becomes the input to the top inverter, which produces a true output, which is Q, and so on. If S is asserted, then the output Q will be asserted and $\bar{Q}$ will be deasserted, while if R is asserted, then the output $\bar{Q}$ will be asserted and Q will be deasserted. When S and R are both deasserted, the last values of Q and $\bar{Q}$ will continue to be stored in the cross-coupled structure. Asserting S and R simultaneously can lead to incorrect operation: depending on how S and R are asserted, the latch may oscillate or become metastable (this is described in more detail in Section A.11).

This cross-coupled structure is the basis for more complex memory elements that allow us to store data signals. These elements contain additional gates used to store signal values and to cause the state to be updated only in conjunction with a clock. The next section shows how these elements are built.

Flip-Flops and Latches

Flip-flops and latches are the simplest memory elements. In both flip-flops and latches, the output is equal to the value of the stored state inside the element. Furthermore, unlike the S-R latch described above, all the latches and flip-flops we will use from this point on are clocked, which means that they have a clock input and the change of state is triggered by that clock. The difference between a flip-flop and a latch is the point at which the clock causes the state to actually change. In a clocked latch, the state is changed whenever the appropriate inputs change and the clock is asserted, whereas in a flip-flop, the state is changed only on a clock edge. Since throughout this text we use an edge-triggered timing methodology where state is only updated on clock edges, we need only use flip-flops. Flip-flops are often built from latches, so we start by describing the operation of a simple clocked latch and then discuss the operation of a flip-flop constructed from that latch.

For computer applications, the function of both flip-flops and latches is to store a signal. A *D latch* or D flip-flop stores the value of its data input signal in the internal memory. Although there are many other types of latch and flip-flop, the D type is the only basic building block that we will need. A D latch has two inputs and two outputs. The inputs are the data value to be stored (called D) and a clock signal (called C) that indicates when the latch should read the value on the D input and store it. The outputs are simply the value of the internal state (Q)

flip-flop A memory element for which the output is equal to the value of the stored state inside the element and for which the internal state is changed only on a clock edge.

latch A memory element in which the output is equal to the value of the stored state inside the element and the state is changed whenever the appropriate inputs change and the clock is asserted.

D flip-flop A flip-flop with one data input that stores the value of that input signal in the internal memory when the clock edge occurs.

and its complement ($\overline{Q}$). When the clock input C is asserted, the latch is said to be *open*, and the value of the output (Q) becomes the value of the input D. When the clock input C is deasserted, the latch is said to be *closed*, and the value of the output (Q) is whatever value was stored the last time the latch was open.

Figure A.8.2 shows how a D latch can be implemented with two additional gates added to the cross-coupled NOR gates. Since when the latch is open the value of Q changes as D changes, this structure is sometimes called a *transparent latch*. Figure A.8.3 shows how this D latch works, assuming that the output Q is initially false and that D changes first.

As mentioned earlier, we use flip-flops as the basic building block, rather than latches. Flip-flops are not transparent: their outputs change *only* on the clock edge. A flip-flop can be built so that it triggers on either the rising (positive) or falling (negative) clock edge; for our designs we can use either type. Figure A.8.4 shows how a falling-edge D flip-flop is constructed from a pair of D latches. In a D flip-flop, the output is stored when the clock edge occurs. Figure A.8.5 shows how this flip-flop operates.

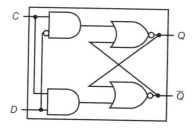

FIGURE A.8.2 A D latch implemented with NOR gates. A NOR gate acts as an inverter if the other input is 0. Thus, the cross-coupled pair of NOR gates acts to store the state value unless the clock input, C, is asserted, in which case the value of input D replaces the value of Q and is stored. The value of input D must be stable when the clock signal C changes from asserted to deasserted.

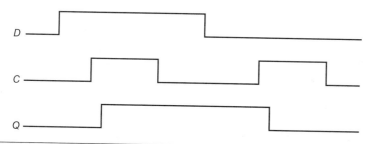

FIGURE A.8.3 Operation of a D latch, assuming the output is initially deasserted. When the clock, C, is asserted, the latch is open and the Q output immediately assumes the value of the D input.

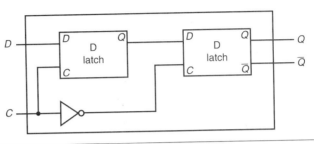

FIGURE A.8.4   A D flip-flop with a falling-edge trigger. The first latch, called the master, is open and follows the input *D* when the clock input, *C*, is asserted. When the clock input, *C*, falls, the first latch is closed, but the second latch, called the slave, is open and gets its input from the output of the master latch.

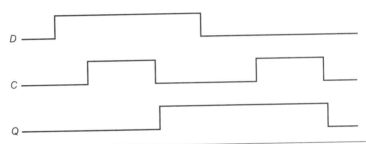

FIGURE A.8.5   Operation of a D flip-flop with a falling-edge trigger, assuming the output is initially deasserted. When the clock input (*C*) changes from asserted to deasserted, the *Q* output stores the value of the *D* input. Compare this behavior to that of the clocked D latch shown in Figure A.8.3. In a clocked latch, the stored value and the output, *Q*, both change whenever *C* is high, as opposed to only when *C* transitions.

Here is a Verilog description of a module for a rising-edge D flip-flop, assuming that C is the clock input and D is the data input:

```
module DFF(clock,D,Q,Qbar);
    input clock, D;
    output reg Q; // Q is a reg since it is assigned in an
always block
    output Qbar;
    assign Qbar= ~ Q; // Qbar is always just the inverse
of Q
    always @(posedge clock) // perform actions whenever the
clock rises
        Q=D;
endmodule
```

setup time The minimum time that the input to a memory device must be valid before the clock edge.

Because the *D* input is sampled on the clock edge, it must be valid for a period of time immediately before and immediately after the clock edge. The minimum time that the input must be valid before the clock edge is called the setup time; the

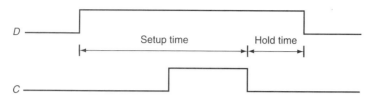

FIGURE A.8.6 Setup and hold time requirements for a D flip-flop with a falling-edge trigger. The input must be stable for a period of time before the clock edge, as well as after the clock edge. The minimum time the signal must be stable before the clock edge is called the setup time, while the minimum time the signal must be stable after the clock edge is called the hold time. Failure to meet these minimum requirements can result in a situation where the output of the flip-flop may not be predictable, as described in Section A.11. Hold times are usually either 0 or very small and thus not a cause of worry.

minimum time during which it must be valid after the clock edge is called the hold time. Thus the inputs to any flip-flop (or anything built using flip-flops) must be valid during a window that begins at time t_{setup} before the clock edge and ends at t_{hold} after the clock edge, as shown in Figure A.8.6. Section A.11 talks about clocking and timing constraints, including the propagation delay through a flip-flop, in more detail.

hold time The minimum time during which the input must be valid after the clock edge.

We can use an array of D flip-flops to build a register that can hold a multibit datum, such as a byte or word. We used registers throughout our datapaths in Chapter 4.

Register Files

One structure that is central to our datapath is a *register file*. A register file consists of a set of registers that can be read and written by supplying a register number to be accessed. A register file can be implemented with a decoder for each read or write port and an array of registers built from D flip-flops. Because reading a register does not change any state, we need only supply a register number as an input, and the only output will be the data contained in that register. For writing a register we will need three inputs: a register number, the data to write, and a clock that controls the writing into the register. In Chapter 4, we used a register file that has two read ports and one write port. This register file is drawn as shown in Figure A.8.7. The read ports can be implemented with a pair of multiplexors, each of which is as wide as the number of bits in each register of the register file. Figure A.8.8 shows the implementation of two register read ports for a 64-bit-wide register file.

Implementing the write port is slightly more complex, since we can only change the contents of the designated register. We can do this by using a decoder to generate a signal that can be used to determine which register to write. Figure A.8.9 shows how to implement the write port for a register file. It is important to remember that the flip-flop changes state only on the clock edge. In Chapter 4, we hooked up write signals for the register file explicitly and assumed the clock shown in Figure A.8.9 is attached implicitly.

What happens if the same register is read and written during a clock cycle? Because the write of the register file occurs on the clock edge, the register will be

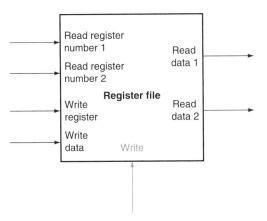

FIGURE A.8.7   A register file with two read ports and one write port has five inputs and two outputs. The control input Write is shown in color.

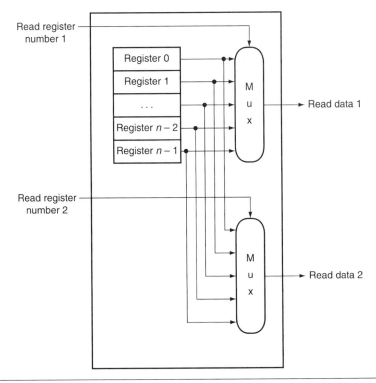

FIGURE A.8.8   The implementation of two read ports for a register file with *n* registers can be done with a pair of *n*-to-1 multiplexors, each 64 bits wide. The register read number signal is used as the multiplexor selector signal. Figure A.8.9 shows how the write port is implemented.

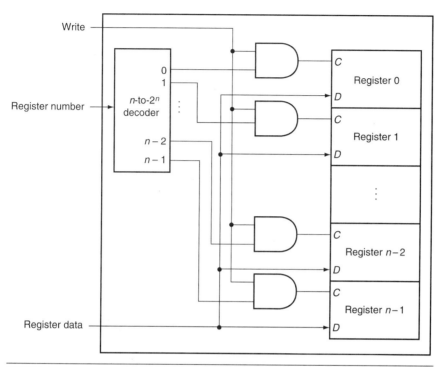

FIGURE A.8.9   The write port for a register file is implemented with a decoder that is used with the write signal to generate the C input to the registers. All three inputs (the register number, the data, and the write signal) will have setup and hold-time constraints that ensure that the correct data are written into the register file.

valid during the time it is read, as we saw earlier in Figure A.7.2. The value returned will be the value written in an earlier clock cycle. If we want a read to return the value currently being written, additional logic in the register file or outside of it is needed. Chapter 4 makes extensive use of such logic.

Specifying Sequential Logic in Verilog

To specify sequential logic in Verilog, we must understand how to generate a clock, how to describe when a value is written into a register, and how to specify sequential control. Let us start by specifying a clock. A clock is not a predefined object in Verilog; instead, we generate a clock by using the Verilog notation #n before a statement; this causes a delay of n simulation time steps before the execution of the statement. In most Verilog simulators, it is also possible to generate a clock as an external input, allowing the user to specify at simulation time the number of clock cycles during which to run a simulation.

The code in Figure A.8.10 implements a simple clock that is high or low for one simulation unit and then switches state. We use the delay capability and blocking assignment to implement the clock.

```
reg clock; // clock is a register
always
#1 clock = 1; #1 clock = 0;
```

FIGURE A.8.10   A specification of a clock.

Next, we must be able to specify the operation of an edge-triggered register. In Verilog, this is done by using the sensitivity list on an always block and specifying as a trigger either the positive or negative edge of a binary variable with the notation posedge or negedge, respectively. Hence, the following Verilog code causes register A to be written with the value b at the positive edge clock:

```
reg [63:0] A;
wire [63:0] b;

always @(posedge clock) A <= b;
```

```
module registerfile (Read1,Read2,WriteReg,WriteData,RegWrite,
Data1,Data2,clock);
    input [5:0] Read1,Read2,WriteReg; // the register numbers
to read or write
    input [63:0] WriteData; // data to write
    input RegWrite, // the write control
      clock; // the clock to trigger write
    output [63:0] Data1, Data2; // the register values read
    reg [63:0] RF [31:0]; // 32 registers each 32 bits long

    assign Data1 = RF[Read1];
    assign Data2 = RF[Read2];

    always begin
        // write the register with new value if Regwrite is
high
        @(posedge clock) if (RegWrite) RF[WriteReg] <=
WriteData;
    end
endmodule
```

FIGURE A.8.11   A LEGv8 register file written in behavioral Verilog. This register file writes on the rising clock edge.

Throughout this chapter and the Verilog sections of Chapter 4, we will assume a positive edge-triggered design. Figure A.8.11 shows a Verilog specification of a LEGv8 register file that assumes two reads and one write, with only the write being clocked.

In the Verilog for the register file in Figure A.8.11, the output ports corresponding to the registers being read are assigned using a continuous assignment, but the register being written is assigned in an `always` block. Which of the following is the reason?

Check Yourself

   a. There is no special reason. It was simply convenient.

   b. Because Data1 and Data2 are output ports and WriteData is an input port.

   c. Because reading is a combinational event, while writing is a sequential event.

A.9   Memory Elements: SRAMs and DRAMs

Registers and register files provide the basic building blocks for small memories, but larger amounts of memory are built using either SRAMs (static random access memories) or *DRAMs* (dynamic random access memories). We first discuss SRAMs, which are somewhat simpler, and then turn to DRAMs.

static random access memory (SRAM) A memory where data are stored statically (as in flip-flops) rather than dynamically (as in DRAM). SRAMs are faster than DRAMs, but less dense and more expensive per bit.

SRAMs

SRAMs are simply integrated circuits that are memory arrays with (usually) a single access port that can provide either a read or a write. SRAMs have a fixed access time to any datum, though the read and write access characteristics often differ. An SRAM chip has a specific configuration in terms of the number of addressable locations, as well as the width of each addressable location. For example, a 4M × 8 SRAM provides 4M entries, each of which is 8 bits wide. Thus it will have 22 address lines (since 4M = 2^{22}), an 8-bit data output line, and an 8-bit single data input line. As with ROMs, the number of addressable locations is often called the *height*, with the number of bits per unit called the *width*. For a variety of technical reasons, the newest and fastest SRAMs are typically available in narrow configurations: × 1 and × 4. Figure A.9.1 shows the input and output signals for a 2M × 16 SRAM.

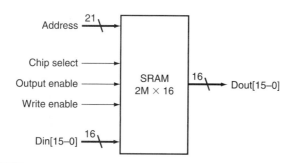

FIGURE A.9.1   A 32K × 8 SRAM showing the 21 address lines (32K = 2^{15}) and 16 data inputs, the three control lines, and the 16 data outputs.

To initiate a read or write access, the Chip select signal must be made active. For reads, we must also activate the Output enable signal that controls whether or not the datum selected by the address is actually driven on the pins. The Output enable is useful for connecting multiple memories to a single-output bus and using Output enable to determine which memory drives the bus. The SRAM read access time is usually specified as the delay from the time that Output enable is true and the address lines are valid until the time that the data are on the output lines. Typical read access times for SRAMs in 2004 varied from about 2–4 ns for the fastest CMOS parts, which tend to be somewhat smaller and narrower, to 8–20 ns for the typical largest parts, which in 2004 had more than 32 million bits of data. The demand for low-power SRAMs for consumer products and digital appliances has grown greatly in the past 5 years; these SRAMs have much lower stand-by and access power, but usually are 5–10 times slower. Most recently, synchronous SRAMs—similar to the synchronous DRAMs, which we discuss in the next section—have also been developed.

For writes, we must supply the data to be written and the address, as well as signals to cause the write to occur. When both the Write enable and Chip select are true, the data on the data input lines are written into the cell specified by the address. There are setup-time and hold-time requirements for the address and data lines, just as there were for D flip-flops and latches. In addition, the Write enable signal is not a clock edge but a pulse with a minimum width requirement. The time to complete a write is specified by the combination of the setup times, the hold times, and the Write enable pulse width.

Large SRAMs cannot be built in the same way we build a register file because, unlike a register file where a 32-to-1 multiplexor might be practical, the 64K-to-1 multiplexor that would be needed for a 64K × 1 SRAM is totally impractical. Rather than use a giant multiplexor, large memories are implemented with a shared output line, called a *bit line*, which multiple memory cells in the memory array can assert. To allow multiple sources to drive a single line, a *three-state buffer* (or *tristate buffer*) is used. A three-state buffer has two inputs—a data signal and an Output enable—and a single output, which is in one of three states: asserted, deasserted, or high impedance. The output of a tristate buffer is equal to the data input signal, either asserted or deasserted, if the Output enable is asserted, and is otherwise in a *high-impedance state* that allows another three-state buffer whose Output enable is asserted to determine the value of a shared output.

Figure A.9.2 shows a set of three-state buffers wired to form a multiplexor with a decoded input. It is critical that the Output enable at most one of the three-state buffers be asserted; otherwise, the three-state buffers may try to set the output line differently. By using three-state buffers in the individual cells of the SRAM, each cell that corresponds to a particular output can share the same output line. The use of a set of distributed three-state buffers is a more efficient implementation than a large centralized multiplexor. The three-state buffers are incorporated into the flip-flops that form the basic cells of the SRAM. Figure A.9.3 shows how a small 4 × 2 SRAM might be built, using D latches with an input called Enable that controls the three-state output.

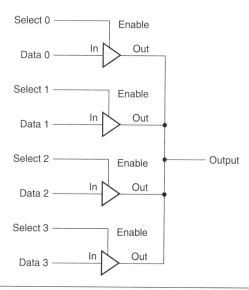

FIGURE A.9.2   Four three-state buffers are used to form a multiplexor. Only one of the four Select inputs can be asserted. A three-state buffer with a deasserted Output enable has a high-impedance output that allows a three-state buffer whose Output enable is asserted to drive the shared output line.

The design in Figure A.9.3 eliminates the need for an enormous multiplexor; however, it still requires a very large decoder and a correspondingly large number of word lines. For example, in a 4M × 8 SRAM, we would need a 22-to-4M decoder and 4M word lines (which are the lines used to enable the individual flip-flops)! To circumvent this problem, large memories are organized as rectangular arrays and use a two-step decoding process. Figure A.9.4 shows how a 4M × 8 SRAM might be organized internally using a two-step decode. As we will see, the two-level decoding process is quite important in understanding how DRAMs operate.

Recently we have seen the development of both synchronous SRAMs (SSRAMs) and synchronous DRAMs (SDRAMs). The key capability provided by synchronous RAMs is the ability to transfer a *burst* of data from a series of sequential addresses within an array or row. The burst is defined by a starting address, supplied in the usual fashion, and a burst length. The speed advantage of synchronous RAMs comes from the ability to transfer the bits in the burst without having to specify additional address bits. Instead, a clock is used to transfer the successive bits in the burst. The elimination of the need to specify the address for the transfers within the burst significantly improves the rate for transferring the block of data. Because of this capability, synchronous SRAMs and DRAMs are rapidly becoming the RAMs of choice for building memory systems in computers. We discuss the use of synchronous DRAMs in a memory system in more detail in the next section and in Chapter 5.

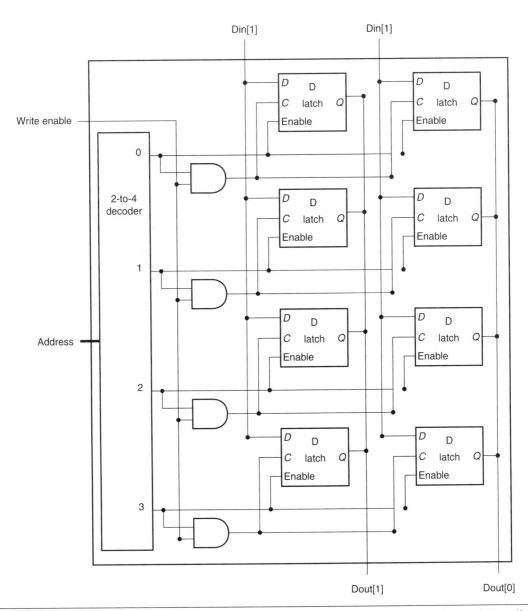

FIGURE A.9.3   The basic structure of a 4 × 2 SRAM consists of a decoder that selects which pair of cells to activate.
The activated cells use a three-state output connected to the vertical bit lines that supply the requested data. The address that selects the cell is sent on one of a set of horizontal address lines, called word lines. For simplicity, the Output enable and Chip select signals have been omitted, but they could easily be added with a few AND gates.

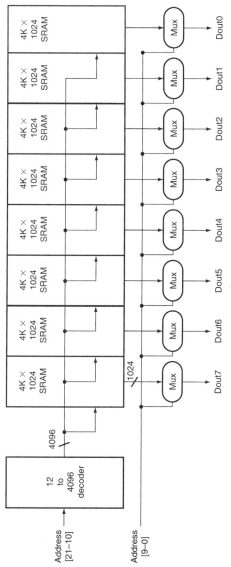

FIGURE A.9.4 Typical organization of a 4M × 8 SRAM as an array of 4K × 1024 arrays. The first decoder generates the addresses for eight 4K × 1024 arrays; then a set of multiplexors is used to select 1 bit from each 1024-bit-wide array. This is a much easier design than a single-level decode that would need either an enormous decoder or a gigantic multiplexor. In practice, a modern SRAM of this size would probably use an even larger number of blocks, each somewhat smaller.

DRAMs

In a static RAM (SRAM), the value stored in a cell is kept on a pair of inverting gates, and as long as power is applied, the value can be kept indefinitely. In a dynamic RAM (DRAM), the value kept in a cell is stored as a charge in a capacitor. A single transistor is then used to access this stored charge, either to read the value or to overwrite the charge stored there. Because DRAMs use only a single transistor per bit of storage, they are much denser and cheaper per bit. By comparison, SRAMs require four to six transistors per bit. Because DRAMs store the charge on a capacitor, it cannot be kept indefinitely and must periodically be *refreshed*. That is why this memory structure is called *dynamic*, as opposed to the static storage in a SRAM cell.

To refresh the cell, we merely read its contents and write it back. The charge can be kept for several milliseconds, which might correspond to close to a million clock cycles. Today, single-chip memory controllers often handle the refresh function independently of the processor. If every bit had to be read out of the DRAM and then written back individually, with large DRAMs containing multiple megabytes, we would constantly be refreshing the DRAM, leaving no time for accessing it. Fortunately, DRAMs also use a two-level decoding structure, and this allows us to refresh an entire row (which shares a word line) with a read cycle followed immediately by a write cycle. Typically, refresh operations consume 1% to 2% of the active cycles of the DRAM, leaving the remaining 98% to 99% of the cycles available for reading and writing data.

Elaboration: How does a DRAM read and write the signal stored in a cell? The transistor inside the cell is a switch, called a *pass transistor*, that allows the value stored on the capacitor to be accessed for either reading or writing. Figure A.9.5 shows how the single-transistor cell looks. The pass transistor acts like a switch: when the signal on the word line is asserted, the switch is closed, connecting the capacitor to the bit line. If the operation is a write, then the value to be written is placed on the bit line. If the value is a 1, the capacitor will be charged. If the value is a 0, then the capacitor will be discharged. Reading is slightly more complex, since the DRAM must detect a very small charge stored in the capacitor. Before activating the word line for a read, the bit line is charged to the voltage that is halfway between the low and high voltage. Then, by activating the word line, the charge on the capacitor is read out onto the bit line. This causes the bit line to move slightly toward the high or low direction, and this change is detected with a sense amplifier, which can detect small changes in voltage.

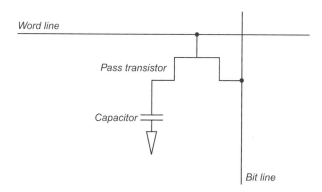

FIGURE A.9.5   A single-transistor DRAM cell contains a capacitor that stores the cell contents and a transistor used to access the cell.

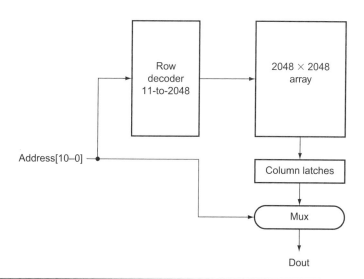

FIGURE A.9.6   A 4M × 1 DRAM is built with a 2048 × 2048 array. The row access uses 11 bits to select a row, which is then latched in 2048 1-bit latches. A multiplexor chooses the output bit from these 2048 latches. The RAS and CAS signals control whether the address lines are sent to the row decoder or column multiplexor.

DRAMs use a two-level decoder consisting of a *row access* followed by a *column access*, as shown in Figure A.9.6. The row access chooses one of a number of rows and activates the corresponding word line. The contents of all the columns in the active row are then stored in a set of latches. The column access then selects the data from the column latches. To save pins and reduce the package cost, the same address lines are used for both the row and column address; a pair of signals called RAS (*Row Access Strobe*) and CAS (*Column Access Strobe*) are used to signal the DRAM that either a row or column address is being supplied. Refresh is performed by simply reading the columns into the column latches and then writing the same values back. Thus, an entire row is refreshed in one cycle. The two-level addressing scheme, combined with the internal circuitry, makes DRAM access times much longer (by a factor of 5–10) than SRAM access times. In 2004, typical DRAM access times ranged from 45 to 65 ns; 256 Mbit DRAMs are in full production, and the first customer samples of 1 GB DRAMs became available in the first quarter of 2004. The much lower cost per bit makes DRAM the choice for main memory, while the faster access time makes SRAM the choice for caches.

You might observe that a 64M × 4 DRAM actually accesses 8K bits on every row access and then throws away all but four of those during a column access. DRAM designers have used the internal structure of the DRAM as a way to provide higher bandwidth out of a DRAM. This is done by allowing the column address to change without changing the row address, resulting in an access to other bits in the column latches. To make this process faster and more precise, the address inputs were clocked, leading to the dominant form of DRAM in use today: synchronous DRAM or SDRAM.

Since about 1999, SDRAMs have been the memory chip of choice for most cache-based main memory systems. SDRAMs provide fast access to a series of bits within a row by sequentially transferring all the bits in a burst under the control of a clock signal. In 2004, DDRRAMs (Double Data Rate RAMs), which are called double data rate because they transfer data on both the rising and falling edge of an externally supplied clock, were the most heavily used form of SDRAMs. As we discuss in Chapter 5, these high-speed transfers can be used to boost the bandwidth available out of main memory to match the needs of the processor and caches.

Error Correction

Because of the potential for data corruption in large memories, most computer systems use some sort of error-checking code to detect possible corruption of data. One simple code that is heavily used is a *parity code*. In a parity code the number of 1s in a word is counted; the word has odd parity if the number of 1s is odd and even otherwise. When a word is written into memory, the parity bit is also written (1 for odd, 0 for even). Then, when the word is read out, the parity bit is read and checked. If the parity of the memory word and the stored parity bit do not match, an error has occurred.

A 1-bit parity scheme can detect at most 1 bit of error in a data item; if there are 2 bits of error, then a 1-bit parity scheme will not detect any errors, since the parity will match the data with two errors. (Actually, a 1-bit parity scheme can detect any odd number of errors; however, the probability of having three errors is much lower than the probability of having two, so, in practice, a 1-bit parity code is limited to detecting a single bit of error.) Of course, a parity code cannot tell which bit in a data item is in error.

A 1-bit parity scheme is an error detection code; there are also *error correction codes* (ECC) that will detect and allow correction of an error. For large main memories, many systems use a code that allows the detection of up to 2 bits of error and the correction of a single bit of error. These codes work by using more bits to encode the data; for example, the typical codes used for main memories require 7 or 8 bits for every 128 bits of data.

error detection code A code that enables the detection of an error in data, but not the precise location and, hence, correction of the error.

Elaboration: A 1-bit parity code is a *distance-2 code*, which means that if we look at the data plus the parity bit, no 1-bit change is sufficient to generate another legal combination of the data plus parity. For example, if we change a bit in the data, the parity will be wrong, and vice versa. Of course, if we change 2 bits (any 2 data bits or 1 data bit and the parity bit), the parity will match the data and the error cannot be detected. Hence, there is a distance of two between legal combinations of parity and data.

To detect more than one error or correct an error, we need a *distance-3 code*, which has the property that any legal combination of the bits in the error correction code and the data has at least 3 bits differing from any other combination. Suppose we have such a code and we have one error in the data. In that case, the code plus data will be one bit away from a legal combination, and we can correct the data to that legal combination. If we have two errors, we can recognize that there is an error, but we cannot correct the errors. Let's look at an example. Here are the data words and a distance-3 error correction code for a 4-bit data item.

Data Word	Code bits	Data	Code bits
0000	000	1000	111
0001	011	1001	100
0010	101	1010	010
0011	110	1011	001
0100	110	1100	001
0101	101	1101	010
0110	011	1110	100
0111	000	1111	111

To see how this works, let's choose a data word, say 0110, whose error correction code is 011. Here are the four 1-bit error possibilities for this data: 1110, 0010, 0100, and 0111. Now look at the data item with the same code (011), which is the entry with the value 0001. If the error correction decoder received one of the four possible data words with an error, it would have to choose between correcting to 0110 or 0001. While these four words with error have only 1 bit changed from the correct pattern of 0110, they each have 2 bits that are different from the alternate correction of 0001. Hence, the error correction mechanism can easily choose to correct to 0110, since a single error is a much higher probability. To see that two errors can be detected, simply notice that all the combinations with 2 bits changed have a different code. The one reuse of the same code is with 3 bits different, but if we correct a 2-bit error, we will correct to the wrong value, since the decoder will assume that only a single error has occurred. If we want to correct 1-bit errors and detect, but not erroneously correct, 2-bit errors, we need a distance-4 code.

Although we distinguished between the code and data in our explanation, in truth, an error correction code treats the combination of code and data as a single word in a larger code (7 bits in this example). Thus, it deals with errors in the code bits in the same fashion as errors in the data bits.

While the above example requires $n - 1$ bits for n bits of data, the number of bits required grows slowly, so that for a distance-3 code, a 64-bit word needs 7 bits and a 128-bit word needs 8. This type of code is called a *Hamming code*, after R. Hamming, who described a method for creating such codes.

A.10 Finite-State Machines

finite-state machine
A sequential logic function consisting of a set of inputs and outputs, a next-state function that maps the current state and the inputs to a new state, and an output function that maps the current state and possibly the inputs to a set of asserted outputs.

next-state function A combinational function that, given the inputs and the current state, determines the next state of a finite-state machine.

As we saw earlier, digital logic systems can be classified as combinational or sequential. Sequential systems contain state stored in memory elements internal to the system. Their behavior depends both on the set of inputs supplied and on the contents of the internal memory, or state of the system. Thus, a sequential system cannot be described with a truth table. Instead, a sequential system is described as a finite-state machine (or often just *state machine*). A finite-state machine has a set of states and two functions, called the next-state function and the *output function*. The set of states corresponds to all the possible values of the internal storage. Thus, if there are n bits of storage, there are $2n$ states. The next-state function is a combinational function that, given the inputs and the current state, determines the next state of the system. The output function produces a set of outputs from current state and the inputs. Figure A.10.1 shows this diagrammatically.

The state machines we discuss here and in Chapter 4 are *synchronous*. This means that the state changes together with the clock cycle, and a new state is computed once every clock. Thus, the state elements are updated only on the clock edge. We use this methodology in this section and throughout Chapter 4, and we do not usually show the clock explicitly. We use state machines throughout Chapter 4 to control the execution of the processor and the actions of the datapath.

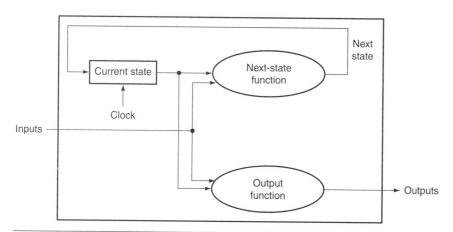

FIGURE A.10.1   A state machine consists of internal storage that contains the state and two combinational functions: the next-state function and the output function. Often, the output function is restricted to take only the current state as its input; this does not change the capability of a sequential machine, but does affect its internals.

To illustrate how a finite-state machine operates and is designed, let's look at a simple and classic example: controlling a traffic light. (Chapters 4 and 5 contain more detailed examples of using finite-state machines to control processor execution.) When a finite-state machine is used as a controller, the output function is often restricted to depend on just the current state. Such a finite-state machine is called a *Moore machine*. This is the type of finite-state machine we use throughout this book. If the output function can depend on both the current state and the current input, the machine is called a *Mealy machine*. These two machines are equivalent in their capabilities, and one can be turned into the other mechanically. The basic advantage of a Moore machine is that it can be faster, while a Mealy machine may be smaller, since it may need fewer states than a Moore machine. In Chapter 5, we discuss the differences in more detail and show a Verilog version of finite-state control using a Mealy machine.

Our example concerns the control of a traffic light at an intersection of a north-south route and an east-west route. For simplicity, we will consider only the green and red lights; adding the yellow light is left for an exercise. We want the lights to cycle no faster than 30 seconds in each direction, so we will use a 0.033-Hz clock so that the machine cycles between states at no faster than once every 30 seconds. There are two output signals:

- *NSlite:* When this signal is asserted, the light on the north-south road is green; when this signal is deasserted, the light on the north-south road is red.

- *EWlite:* When this signal is asserted, the light on the east-west road is green; when this signal is deasserted, the light on the east-west road is red.

In addition, there are two inputs:

- *NScar:* Indicates that a car is over the detector placed in the roadbed in front of the light on the north-south road (going north or south).

- *EWcar:* Indicates that a car is over the detector placed in the roadbed in front of the light on the east-west road (going east or west).

The traffic light should change from one direction to the other only if a car is waiting to go in the other direction; otherwise, the light should continue to show green in the same direction as the last car that crossed the intersection.

To implement this simple traffic light we need two states:

- *NSgreen:* The traffic light is green in the north-south direction.

- *EWgreen:* The traffic light is green in the east-west direction.

We also need to create the next-state function, which can be specified with a table:

	Inputs		
	NScar	**EWcar**	**Next state**
NSgreen	0	0	NSgreen
NSgreen	0	1	EWgreen
NSgreen	1	0	NSgreen
NSgreen	1	1	EWgreen
EWgreen	0	0	EWgreen
EWgreen	0	1	EWgreen
EWgreen	1	0	NSgreen
EWgreen	1	1	NSgreen

Notice that we didn't specify in the algorithm what happens when a car approaches from both directions. In this case, the next-state function given above changes the state to ensure that a steady stream of cars from one direction cannot lock out a car in the other direction.

The finite-state machine is completed by specifying the output function.

Before we examine how to implement this finite-state machine, let's look at a graphical representation, which is often used for finite-state machines. In this representation, nodes are used to indicate states. Inside the node we place a list of the outputs that are active for that state. Directed arcs are used to show the next-state function, with labels on the arcs specifying the input condition as logic functions. Figure A.10.2 shows the graphical representation for this finite-state machine.

	Outputs	
	NSlite	**EWlite**
NSgreen	1	0
EWgreen	0	1

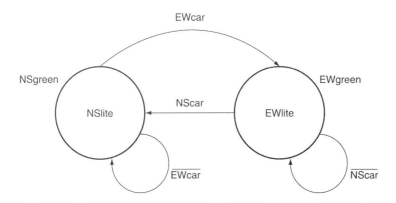

FIGURE A.10.2 The graphical representation of the two-state traffic light controller. We simplified the logic functions on the state transitions. For example, the transition from NSgreen to EWgreen in the next-state table is ($\overline{\text{NScar}}$ · EWcar) + (NScar · EWcar), which is equivalent to EWcar.

A finite-state machine can be implemented with a register to hold the current state and a block of combinational logic that computes the next-state function and the output function. Figure A.10.3 shows how a finite-state machine with 4 bits of state, and thus up to 16 states, might look. To implement the finite-state machine in this way, we must first assign state numbers to the states. This process is called *state assignment*. For example, we could assign NSgreen to state 0 and EWgreen to state 1. The state register would contain a single bit. The next-state function would be given as

$$\text{NextState} = (\overline{\text{CurrentState} \cdot \text{EWcar}}) + (\text{CurrentState} \cdot \overline{\text{NScar}})$$

where CurrentState is the contents of the state register (0 or 1) and NextState is the output of the next-state function that will be written into the state register at the end of the clock cycle. The output function is also simple:

$$\text{NSlite} = \overline{\text{CurrentState}}$$
$$\text{EWlite} = \text{CurrentState}$$

The combinational logic block is often implemented using structured logic, such as a PLA. A PLA can be constructed automatically from the next-state and output function tables. In fact, there are *computer-aided design* (CAD) programs

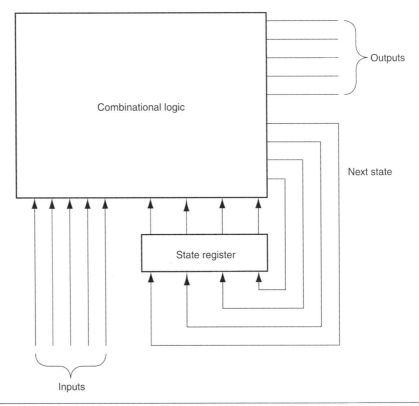

FIGURE A.10.3   A finite-state machine is implemented with a state register that holds the current state and a combinational logic block to compute the next state and output functions. The latter two functions are often split apart and implemented with two separate blocks of logic, which may require fewer gates.

that take either a graphical or textual representation of a finite-state machine and produce an optimized implementation automatically. In Chapters 4 and 5, finite-state machines were used to control processor execution. Appendix C discusses the detailed implementation of these controllers with both PLAs and ROMs.

To show how we might write the control in Verilog, Figure A.10.4 shows a Verilog version designed for synthesis. Note that for this simple control function, a Mealy machine is not useful, but this style of specification is used in Chapter 5 to implement a control function that is a Mealy machine and has fewer states than the Moore machine controller.

```
module TrafficLite (EWCar,NSCar,EWLite,NSLite,clock);

   input EWCar, NSCar,clock;
output EWLite,NSLite;

reg state;

initial state=0;  //set initial state

//following two assignments set the output, which is based
only on the state variable
assign NSLite = ~ state; //NSLite on if state = 0;
assign EWLite = state; //EWLite on if state = 1

always @(posedge clock) // all state updates on a positive
clock edge
   case (state)
      0: state = EWCar; //change state only if EWCar

      1: state = NSCar; //change state only if NSCar

   endcase
endmodule
```

FIGURE A.10.4 A Verilog version of the traffic light controller.

What is the smallest number of states in a Moore machine for which a Mealy machine could have fewer states?

a. Two, since there could be a one-state Mealy machine that might do the same thing.

b. Three, since there could be a simple Moore machine that went to one of two different states and always returned to the original state after that. For such a simple machine, a two-state Mealy machine is possible.

c. You need at least four states to exploit the advantages of a Mealy machine over a Moore machine.

A.11 Timing Methodologies

Throughout this appendix and in the rest of the text, we use an edge-triggered timing methodology. This timing methodology has an advantage in that it is simpler to explain and understand than a level-triggered methodology. In this section, we explain this timing methodology in a little more detail and also introduce level-sensitive clocking. We conclude this section by briefly discussing the issue of asynchronous signals and synchronizers, an important problem for digital designers.

The purpose of this section is to introduce the major concepts in clocking methodology. The section makes some important simplifying assumptions; if you are interested in understanding timing methodology in more detail, consult one of the references listed at the end of this appendix.

We use an edge-triggered timing methodology because it is simpler to explain and has fewer rules required for correctness. In particular, if we assume that all clocks arrive at the same time, we are guaranteed that a system with edge-triggered registers between blocks of combinational logic can operate correctly without races if we simply make the clock long enough. A *race* occurs when the contents of a state element depend on the relative speed of different logic elements. In an edge-triggered design, the clock cycle must be long enough to accommodate the path from one flip-flop through the combinational logic to another flip-flop where it must satisfy the setup-time requirement. Figure A.11.1 shows this requirement for a system using rising edge-triggered flip-flops. In such a system the clock period (or cycle time) must be at least as large as

$$t_{\mathrm{prop}} + t_{\mathrm{combinational}} + t_{\mathrm{setup}}$$

for the worst-case values of these three delays, which are defined as follows:

- t_{prop} is the time for a signal to propagate through a flip-flop; it is also sometimes called clock-to-Q.

- $t_{\mathrm{combinational}}$ is the longest delay for any combinational logic (which by definition is surrounded by two flip-flops).

- t_{setup} is the time before the rising clock edge that the input to a flip-flop must be valid.

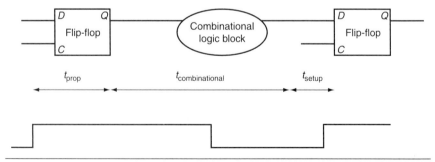

FIGURE A.11.1 In an edge-triggered design, the clock must be long enough to allow signals to be valid for the required setup time before the next clock edge. The time for a flip-flop input to propagate to the flip-flip outputs is t_{prop}; the signal then takes $t_{\mathrm{combinational}}$ to travel through the combinational logic and must be valid t_{setup} before the next clock edge.

We make one simplifying assumption: the hold-time requirements are satisfied, which is almost never an issue with modern logic.

One additional complication that must be considered in edge-triggered designs is clock skew. Clock skew is the difference in absolute time between when two state elements see a clock edge. Clock skew arises because the clock signal will often use two different paths, with slightly different delays, to reach two different state elements. If the clock skew is large enough, it may be possible for a state element to change and cause the input to another flip-flop to change before the clock edge is seen by the second flip-flop.

clock skew The difference in absolute time between the times when two state elements see a clock edge.

Figure A.11.2 illustrates this problem, ignoring setup time and flip-flop propagation delay. To avoid incorrect operation, the clock period is increased to allow for the maximum clock skew. Thus, the clock period must be longer than

$$t_{prop} + t_{combinational} + t_{setup} + t_{skew}$$

With this constraint on the clock period, the two clocks can also arrive in the opposite order, with the second clock arriving t_{skew} earlier, and the circuit will work

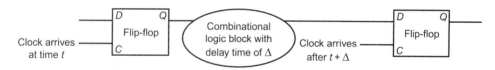

FIGURE A.11.2 Illustration of how clock skew can cause a race, leading to incorrect operation. Because of the difference in when the two flip-flops see the clock, the signal that is stored into the first flip-flop can race forward and change the input to the second flip-flop before the clock arrives at the second flip-flop.

correctly. Designers reduce clock-skew problems by carefully routing the clock signal to minimize the difference in arrival times. In addition, smart designers also provide some margin by making the clock a little longer than the minimum; this allows for variation in components as well as in the power supply. Since clock skew can also affect the hold-time requirements, minimizing the size of the clock skew is important.

Edge-triggered designs have two drawbacks: they require extra logic and they may sometimes be slower. Just looking at the D flip-flop versus the level-sensitive latch that we used to construct the flip-flop shows that edge-triggered design requires more logic. An alternative is to use level-sensitive clocking. Because state changes in a level-sensitive methodology are not instantaneous, a level-sensitive scheme is slightly more complex and requires additional care to make it operate correctly.

level-sensitive clocking A timing methodology in which state changes occur at either high or low clock levels but are not instantaneous as such changes are in edge-triggered designs.

Level-Sensitive Timing

In level-sensitive timing, the state changes occur at either high or low levels, but they are not instantaneous as they are in an edge-triggered methodology. Because of the noninstantaneous change in state, races can easily occur. To ensure that a level-sensitive design will also work correctly if the clock is slow enough, designers use *two-phase clocking*. Two-phase clocking is a scheme that makes use of two nonoverlapping clock signals. Since the two clocks, typically called ϕ_1 and ϕ_2, are nonoverlapping, at most one of the clock signals is high at any given time, as Figure A.11.3 shows. We can use these two clocks to build a system that contains level-sensitive latches but is free from any race conditions, just as the edge-triggered designs were.

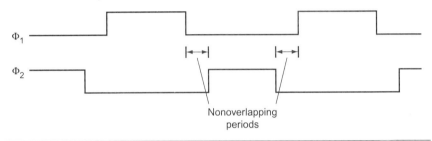

FIGURE A.11.3   A two-phase clocking scheme showing the cycle of each clock and the nonoverlapping periods.

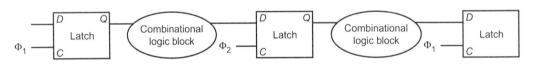

FIGURE A.11.4   A two-phase timing scheme with alternating latches showing how the system operates on both clock phases. The output of a latch is stable on the opposite phase from its C input. Thus, the first block of combinational inputs has a stable input during ϕ_2, and its output is latched by ϕ_2. The second (rightmost) combinational block operates in just the opposite fashion, with stable inputs during ϕ_1. Thus, the delays through the combinational blocks determine the minimum time that the respective clocks must be asserted. The size of the nonoverlapping period is determined by the maximum clock skew and the minimum delay of any logic block.

One simple way to design such a system is to alternate the use of latches that are open on ϕ_1 with latches that are open on ϕ_2. Because both clocks are not asserted at the same time, a race cannot occur. If the input to a combinational block is a ϕ_1 clock, then its output is latched by a ϕ_2 clock, which is open only during ϕ_2 when the input latch is closed and hence has a valid output. Figure A.11.4 shows how a system with two-phase timing and alternating latches operates. As in an edge-triggered design, we must pay attention to clock skew, particularly between the two clock phases. By increasing the amount of nonoverlap between the two phases, we can reduce the potential margin of error. Thus, the system is guaranteed to operate correctly if each phase is long enough and if there is large enough nonoverlap between the phases.

Asynchronous Inputs and Synchronizers

By using a single clock or a two-phase clock, we can eliminate race conditions if clock-skew problems are avoided. Unfortunately, it is impractical to make an entire system function with a single clock and still keep the clock skew small. While the CPU may use a single clock, I/O devices will probably have their own clock. An asynchronous device may communicate with the CPU through a series of handshaking steps. To translate the asynchronous input to a synchronous signal that can be used to change the state of a system, we need to use a *synchronizer*, whose inputs are the asynchronous signal and a clock and whose output is a signal synchronous with the input clock.

Our first attempt to build a synchronizer uses an edge-triggered D flip-flop, whose D input is the asynchronous signal, as Figure A.11.5 shows. Because we communicate with a handshaking protocol, it does not matter whether we detect the asserted state of the asynchronous signal on one clock or the next, since the signal will be held asserted until it is acknowledged. Thus, you might think that this simple structure is enough to sample the signal accurately, which would be the case except for one small problem.

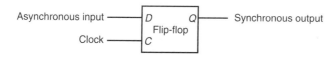

FIGURE A.11.5   A synchronizer built from a D flip-flop is used to sample an asynchronous signal to produce an output that is synchronous with the clock. This "synchronizer" will *not* work properly!

The problem is a situation called metastability. Suppose the asynchronous signal is transitioning between high and low when the clock edge arrives. Clearly, it is not possible to know whether the signal will be latched as high or low. That problem we could live with. Unfortunately, the situation is worse: when the signal that is sampled is not stable for the required setup and hold times, the flip-flop may go into a *metastable* state. In such a state, the output will not have a legitimate high or low value, but will be in the indeterminate region between them. Furthermore,

metastability
A situation that occurs if a signal is sampled when it is not stable for the required setup and hold times, possibly causing the sampled value to fall in the indeterminate region between a high and low value.

the flip-flop is not guaranteed to exit this state in any bounded amount of time. Some logic blocks that look at the output of the flip-flop may see its output as 0, while others may see it as 1. This situation is called a synchronizer failure.

In a purely synchronous system, synchronizer failure can be avoided by ensuring that the setup and hold times for a flip-flop or latch are always met, but this is impossible when the input is asynchronous. Instead, the only solution possible is to wait long enough before looking at the output of the flip-flop to ensure that its output is stable, and that it has exited the metastable state, if it ever entered it. How long is long enough? Well, the probability that the flip-flop will stay in the metastable state decreases exponentially, so after a very short time the probability that the flip-flop is in the metastable state is very low; however, the probability never reaches 0! So designers wait long enough such that the probability of a synchronizer failure is very low, and the time between such failures will be years or even thousands of years.

For most flip-flop designs, waiting for a period that is several times longer than the setup time makes the probability of synchronization failure very low. If the clock rate is longer than the potential metastability period (which is likely), then a safe synchronizer can be built with two D flip-flops, as Figure A.11.6 shows. If you are interested in reading more about these problems, look into the references.

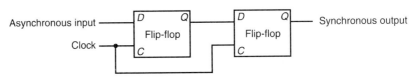

FIGURE A.11.6   This synchronizer will work correctly if the period of metastability that we wish to guard against is less than the clock period. Although the output of the first flip-flop may be metastable, it will not be seen by any other logic element until the second clock, when the second D flip-flop samples the signal, which by that time should no longer be in a metastable state.

Check Yourself

Suppose we have a design with very large clock skew—longer than the register propagation time. Is it always possible for such a design to slow the clock down enough to guarantee that the logic operates properly?

a. Yes, if the clock is slow enough the signals can always propagate and the design will work, even if the skew is very large.

b. No, since it is possible that two registers see the same clock edge far enough apart that a register is triggered, and its outputs propagated and seen by a second register with the same clock edge.

A.12 Field Programmable Devices

Within a custom or semicustom chip, designers can make use of the flexibility of the underlying structure to easily implement combinational or sequential logic. How can a designer who does not want to use a custom or semicustom IC implement a complex piece of logic taking advantage of the very high levels of integration available? The most popular component used for sequential and combinational logic design outside of a custom or semicustom IC is a field programmable device (FPD). An FPD is an integrated circuit containing combinational logic, and possibly memory devices, that are configurable by the end user.

FPDs generally fall into two camps: programmable logic devices (PLDs), which are purely combinational, and field programmable gate arrays (FPGAs), which provide both combinational logic and flip-flops. PLDs consist of two forms: simple PLDs (SPLDs), which are usually either a PLA or a programmable array logic (PAL), and complex PLDs, which allow more than one logic block as well as configurable interconnections among blocks. When we speak of a PLA in a PLD, we mean a PLA with user programmable and-plane and or-plane. A PAL is like a PLA, except that the or-plane is fixed.

Before we discuss FPGAs, it is useful to talk about how FPDs are configured. Configuration is essentially a question of where to make or break connections. Gate and register structures are static, but the connections can be configured. Notice that by configuring the connections, a user determines what logic functions are implemented. Consider a configurable PLA: by determining where the connections are in the and-plane and the or-plane, the user dictates what logical functions are computed in the PLA. Connections in FPDs are either permanent or reconfigurable. Permanent connections involve the creation or destruction of a connection between two wires. Current FPLDs all use an antifuse technology, which allows a connection to be built at programming time that is then permanent. The other way to configure CMOS FPLDs is through a SRAM. The SRAM is downloaded at power-on, and the contents control the setting of switches, which in turn determines which metal lines are connected. The use of SRAM control has the advantage in that the FPD can be reconfigured by changing the contents of the SRAM. The disadvantages of the SRAM-based control are two-fold: the configuration is volatile and must be reloaded on power-on, and the use of active transistors for switches slightly increases the resistance of such connections.

FPGAs include both logic and memory devices, usually structured in a two-dimensional array with the corridors dividing the rows and columns used for

---

field programmable devices (FPD) An integrated circuit containing combinational logic, and possibly memory devices, that are configurable by the end user.

programmable logic device (PLD) An integrated circuit containing combinational logic whose function is configured by the end user.

field programmable gate array (FPGA) A configurable integrated circuit containing both combinational logic blocks and flip-flops.

simple programmable logic device (SPLD) Programmable logic device, usually containing either a single PAL or PLA.

programmable array logic (PAL) Contains a programmable and-plane followed by a fixed or-plane.

antifuse A structure in an integrated circuit that when programmed makes a permanent connection between two wires.

global interconnect between the cells of the array. Each cell is a combination of gates and flip-flops that can be programmed to perform some specific function. Because they are basically small, programmable RAMs, they are also called lookup tables (LUTs). Newer FPGAs contain more sophisticated building blocks such as pieces of adders and RAM blocks that can be used to build register files. A few large FPGAs even contain 32-bit RISC cores!

In addition to programming each cell to perform a specific function, the interconnections between cells are also programmable, allowing modern FPGAs with hundreds of blocks and hundreds of thousands of gates to be used for complex logic functions. Interconnect is a major challenge in custom chips, and this is even more true for FPGAs, because cells do not represent natural units of decomposition for structured design. In many FPGAs, 90% of the area is reserved for interconnect and only 10% is for logic and memory blocks.

Just as you cannot design a custom or semicustom chip without CAD tools, you also need them for FPDs. Logic synthesis tools have been developed that target FPGAs, allowing the generation of a system using FPGAs from structural and behavioral Verilog.

A.13   Concluding Remarks

This appendix introduces the basics of logic design. If you have digested the material in this appendix, you are ready to tackle the material in Chapters 4 and 5, both of which use the concepts discussed in this appendix extensively.

Further Reading

There are a number of good texts on logic design. Here are some you might like to look into.

Ciletti, M. D. [2002]. *Advanced Digital Design with the Verilog HDL*, Englewood Cliffs, NJ: Prentice Hall.
A thorough book on logic design using Verilog.

Katz, R. H. [2004]. *Modern Logic Design*, 2nd ed., Reading, MA: Addison-Wesley.
A general text on logic design.

Wakerly, J. F. [2000]. *Digital Design: Principles and Practices*, 3rd ed., Englewood Cliffs, NJ: Prentice Hall.
A general text on logic design.

Exercises

A.1 [10] < §A.2> In addition to the basic laws we discussed in this section, there are two important theorems, called DeMorgan's theorems:

$$\overline{A + B} = \overline{A} \cdot \overline{B} \quad \text{and} \quad \overline{A \cdot B} = \overline{A} + \overline{B}$$

Prove DeMorgan's theorems with a truth table of the form

A	B	$\overline{A}$	$\overline{B}$	$\overline{A + B}$	$\overline{A} \cdot \overline{B}$	$\overline{A \cdot B}$	$\overline{A} + \overline{B}$
0	0	1	1	1	1	1	1
0	1	1	0	0	0	1	1
1	0	0	1	0	0	1	1
1	1	0	0	0	0	0	0

A.2 [15] < §A.2> Prove that the two equations for E in the example starting on page A-7 are equivalent by using DeMorgan's theorems and the axioms shown on page A-7.

A.3 [10] < §A.2> Show that there are 2^n entries in a truth table for a function with n inputs.

A.4 [10] < §A.2> One logic function that is used for a variety of purposes (including within adders and to compute parity) is *exclusive OR*. The output of a two-input exclusive OR function is true only if exactly one of the inputs is true. Show the truth table for a two-input exclusive OR function and implement this function using AND gates, OR gates, and inverters.

A.5 [15] < §A.2> Prove that the NOR gate is universal by showing how to build the AND, OR, and NOT functions using a two-input NOR gate.

A.6 [15] < §A.2> Prove that the NAND gate is universal by showing how to build the AND, OR, and NOT functions using a two-input NAND gate.

A.7 [10] < §§A.2, A.3> Construct the truth table for a four-input odd-parity function (see page A-65 for a description of parity).

A.8 [10] < §§A.2, A.3> Implement the four-input odd-parity function with AND and OR gates using bubbled inputs and outputs.

A.9 [10] < §§A.2, A.3> Implement the four-input odd-parity function with a PLA.

A.10 [15] < §§A.2, A.3> Prove that a two-input multiplexor is also universal by showing how to build the NAND (or NOR) gate using a multiplexor.

A.11 [5] < §§4.2, A.2, A.3> Assume that X consists of 3 bits, x2 x1 x0. Write four logic functions that are true if and only if

- X contains only one 0

- X contains an even number of 0s

- X when interpreted as an unsigned binary number is less than 4

- X when interpreted as a signed (two's complement) number is negative

A.12 [5] < §§4.2, A.2, A.3> Implement the four functions described in Exercise A.11 using a PLA.

A.13 [5] < §§4.2, A.2, A.3> Assume that X consists of 3 bits, x2 x1 x0, and Y consists of 3 bits, y2 y1 y0. Write logic functions that are true if and only if

- X < Y, where X and Y are thought of as unsigned binary numbers

- X < Y, where X and Y are thought of as signed (two's complement) numbers

- X = Y

Use a hierarchical approach that can be extended to larger numbers of bits. Show how can you extend it to 6-bit comparison.

A.14 [5] < §§A.2, A.3> Implement a switching network that has two data inputs (*A* and *B*), two data outputs (*C* and *D*), and a control input (*S*). If *S* equals 1, the network is in pass-through mode, and *C* should equal *A*, and *D* should equal *B*. If *S* equals 0, the network is in crossing mode, and *C* should equal *B*, and *D* should equal *A*.

A.15 [15] < §§A.2, A.3> Derive the product-of-sums representation for *E* shown on page A-11 starting with the sum-of-products representation. You will need to use DeMorgan's theorems.

A.16 [30] < §§A.2, A.3> Give an algorithm for constructing the sum-of-products representation for an arbitrary logic equation consisting of AND, OR, and NOT. The algorithm should be recursive and should not construct the truth table in the process.

A.17 [5] < §§A.2, A.3> Show a truth table for a multiplexor (inputs *A*, *B*, and *S*; output *C*), using don't cares to simplify the table where possible.

A.18 [5] < §A.3> What is the function implemented by the following Verilog modules:

```
module FUNC1 (I0, I1, S, out);
      input I0, I1;
      input S;
      output out;
      out = S? I1: I0;
endmodule

module FUNC2 (out,ctl,clk,reset);
      output [7:0] out;
      input ctl, clk, reset;
      reg [7:0] out;
      always @(posedge clk)
      if (reset) begin
                  out <= 8'b0 ;
      end
      else if (ctl) begin
                  out <= out + 1;
      end
      else begin
                  out <= out - 1;
      end
endmodule
```

A.19 [5] < §A.4> The Verilog code on page A-53 is for a D flip-flop. Show the Verilog code for a D latch.

A.20 [10] < §§A.3, A.4> Write down a Verilog module implementation of a 2-to-4 decoder (and/or encoder).

A.21 [10] < §§A.3, A.4> Given the following logic diagram for an accumulator, write down the Verilog module implementation of it. Assume a positive edge-triggered register and asynchronous Rst.

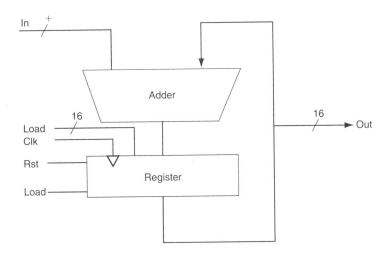

A.22 [20] < §§B3, A.4, A.5> Section 3.3 presents basic operation and possible implementations of multipliers. A basic unit of such implementations is a shift-and-add unit. Show a Verilog implementation for this unit. Show how can you use this unit to build a 32-bit multiplier.

A.23 [20] < §§B3, A.4, A.5> Repeat Exercise A.22, but for an unsigned divider rather than a multiplier.

A.24 [15] < §A.5> The ALU supported set on less than (slt) using just the sign bit of the adder. Let's try a set on less than operation using the values -7_{ten} and 6_{ten}. To make it simpler to follow the example, let's limit the binary representations to 4 bits: 1001_{two} and 0110_{two}.

$$1001_{two} - 0110_{two} = 1001_{two} + 1010_{two} = 0011_{two}$$

This result would suggest that $-7 > 6$, which is clearly wrong. Hence, we must factor in overflow in the decision. Modify the 1-bit ALU in Figure A.5.10 on page A-33 to handle slt correctly. Make your changes on a photocopy of this figure to save time.

A.25 [20] < §A.6> A simple check for overflow during addition is to see if the CarryIn to the most significant bit is not the same as the CarryOut of the most significant bit. Prove that this check is the same as in Figure 3.2.

A.26 [5] < §A.6> Rewrite the equations on page A-44 for a carry-lookahead logic for a 16-bit adder using a new notation. First, use the names for the CarryIn signals of the individual bits of the adder. That is, use c4, c8, c12, ... instead of C1, C2, C3, ... In addition, let Pi,j; mean a propagate signal for bits i to j, and Gi,j; mean a generate signal for bits i to j. For example, the equation

$$C2 = G1 + (P1 \cdot G0) + (P1 \cdot P0 \cdot c0)$$

can be rewritten as

$$c8 = G_{7,4} + (P_{7,4} \cdot G_{3,0}) + (P_{7,4} \cdot P_{3,0} \cdot c0)$$

This more general notation is useful in creating wider adders.

A.27 [15] < §A.6> Write the equations for the carry-lookahead logic for a 64-bit adder using the new notation from Exercise A.26 and using 16-bit adders as building blocks. Include a drawing similar to Figure A.6.3 in your solution.

A.28 [10] < §A.6> Now calculate the relative performance of adders. Assume that hardware corresponding to any equation containing only OR or AND terms, such as the equations for pi and gi on page A-40, takes one time unit T. Equations that consist of the OR of several AND terms, such as the equations for $c1$, $c2$, $c3$, and $c4$ on page A-40, would thus take two time units, 2T. The reason is it would take T to produce the AND terms and then an additional T to produce the result of the OR. Calculate the numbers and performance ratio for 4-bit adders for both ripple carry and carry lookahead. If the terms in equations are further defined by other equations, then add the appropriate delays for those intermediate equations, and continue recursively until the actual input bits of the adder are used in an equation. Include a drawing of each adder labeled with the calculated delays and the path of the worst-case delay highlighted.

A.29 [15] < §A.6> This exercise is similar to Exercise A.28, but this time calculate the relative speeds of a 16-bit adder using ripple carry only, ripple carry of 4-bit groups that use carry lookahead, and the carry-lookahead scheme on page A-39.

A.30 [15] < §A.6> This exercise is similar to Exercises A.28 and A.29, but this time calculate the relative speeds of a 64-bit adder using ripple carry only, ripple carry of 4-bit groups that use carry lookahead, ripple carry of 16-bit groups that use carry lookahead, and the carry-lookahead scheme from Exercise A.27.

A.31 [10] < §A.6> Instead of thinking of an adder as a device that adds two numbers and then links the carries together, we can think of the adder as a hardware device that can add three inputs together (ai, bi, ci) and produce two outputs (s, $ci + 1$). When adding two numbers together, there is little we can do with this observation. When we are adding more than two operands, it is possible to reduce the cost of the carry. The idea is to form two independent sums, called S' (sum bits) and C' (carry bits). At the end of the process, we need to add C' and S' together using a normal adder. This technique of delaying carry propagation until the end of a sum of numbers is called *carry save addition*. The block drawing on the lower right of Figure A.14.1 (see below) shows the organization, with two levels of carry save adders connected by a single normal adder.

Calculate the delays to add four 16-bit numbers using full carry-lookahead adders versus carry save with a carry-lookahead adder forming the final sum. (The time unit T in Exercise A.28 is the same.)

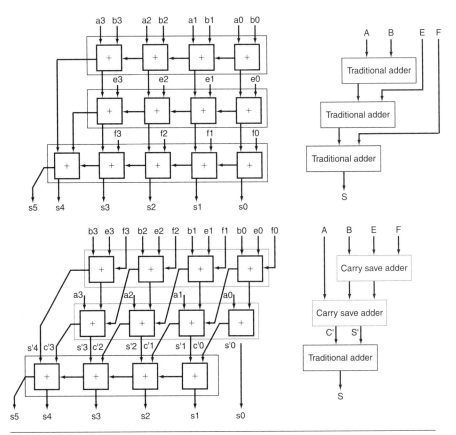

FIGURE A.14.1   Traditional ripple carry and carry save addition of four 4-bit numbers. The details are shown on the left, with the individual signals in lowercase, and the corresponding higher-level blocks are on the right, with collective signals in upper case. Note that the sum of four *n*-bit numbers can take *n* + 2 bits.

A.32 [20] < §A.6> Perhaps the most likely case of adding many numbers at once in a computer would be when trying to multiply more quickly by using many adders to add many numbers in a single clock cycle. Compared to the multiply algorithm in Chapter 3, a carry save scheme with many adders could multiply more than 10 times faster. This exercise estimates the cost and speed of a combinational multiplier to multiply two positive 16-bit numbers. Assume that you have 16 intermediate terms M15, M14, …, M0, called *partial products*, that contain the multiplicand ANDed with multiplier bits m15, m14, …, m0. The idea is to use carry save adders to reduce the *n* operands into 2*n*/3 in parallel groups of three, and do this repeatedly until you get two large numbers to add together with a traditional adder.

First, show the block organization of the 16-bit carry save adders to add these 16 terms, as shown on the right in Figure A.14.1. Then calculate the delays to add these 16 numbers. Compare this time to the iterative multiplication scheme in Chapter 3 but only assume 16 iterations using a 16-bit adder that has full carry lookahead whose speed was calculated in Exercise A.29.

A.33 [10] < §A.6> There are times when we want to add a collection of numbers together. Suppose you wanted to add four 4-bit numbers (A, B, E, F) using 1-bit full adders. Let's ignore carry lookahead for now. You would likely connect the 1-bit adders in the organization at the top of Figure A.14.1. Below the traditional organization is a novel organization of full adders. Try adding four numbers using both organizations to convince yourself that you get the same answer.

A.34 [5] < §A.6> First, show the block organization of the 16-bit carry save adders to add these 16 terms, as shown in Figure A.14.1. Assume that the time delay through each 1-bit adder is 2T. Calculate the time of adding four 4-bit numbers to the organization at the top versus the organization at the bottom of Figure A.14.1.

A.35 [5] < §A.8> Quite often, you would expect that given a timing diagram containing a description of changes that take place on a data input D and a clock input C (as in Figures A.8.3 and A.8.6 on pages A-52 and A-54, respectively), there would be differences between the output waveforms (Q) for a D latch and a D flip-flop. In a sentence or two, describe the circumstances (e.g., the nature of the inputs) for which there would not be any difference between the two output waveforms.

A.36 [5] < §A.8> Figure A.8.8 on page A-55 illustrates the implementation of the register file for the LEGv8 datapath. Pretend that a new register file is to be built, but that there are only two registers and only one read port, and that each register has only 2 bits of data. Redraw Figure A.8.8 so that every wire in your diagram corresponds to only 1 bit of data (unlike the diagram in Figure A.8.8, in which some wires are 5 bits and some wires are 32 bits). Redraw the registers using D flip-flops. You do not need to show how to implement a D flip-flop or a multiplexor.

A.37 [10] < §A.10> A friend would like you to build an "electronic eye" for use as a fake security device. The device consists of three lights lined up in a row, controlled by the outputs Left, Middle, and Right, which, if asserted, indicate that a light should be on. Only one light is on at a time, and the light "moves" from left to right and then from right to left, thus scaring away thieves who believe that the device is monitoring their activity. Draw the graphical representation for the finite-state machine used to specify the electronic eye. Note that the rate of the eye's movement will be controlled by the clock speed (which should not be too great) and that there are essentially no inputs.

A.38 [10] < §A.10> Assign state numbers to the states of the finite-state machine you constructed for Exercise A.37 and write a set of logic equations for each of the outputs, including the next-state bits.

A.39 [15] < §§A.2, A.8, A.10>   Construct a 3-bit counter using three D flip-flops and a selection of gates. The inputs should consist of a signal that resets the counter to 0, called *reset*, and a signal to increment the counter, called *inc*. The outputs should be the value of the counter. When the counter has value 7 and is incremented, it should wrap around and become 0.

A.40 [20] < §A.10>   A *Gray code* is a sequence of binary numbers with the property that no more than 1 bit changes in going from one element of the sequence to another. For example, here is a 3-bit binary Gray code: 000, 001, 011, 010, 110, 111, 101, and 100. Using three D flip-flops and a PLA, construct a 3-bit Gray code counter that has two inputs: *reset*, which sets the counter to 000, and *inc*, which makes the counter go to the next value in the sequence. Note that the code is cyclic, so that the value after 100 in the sequence is 000.

A.41 [25] < §A.10>   We wish to add a yellow light to our traffic light example on page A-68. We will do this by changing the clock to run at 0.25 Hz (a 4-second clock cycle time), which is the duration of a yellow light. To prevent the green and red lights from cycling too fast, we add a 30-second timer. The timer has a single input, called *TimerReset*, which restarts the timer, and a single output, called *TimerSignal*, which indicates that the 30-second period has expired. Also, we must redefine the traffic signals to include yellow. We do this by defining two output signals for each light: green and yellow. If the output NSgreen is asserted, the green light is on; if the output NSyellow is asserted, the yellow light is on. If both signals are off, the red light is on. Do *not* assert both the green and yellow signals at the same time, since American drivers will certainly be confused, even if European drivers understand what this means! Draw the graphical representation for the finite-state machine for this improved controller. Choose names for the states that are *different* from the names of the outputs.

A.42 [15] < §A.10>   Write down the next-state and output-function tables for the traffic light controller described in Exercise A.41.

A.43 [15] < §§A.2, A.10>   Assign state numbers to the states in the traffic light example of Exercise A.41 and use the tables of Exercise A.42 to write a set of logic equations for each of the outputs, including the next-state outputs.

A.44 [15] < §§A.3, A.10>   Implement the logic equations of Exercise A.43 as a PLA.

Answers to
Check Yourself

§A.2, page A-8: No. If $A = 1$, $C = 1$, $B = 0$, the first is true, but the second is false.
§A.3, page A-20: C.
§A.4, page A-22: They are all exactly the same.
§A.4, page A-26: A $= 0$, B $= 1$.
§A.5, page A-38: 2.
§A.6, page A-47: 1.
§A.8, page A-58: c.
§A.10, page A-72: b.
§A.11, page A-77: b.

Index

Note: Online information is listed by chapter and section number followed by page numbers (OL3.11-7). Page references preceded by a single letter with hyphen refer to appendices.